The Subnuclear Series • Volume 47

Proceedings of the International School of Subnuclear Physics

THE MOST UNEXPECTED AT LHC AND THE STATUS OF HIGH ENERGY FRONTIER

THE SUBNUCLEAR SERIES*

Series Editor: ANTONINO ZICHICHI, *European Physical Society, Geneva, Switzerland*

16. 1978 THE NEW ASPECTS OF SUBNUCLEAR PHYSICS
17. 1979 POINTLIKE STRUCTURES INSIDE AND OUTSIDE HADRONS
18. 1980 THE HIGH-ENERGY LIMIT
19. 1981 THE UNITY OF THE FUNDAMENTAL INTERACTIONS
20. 1982 GAUGE INTERACTIONS: Theory and Experiment
21. 1983 HOW FAR ARE WE FROM THE GAUGE FORCES?
22. 1984 QUARKS, LEPTONS, AND THEIR CONSTITUENTS
23. 1985 OLD AND NEW FORCES OF NATURE
24. 1986 THE SUPERWORLD I
25. 1987 THE SUPERWORLD II
26. 1988 THE SUPERWORLD III
27. 1989 THE CHALLENGING QUESTIONS
28. 1990 PHYSICS UP TO 200 TeV
29. 1991 PHYSICS AT THE HIGHEST ENERGY AND LUMINOSITY: To Understand the Origin of Mass
30. 1992 FROM SUPERSTRINGS TO THE REAL SUPERWORLD
31. 1993 FROM SUPERSYMMETRY TO THE ORIGIN OF SPACE–TIME
32. 1994 FROM SUPERSTRING TO PRESENT-DAY PHYSICS
33. 1995 VACUUM AND VACUA: The Physics of Nothing
34. 1996 EFFECTIVE THEORIES AND FUNDAMENTAL INTERACTIONS
35. 1997 HIGHLIGHTS OF SUBNUCLEAR PHYSICS: 50 Years Later
36. 1998 FROM THE PLANCK LENGTH TO THE HUBBLE RADIUS
37. 1999 BASICS AND HIGHLIGHTS IN FUNDAMENTAL PHYSICS
38. 2000 THEORY AND EXPERIMENT HEADING FOR NEW PHYSICS
39. 2001 NEW FIELDS AND STRINGS IN SUBNUCLEAR PHYSICS
40. 2002 FROM QUARKS AND GLUONS TO QUANTUM GRAVITY
41. 2003 FROM QUARKS TO BLACK HOLES: PROGRESS IN UNDERSTANDING THE LOGIC OF NATURE
42. 2004 HOW AND WHERE TO GO BEYOND THE STANDARD MODEL
43. 2005 TOWARDS NEW MILESTONES IN OUR QUEST TO GO BEYOND THE STANDARD MODEL
44. 2006 THE LOGIC OF NATURE, COMPLEXITY AND NEW PHYSICS: FROM QUARK-GLUON PLASMA TO SUPERSTRINGS, QUANTUM GRAVITY AND BEYOND
45. 2007 SEARCH FOR THE "TOTALLY UNEXPECTED" IN THE LHC ERA
46. 2009 PREDICTED AND TOTALLY UNEXPECTED IN THE ENERGY FRONTIER OPENED BY LHC
47. 2010 THE MOST UNEXPECTED AT LHC AND THE STATUS OF HIGH ENERGY FRONTIER

Volume 1 was published by W. A. Benjamin, Inc., New York; 2–8 and 11–12 by Academic Press, New York and London; 9–10 by Editrice Compositori, Bologna; 13–29 by Plenum Press, New York and London; 30–47 by World Scientific, Singapore.

*For the complete list of titles in this series, please go to

http://www.worldscibooks.com/series/ss_series.shtml

The Subnuclear Series • Volume 47

Proceedings of the International School of Subnuclear Physics

THE MOST UNEXPECTED AT LHC AND THE STATUS OF HIGH ENERGY FRONTIER

Edited by

Antonino Zichichi

European Physical Society
Geneva, Switzerland

World Scientific

NEW JERSEY • LONDON • SINGAPORE • BEIJING • SHANGHAI • HONG KONG • TAIPEI • CHENNAI

Published by

World Scientific Publishing Co. Pte. Ltd.
5 Toh Tuck Link, Singapore 596224
USA office: 27 Warren Street, Suite 401-402, Hackensack, NJ 07601
UK office: 57 Shelton Street, Covent Garden, London WC2H 9HE

British Library Cataloguing-in-Publication Data
A catalogue record for this book is available from the British Library.

The Subnuclear Series — Vol. 47
THE MOST UNEXPECTED AT LHC AND THE STATUS OF HIGH ENERGY FRONTIER
Proceedings of the International School of Subnuclear Physics

ISBN-13 978-981-4374-11-8
ISBN-10 981-4374-11-3

Printed in Singapore by World Scientific Printers.

PREFACE

In August/September 2009, a group of 78 physicists from 18 countries, working in 67 Universities, Laboratories and Centers, located in 16 countries, met in Erice for the 47th Course of the International School of Subnuclear Physics. The countries represented by the participants were: Armenia, France, Georgia, Germany, Greece, Hungary, India, Israel, Italy, Japan, Mexico, The Netherlands, Pakistan, Poland, Russia, Ukraine, United Kingdom and the United States of America.

The School was sponsored by the Academies of Sciences of Estonia, Georgia, Lithuania, Russia and Ukraine; the Chinese Academy of Sciences; the Commission of the European Communities; the European Physical Society; the Italian Ministry of Education, University and Scientific Research; the Sicilian Regional Government; the Weizmann Institute of Science; the World Federation of Scientists and the World Laboratory.

Purpose of the School was to focus attention on "The most Unexpected at LHC and the Status of High Energy Frontier", the key point being that we should not forget the past of our Physics; i.e. the unexpected events with enormous consequences, such as the discovery of the strange particles and of Δ^{++}. Experimental highlights from the most relevant sources of new data were presented and discussed together with the frontiers of theoretical studies.

An original feature of the School, introduced in 1996, is a series of special sessions devoted to "New Talents", with the aim of encouraging and promoting young physicists to achieve recognition at an international level. This is a serious problem in Experimental Physics, where collaborations count several hundreds of participants and it is almost impossible for young fellows to be noticed. Even if with much less emphasis, the problem exists also in Theoretical Physics. We therefore decided to offer young fellows the possibility to give an open presentation of the results of their studies, followed by a discussion. Among them, four best "New Talents" have been selected for publication in this volume, and were given an award.

During the organization and the running of the Course, I enjoyed the collaboration of my colleague and friend, Gerardus 't Hooft, who shared with me the Directorship of the School. I would like to thank him, together with the group of invited scientists and all the people who have contributed to the success of the Course.

I hope the reader will enjoy the book as much as the students attending the lectures and the discussion sessions, which, as every year, have been the focal point of our School activity. Thanks to the work of the Scientific Secretaries, the discussions have been transcribed as faithfully as possible. At various stages of my work, I have enjoyed the collaboration of many friends whose contributions have been extremely important for the School and are highly appreciated. I thank them most warmly. A final acknowledgement to all those in Erice, Bologna and Geneva who have helped me in so many occasions and to whom I feel very much indebted.

Antonino Zichichi

Geneva, May 2011

CONTENTS

Ultraviolet Behavior of $\mathcal{N} = 8$ Supergravity[1]

Lance J. Dixon

SLAC National Accelerator Laboratory
Stanford University
Stanford, CA 94309, USA

Abstract

In these lectures I describe the remarkable ultraviolet behavior of $\mathcal{N} = 8$ supergravity, which through four loops is no worse than that of $\mathcal{N} = 4$ super-Yang-Mills theory (a finite theory). I also explain the computational tools that allow multi-loop amplitudes to be evaluated in this theory — the KLT relations and the unitarity method — and sketch how ultraviolet divergences are extracted from the amplitudes.

1 Introduction

Quantum gravity is nonrenormalizable by power counting, because the coupling, Newton's constant, $G_N = 1/M_{\rm Pl}^2$, is dimensionful. In contrast, the coupling constants of unbroken gauge theories, such as α for QED and α_s for QCD, are dimensionless. At each loop order in perturbation theory, ultraviolet divergences in quantum gravity should get worse and worse, in comparison with gauge theory: There are two more powers of loop momentum in each loop, to compensate dimensionally for the two powers of the Planck mass in the denominator of the gravitational coupling. Instead of the logarithmic divergences of gauge theory, which are renormalizable using a finite set of counterterms, quantum gravity should possess an infinite set of counterterms.

String theory is well known to cure the divergences of quantum gravity. It does so by introducing a new length scale, related to the string tension, at which particles are no longer pointlike. The question these lectures will address is: Is this necessary? Or could enough symmetry allow a point-particle theory of quantum gravity to be perturbatively finite in the ultraviolet? If the latter is true, even if in a "toy model", it would have a big impact on how we think about quantum gravity. While string theory makes quantum gravity finite, it does so at the price of having a huge number of ground state vacua, perhaps 10^{500}. Is it possible that there are consistent theories of quantum gravity with fewer degrees of freedom, and fewer ground states?

The particular approach we will take in these lectures is to see whether the divergences generic to point-like theories of quantum gravity can be cured using (in part) a large amount of symmetry, particularly supersymmetry. The maximal supersymmetry available for a theory with a maximal spin of two is $\mathcal{N} = 8$. In $\mathcal{N} = 8$ supergravity, eight applications of the spin-1/2 supercharges Q_i, $i = 1, 2, \ldots, 8$, connect the helicity $h = +2$ graviton state with its CPT conjugate state with helicity $h = 2 - 8 \times 1/2 = -2$. Each anti-commuting supercharge can be applied either once or not at all, so the total

[1]Presented at International School of Subnuclear Physics, 47th Course, Erice Sicily, August 29-September 7, 2009.

number of massless states is $2^8 = 256$. The complete Lagrangian for this theory was obtained by Cremmer and Julia [1] in the late 1970s, following earlier work by de Wit and Freedman [2] and by Cremmer, Julia and Scherk [3]. While this theory has maximal supersymmetry, it seems unlikely that supersymmetry alone can render it finite to all orders in perturbation theory; other symmetries or dynamical principles may have to come in to play.

Many other approaches to quantum gravity have been considered. For example, the asymptotic safety program [4] posits that there is a nontrivial ultraviolet fixed point in the exact renormalization group. Hořava [5] has also proposed a renormalization group flow solution, in which the ultraviolet fixed point is not Lorentz symmetric; space and time scale differently at the fixed point. These proposals are very interesting, but usually require truncation of the number of operators or other assumptions. In contrast, here we will work in a conventional perturbative framework, with an action that is quadratic in derivatives.

The perturbative ultraviolet behavior of gravity theories in general, and $\mathcal{N} = 8$ supergravity in particular, has been under investigation for a few decades. Through about 1998, the general suspicion was that $\mathcal{N} = 8$ supergravity in four space-time dimensions would first diverge at three loops, based on the existence of an $\mathcal{N} = 8$ supersymmetric local counterterm at this order [6, 7, 8, 9, 10]. However, in 1998 the two-loop four-graviton amplitude was computed [11], and its ultraviolet behavior was found to be better than expected, leading to the speculation that the first divergence might occur at five loops. In 2002, Howe and Stelle speculated [12], based on the possible existence of a superspace formalism realizing seven of the eight supersymmetries, that the divergence might be delayed until six loops. However, a more recent analysis by Bossard, Howe and Stelle [13] suggests a five-loop divergence, unless additional cancellation mechanisms are present.

In the early 1980s, a seven-loop $\mathcal{N} = 8$ supersymmetric counterterm was constructed by Howe and Lindström [9]. Also at that time, an eight-loop counterterm was presented [9, 10], which is manifestly invariant under a continuous noncompact coset symmetry possessed by $\mathcal{N} = 8$ supergravity, referred to as $E_{7(7)}$. The status of lower-loop counterterms with respect to $E_{7(7)}$ is not totally clear. For example, the volume of the on-shell $\mathcal{N} = 8$ superspace would appear at seven loops, and is invariant under $E_{7(7)}$, but it might also vanish [14].

In the last few years, a variety of arguments have highlighted the excellent ultraviolet properties of $\mathcal{N} = 8$ supergravity, mostly suggesting finiteness until at least seven loops, although some arguments go much further. Based on multi-loop superstring results obtained by Berkovits using the pure spinor formalism [15], Green, Russo and Vanhove [16] argued that the theory should be finite through nine loops. However, a recent re-analysis by the same authors [17] (see also refs. [18, 19]) indicates that technical issues with the pure spinor formalism beginning at five loops invalidate this argument, and suggest a first divergence at seven loops. Note that refs. [9, 14, 20] also point to seven loops for a possible first divergence.

There are also arguments based on M-theory dualities which suggest finiteness to all orders [21]. On the other hand, $\mathcal{N} = 8$ supergravity is a point-particle theory in four dimensions containing only massless particles. String theory and M theory are quite different. Perturbatively, they contain infinite towers of massive excited states (string and/or Kaluza-Klein excitations); nonperturbatively, they contain additional extended objects. It is not at all obvious that results found in those theories can be applied directly to $\mathcal{N}=8$ supergravity, because of issues in decoupling the infinite towers of states [22]. The safest approach to determining the ultraviolet properties of $\mathcal{N}=8$ supergravity is to work directly in the field theory.

The purpose of these lectures is to describe some of the main methods that have been used to determine multi-loop amplitudes in $\mathcal{N}=8$ supergravity, and to extract from the amplitudes the ultraviolet behavior of the theory. Another review which overlaps this one in subject matter has been written recently by Bern, Carrasco and Johansson [23].

Multi-loop computations in $\mathcal{N}=8$ supergravity are feasible thanks to two key ideas that work hand in hand with each other: the unitarity method [24], which allows loop computations to be reduced to tree computations; and the Kawai-Lewellen-Tye (KLT) relations [25], which allow $\mathcal{N}=8$ supergravity tree amplitudes to be written in terms of tree amplitudes in a gauge theory, $\mathcal{N}=4$ super-Yang-Mills theory ($\mathcal{N}=4$ SYM). Both of these ideas have more modern refinements, to be described later, which are useful for pushing the results to the maximal number of loops. Using these methods, as I will explain in more detail below, the four-graviton amplitude in $\mathcal{N}=8$ supergravity was computed at two loops in 1998 [11], and eventually at three [26, 27] and four loops [28]. The basic idea is to first compute the four-gluon amplitude in $\mathcal{N}=4$ SYM at the same number of loops, and then use this information, along with unitarity and the KLT relations, to reconstruct the four-graviton amplitude in $\mathcal{N}=8$ supergravity.

Much of the motivation for these multi-loop efforts came from the results of one-loop computations with a large number of external gravitons [29, 30, 31]. The one-loop results all had the property (dubbed the "no-triangle hypothesis" [31]) that the ultraviolet behavior of n-graviton amplitudes in $\mathcal{N}=8$ supergravity was *never any worse* than that of the corresponding n-gluon amplitudes in $\mathcal{N}=4$ SYM. Both sets of one-loop amplitudes are finite in four dimensions. Considered as theories with the same number of supercharges in a higher space-time dimension, they first begin to diverge in eight dimensions. It is possible to use unitarity to argue that the one-loop behavior also implies large classes of multi-loop cancellations [32], although it clearly does not control the complete multi-loop behavior.

Remarkably, the same property, that $\mathcal{N}=8$ supergravity is no worse behaved than $\mathcal{N}=4$ SYM, has also been found to hold for all the multi-loop four-graviton results that have been computed to date, all the way through four loops. Now $\mathcal{N}=4$ SYM is well known to be an ultraviolet-finite theory to all orders in perturbation theory [33]; indeed, it is the prototypical four-dimensional conformal field theory. As long as $\mathcal{N}=8$ supergravity is as well-behaved as $\mathcal{N}=4$ SYM is in the ultraviolet, it must also remain finite. It is still an open question as to whether this property holds to all orders in perturbation theory. It has been suggested [19] that it might fail as soon as the next uncomputed amplitude, namely at five loops. (Such a failure at five loops could indicate [19] a four-dimensional divergence as early as seven loops, a loop order also suggested in refs. [9, 14, 20].) However, at present it seems that only a complete computation of the five-loop amplitude can definitively answer this question.

The remainder of this article is organized as follows. In section 2 the general structure of divergences in quantum gravity and supergravity is outlined. In section 3 we describe the KLT relations and the unitarity method. We discuss the method of maximal cuts, and the rung rule for $\mathcal{N}=4$ super-Yang-Mills theory. In section 4 we explain how the KLT relations allow generalized cuts in $\mathcal{N}=8$ supergravity to be obtained from two copies of the cuts in $\mathcal{N}=4$ super-Yang-Mills theory, and we discuss the ultraviolet behavior of typical contributions. Section 5 discusses the four-point scattering amplitude and ultraviolet behavior of $\mathcal{N}=8$ supergravity at three loops, and section 6 does the same at four loops. In section 7 conclusions are presented, along with an outlook for the future.

2 Divergences in quantum gravity

In general, divergences in quantum field theory are associated with local operators, or *counterterms*. Divergences in off-shell quantities, such as off-shell two-point functions, can depend on the gauge. However, the ultraviolet divergences in on-shell scattering amplitudes, which we shall be concerned with here, are gauge invariant, *i.e.* diffeomorphism invariant. Therefore they should be constructed from tensors that transform homogeneously under coordinate transformations. For scattering amplitudes containing only external gravitons, counterterms should be built from products of the Riemann tensor $R_{\mu\nu\sigma\rho}$, including contractions of it such as the Ricci tensor $R_{\mu\nu}$ and the Ricci scalar R.

The available counterterms are also constrained by dimensional analysis. The loop-counting parameter G_N has mass dimension -2, so at each successive loop order a logarithmic divergence must be associated with a counterterm with a dimension two units greater than in the previous loop. The Riemann tensor has mass dimension 2: $R^{\mu}_{\ \nu\sigma\rho} \sim \partial_\rho \Gamma^{\mu}_{\ \nu\sigma} \sim g^{\mu\kappa}\partial_\rho\partial_\nu g_{\kappa\sigma}$. Because the (tree-level) Einstein Lagrangian R also has dimension 2, an L-loop counterterm should have the schematic form $\mathcal{D}^{2k} R_{\dots}^{L+1-k}$, where $\mathcal{D}$ is a covariant derivative and $R_{\dots}$ is a generic Riemann tensor.

Even for on-shell divergences, there is still an ambiguity in associating counterterms, because of the freedom to perform (nonlinear) field redefinitions on the action. For example, in the case of pure Einstein gravity, with action $S=\int d^4x\sqrt{-g}R$, one can redefine the metric by $g_{\mu\nu}\to f(g)g_{\mu\nu}$. Because the variation $\delta S/\delta g_{\mu\nu}$ is proportional to the Ricci tensor $R_{\mu\nu}$, one can adjust in this way the coefficient of any potential counterterm in pure Einstein gravity that contains $R_{\mu\nu}$. At one-loop, the potential counterterm $R_{\mu\nu\sigma\rho}R^{\mu\nu\sigma\rho}$ is thus equivalent to $R^2-4R_{\mu\nu}R^{\mu\nu}+R_{\mu\nu\sigma\rho}R^{\mu\nu\sigma\rho}$, which is the Gauss-Bonnet term, a total derivative in four dimensions. Total derivatives cannot be produced in perturbation theory. For this reason, pure Einstein gravity is ultraviolet-finite at one loop [34]. On the other hand, if matter is coupled to gravity, there are generically counterterms of the form $T_{\mu\nu}T^{\mu\nu}$, where $T^{\mu\nu}$ is the matter stress-energy tensor, and the theory is nonrenormalizable at one loop [34].

At two loops, there is a nontrivial counterterm for pure gravity, which cannot be related to a total derivative. It has the form,

$$R^3 \equiv R^{\lambda\rho}_{\mu\nu}R^{\mu\nu}_{\sigma\tau}R^{\sigma\tau}_{\lambda\rho}\,. \tag{1}$$

Two independent two-loop computations using Feynman diagrams, by Goroff and Sagnotti [35] and by van de Ven [36], showed that the coefficient of this counterterm is nonzero, *i.e.* that pure gravity diverges at two loops.

In any supergravity, even $\mathcal{N} = 1$, if all states are in the same multiplet with the graviton (pure supergravity), then it is possible to show that there are no divergences until at least three loops [37, 6, 38]. This property can be understood by computing the tree-level matrix element of the operator R^3 between four on-shell graviton states with outgoing helicity $(-{+}{+}{+})$ [37]. On the one hand, the matrix element is nonzero [39]. On the other hand, such a helicity configuration is forbidden by supersymmetric Ward identities acting on the S matrix [40] (see section 3.4), which require at least two negative and two positive helicities. Thus R^3 is unavailable as a potential two-loop counterterm. In a pure supergravity, all four-point amplitudes can be rotated by supersymmetry into four-graviton amplitudes, so they must also be finite.

At three loops, the first potential counterterm in pure supergravity arises,

$$R^4 \equiv t_8^{\mu_1\nu_1\ldots\mu_4\nu_4} t_8^{\rho_1\sigma_1\ldots\rho_4\sigma_4} R_{\mu_1\nu_1\rho_1\sigma_1} R_{\mu_2\nu_2\rho_2\sigma_2} R_{\mu_3\nu_3\rho_3\sigma_3} R_{\mu_4\nu_4\rho_4\sigma_4} \,, \tag{2}$$

where the tensor t_8 is defined in ref. [41]. This particular operator [6, 7, 8, 9, 10] is also known as the square of the Bel-Robinson tensor [42]. It is compatible with supersymmetry, not just $\mathcal{N} = 1$ but all the way up to maximal $\mathcal{N} = 8$ supersymmetry. This latter property follows from the appearance of R^4 in the low-energy effective action of the $\mathcal{N} = 8$ supersymmetric closed superstring [43]; indeed, it represents the first correction term beyond the limit of $\mathcal{N} = 8$ supergravity [44], appearing at order α'^3. More precisely, the matrix element of this operator on four-particle states takes the form,

$$\langle R^4 \rangle |_{4\text{-point}} = stu\, M_4^{\text{tree}}(1,2,3,4)\,, \tag{3}$$

where the momentum invariants are $s = (k_1 + k_2)^2$, $t = (k_2 + k_3)^2$, $u = (k_1 + k_3)^2$, and M_4^{tree} stands for any of the 256^4 four-point amplitudes in $\mathcal{N} = 8$ supergravity (stripped of the gravitational coupling constant).

As mentioned earlier, it was generally believed prior to 1998 that the R^4 counterterm would control the first divergence in $\mathcal{N} = 8$ supergravity. However, when the two-loop four-graviton amplitude was computed [11], it was found to have two additional powers of the momentum-invariants emerging from the integrand, so that the amplitude had the schematic form,

$$M_4^{2\text{-loop}}(1,2,3,4) = stu\, M_4^{\text{tree}}(1,2,3,4) \times \Big[s^2\, I_4^X(s,t) \;+\; \text{permutations} \Big]\,, \tag{4}$$

where $I_4^X(s,t)$ denotes a scalar double box integral. The two extra powers of s in eq. (4), when written in position space, correspond to a counterterm with four covariant derivatives, of the form $\mathcal{D}^4 R^4$, where we have not specified the precise index contractions. This operator has a dimension appropriate for a four-dimensional divergence at five loops, not three loops. Ref. [11] also investigated a few different classes of higher-loop contributions, and found that in each case, two extra powers of momentum-invariants continued to emerge from the loops. As it later transpired, the true power-counting for the three- and four-loop amplitudes is considerably better, due to cancellations between different contributions. As we shall see, the actual behavior observed so far is $\mathcal{D}^6 R^4$ at three loops [26, 27, 19] and $\mathcal{D}^8 R^4$ at four loops [28, 19].

For simplicity, I have represented the ultraviolet behavior at a given loop order in terms of local counterterms, even when the theory is finite in four dimensions. This terminology corresponds to considering $\mathcal{N}=8$ supergravity generalized to $D \geq 4$ dimensions, while keeping the same number of supercharges as in four dimensions (thirty-two). The same 256 physical states circulate in the loop for any value of D, but the virtual loop integration measure $d^D\ell$ leads to poorer ultraviolet behavior as D increases. At each loop order L, there is a critical dimension $D_c(L)$ at which the four-point amplitude first diverges in the ultraviolet, and this can be related to a counterterm of the form $\mathcal{D}^{2k}R^4$. The gravitational coupling κ is related to Newton's constant by $\kappa^2 = 32\pi^2 G_N$. In D dimensions, κ^2 (the loop-counting parameter) has a mass dimension of $-(D-2)$. A counterterm of the form $\mathcal{D}^{2k}R^4$ corresponds to a logarithmic amplitude divergence of the form

$$\mathcal{M}_4^{L\text{-loop}} \sim \kappa^{2L} \times stu\mathcal{M}_4^{\text{tree}} \times s^k \times \left[\frac{1}{\epsilon} \text{ or } \ln\Lambda\right], \tag{5}$$

when using dimensional regularization ($D \to D-2\epsilon$) or a dimensionful cutoff Λ. Dimensional analysis on eq. (5) then implies that

$$2k+6 = L(D_c-2). \tag{6}$$

Suppose $k=2$ at each loop order beyond $L=1$, the behavior suggested in ref. [11]. (The one-loop case is special, as no extra powers of s emerge at this order [44], giving $D_c(1)=8$.) This behavior would imply

$$D_c(L) = 2+\frac{10}{L} \qquad (L>1), \tag{7}$$

yielding a divergence in four dimensions at $L=5$. On the other hand, the behavior $k=L$ is seen through four loops, and this would imply, if valid to all loops,

$$D_c(L) = 4+\frac{6}{L} \qquad (L>1). \tag{8}$$

Because this formula gives a critical dimension above 4 for all L, its validity would result in the perturbative finiteness of $\mathcal{N}=8$ supergravity.

It is convenient to consider simultaneously the ultraviolet behavior of $\mathcal{N}=4$ SYM generalized to $D \geq 4$. Ref. [11] studied the four-gluon amplitude in this theory [45], the complete amplitude at two loops (see also ref. [46]), and classes of higher-loop contributions. In this case, a single additional power of s emerged, beyond that at one loop, leading to the generic expected behavior,

$$\mathcal{A}_4^{L\text{-loop}} \sim g^{2L} \times st\mathcal{A}_4^{\text{tree}} \times s \times \left[\frac{1}{\epsilon} \text{ or } \ln\Lambda\right]. \tag{9}$$

In D dimensions, g^2 (the gauge theory loop-counting parameter) has a mass dimension of $-(D-4)$. Dimensional analysis on eq. (9) then implies $4+2 = L(D_c-4)$, or $D_c(L) = 4+6/L$, the same behavior as in eq. (8). The associated higher-dimensional counterterms have the form $\mathcal{D}^2F^4$, where $F=F^{\mu\nu}$ is the gauge field strength and $\mathcal{D}$ a gauge covariant derivative.

This multi-loop behavior of $\mathcal{N}=4$ SYM follows from an argument by Howe and Stelle [12] using $\mathcal{N}=3$ harmonic superspace [47]. In the limit of a large number of colors, *i.e.* for gauge group $SU(N_c)$ for large N_c, the behavior has been confirmed by explicit computation through five loops, using the planar amplitudes in refs. [48, 49, 50]. In other words, at the four-point level, at each loop order, eq. (8) is equivalent to the statement that the ultraviolet behavior of $\mathcal{N}=8$ supergravity is no worse than that of $\mathcal{N}=4$ super-Yang-Mills theory.

There have been suggestions [19, 18] recently that the four-point five-loop amplitude in $\mathcal{N}=8$ supergravity will behave as $\mathcal{D}^8R^4$, not $\mathcal{D}^{10}R^4$. This behavior would be worse than that of $\mathcal{N}=4$ super-Yang-Mills theory. Although it would not directly imply a four-dimensional divergence at five loops, it would have important implications for the possibility of finiteness to all orders. However, it seems that only a complete computation of the amplitude can definitively answer this question.

I have not been very specific about the precise index contractions for the various counterterms, nor about the number of independent counterterms at each dimension (or value of k in $\mathcal{D}^{2k}R^4$). Also, what about the possibility of counterterms that do not show up at the four-point level at all? Recently, Elvang, Freedman and Kiermaier used $\mathcal{N}=8$ supersymmetry and locality to heavily constrain the possibilities [51]. At the four-point level, the constraints on operators of the form $\mathcal{D}^{2k}R^4$ are that their matrix elements should generalize eq. (3) as follows,

$$\langle \mathcal{D}^{2k}R^4\rangle|_{4-\text{point}} = P_k(s,t,u)\, stu\, M_4^{\text{tree}}(1,2,3,4)\,, \tag{10}$$

where $P_k(s,t,u)$ must be a Bose symmetric polynomial of degree k, subject to the on-shell constraint $s+t+u=0$. It is easy to count the independent polynomials, and hence the number of four-point counterterms allowed by $\mathcal{N}=8$ supersymmetry, at loop order $L=k+3$. There is one such operator for $k=0$; none for $k=1$ (because $s+t+u=0$); one each for $k=2,3,4,5$; two for $k=6$; and so on [51]. More impressively, ref. [51] also showed that *beyond* the four-point level very few operators are allowed at low loop order; in fact, there are no such operators until $L=7$. (The four-loop case was analyzed earlier [52, 53].) Thus through six loops, all ultraviolet divergences must appear in the four-graviton amplitude.

3 Computational tools

In this section we outline the two major tools that have been used to compute complete multi-loop amplitudes in $\mathcal{N}=8$ supergravity: the KLT relations and the unitarity method. We then discuss generalized unitarity at the multi-loop level, the method of maximal cuts, and the rung rule for $\mathcal{N}=4$ super-Yang-Mills theory.

3.1 KLT relations

The $2^8=256$ massless states of $\mathcal{N}=8$ supergravity are tabulated in table 1. The multiplicity for helicity h is given by $\binom{8}{p}$, where $p=2(2-h)$ is the number of applications of the spin-1/2 supersymmetry generators Q_i needed to reach that helicity from the $h=+2$ graviton state h^+. Alternatively, the multiplicities can be read off from the 8th row of Pascal's triangle, or the coefficients in the binomial expansion of $(x+y)^8$. Also tabulated in table 1 are the multiplicities for the $2^4=16$ massless states of $\mathcal{N}=4$ SYM.

$\mathcal{N}=8$ supergravity									
h	-2	$-\frac{3}{2}$	-1	$-\frac{1}{2}$	0	$\frac{1}{2}$	1	$\frac{3}{2}$	2
# of states	1	8	28	56	70	56	28	8	1
field	h^-	ψ_i^-	v_{ij}^-	χ_{ijk}^-	s_{ijkl}	χ_{ijk}^+	v_{ij}^+	ψ_i^+	h^+
$\mathcal{N}=4$ super-Yang-Mills									
h			-1	$-\frac{1}{2}$	0	$\frac{1}{2}$	1		
# of states			1	4	6	4	1		
field			g^-	λ_A^-	ϕ_{AB}	λ_A^+	g^+		

Table 1: Table of state multiplicities, as a function of helicity h, for the $2^8 = 256$ states in $\mathcal{N}=8$ supergravity and for the $2^4 = 16$ states in $\mathcal{N}=4$ super-Yang-Mills theory.

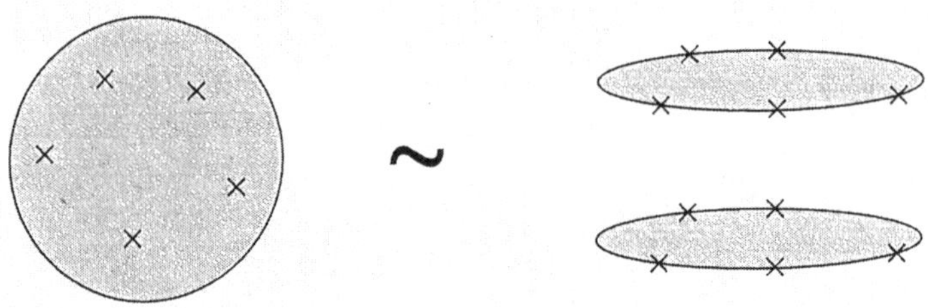

Figure 1: Schematic depiction of the KLT relations. The closed-string world-sheet, a sphere, can be thought of as two copies of the open-string world-sheet, a disk. Vertex operator insertions are marked with $\times$'s.

They are given by $\binom{4}{p}$ with $p = 2(1-h)$, or as the coefficients in the binomial expansion of $(x+y)^4$. Both $\mathcal{N}=8$ supergravity and $\mathcal{N}=4$ SYM arise as the low-energy limit of string theories [44], the type II closed superstring and the type I open superstring, respectively. The closed superstring contains both left- and right-moving modes, which are each in correspondence with one copy of the the open superstring modes. Therefore the $\mathcal{N}=8$ supergravity Fock space can be written as the tensor product of two copies of the $\mathcal{N}=4$ SYM Fock space,

$$[\mathcal{N}=8] \;=\; [\mathcal{N}=4]_L \otimes [\mathcal{N}=4]_R \,. \tag{11}$$

The fact that the multiplicities work out is a simple consequence of the identity $(x+y)^8 = [(x+y)^4]^2$.

Kawai, Lewellen and Tye [25] first observed that closed and open string amplitudes are very closely related at tree level. As shown in fig. 1, the closed-string world-sheet is a sphere, and the emission of a particular state is described by inserting a closed-string vertex operator $V^{\rm closed}(z,\bar{z})$ somewhere on the sphere. In contrast, the open-string world-sheet is a disk, and open string states are emitted off the boundary, with a vertex operator $V^{\rm open}(x)$. The KLT relations derive from the fact that the closed-string vertex operator is a product of two open-string vertex operators,

$$V^{\rm closed}(z_i,\bar{z}_i) = V_L^{\rm open}(z_i) \,\times\, \overline{V}_R^{\rm open}(\bar{z}_i)\,, \tag{12}$$

one for the left-movers and one for the right-movers. The left and right string oscillators appearing in $V_L^{\rm open}$ and $\overline{V}_L^{\rm open}$ are distinct, but the zero-mode momentum is shared.

Closed-string tree amplitudes are given by integrating correlation functions of the form

$$\langle V_1^{\text{closed}}(z_1, \bar{z}_1) \cdots V_n^{\text{closed}}(z_n, \bar{z}_n) \rangle \tag{13}$$

over $n-3$ copies of the sphere. KLT noticed that the integrand (13) was just the product of corresponding open-string integrands. They expressed the closed-string complex integrations as products of z and $\bar{z}$ contour integrals, and then deformed the contours until they were equivalent to open-string integrals, over real variables x_i, multiplied by momentum-dependent phase factors arising from branch cuts in the integrand. In this way, arbitrary closed-string tree amplitudes were expressed as quadratic combinations of open-string tree amplitudes.

Taking the low-energy limit of the KLT relations for string theory gives corresponding relations at the field-theory level, relating n-point tree amplitudes M_n^{tree} in $\mathcal{N}=8$ supergravity to quadratic combinations of tree amplitudes A_n^{tree} in $\mathcal{N}=4$ super-Yang-Mills theory. In the low-energy limit, the momentum-dependent phase factors generate powers of momentum-invariants. Strictly speaking, the gauge theory tree amplitudes that appear are those from which the Chan-Paton factors have been removed (in string terminology), or color-ordered subamplitudes (in QCD terminology). (For a review, see ref. [54].) We write the full tree amplitude as

$$\mathcal{A}_n^{\text{tree}}(\{k_i, a_i\}) = g^{n-2} \sum_{\rho \in S_n/\mathbb{Z}_n} \text{Tr}(T^{a_{\rho(1)}} T^{a_{\rho(2)}} \ldots T^{a_{\rho(n)}})\, A_n^{\text{tree}}(\rho(1), \rho(2), \ldots, \rho(n))\,, \tag{14}$$

where g is the gauge coupling, a_i is an adjoint index, T^{a_i} is a generator matrix in the fundamental representation of $SU(N_c)$, the sum is over all $(n-1)!$ inequivalent (non-cyclic) permutations ρ of n objects, and the argument i of A_n^{tree} labels both the momentum k_i and state information (helicity h_i, *etc.*). In the case of supergravity amplitudes, we only strip off powers of the coupling κ, defining M_n^{tree} by

$$\mathcal{M}_n^{\text{tree}}(\{k_i\}) = \left(\frac{\kappa}{2}\right)^{n-2} M_n^{\text{tree}}(1, 2, \ldots, n)\,. \tag{15}$$

Then the first few KLT relations have the form,

$$M_3^{\text{tree}}(1,2,3) = i\, A_3^{\text{tree}}(1,2,3)\tilde{A}_3^{\text{tree}}(1,2,3)\,, \tag{16}$$

$$M_4^{\text{tree}}(1,2,3,4) = -is_{12}\, A_4^{\text{tree}}(1,2,3,4)\, \tilde{A}_4^{\text{tree}}(1,2,4,3)\,, \tag{17}$$

$$M_5^{\text{tree}}(1,2,3,4,5) = is_{12}s_{34}\, A_5^{\text{tree}}(1,2,3,4,5)\tilde{A}_5^{\text{tree}}(2,1,4,3,5) + \mathcal{P}(2,3)\,, \tag{18}$$

$$\begin{aligned} M_6^{\text{tree}}(1,2,3,4,5,6) &= -is_{12}s_{45}\, A_6^{\text{tree}}(1,2,3,4,5,6) \\ &\quad \times \left[s_{35}\, \tilde{A}_6^{\text{tree}}(2,1,5,3,4,6) + (s_{34}+s_{35})\, \tilde{A}_6^{\text{tree}}(2,1,5,4,3,6) \right] \\ &\quad + \mathcal{P}(2,3,4)\,, \end{aligned} \tag{19}$$

where $s_{ij} \equiv (k_i + k_j)^2$, and "$+\,\mathcal{P}$" indicates a sum over the $m!$ permutations of the m arguments of $\mathcal{P}$. Here A_n^{tree} indicates a tree amplitude for which the external states are drawn from the left-moving Fock space $[\mathcal{N}=4]_L$ in the tensor product (11), while $\tilde{A}_n^{\text{tree}}$ denotes an amplitude from the right-moving copy $[\mathcal{N}=4]_R$.

The KLT relations are quite general, in the sense that left- and right-movers apparently do not need to be drawn from the particular gauge theory $\mathcal{N}=4$ SYM (or any truncation of it) [55, 56]. Furthermore, a general 'double-copy' formula for gravity amplitudes was proposed recently [57, 58, 56], which is consistent with the KLT relations, but generates additional novel representations of gravity amplitudes. It starts with a representation [57] of the full color-dressed gauge tree amplitudes in terms of cubic graphs labelled by i,

$$\frac{\mathcal{A}_n^{\text{tree}}}{g^{n-2}} = \sum_i \frac{n_i c_i}{(\prod_j p_j^2)_i}, \qquad \frac{\tilde{\mathcal{A}}_n^{\text{tree}}}{g^{n-2}} = \sum_i \frac{\tilde{n}_i c_i}{(\prod_j p_j^2)_i}, \tag{20}$$

where $1/p_j^2$ are scalar propagators, and $n_i, \tilde{n}_i$ are kinematical numerator factors for the left- and right-moving theory. The c_i are color factors, taken for convenience to be in a theory with adjoint particles only, so that they are specific products of the structure constants f^{abc}, one for each cubic vertex, and contracted according to the topology of the graph. The Jacobi identity,

$$f^{abe} f^{cde} + f^{cae} f^{bde} + f^{bce} f^{ade} = 0, \tag{21}$$

induces relations between the color factors for n-point amplitudes, namely

$$c_i + c_j + c_k = 0, \tag{22}$$

where the three graphs i, j, k are identical except for exchanging the connections between two of the vertices, following the relation (21).

While there is considerable freedom in choosing the numerator factors, it seems that it is always possible to choose them to satisfy kinematic analogs of the color Jacobi identity [57],

$$n_i + n_j + n_k = 0, \tag{23}$$

for all triplets (i, j, k) for which the color Jacobi identity (22) holds. With such a choice made, the double-copy formula for the gravity amplitudes is then

$$M_n^{\text{tree}} = i \sum_i \frac{n_i \tilde{n}_i}{(\prod_j p_j^2)_i}. \tag{24}$$

Although much of the multi-loop progress in $\mathcal{N} = 8$ supergravity to date has been based on the KLT relations, it is quite likely that representations such as eq. (24) will play a key role in the future.

These relations between gravity and gauge theory are reminiscent of the AdS/CFT duality [59]. Of course, the details are very different: AdS/CFT relates a weakly-coupled gravitational theory to a strongly-coupled gauge theory, whereas KLT and associated relations related a weakly-coupled gravitational theory to the *square* of a weakly-coupled gauge theory. AdS/CFT is tied to the notion of holography. Similarly, there is undoubtedly a deep principle attached to KLT-like dualities, but its full nature has not yet been unraveled.

3.2 Unitarity method

The scattering matrix is a unitary operator between in and out states. That is, $S^\dagger S = 1$, or in terms of the more standard "off-forward" part of the S matrix, $T \equiv (S-1)/i$, we have

$$2\,\text{Disc}\,T = T^\dagger T\,, \tag{25}$$

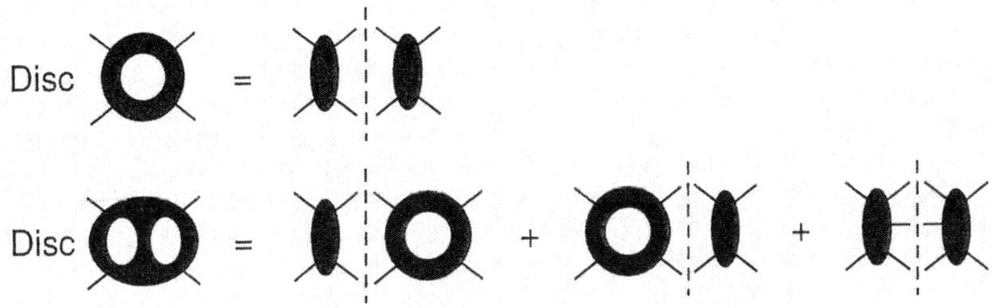

Figure 2: Unitarity relations for the four-point amplitude at one and two loops. The number of holes in a blob indicates the number of loops in the corresponding amplitude.

where $\mathrm{Disc}\, T \equiv (T - T^\dagger)/2i$. This simple relation gives rise to the well-known unitarity relations, or cutting rules [60], for the discontinuities (or absorptive parts) of perturbative amplitudes. If one inserts a perturbative expansion for T into eq. (25), say

$$T_4 = g^2\, T_4^{\text{tree}} + g^4\, T_4^{\text{1-loop}} + g^6\, T_4^{\text{2-loop}} + \ldots\,, \tag{26}$$
$$T_5 = g^3\, T_4^{\text{tree}} + g^5\, T_5^{\text{1-loop}} + g^7\, T_5^{\text{2-loop}} + \ldots\,, \tag{27}$$

for the four- and five-point amplitudes, then one obtains the unitarity relations shown in fig. 2.

At order g^4, the discontunity in the one-loop four-point amplitude is given by the product of two order g^2 four-point tree amplitudes. The product must be summed over all possible intermediate states crossing the cut (indicated by the dashed line), and integrated over all possible intermediate momenta. At two loops, or order g^6, there are two possible types of cuts: the product of a tree-level and a one-loop four-point amplitude ($g^2 \times g^4$), and the product of two tree-level five-point amplitudes ($g^3 \times g^3$).

To get the complete scattering amplitude, not just the absorptive part, one might attempt to reconstruct the real part via a dispersion relation. However, in the context of perturbation theory, an easier method is available, because one knows that the amplitude could have been calculated in terms of Feynman diagrams. Therefore it can be expressed as a linear combination of appropriate Feynman integrals, with coefficients that are rational functions of the kinematics. The unitarity method [24] matches the information coming from the cuts against the set of available loop integrals in order to determine these rational coefficients. There are also additive rational terms in the amplitude, terms which have no cuts in four dimensions. There are two general ways to determine these terms: one can use unitarity in $D - 4 - 2\epsilon$ dimensions [61, 62], or one can exploit factorization information to relate the rational terms for an n-point amplitude to those for amplitudes with $(n-1)$ or fewer legs [63, 64, 65].

At one loop, there have been many recent refinements to these on-shell methods, allowing for automation and numerical implementation by several groups [66, 67, 68] (as reviewed recently in ref. [69]). These results have led in turn to state-of-the-art results for next-to-leading order QCD cross sections, providing precise predictions for important Standard Model backgrounds at the LHC.

In particular, *generalized unitarity* [70], which corresponds at one loop to cutting more than two lines, can be used to simplify the information required for computing many terms in the amplitude [63, 71, 72]. Fig. 3(a) depicts an ordinary cut for a one-

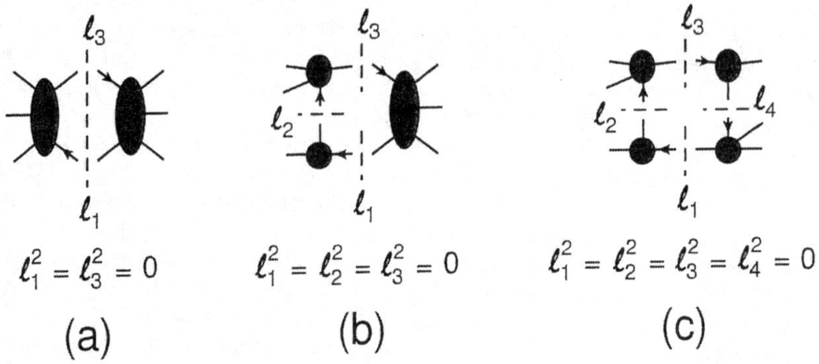

Figure 3: Generalized unitarity at one loop: (a) the ordinary two-particle cut imposes two constraints on the loop-momentum; (b) the triple cut imposes three constraints and is sensitive to triangle coefficients; (c) the quadruple cut imposes four constraints (freezing all four components of ℓ_1^μ) and is sensitive to box coefficients.

loop six-point amplitude in four space-time dimensions. The two lines crossing the cut, with momenta ℓ_1^μ and ℓ_3^μ, are on shell, so that $\ell_1^2 = \ell_3^2 = 0$. Of course ℓ_1^μ and ℓ_3^μ are not independent (they differ by the fixed external momenta), so these conditions represent two equations for the four components of ℓ_1^μ. One can impose up to two more equations, leading to the more restrictive cut kinematics in fig. 3(b) (the triple cut), and finally fig. 3(c) (the quadruple cut). For the latter condition, $\ell_1^2 = \ell_2^2 = \ell_3^2 = \ell_4^2 = 0$, the number of equations equals the number of unknowns, and so there are generically two discrete solutions. The coefficient d_j of a scalar box integral with the indicated topology can be found [72] simply by summing over the product of the four tree amplitudes, evaluated for the two solutions labeled by σ,

$$\begin{aligned} d_j &= \frac{1}{2}\sum_{\sigma=\pm} A_{n_1}^{\text{tree}}(\ldots,-\ell_1^\sigma,\ell_2^\sigma)\, A_{n_2}^{\text{tree}}(\ldots,-\ell_2^\sigma,\ell_3^\sigma) \\ &\qquad\quad \times A_{n_3}^{\text{tree}}(\ldots,-\ell_3^\sigma,\ell_4^\sigma)\, A_{n_4}^{\text{tree}}(\ldots,-\ell_4^\sigma,\ell_1^\sigma)\,, \end{aligned} \tag{28}$$

where the ellipses stand for the external momenta.

From fig. 3 it is clear that as one imposes more constraints, the required tree amplitudes have fewer legs, so they will become simpler. On the other hand, the kinematic conditions for generalized cuts are more constraining, and often cannot be satisfied by real Minkowski momenta. The limiting case of this is the three-point amplitude. When all three legs are massless, the only solution to

$$k_1^2 = k_2^2 = k_3^2 = 0, \qquad k_1^\mu + k_2^\mu + k_3^\mu = 0, \tag{29}$$

with real momenta k_i^μ, is for all three momenta to be parallel, such that $k_1^\mu = -\alpha k_3^\mu$, $k_2^\mu = -(1-\alpha)k_3^\mu$, for some real number α. This configuration is pathological because all kinematic invariants vanish. Clearly, the momentum-invariants s_{ij} all vanish: $s_{12} = (k_1+k_2)^2 = k_3^2 = 0$, and similarly $s_{23} = s_{13} = 0$.

One can also construct kinematic invariants from Weyl spinors based on the particle momenta. Defining $\lambda_i^\alpha \equiv |k_i^+\rangle \equiv \frac{1}{2}(1+\gamma_5)u(k_i)$ and $\tilde\lambda_i^{\dot\alpha} \equiv |k_i^-\rangle \equiv \frac{1}{2}(1-\gamma_5)u(k_i)$, the spinor products are

$$\langle i\, j\rangle \;=\; \varepsilon_{\alpha\beta}\lambda_i^{\alpha}\lambda_j^{\beta} = \langle k_i^-|k_j^+\rangle\,, \tag{30}$$

$$[i\, j] \;=\; \varepsilon_{\dot\alpha\dot\beta}\tilde\lambda_i^{\dot\alpha}\tilde\lambda_j^{\dot\beta} = \langle k_i^+|k_j^-\rangle\,. \tag{31}$$

They satisfy

$$[i\, j]\,\langle j\, i\rangle = \mathrm{Tr}[\tfrac{1}{2}(1+\gamma_5)\not{k}_i\not{k}_j] = s_{ij}\,. \tag{32}$$

For real momenta, $\langle i\, j\rangle$ and $[i\, j]$ are complex conjugates of each other. Therefore they are complex square roots of s_{ij},

$$\langle i\, j\rangle = e^{i\phi_{ij}}\sqrt{s_{ij}}\,, \qquad [i\, j] = -e^{-i\phi_{ij}}\sqrt{s_{ij}}\,, \tag{33}$$

for some phase angle ϕ_{ij}, which means that they too vanish for three-point kinematics.

However, for complex momenta k_i there is another type of solution: If we choose all three negative-helicity two-component spinors to be proportional,

$$\tilde\lambda_1^{\dot\alpha} \propto \tilde\lambda_2^{\dot\alpha} \propto \tilde\lambda_3^{\dot\alpha}\,, \tag{34}$$

then according to eq. (31) we have $[1\,2] = [2\,3] = [1\,3] = 0$, but the other three spinor products, $\langle 1\,2\rangle$, $[2\,3]$, and $[1\,3]$, are allowed to be nonzero (consistent with eq. (32)). Hence an amplitude built solely from $\langle i\, j\rangle$ is nonzero and finite for this choice of kinematics. The three-gluon amplitude with two negative and one positive helicity is such an object,

$$A_3^{\rm tree}(1^-,2^-,3^+) = i\,\frac{\langle 1\,2\rangle^4}{\langle 1\,2\rangle\,\langle 2\,3\rangle\,\langle 3\,1\rangle}\,, \tag{35}$$

where we assign helicities with an all-outgoing convention (*i.e.* if a particle is incoming, it has a physical helicity opposite from its helicity label). In contrast, the parity conjugate of this amplitude,

$$A_3^{\rm tree}(1^+,2^+,3^-) = -i\,\frac{[1\,2]^4}{[1\,2]\,[2\,3]\,[3\,1]}\,, \tag{36}$$

vanishes for the kinematics (34), but is nonzero and finite for the conjugate kinematics satisfying,

$$\lambda_1^{\alpha} \propto \lambda_2^{\alpha} \propto \lambda_3^{\alpha}\,. \tag{37}$$

3.3 Multi-loop generalized unitarity and maximal cuts

At the multi-loop level, generalized unitarity is also extremely powerful [62, 49, 26, 50, 73]. It allows one to avoid cuts like the ones shown in fig. 2, in which a loop amplitude appears on one side of the cut. One can impose additional cuts in order to chop such loops further into trees. Doing so is essential in order to make use of the KLT relations, which hold only for tree amplitudes. Fig. 4 illustrates this, starting with an ordinary three-particle cut for the three-loop four-point amplitude. The information in this cut can be extracted more easily by cutting the one-loop five-point amplitude on the right-hand side of the cut, decomposing it into the product of a four-point tree and a five-point tree; as illustrated, there are three inequivalent ways to do this.

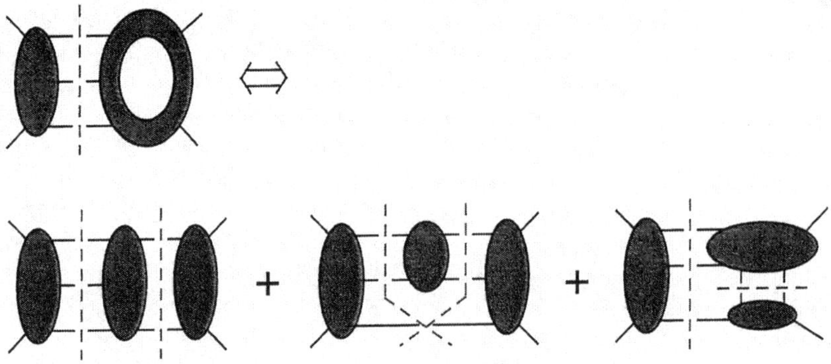

Figure 4: An example of multi-loop generalized unitarity. The one-loop five-point amplitude, appearing on the right side of the ordinary cut, is further cut into products of trees, in three inequivalent ways.

If one finds a representation of the amplitude that reproduces all the generalized cuts (in D dimensions), then that representation is guaranteed to be correct. The reason is that the generalized cuts simply provide a way of efficiently sorting all the Feynman diagrams contributing to an amplitude, and every Feynman diagram has a cut in some channel. This statement assumes that all particles are massless — Feynman diagrams for external wave-function corrections do not contain cuts, but they also vanish in dimensional regularization in the massless case, because there is no scale on which they can depend. The reason D dimensions is required is that some cuts may vanish as $D \to 4$. An alternate but equivalent argument uses dimensional analysis: An L-loop amplitude in $D = 4 - 2\epsilon$ carries a fractional mass dimension of $-2\epsilon L$, from the integration measure $\sim (\int d^{4-2\epsilon}\ell)^L$. If there are no particle masses, then all dimensions are carried by momentum invariants, corresponding to channels that can be cut through. A rational function $R(s,t)$ that is present in the amplitude in the limit $\epsilon \to 0$ must actually have the form in $D = 4 - 2\epsilon$ of $R(s,t) \times (-s)^{-\epsilon L} \approx R(s,t) \times (1 - \epsilon L \ln(-s))$. The logarithm indicates that the rational function is visible in the cuts at the next order in ϵ.

Fig. 4 illustrates a particular type of generalized unitarity in which all cut momenta are real. As at one loop, it is profitable to allow for complex momenta, and slice the tree amplitudes into yet smaller ones. The method of *maximal cuts* [50, 27] starts with the limiting case in which all tree amplitudes are three-point ones. Fig. 5 shows how one of the real-momentum configurations in fig. 4 spawns several maximal cuts. The maximal cuts are simply enumerated by drawing all cubic graphs. Their evaluation is also simple, because the three-point tree amplitudes are so compact. In the case of gauge theory, they are given by eqs. (35) and (36), and by other formulae related by supersymmetry; for gravity, they are obtained by squaring the gauge amplitudes, using eq. (16).

Even though the maximal cuts are maximally simple, they give an excellent starting point for constructing the full amplitude. For example, for the four-gluon amplitude in $\mathcal{N} = 4$ SYM, they detect all terms in the complete answer through two loops [45, 11], and all terms in the planar (leading in N_c) contribution through three loops [48]. The terms they do not detect can be expressed as "contact terms". Suppose we take the Feynman integral associated with a cubic graph from scalar ϕ^3 theory, and insert into the numerator

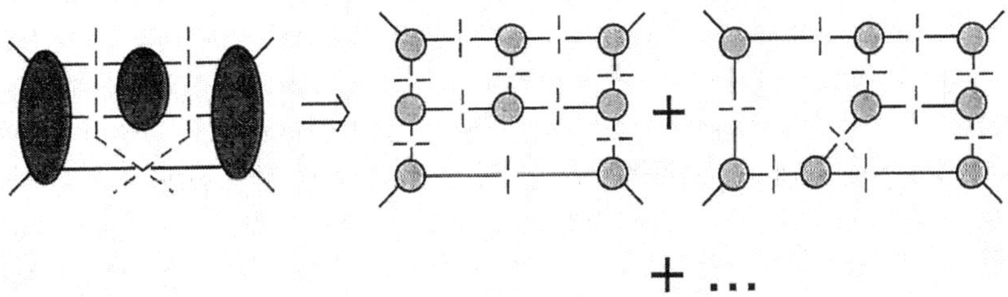

Figure 5: Example of a real-momentum generalized cut generating several maximal cuts; the latter contain only three-point tree amplitudes.

one power of an inverse propagator ℓ_i^2 for some loop momentum ℓ_i. This insertion cancels a propagator in the ϕ^3 graph, which corresponds to deleting one of the cut lines in the corresponding maximal cut, and merging two of the three-point amplitudes in that cut into a four-point amplitude. By definition, such a contribution is not detectable in the maximal cut, which assumes $\ell_i^2 = 0$. However, it, and indeed all remaining terms in the amplitude, can be found systematically, by considering the *near-maximal cuts*, which are found by collapsing one or more propagators in each maximal cut. Starting with an ansatz based on the maximal cuts, one adds contact terms to it, fixing their coefficients by requiring that the new ansatz reproduces the near-maximal cuts. The procedure is then iterated, in the number of cancelled propagators, until all cuts are successfully reproduced. In the case of $\mathcal{N} = 4$ SYM at L loops, the procedure converges after only $(L-2)$ cancelled propagators, because of the excellent ultraviolet behavior of this theory, eq. (8), which corresponds to only allowing $2(L-2)$ powers of loop momentum in the numerator of each cubic loop integral.

3.4 Supersymmetric Ward identities and the rung rule

Although the maximal cuts are quite simple, there is an even simpler subclass of contributions for maximally supersymmetric theories, those which contain iterated two-particle cuts, *i.e.* graphs that can be reduced to tree amplitudes by a succession of two-particle cuts. The reason they are so simple is two-fold: (1) at each stage one encounters only four-point amplitudes, whose dependence on the external states is completely dictated by supersymmetry, at any loop order; and (2) the sum over intermediate states for this case can be performed simply, once and for all.

The first statement follows from supersymmetric Ward identities (SWI) for the S matrix [40]. These identities are derived by requiring that supercharges annihilate the vacuum. The $\mathcal{N} = 1$ SWI can be found easily by letting $Q = Q^\alpha \eta_\alpha$, where η is a Grassmann parameter, and writing

$$0 = \langle 0|[Q, \Phi_1\Phi_2 \cdots \Phi_n]|0\rangle = \sum_{i=1}^{n} \langle 0|\Phi_1 \cdots [Q, \Phi_i] \cdots \Phi_n]|0\rangle \,, \tag{38}$$

where Φ_i are fields making the n external states for the amplitude. The commutators $[Q, \Phi_i] \equiv \tilde{\Phi}_i$ make the corresponding $\mathcal{N} = 1$ superpartner states. If the Φ_i are chosen to make helicity eigenstates, and in particular if many of the states are gluons with the same helicity, then many terms in the SWI vanish [40, 54].

For an amplitude containing only gluons, in which all, or all but one, of the gluons have positive helicity, one can arrange that there is only one term in the SWI, so that the amplitude itself vanishes,

$$A_n^{\rm SUSY}(1^\pm, 2^+, 3^+, \ldots, n^+) = 0\,. \tag{39}$$

(The same argument applied to gravity implies the vanishing of the four-point amplitude $M_4^{\rm SUSY}(-{+}{+}{+})$; this fact was used the argument in section 2 for the absence of the R^3 counterterm in pure supergravity.) A second type of vanishing amplitude contains $(n-2)$ gluons, plus a single pair of states $(\overline{P}, P)$:

$$A_n^{\rm SUSY}(1_{\overline{P}}^{-h_P}, 2_P^{h_P}, 3^+, \ldots, n^+) = 0\,. \tag{40}$$

Here P may be a scalar ϕ, a gluino λ, or a gluon g (reverting to the previous case), with helicity $h_P = 0, \pm 1/2, \pm 1$ respectively:

The first nonvanishing class of n-point amplitudes are called maximally-helicity-violating (MHV). They include the pure-gluon amplitudes with precisely two negative helicities (labeled by i and j), which were first written down at tree level by Parke and Taylor [74],

$$A_n^{\rm tree}(1^+, \ldots, i^-, \ldots, j^-, \ldots, n^+) = i\,\frac{\langle i\,j\rangle^4}{\langle 1\,2\rangle\,\langle 2\,3\rangle \cdots \langle n\,1\rangle}\,. \tag{41}$$

Other MHV amplitudes include those with a pair of states $(\overline{P}, P)$ as above, plus exactly one negative-helicity gluon (labelled by j). The SWI relate all such amplitudes to each other, according to

$$A_n^{\rm SUSY}(1_{\overline{P}}^{-h_P}, 2_P^{h_P}, 3^+, \ldots, j^-, \ldots, n^+) = \left(\frac{\langle 1\,j\rangle}{\langle 2\,j\rangle}\right)^{2h_P} A_n^{\rm tree}(1_{\bar\phi}, 2_\phi, 3^+, \ldots, j^-, \ldots, n^+)\,. \tag{42}$$

Comparing the two cases in which P is a gluon, $h_P = \pm 1$, and repeating for other labelings, eq. (42) implies that the dependence of the MHV pure-gluon amplitudes on the location of the negative helicities is trivial,

$$A_n^{\mathcal{N}=4\ \rm SYM}(1^+, \ldots, i^-, \ldots, j^-, \ldots, n^+) = \langle i\,j\rangle^4 \times f_n(s_{m,m+1})\,, \tag{43}$$

where $f_n(s_{m,m+1})$ is invariant under cyclic permutations, and is completely independent of i and j. Equation (43) is clearly satisfied by the Parke-Taylor tree amplitudes (41), but it holds to arbitrary loop order in $\mathcal{N} = 4$ SYM. At the four-point level, all nonvanishing amplitudes are MHV, and so they are all related by $\mathcal{N} = 4$ supersymmetry. The same factor $\langle i\,j\rangle^4$ occurs at every loop order, and hence the ratio $A_n^{(L)}/A_n^{\rm tree}$ is independent of i and j in $\mathcal{N} = 4$ SYM.

With this information in hand, we consider the simplest unitarity cut in $\mathcal{N} = 4$ SYM, the cut in the s channel of the one-loop four-point amplitude. We use eq. (43) to move the two negative helicities, so that they are in locations 1 and 2; *i.e.* we consider $A_4^{\text{1-loop}}(1^-, 2^-, 3^+, 4^+)$. Its s-channel cut is depicted in fig. 6(a), and is given by,

$$\begin{aligned} C_s &\equiv \operatorname{Disc}_s A_4^{\text{1-loop}}(1^-, 2^-, 3^+, 4^+) \\ &= \sum_{P \in \mathcal{N}=4} \int d\text{LIPS}(\ell_1, \ell_3)\, A_4^{\rm tree}(1^-, 2^-, \ell_{3,P}, -\ell_{1,P})\, A_4^{\rm tree}(\ell_{1,P}, -\ell_{3,P}, 3^+, 4^+)\,, \end{aligned} \tag{44}$$

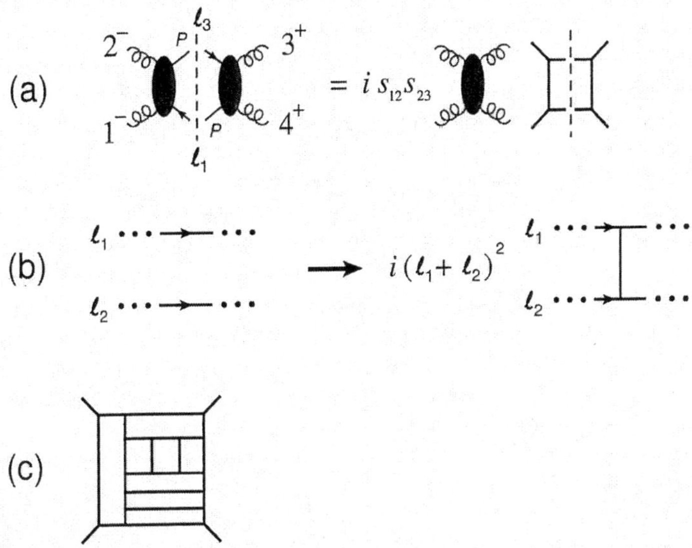

Figure 6: (a) The s-channel cut of the one-loop four-gluon amplitude in $\mathcal{N} = 4$ SYM is expressed in terms of the tree amplitude, the cut scalar box integral, and external momentum invariants. (b) The rung rule for generating numerator factors using iterated two-particle cuts. (c) Example of a graph whose numerator factor can be determined by the rung rule.

where $\text{dLIPS}(\ell_1, \ell_3)$ stands for the intermediate phase-space measure. In principle, we have to sum over all 16 states P in the multiplet. However, the SWI (39) and (40) imply that there is only one nonvanishing configuration, the one in which two identical-helicity gluons cross the cut,

$$C_s = \int \text{dLIPS}(\ell_1, \ell_3)\, A_4^{\text{tree}}(1^-, 2^-, \ell_3^+, -\ell_1^+) A_4^{\text{tree}}(\ell_1^-, -\ell_3^-, 3^+, 4^+)\,. \tag{45}$$

Using eq. (33), the four-point tree amplitude (41) can be written as

$$A_4^{\text{tree}}(1^-, 2^-, 3^+, 4^+) = \frac{\langle 1\,2\rangle^4}{\langle 1\,2\rangle\,\langle 2\,3\rangle\,\langle 3\,4\rangle\,\langle 4\,1\rangle} = \text{phase} \times \frac{s_{12}}{s_{23}}\,. \tag{46}$$

Substituting for $1/s_{23}$ the appropriate kinematic variable inside the cut, eq. (45) becomes

$$\begin{aligned} C_s &= \text{phase} \times \int \text{dLIPS}(\ell_1, \ell_3)\, \frac{s_{12}}{(\ell_1 - k_1)^2} \frac{s_{12}}{(\ell_3 - k_3)^2} \\ &= i\, s_{12} s_{23}\, A_4^{\text{tree}}(1^-, 2^-, 3^+, 4^+) \int \text{dLIPS}(\ell_1, \ell_3)\, \frac{1}{(\ell_1 - k_1)^2\,(\ell_3 - k_3)^2}\,. \end{aligned} \tag{47}$$

This result is represented diagramatically in fig. 6(a). The last factor in eq. (47) is simply the s-channel cut of the scalar one-loop box integral,

$$I_4^D(s_{12}, s_{23}) = \int \frac{d^D \ell_1}{(2\pi)^D} \frac{1}{\ell_1^2\,(\ell_1 - k_1)^2\,\ell_3^2\,(\ell_3 - k_3)^2}\,, \tag{48}$$

because the two propagators with momenta ℓ_1 and $\ell_3 = \ell_1 - k_1 - k_2$ are replaced by delta functions in the cut. Thus an expression for the one-loop four-point amplitude that matches the s-channel cut is,

$$A_4^{\text{1-loop}}(1^-, 2^-, 3^+, 4^+) = i\, s_{12} s_{23}\, A_4^{\text{tree}}(1^-, 2^-, 3^+, 4^+)\, I_4^D(s_{12}, s_{23})\,. \tag{49}$$

Because of eq. (43), the ratio $A_4^{\text{1-loop}}/A_4^{\text{tree}}$ is independent of the location of the negative helicities. So we know that eq. (49) must also match the t-channel cut. Here we only computed the cut in four dimensions, using helicity states to perform the intermediate state sum. However, the same expression (49) matches the full D-dimensional cuts (when a supersymmetric regulator is used [75]), and therefore it must be the correct answer for the full amplitude, including the dispersive part [44].

Furthermore, the two-particle cut can be iterated to higher loops [45]. In the next iteration, there is an additional numerator factor of s_{12} from sewing on the next tree, while the factor of $1/s_{23}$ becomes an additional propagator. This result has been abstracted to give the *rung rule* [45] shown in fig. 6(b), which generates numerator factors for certain contributions to the $(L+1)$-loop amplitudes, in terms of those at L loops. Whenever a rung is sewn on perpendicular to two lines carrying loop momenta ℓ_1 and ℓ_2, an extra factor of $i(\ell_1 + \ell_2)^2$ is generated in the numerator. An example of a graph with iterated two-particle cuts, whose numerator can be determined by the rung rule, is given in fig. 6(c).

4 KLT copying and rung-rule behavior

Suppose that we have a representation of the L-loop four-point amplitude in $\mathcal{N}=4$ SYM, and we want to compute the L-loop four-point amplitude in $\mathcal{N}=8$ supergravity. The KLT relations give us an efficient way of exporting the information from the first theory to the second. An arbitrary generalized cut in $\mathcal{N}=8$ supergravity is given in terms of $\mathcal{N}=8$ supergravity tree amplitudes, summed over all intermediate states. We rewrite each tree using KLT in terms of two copies of $\mathcal{N}=4$ SYM trees. The net result is a sum over products of *two copies* of the $\mathcal{N}=4$ SYM cuts. The sum over intermediate states in $\mathcal{N}=8$ supergravity is automatically carried out as a double sum over the $\mathcal{N}=4$ SYM states, $\sum_{\mathcal{N}=8} = \sum_{\mathcal{N}=4}\sum_{\mathcal{N}=4}$. Because the KLT relations contain different cyclic orderings of the $\mathcal{N}=4$ SYM tree amplitudes, we need both planar (leading-in-N_c) and non-planar (subleading-in-N_c) terms in the L-loop $\mathcal{N}=4$ SYM amplitude. This is not a surprise; the gravitational amplitude has no notion of color ordering.

Fig. 7 shows an example of KLT copying at three loops. The $\mathcal{N}=8$ supergravity cut contains one four-point tree amplitude and two five-point ones. We use eqs. (17) and (18). It is convenient to rewrite them as

$$\begin{aligned}
M_4^{\text{tree}}(\ell_1,\ell_2,\ell_3,\ell_4) &= -i\frac{s_{\ell_1\ell_2}s_{\ell_2\ell_3}}{s_{\ell_1\ell_3}}\, A_4^{\text{tree}}(\ell_1,\ell_2,\ell_3,\ell_4)\,\tilde{A}_4^{\text{tree}}(\ell_1,\ell_2,\ell_3,\ell_4)\,, \\
M_5^{\text{tree}}(1,2,\ell_2,\ell_1,\ell_5) &= -is_{\ell_5 1}s_{2\ell_2}\, A_5^{\text{tree}}(1,2,\ell_2,\ell_1,\ell_5)\,\tilde{A}_5^{\text{tree}}(1,\ell_1,2,\ell_2,\ell_5) + \mathcal{P}(1,2)\,, \\
M_5^{\text{tree}}(4,3,\ell_3,\ell_4,5) &= -is_{\ell_5 4}s_{3\ell_3}\, A_5^{\text{tree}}(4,3,\ell_3,\ell_4,\ell_5)\,\tilde{A}_5^{\text{tree}}(4,\ell_4,3,\ell_3,\ell_5) + \mathcal{P}(3,4)\,.
\end{aligned} \tag{50}$$

In this way, both occurrences of the four-point $\mathcal{N}=4$ SYM amplitude carry the same cyclic ordering as the $\mathcal{N}=8$ supergravity one, as shown in the figure. One of the two

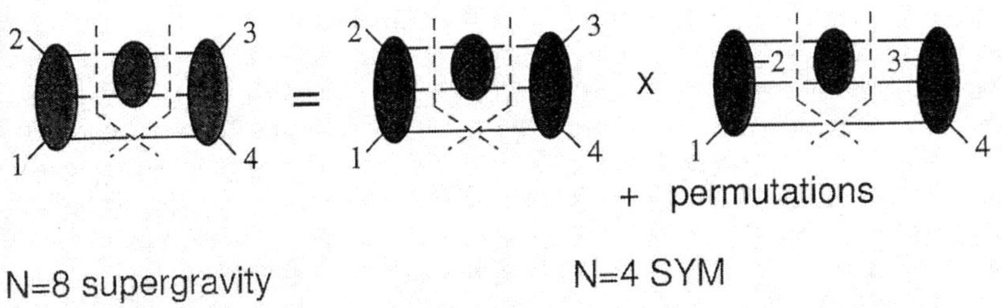

Figure 7: Evaluation of a generalized cut in $\mathcal{N}=8$ supergravity at three loops, in terms of planar and non-planar cuts in $\mathcal{N}=4$ SYM.

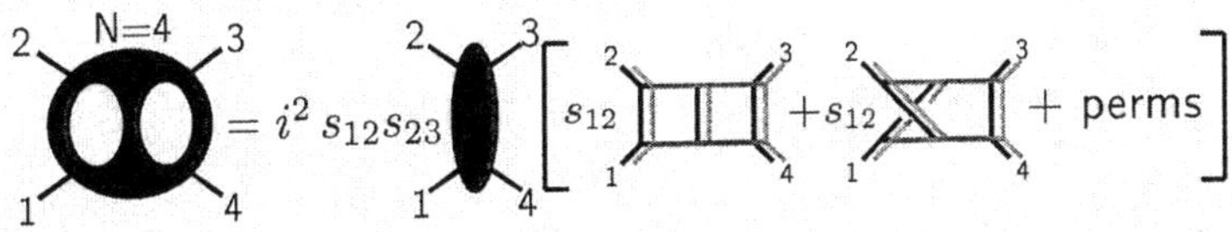

Figure 8: The two-loop amplitude in $\mathcal{N}=4$ SYM. The blob on the right represents the color-ordered tree amplitude A_4^{tree}. In the the brackets, black lines are kinematic $1/p^2$ propagators, with scalar (ϕ^3) vertices. Green lines are color δ^{ab} propagators, with structure constant (f^{abc}) vertices.

five-point amplitudes carries the same ordering, as shown in the left copy, but the other one is twisted, leading to the right copy. A reflection symmetry under the permutation $(1 \leftrightarrow 4,\ 2 \leftrightarrow 3)$ is preserved by this representation. The two-fold permutation sum in M_5^{tree} in eq. (50) leads to a four-fold permutation sum in the figure; one must add the permutations $(1 \leftrightarrow 2)$, $(3 \leftrightarrow 4)$ and $(1 \leftrightarrow 2,\ 3 \leftrightarrow 4)$.

KLT copying has a simple consequence for terms that are detected in the maximal cuts, because of the simple relation between gravity and gauge three-point amplitudes, eq. (16): the numerator factors are simply squared.

For example, as mentioned above, the maximal cuts capture the full two-loop four-point amplitude in $\mathcal{N}=4$ SYM, which is given by

$$\begin{aligned} A_4^{\text{2-loop}} &= -s_{12}s_{23}A_4^{\text{tree}}\Big[C_{1234}^{\text{P}}\, s_{12}\,\mathcal{I}_4^{\text{2-loop,P}}(s_{12},s_{23}) + C_{1234}^{\text{NP}}\, s_{12}\,\mathcal{I}_4^{\text{2-loop,NP}}(s_{12},s_{23}) \\ &\qquad + \mathcal{P}(2,3,4)\Big], \end{aligned} \tag{51}$$

where $\mathcal{I}_4^{\text{2-loop,(P,NP)}}$ are the scalar planar and non-planar double box integrals shown in fig. 8, and $C_{1234}^{\text{(P,NP)}}$ are color factors constructed from structure constant vertices, with the same graphical structure as the corresponding integral. The quantity $s_{12}s_{23}A_4^{\text{tree}}$ is totally symmetric under gluon interchange, and its square is the R^4 matrix element (3), up to a factor of i. Therefore squaring the prefactors in eq. (51) (and removing the color factors, as appropriate for gravity) gives the complete two-loop four-point amplitude in

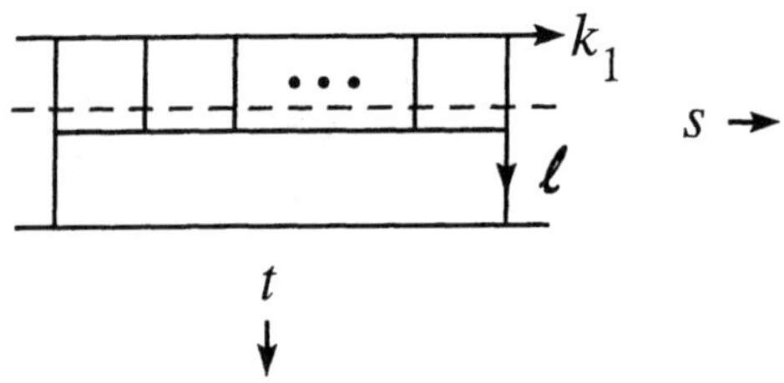

Figure 9: Example of a rung-rule contribution with worse ultraviolet behavior for $\mathcal{N}=8$ supergravity than for $\mathcal{N}=4$ SYM. The dashed line shows a unitarity cut which exposes a one-loop $(L+2)$-point amplitude that violates the no-triangle property.

$\mathcal{N}=8$ supergravity,

$$\begin{aligned} M_4^{\text{2-loop}} &= -i(s_{12}s_{23}A_4^{\text{tree}})^2\Big[s_{12}^2\,\mathcal{I}_4^{\text{2-loop,P}}(s_{12},s_{23}) + s_{12}^2\,\mathcal{I}_4^{\text{2-loop,NP}}(s_{12},s_{23}) + \mathcal{P}(2,3,4)\Big] \\ &= s_{12}s_{23}s_{13}M_4^{\text{tree}}\Big[s_{12}^2\,\mathcal{I}_4^{\text{2-loop,P}}(s_{12},s_{23}) + s_{12}^2\,\mathcal{I}_4^{\text{2-loop,NP}}(s_{12},s_{23}) + \mathcal{P}(2,3,4)\Big]. \end{aligned} \tag{52}$$

Because the loop integrals appearing in the two amplitudes, eqs. (51) and (52), are precisely the same, the critical dimension D_c is automatically the same for both theories at two loops. This value is $D_c = 7$, the dimension in which the two-loop, seven-propagator integrals, $\sim \int d^{2D}\ell/(\ell^2)^7$, are log divergent. As mentioned in section 2, the divergence is associated with a counterterm of the form $\mathcal{D}^4R^4$ in $D=7$.

Another class of diagrams detectable by maximal cuts are the L-loop ladder diagrams. Their numerator factors are given by the rung rule,

$$stA_4^{\text{tree}} \times s^{L-1} \qquad (\mathcal{N}=4 \text{ SYM}), \tag{53}$$
$$stuM_4^{\text{tree}} \times s^{2(L-1)} \qquad (\mathcal{N}=8 \text{ supergravity}), \tag{54}$$

where the second result was obtained by squaring the first one. The extra factor of s per loop for gravity corresponds, in the semi-classical high-energy limit $s \gg t$, to the fact that the gravitational "charge" is the energy in the center of mass, and it appears squared for each additional rung exchange, That is, g^2 in gauge theory is replaced by $\kappa^2 E_{\text{CM}}^2 = \kappa^2 s$ in gravity.

If one now takes an $(L-1)$-loop ladder diagram, and sews two opposite ends together, one gets the L-loop graph shown in fig. 9. The respective rung-rule numerator factors are now

$$stA_4^{\text{tree}} \times t \times [(\ell+k_1)^2]^{L-2} \qquad (\mathcal{N}=4 \text{ SYM}), \tag{55}$$
$$stuM_4^{\text{tree}} \times t^2 \times [(\ell+k_1)^2]^{2(L-2)} \qquad (\mathcal{N}=8 \text{ supergravity}). \tag{56}$$

The extra factors of the external momentum invariant s in the ladder graph have become factors of $(\ell+k_1)^2$, where ℓ is the loop momentum shown in the figure. This particular contribution, with $3L+1$ propagators, behaves in the ultraviolet as

$$\mathcal{I} \sim \int d^{DL}\ell \, \frac{(\ell^2)^{p(L-2)}}{(\ell^2)^{3L+1}}, \tag{57}$$

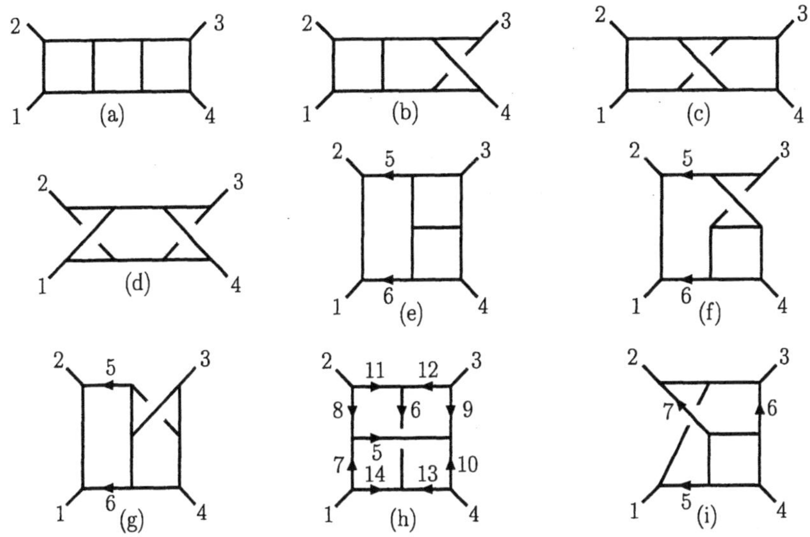

Figure 10: The different cubic four-point graphs in terms of which four-point three-loop amplitudes may be expressed.

which is worse for $\mathcal{N}=8$ supergravity ($p=2$) than for $\mathcal{N}=4$ SYM ($p=1$). The critical dimension for this integral satisfies $D_C(L)\times L/2+p(L-2)=3L+1$. For $p=1$, this gives the standard answer for $\mathcal{N}=4$ SYM, $D_c(L)=4+6/L$, eq. (8). For $p=2$, if there are no additional cancellations, it gives $D_c(L)=2+10/L$, eq. (7), suggesting a divergence in $D=4$ at $L=5$.

However, there were reasons to expect additional cancellations beginning at $L=3$ [32], which were revealed by a particular cut of fig. 9, shown as the dashed line slicing through all the rungs of the ladder. This L-particle cut has a one-loop $(L+2)$-point amplitude on one side of it with $4(L-2)$ factors of L in the numerator. This one-loop integral would have generated, after reduction to a basis of scalar integrals, nonvanishing coefficients for triangle integrals, in contradiction with the no-triangle hypothesis [31]. This hypothesis was a known fact [29] already for the five-point amplitude encountered at $L=3$, and would later be proved for an arbitrary number of legs [76, 77]. Therefore other graphs had to combine to cancel its large ℓ behavior, integrals that were not detectable in the two-particle cuts. This result inspired the computation of the complete four-graviton amplitude in $\mathcal{N}=8$ supergravity at three, and later four, loops.

5 $\mathcal{N}=8$ supergravity at three loops

Fig. 10 shows the nine cubic four-point graphs at three loops that do not contain three-point subdiagrams. Both the $\mathcal{N}=4$ SYM and $\mathcal{N}=8$ supergravity amplitudes can be described by giving the loop-momentum numerator polynomials $N^{(x)}$ for these graphs. (A recent, alternate representation [58] makes use of the graphs with three-point subdiagrams, but they are not necessary.) In addition, the $\mathcal{N}=4$ SYM graphs are to be multiplied by the corresponding color structure, as in fig. 8. Notice that seven of the nine

Table 2: The numerator factors $N^{(x)}$ for the integrals $I^{(x)}$ in fig. 10 for $\mathcal{N}=4$ super-Yang-Mills theory. The first column labels the integral, the second column the relative numerator factor. An overall factor of $s_{12}s_{23}A_4^{\rm tree}$ has been removed. The invariants s_{ij} and τ_{ij} are defined in eq. (58).

Integral $I^{(x)}$	$N^{(x)}$ for $\mathcal{N}=4$ super-Yang-Mills
(a)–(d)	s_{12}^2
(e)–(g)	$s_{12}\, s_{46}$
(h)	$s_{12}(\tau_{26}+\tau_{36})+s_{23}(\tau_{15}+\tau_{25})+s_{12}s_{23}$
(i)	$s_{12}s_{45}-s_{23}s_{46}-\frac{1}{3}(s_{12}-s_{23})\ell_7^2$

Table 3: Numerator factors $N^{(x)}$ for $\mathcal{N}=8$ supergravity. The first column labels the integral, the second column the relative numerator factor. An overall factor of $s_{12}s_{13}s_{14}M_4^{\rm tree}$ has been removed.

Integral $I^{(x)}$	$N^{(x)}$ for $\mathcal{N}=8$ supergravity
(a)–(d)	$[s_{12}^2]^2$
(e)–(g)	$s_{12}^2\,\tau_{35}\,\tau_{46}$
(h)	$(s_{12}(\tau_{26}+\tau_{36})+s_{23}(\tau_{15}+\tau_{25})+s_{12}s_{23})^2$ $+\,(s_{12}^2(\tau_{26}+\tau_{36})-s_{23}^2(\tau_{15}+\tau_{25}))(\tau_{17}+\tau_{28}+\tau_{39}+\tau_{4,10})$ $+\,s_{12}^2(\tau_{17}\tau_{28}+\tau_{39}\tau_{4,10})+s_{23}^2(\tau_{28}\tau_{39}+\tau_{17}\tau_{4,10})+s_{13}^2(\tau_{17}\tau_{39}+\tau_{28}\tau_{4,10})$
(i)	$(s_{12}\,\tau_{45}-s_{23}\,\tau_{46})^2-\tau_{27}(s_{12}^2\,\tau_{45}+s_{23}^2\,\tau_{46})-\tau_{15}(s_{12}^2\,\tau_{47}+s_{13}^2\,\tau_{46})$ $-\,\tau_{36}(s_{23}^2\,\tau_{47}+s_{13}^2\,\tau_{45})+\ell_5^2\,s_{12}^2\,s_{23}+\ell_6^2\,s_{12}\,s_{23}^2-\frac{1}{3}\ell_7^2\,s_{12}\,s_{13}\,s_{23}$

graphs, (a)–(g), have iterated two-particle cuts, so they can be computed via the rung rule. Only (h) and (i) require additional work.

Table 2 gives the values of $N^{(x)}$ for $\mathcal{N}=4$ SYM in terms of the following invariants,

$$\begin{aligned} s_{ij} &= (k_i+k_j)^2\,, & & & (i,j\le 4) \\ s_{ij} &= (k_i+\ell_j)^2\,, & \tau_{ij} &= 2k_i\cdot\ell_j\,, & (i\le 4,\quad j\ge 5) \\ s_{ij} &= (\ell_i+\ell_j)^2\,. & & & (i,j\ge 5) \end{aligned} \qquad (58)$$

Note that s_{ij} is quadratic in the loop momenta ℓ_j, if $j>4$, but τ_{ij} is linear. Every $N^{(x)}$ is manifestly quadratic in the loop momenta, consistent with $D_c(L)=4+6/L$ for $L=3$. It is easy to verify the numerators for graphs (a)–(g) using the rung rule.

Table 3 gives the values of $N^{(x)}$ for $\mathcal{N}=8$ supergravity, in a form [27] which is also manifestly quadratic in the loop momenta. (In an earlier version of the amplitude [26], the quadratic nature was not yet manifest.) Comparing the two sets of numerators, we see that the $\mathcal{N}=8$ supergravity ones are the squares of the $\mathcal{N}=4$ SYM ones, up to contact terms, as expected from the KLT relations. For example, in graphs (e)–(g), $s_{46}=\tau_{46}+\ell_6^2=\tau_{35}+\ell_5^2$, so $s_{12}^2\tau_{35}\tau_{46}\approx[s_{12}s_{46}]^2$ (modulo ℓ_i^2 terms).

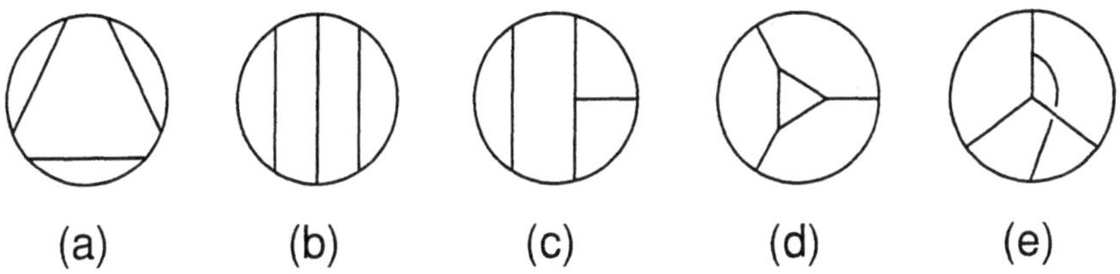

Figure 11: Cubic vacuum graphs at four loops.

Because the numerator factors for both $\mathcal{N}=8$ supergravity and $\mathcal{N}=4$ SYM are manifestly quadratic in the loop momenta, the critical dimension D_c at three loops should continue to obey eq. (8), *i.e.* $D_c = 6$. Indeed, when the ultraviolet poles in the nine integrals for $\mathcal{N}=8$ supergravity are evaluated, no further cancellation is found, and the resulting pole is

$$M_4^{(3),D=6-2\epsilon}\Big|_{\text{pole}} = \frac{1}{\epsilon}\frac{5\zeta_3}{(4\pi)^9}(s_{12}s_{23}s_{13})^2 M_4^{\text{tree}}, \tag{59}$$

corresponding to a counterterm of the form $\mathcal{D}^6R^4$ in $D=6$. (The form of this divergence was recently reproduced from duality arguments in string theory [17]; however, the rational number does not agree with eq. (59) as of this writing. Whether or not this indicates an issue in decoupling massive states [22] remains unclear.)

6 $\mathcal{N}=8$ supergravity at four loops

The general strategy [28, 78] used to compute the four-loop four-point amplitude in $\mathcal{N}=8$ supergravity is the same as at three loops; however, the bookkeeping issues are considerably greater. One can start by classifying the cubic vacuum graphs. At three loops there were only two; at four loops there are five, shown in fig. 11.

The next step is to decorate the five vacuum graphs with four external legs to get the cubic four-point graphs. As at lower loops, graphs containing triangles (three propagators or fewer on a loop) or other three point subgraphs can be dropped. Fig. 11(a) only gives rise to triangle-containing graphs, so it can be neglected henceforth. Altogether there are 50 cubic four-point graphs with nonvanishing numerators. Graphs (b) and (c) do generate no-triangle four-point graphs, but the numerators for all such graphs can be determined, up to possible contact terms, using the rung rule. For this reason, their associated numerator polynomials are the simplest. Graphs (d), and particularly (e), give rise to the most complex numerators.

The numerators are first determined for $\mathcal{N}=4$ SYM using the method of maximal cuts. At four loops the maximal cuts have 13 cut conditions $\ell_i^2=0$. Then near-maximal cuts with only 12 cut conditions are considered, followed by ones with 11. At this point the $\mathcal{N}=4$ SYM ansatz is complete; no more terms need to be added. The result is verified by comparison against many of the generalized cuts with real momenta, shown in fig. 12. Cases (a) and (i) involve six- and seven-point next-to-MHV amplitudes, for which the sums over super-multiplets crossing the cuts are more intricate than when all amplitudes are MHV. These cuts were evaluated using compact results for the super-sums

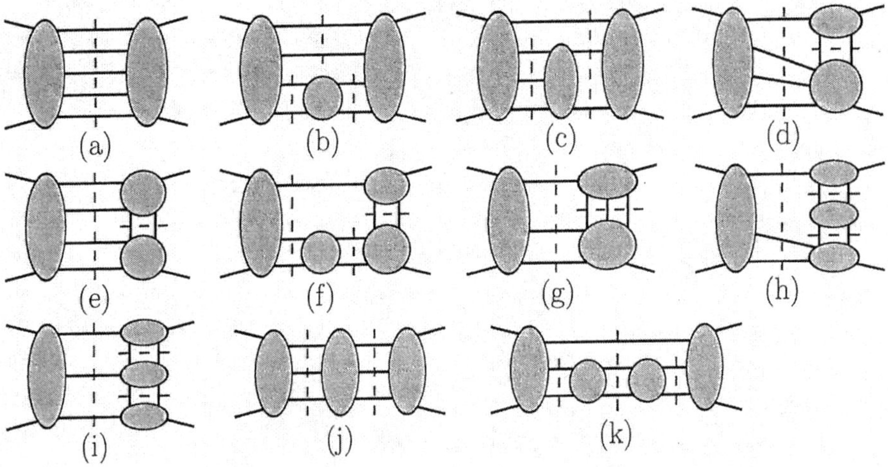

Figure 12: Generalized cuts which, along with iterated two-particle cuts, suffice to determine any massless four-point four-loop amplitude.

obtained by Elvang, Freedman and Kiermaier [79].

The 50 numerator polynomials for $\mathcal{N} = 8$ supergravity are then constructed using information provided by the KLT relations. The results are quite lengthy, but are provided as MATHEMATICA readable files in ref. [28], along with some tools for manipulating them.

From the numerator polynomials, the ultraviolet behavior of the amplitude can be extracted. One has to expand the integrals in the limit of small external momenta, relative to the loop momenta [80]. If the ultraviolet behavior is manifest, as at three loops with the representation found in ref. [27], then only the first term in this expansion is required. At four loops, it was necessary to go to third order to see all observed cancellations. More concretely, the numerator polynomials, omitting an overall factor of $stuM_4^{\text{tree}}$, have a mass dimension of 12, *i.e.* each term is of the form $k^{12-m}\ell^m$, where k and ℓ stand respectively for external and loop momenta. The maximum value of m turns out to be 8, for every integral. The integrals all have 13 propagators, so they have the form $\mathcal{I} \sim \int d^{4D}\ell\, \ell^{8-26}$. The amplitude is manifestly finite in $D = 4$, because $4\times 4+8-26 < 0$. (This result is not unexpected, given the absence of a $\mathcal{D}^2R^4$ counterterm [52, 53].) The amplitude is not manifestly finite in $D = 5$; to see that requires cancellation of the $k^4\ell^8$, $k^5\ell^7$ and $k^6\ell^6$ terms, after expansion around small k.

The cancellation of the $k^4\ell^8$ terms is relatively simple; one can set the external momenta k_i to zero, and collect the coefficients of the two resulting vacuum graphs, (d) and (e) in fig. 11, observing that the (d) and (e) coefficients both vanish. The cancellation of the $k^5\ell^7$ terms (and the $k^7\ell^5$ terms) is trivial: Using dimensional regularization, with no dimensionful parameter, Lorentz invariance does not allow an odd-power divergence. The most intricate cancellation is of the $k^6\ell^6$ terms, corresponding to the vanishing of the coefficient of the potential counterterm $\mathcal{D}^6R^4$ in $D = 5$. In the expansion of the integrals to the second subleading order as $k_i \to 0$, 30 different four-loop vacuum integrals are generated. However, there are consistency relations between the integrals, corresponding to the ability to shift the loop momenta by external momenta before expanding around $k_i = 0$. These consistency relations are powerful enough to imply the cancellation of the ultraviolet pole in $D = 5 - 2\epsilon$. As a check, we evaluated all the ultraviolet poles directly, with the same conclusion.

In summary, the four-loop four-point amplitude of $\mathcal{N} = 8$ supergravity is ultraviolet finite for $D < 11/2$ [28], the same critical dimension as for $\mathcal{N} = 4$ super-Yang-Mills theory. Finiteness in $5 \leq D < 11/2$ is a consequence of nontrivial cancellations, beyond those already found at three loops [26, 27]. These results provide the strongest direct support to date for the possibility that $\mathcal{N} = 8$ supergravity might be a perturbatively finite quantum theory of gravity.

7 Conclusions and Outlook

What are the prospects for going beyond four loops? As mentioned in the Introduction, there are indications of a problem [17, 18, 19] with the pure spinor formalism [15] at five loops, which could lead to a deviation from the critical dimension formula (8) at this order. It would be very interesting to clarify the situation by computing the complete amplitude. This would be a major undertaking, but recent developments might provide some assistance [57, 58, 56]. It is also worth noting that the unitarity method generically implies sets of higher-loop cancellations as a consequence of lower-loop ones [32], by finding the latter contributions as subgraphs exposed by unitarity cuts. The fact that certain higher-loop cancellations were implied by the no-triangle behavior of one-loop n-point amplitudes was discussed in section 4, but the excellent ultraviolet behavior of the two-, three- and four-loop four-point amplitudes has similar implications. Combining this information properly is still an issue, however.

The unitarity method works with on-shell objects, so it can maintain more supersymmetry than is generally possible with off-shell Feynman diagrams. However, not all symmetries of $\mathcal{N} = 8$ supergravity are kept manifest in the current approach. In particular the theory has a continuous noncompact coset symmetry, $E_{7(7)}$ [1]. The 70 massless scalars parametrize the coset space $E_{7(7)}/SU(8)$, and the non-$SU(8)$ part of the symmetry is realized nonlinearly. When using the KLT relations, there is an $SU(4)$ R symmetry associated with each $\mathcal{N} = 4$ supersymmetry, so only an $SU(4) \times SU(4)$ subgroup of the linearly-realized $SU(8)$ is kept manifest, and none of the $E_{7(7)}/SU(8)$. There has been some recent progress on the role of $E_{7(7)}$ in $\mathcal{N} = 8$ supergravity: using it to construct terms in the light-cone superspace Hamiltonian [81]; describing its explicit action on covariant fields [82]; exploring its implications for tree-level S matrix elements [83, 77, 84], and also at loop level [85, 86]; and assessing the $E_{7(7)}$ invariance of the R^4 counterterm [87]. However, a deeper understanding of the implications of $E_{7(7)}$ invariance would certainly be welcome.

Suppose that $\mathcal{N} = 8$ supergravity is finite to all loop orders. This still would not prove that it is a nonperturbatively consistent theory of quantum gravity. There are at least two reasons to think that it might need a nonperturbative ultraviolet completion:

1. The (likely) $L!$ or worse growth of the coefficients of the order L terms in the perturbative expansion, which for fixed-angle scattering, means a non-convergent behavior $\sim L!\,(s/M_{\rm Pl}^2)^L$.

2. The fact that the perturbative series seems to be $E_{7(7)}$ invariant, while the mass

spectrum of black holes is non-invariant (see *e.g.* ref. [88] for recent discussions).

QED is an example of a perturbatively well-defined theory that needs an ultraviolet completion; it also has factorial growth in its perturbative coefficients, $\sim L!\,\alpha^L$, due to ultraviolet renormalons associated with the Landau pole. Of course, for small values of α, QED works extremely well; for example, it predicts the anomalous magnetic moment of the electron to 10 digits of accuracy. Also, we know of many pointlike nonperturbative ultraviolet completions for QED, namely asymptotically free grand unified theories. Are there any imaginable pointlike completions for $\mathcal{N}=8$ supergravity? Maybe the only completion is string theory; or maybe this cannot happen because of the impossibility of decoupling nonperturbative states [22].

In any event, it is clear that the remarkably good perturbative ultraviolet behavior of $\mathcal{N}=8$ supergravity has provided many surprises to date. Although the theory may not be of direct phenomenological interest, perhaps it could some day point the way to other, more realistic finite theories. As a "toy model" for a pointlike theory of quantum gravity, it has been extremely instructive, and is still well worth exploring further.

Acknowledgments

I am grateful to Zvi Bern, John Joseph Carrasco, Henrik Johansson, David Kosower and Radu Roiban for collaboration on many of the topics described here, and to Zvi Bern for comments on the manuscript. I thank the organizers of the 47th International School of Subnuclear Physics for the opportunity to present these lectures in such a stimulating and pleasant environment. This work was supported by the US Department of Energy under contract DE-AC02-76SF00515.

References

[1] E. Cremmer and B. Julia, Phys. Lett. B **80**, 48 (1978); Nucl. Phys. B **159**, 141 (1979).

[2] B. de Wit and D. Z. Freedman, Nucl. Phys. B **130**, 105 (1977).

[3] E. Cremmer, B. Julia and J. Scherk, Phys. Lett. B **76**, 409 (1978).

[4] S. Weinberg, in *Understanding the Fundamental Constituents of Matter*, ed. A. Zichichi (Plenum Press, New York, 1977); S. Weinberg, in *General Relativity*, eds. S. W. Hawking and W. Israel (Cambridge University Press, 1979), p. 700;
M. Niedermaier and M. Reuter, Living Rev. Rel. **9**, 5 (2006).

[5] P. Hořava, Phys. Rev. D **79**, 084008 (2009) [0901.3775 [hep-th]].

[6] S. Deser, J. H. Kay and K. S. Stelle, Phys. Rev. Lett. **38**, 527 (1977).

[7] S. Ferrara and B. Zumino, Nucl. Phys. B **134**, 301 (1978).

[8] S. Deser and J. H. Kay, Phys. Lett. B **76**, 400 (1978).

[9] P. S. Howe and U. Lindström, Nucl. Phys. B **181**, 487 (1981).

[10] R. E. Kallosh, Phys. Lett. B **99**, 122 (1981).

[11] Z. Bern, L. J. Dixon, D. C. Dunbar, M. Perelstein and J. S. Rozowsky, Nucl. Phys. B **530**, 401 (1998) [hep-th/9802162].

[12] P. S. Howe and K. S. Stelle, Phys. Lett. B **554**, 190 (2003) [hep-th/0211279].

[13] G. Bossard, P. S. Howe and K. S. Stelle, Gen. Rel. Grav. **41**, 919 (2009) [0901.4661 [hep-th]].

[14] G. Bossard, P. S. Howe and K. S. Stelle, Phys. Lett. B **682**, 137 (2009) [0908.3883 [hep-th]].

[15] N. Berkovits, Phys. Rev. Lett. **98**, 211601 (2007) [hep-th/0609006].

[16] M. B. Green, J. G. Russo and P. Vanhove, Phys. Rev. Lett. **98**, 131602 (2007) [hep-th/0611273].

[17] M. B. Green, J. G. Russo and P. Vanhove, 1002.3805 [hep-th].

[18] P. Vanhove, 1004.1392 [hep-th].

[19] J. Björnsson and M. B. Green, 1004.2692 [hep-th].

[20] R. Kallosh, Phys. Rev. D **80**, 105022 (2009) [0903.4630 [hep-th]].

[21] G. Chalmers, hep-th/0008162;
M. B. Green, J. G. Russo and P. Vanhove, JHEP **0702**, 099 (2007) [hep-th/0610299].

[22] M. B. Green, H. Ooguri and J. H. Schwarz, Phys. Rev. Lett. **99**, 041601 (2007) [0704.0777 [hep-th]].

[23] Z. Bern, J. J. M. Carrasco, H. Johansson, to appear in the Proceedings of the 46th International School of Subnuclear Physics [0902.3765 [hep-th]].

[24] Z. Bern, L. J. Dixon, D. C. Dunbar and D. A. Kosower, Nucl. Phys. B **425**, 217 (1994) [hep-ph/9403226]; Nucl. Phys. B **435**, 59 (1995) [hep-ph/9409265].

[25] H. Kawai, D. C. Lewellen and S. H. H. Tye, Nucl. Phys. B **269**, 1 (1986).

[26] Z. Bern, J. J. Carrasco, L. J. Dixon, H. Johansson, D. A. Kosower and R. Roiban, Phys. Rev. Lett. **98**, 161303 (2007) [hep-th/0702112].

[27] Z. Bern, J. J. M. Carrasco, L. J. Dixon, H. Johansson and R. Roiban, Phys. Rev. D **78**, 105019 (2008) [0808.4112 [hep-th]].

[28] Z. Bern, J. J. Carrasco, L. J. Dixon, H. Johansson and R. Roiban, Phys. Rev. Lett. **103**, 081301 (2009) [0905.2326 [hep-th]].

[29] Z. Bern, L. J. Dixon, M. Perelstein and J. S. Rozowsky, Nucl. Phys. B **546**, 423 (1999) [hep-th/9811140].

[30] Z. Bern, N. E. J. Bjerrum-Bohr and D. C. Dunbar, JHEP **0505**, 056 (2005) [hep-th/0501137];
N. E. J. Bjerrum-Bohr, D. C. Dunbar and H. Ita, Phys. Lett. B **621**, 183 (2005) [hep-th/0503102].

[31] N. E. J. Bjerrum-Bohr, D. C. Dunbar, H. Ita, W. B. Perkins and K. Risager, JHEP **0612**, 072 (2006) [hep-th/0610043].

[32] Z. Bern, L. J. Dixon and R. Roiban, Phys. Lett. B **644**, 265 (2007) [hep-th/0611086].

[33] S. Mandelstam, Nucl. Phys. B **213**, 149 (1983);
P. S. Howe, K. S. Stelle and P. K. Townsend, Nucl. Phys. B **214**, 519 (1983);
L. Brink, O. Lindgren and B. E. W. Nilsson, Phys. Lett. B **123**, 323 (1983).

[34] G. 't Hooft and M. J. G. Veltman, Annales Poincare Phys. Theor. A **20**, 69 (1974).

[35] M. H. Goroff and A. Sagnotti, Phys. Lett. B **160**, 81 (1985); Nucl. Phys. B **266**, 709 (1986).

[36] A. E. M. van de Ven, Nucl. Phys. B **378**, 309 (1992).

[37] M. T. Grisaru, Phys. Lett. B **66**, 75 (1977).

[38] E. Tomboulis, Phys. Lett. B **67**, 417 (1977).

[39] P. van Nieuwenhuizen and C. C. Wu, J. Math. Phys. **18**, 182 (1977).

[40] M. T. Grisaru, H. N. Pendleton and P. van Nieuwenhuizen, Phys. Rev. D **15**, 996 (1977);
M. T. Grisaru and H. N. Pendleton, Nucl. Phys. B **124**, 81 (1977).

[41] J. H. Schwarz, Phys. Rept. **89**, 223 (1982).

[42] I. Robinson, unpublished;
L. Bel, Acad. Sci. Paris, Comptes Rend. **247**, 1094 (1958) and **248**, 1297 (1959).

[43] D. J. Gross and E. Witten, Nucl. Phys. B **277**, 1 (1986).

[44] M. B. Green, J. H. Schwarz and L. Brink, Nucl. Phys. B **198**, 474 (1982).

[45] Z. Bern, J. S. Rozowsky and B. Yan, Phys. Lett. B **401**, 273 (1997) [hep-ph/9702424].

[46] N. Marcus and A. Sagnotti, Nucl. Phys. B **256**, 77 (1985).

[47] A. Galperin, E. Ivanov, S. Kalitsyn, V. Ogievetsky and E. Sokatchev, Class. Quant. Grav. **1**, 469 (1984); Class. Quant. Grav. **2**, 155 (1985).

[48] Z. Bern, L. J. Dixon and V. A. Smirnov, Phys. Rev. D **72**, 085001 (2005) [hep-th/0505205].

[49] Z. Bern, M. Czakon, L. J. Dixon, D. A. Kosower and V. A. Smirnov, Phys. Rev. D **75**, 085010 (2007) [hep-th/0610248].

[50] Z. Bern, J. J. M. Carrasco, H. Johansson and D. A. Kosower, Phys. Rev. D **76**, 125020 (2007) [0705.1864 [hep-th]].

[51] H. Elvang, D. Z. Freedman and M. Kiermaier, 1003.5018 [hep-th].

[52] J. M. Drummond, P. J. Heslop, P. S. Howe and S. F. Kerstan, JHEP **0308**, 016 (2003) [hep-th/0305202].

[53] R. Kallosh, JHEP **0909**, 116 (2009) [0906.3495 [hep-th]].

[54] L. J. Dixon, in *QCD & Beyond: Proceedings of TASI '95*, ed. D. E. Soper (World Scientific, 1996) [hep-ph/9601359].

[55] Z. Bern, A. De Freitas and H. L. Wong, Phys. Rev. Lett. **84**, 3531 (2000) [hep-th/9912033];
N. E. J. Bjerrum-Bohr and K. Risager, Phys. Rev. D **70**, 086011 (2004) [hep-th/0407085];
N. E. J. Bjerrum-Bohr and O. T. Engelund, 1002.2279 [hep-th].

[56] Z. Bern, T. Dennen, Y. t. Huang and M. Kiermaier, 1004.0693 [hep-th].

[57] Z. Bern, J. J. M. Carrasco and H. Johansson, Phys. Rev. D **78**, 085011 (2008) [0805.3993 [hep-ph]].

[58] Z. Bern, J. J. M. Carrasco and H. Johansson, 1004.0476 [hep-th].

[59] J. M. Maldacena, Adv. Theor. Math. Phys. **2**, 231 (1998) [Int. J. Theor. Phys. **38**, 1113 (1999)] [hep-th/9711200];
S. S. Gubser, I. R. Klebanov and A. M. Polyakov, Phys. Lett. B **428**, 105 (1998) [hep-th/9802109];
O. Aharony, S. S. Gubser, J. M. Maldacena, H. Ooguri and Y. Oz, Phys. Rept. **323**, 183 (2000) [hep-th/9905111].

[60] L. D. Landau, Nucl. Phys. **13**, 181 (1959);
S. Mandelstam, Phys. Rev. **115**, 1741 (1959);
R. E. Cutkosky, J. Math. Phys. **1**, 429 (1960).

[61] Z. Bern and A. G. Morgan, Nucl. Phys. B **467**, 479 (1996) [hep-ph/9511336];
Z. Bern, L. J. Dixon, D. C. Dunbar and D. A. Kosower, Phys. Lett. B **394**, 105 (1997) [hep-th/9611127].

[62] Z. Bern, L. J. Dixon and D. A. Kosower, JHEP **0001**, 027 (2000) [hep-ph/0001001];
JHEP **0408**, 012 (2004) [hep-ph/0404293].

[63] Z. Bern, L. J. Dixon and D. A. Kosower, Nucl. Phys. B **513**, 3 (1998) [hep-ph/9708239].

[64] R. Britto, F. Cachazo and B. Feng, Nucl. Phys. B **715**, 499 (2005) [hep-th/0412308];
R. Britto, F. Cachazo, B. Feng and E. Witten, Phys. Rev. Lett. **94**, 181602 (2005) [hep-th/0501052].

[65] Z. Bern, L. J. Dixon and D. A. Kosower, Phys. Rev. D **73**, 065013 (2006) [hep-ph/0507005];
C. F. Berger, Z. Bern, L. J. Dixon, D. Forde and D. A. Kosower, Phys. Rev. D **74**, 036009 (2006) [hep-ph/0604195].

[66] G. Ossola, C. G. Papadopoulos and R. Pittau, Nucl. Phys. B **763**, 147 (2007) [hep-ph/0609007]; JHEP **0803**, 042 (2008) [0711.3596 [hep-ph]]; JHEP **0805**, 004 (2008) [0802.1876 [hep-ph]].

[67] R. K. Ellis, W. T. Giele and Z. Kunszt, JHEP **0803**, 003 (2008) [0708.2398 [hep-ph]];
W. T. Giele, Z. Kunszt and K. Melnikov, JHEP **0804**, 049 (2008) [0801.2237 [hep-ph]];
W. T. Giele and G. Zanderighi, JHEP **0806**, 038 (2008) [0805.2152 [hep-ph]].

[68] C. F. Berger *et al.*, Phys. Rev. D **78**, 036003 (2008) [0803.4180 [hep-ph]].

[69] C. F. Berger and D. Forde, 0912.3534 [hep-ph].

[70] R. J. Eden, P. V. Landshoff, D. I. Olive and J. C. Polkinghorne, *The Analytic S Matrix* (Cambridge University Press, 1966).

[71] Z. Bern, V. Del Duca, L. J. Dixon and D. A. Kosower, Phys. Rev. D **71**, 045006 (2005) [hep-th/0410224].

[72] R. Britto, F. Cachazo and B. Feng, Nucl. Phys. B **725**, 275 (2005) [hep-th/0412103].

[73] E. I. Buchbinder and F. Cachazo, JHEP **0511**, 036 (2005) [hep-th/0506126];
F. Cachazo and D. Skinner, 0801.4574 [hep-th];
F. Cachazo, 0803.1988 [hep-th];
F. Cachazo, M. Spradlin and A. Volovich, Phys. Rev. D **78**, 105022 (2008) [0805.4832 [hep-th]];
M. Spradlin, A. Volovich and C. Wen, Phys. Rev. D **78**, 085025 (2008) [0808.1054 [hep-th]].

[74] S. J. Parke and T. R. Taylor, Phys. Rev. Lett. **56**, 2459 (1986).

[75] W. Siegel, Phys. Lett. B **84**, 193 (1979);
Z. Bern and D. A. Kosower, Nucl. Phys. B **379**, 451 (1992);
Z. Bern, A. De Freitas, L. J. Dixon and H. L. Wong, Phys. Rev. D **66**, 085002 (2002) [hep-ph/0202271].

[76] N. E. J. Bjerrum-Bohr and P. Vanhove, JHEP **0804**, 065 (2008) [0802.0868 [hep-th]]; JHEP **0810**, 006 (2008) [0805.3682 [hep-th]]; Fortsch. Phys. **56**, 824 (2008) [0806.1726 [hep-th]].

[77] N. Arkani-Hamed, F. Cachazo and J. Kaplan, 0808.1446 [hep-th].

[78] Z. Bern, J. J. M. Carrasco, L. J. Dixon, H. Johansson and R. Roiban, to appear.

[79] H. Elvang, D. Z. Freedman and M. Kiermaier, JHEP **0904**, 009 (2009) [0808.1720 [hep-th]]; and private communication.

[80] A. A. Vladimirov, Theor. Math. Phys. **43**, 417 (1980) [Teor. Mat. Fiz. **43**, 210 (1980)];
N. Marcus and A. Sagnotti, Nuovo Cim. A **87**, 1 (1985).

[81] L. Brink, S. S. Kim and P. Ramond, JHEP **0806**, 034 (2008) [0801.2993 [hep-th]].

[82] R. Kallosh and M. Soroush, Nucl. Phys. B **801**, 25 (2008) [0802.4106 [hep-th]].

[83] M. Bianchi, H. Elvang and D. Z. Freedman, JHEP **0809**, 063 (2008) [0805.0757 [hep-th]].

[84] R. Kallosh and T. Kugo, JHEP **0901** (2009) 072 [0811.3414 [hep-th]].

[85] R. Kallosh, C. H. Lee and T. Rube, JHEP **0902** (2009) 050 [0811.3417 [hep-th]].

[86] S. He and H. Zhu, 0812.4533 [hep-th].

[87] J. Brödel and L. J. Dixon, JHEP **1005**, 003 (2010) [0911.5704 [hep-th]].

[88] M. Bianchi, S. Ferrara and R. Kallosh, 0910.3674 [hep-th]; JHEP **1003**, 081 (2010) [0912.0057 [hep-th]].

CHAIRMAN: L. J. DIXON

Scientific Secretaries: V. Pauk, A. Rebesh

DISCUSSION I

- *Burda:*

If I understood it properly, it is a long-standing dream to find out whether N=8 supergravity is finite or not. And so I have two questions. Now you are at the four loop level, correct? What will be the next step? Is there a general program, or idea of how we can finally prove finiteness or something like this? That is the first question. And the second one is: suppose we have proved that N=8 supergravity is finite. Then, how will this knowledge affect our understanding of nature?

- *Dixon:*

The simple answer is that the next step after 4 loops is 5 loops, but things definitely get more complicated. So, one possibility is to try to do the sort of things that we are doing, but simplify them to get just the divergent part of the amplitude. Right now we try to construct the complete amplitude. Then we get lots and lots of cross checks. However, only afterwards, as I explain more in the next lecture, we do extract the ultraviolet divergence. Maybe we can just limit ourselves to terms that would affect the divergences and that might simplify things. Ideally we would like to have some kind of understanding that is even simpler, and that allows us to go to high loop orders. There are some alternative approaches from string theory. Some of them are very different from field theory, but maybe some of them will give us some clues how to see cancellations more directly in field theory. But the answer is that we don't have a great plan for all loop orders at the moment. As to the second question: Suppose that the theory is finite to all orders in perturbation theory, then what? (Actually I have some comments on this topic in the next lecture, so I will probably defer most of this question until then.) I want to give you a warning that just because you have a finite perturbation theory, it doesn't mean that the perturbative series converges. If it doesn't converge then you can speculate on what nonperturbative properties you might need to add in order to complete the theory so that it converges.

- *Mastrolia:*

You said that you use maximal cuts, and also near-to-maximal cuts. Do you need to go further down in the number of cuts that you get or is it enough for the reconstruction of the loop amplitudes to do the near-to-maximal cuts?

- *Dixon:*

You have to go down two or three "layers" when constructing the amplitude, where each layer corresponds to an additional power of inverse propagators (factors of l_i^2). Let me remind you that the maximal cuts were defined as cuts that have only three-point vertices. The first layer below that, or the near-to-maximal cuts, allows for one power of an inverse propagator, which corresponds to collapsing one propagator in some maximal cut, or in other words merging two 3-point vertices into a single 4-point vertex. The next layer after that contains two collapsed propagators, leading to either two 4-point vertices or one 5-point vertex. The depth in layers that we go to in constructing the amplitude is simply the point at which, when we go to the next layer, we find we don't have to add any more terms, because the cuts are all consistent. Then we go on to check that our ansatz is consistent with the whole set of cuts, in particular those cuts which can be realized using real (not complex) momenta in the cuts.

- *Mastrolia:*

How do you distinguish the contributions coming from different "master integrals" which share the same set of denominators but have different numerators?

- *Dixon:*

First I would like to briefly explain the notion of "master integral" to people who haven't worked with it before. At one loop, in the numerator of a Feynman integral, the Lorentz product of a loop momentum l with any external momentum p_i can be written as a difference of two inverse propagators [$2l.p_i = l^2 - (l-p_i)^2 + p_i^2$]. The inverse propagators cancel Feynman propagators and reduce the integral to one with one fewer leg. For example, a hexagon with $l.p_i$ in the numerator can be reduced to a couple of scalar pentagons and a scalar hexagon. But when you go to the two-loop level you can no longer cancel all such tensors. There are numerator factors which are irreducible from the point of view of this kind of reduction. But that fact doesn't really matter too much to us. We are just trying to match the loop-momentum polynomials in the numerator against the cuts. In any case, there is always an ambiguity in how you represent the polynomials. For example, even at tree level if we took a 4-point tree graph with a propagator in the s-channel, and if it had a numerator polynomial of s, then they would cancel to give a 4-point contact interaction. But we could rewrite this interaction equivalently as a 4-point graph with a propagator in the t-channel, and with a numerator polynomial of t. Now you can plug this simple tree-level relation in at the loop level to get simple examples of the ambiguity I was mentioning. So there are always ambiguities, but as long as we find a solution that matches all the cuts, that is sufficient for our purposes.

- *Douglas:*

In supergravity the coupling is dimensionful, and the effective loop expansion parameter is E^2/M_{pl}^2 for a scattering at energy E. So the theory is strongly coupled at the Planck scale. How do we think about it?

- *Dixon:*

This is related to the question by P. Burda that I was going to defer. You are right, for fixed-angle scattering the expansion parameter is E^2/M_{pl}^2. The answer depends on how fast the coefficients of the perturbative expansion grow with the loop order. It is quite likely that this growth will be factorial, that is the typical thing you get. (In large N_c gauge theory, or planar graphs, it may be milder). Because of the factorial growth, the expansion will be asymptotic one. It should be very accurate at small energies, but we already know that the answer is incredibly tiny in this case. We are really interested in what happens as you come up in energy from well below the Planck scale. Is it possible to get any predictions at energies near, but below the Planck scale? In other words, can we resum the series even if it is an asymptotic one? In some cases the series may be asymptotic, but you can resum it by using, for example, a Borel resummation. Then you can get an estimate of the error by looking in the plane of the Borel parameter, and see where the first singularity is. If the first singularity is at the origin, then you cannot resum it at all. But basically that would be the goal, to try to understand how we can push up out of the perturbative regime, and what sorts of contributions that forces us to include. Do we get string theory back in the end? Or is there a kind of pointlike completion using nonperturbative states in a pointlike theory?

- *Smirnov:*

Starting from 2005, several results about the cancellation of the UV divergences in 4-dimensional SUGRA up to seven or even nine loops were suggested. What is the difference between your five-loop cancellation results and their works? Even if the UV cancellation were proved up to nine loops, to what extent can we believe in finiteness in all orders?

- *Dixon:*

Most of the papers you referred to discuss the constraints on potential counterterms that you can write down, are based on assumptions about what amount of supersymmetry can be maintained off shell, e.g. in superspace Feynman rules. The case of N=8 is complicated because you cannot maintain all the supersymmetries on shell. In gauge theory, wc know how to write down N=1 superspace and so we can write Feynman rules in superspace and conclude how supersymmetry restricts the available counterterms. We can also do N=2 superspace but there is no N=4 supersymmetry on-shell for gauge theory. So, you have to build your argument for N=4 super-Yang-Mills theory based on the low amount of supersymmetry, for example N=2, and this is usually enough. When you come to gravity you face a similar problem: you can't maintain the full N=8 off-shell, but (easily) only N=4. There are also arguments (due to Howe and Stelle) about how the counterterms can be further restricted by considering what happens when you go from D=4 up to higher dimensions. But in short, different assumptions about the superspace properties give you different conclusions about the loop order to which finiteness is protected. These arguments typically give finiteness through L=5 or L=6. The value of L=9 that you referred to comes from the Berkovits (pure spinor) formalism in string theory, as analyzed by Green, Russo and Vanhove. In this case, there is a question how you go from an argument in D=10 string theory, to D=4 field theory. The massless sector of the string is N=8 supergravity, but above it you have all these massive states. They come from string excitations, and from Kaluza-Klein modes from compactifying six dimensions. In principle there could be cancellations. That is, loops containing the massless modes could diverge, and loops containing the massive modes diverge, but the sum (in string theory) is finite. So far, however, things seem to agree, in that the critical dimension $D_c(L)$ for a divergence at L loops follows the same formula in both cases through four loops. The second part of your question was, what about beyond eight or nine loops? The answer is related to P. Burda's question about how we try to make these arguments to all orders, and right now we don't have the answer.

- *Douglas:*

Would it be any easier to prove the theory is finite in 3 dimensions?

- *Dixon:*

Yes, it might be. Two dimensions is even easier. But its gravitons (by themselves) aren't dynamical.

- *Douglas:*

But it is still a highly interacting theory. And in three dimensions it still has a dimensionful coupling.

- *Dixon:*

Yes, you are right. In the next lecture I will show you that when we construct the N=8 supergravity amplitudes, the generic term for L loops looks like it diverges in the critical dimension $D_c = 2+10/L$. On the other hand, in the N=4 case, the formula that we find at each loop order is $D_c = 4+6/L$. This second formula has the nice feature that it is always greater than 4. The supergravity one is always than greater than 2. So, if you just naively take the formula in 2 dimensions (it is naive because you can't define a general scattering angle in 2 dimensions), then maybe you could argue for finiteness to all orders. But in three dimensions, like four, you would still have to find additional kinds of cancellations in order to show finiteness to all orders.

- *Douglas:*

The three-dimensional gravity case will be more interesting and exciting than two dimensions.

- *Dixon:*

Yes, I agree, but we would still have to find the cancellation mechanism, because of the structure of the formula $D_c = 2 + 10/L$.

- *Davey:*

Can I just ask you a pretty basic question? In your personal opinion how likely is it that N=8 supergravity is finite: fifty-fifty or is that too optimistic?

- *Dixon:*

It either is finite or it isn't. So, I don't know how to assign a probability to it.

- *Davey:*

I'm just after your personal opinion.

- *Dixon:*

I am not really sure. However, my working attitude is that I consider it to be finite until the first time that I see it behave worse then N=4 super-Yang-Mills. So my default is that it is finite, but I am prepared to be reversed (and I always do a "blind analysis" at every loop order). So, my probability is "100%" right now, but I am prepared to change the odds at any time.

- *Korthals-Altes:*

So, you have these Kawai-Lewellen-Tye relations, and they are valid for any set of off-shell momenta, is that correct?

- *Dixon:*

No, the KLT relations are only valid for on-shell momenta. The external states all have to be on their mass shell. In principle they could be excited string states with masses greater then zero, but in our application we have always taken them to be massless states, so, they obey $k_i^2 = 0$.

- *Korthals-Altes:*

So, your method is really limited to computing on-shell amplitudes, right?

- *Dixon:*

Yes, we used the unitarity approach in order to use on-shell information, because that information has the most gauge independence. Also, you can typically maintain the most supersymmetry in intermediate steps: we can take advantage of supersymmetric relations between on-shell tree amplitudes.

- *Korthals-Altes:*

And once the theory is finite for on-shell amplitudes, you cannot expect anything more?

- *Dixon:*

Yes, I don't think you can expect much more than that. I think that all the real physical information is in the on-shell S-matrix. Maybe you have to factorize it in certain complicated ways to extract other physical information, but it should be possible to do so.

- *Zichichi:*

What is the probability that you are going to get a theorem establishing that N=8 supergravity is UV finite?

- *Dixon:*

It is lower then my confidence of 100% (at the moment) that it is finite. It is somewhat smaller then that. We need some new ideas in order to construct a proof.

CHAIRMAN: L. J. DIXON

Scientific Secretaries: V. Iglovikov, P.P. Mastrolia

DISCUSSION II

- *McLerran:*

Why don't you believe the result of Kallosh that the theory is finite in 7 loops?

- *Dixon:*

Kallosh's results are obtained by using light-cone gauge and contain singular denominators in intermediate steps, which might lead to singular denominators in the final answer that would invalidate the argument. Nevertheless at this stage I cannot say that her argument is wrong; it might well be right.

- *Mastrolia:*

You have discussed about the finiteness of the 4-graviton scattering up to the 4-loop order. What about the 5-graviton (and more) case? Are we guaranteed they would be finite?

- *Dixon:*

Because the 4-graviton scattering amplitude is now known through 4 loops, and is finite through that order, it is interesting to investigate the 5–graviton case. We don't expect any divergence in 5–graviton scattering until 4 loops. This is because from 3-loop 4-graviton scattering, the coefficient of R^4 is zero, and the next available type of counterterm, R^5, could only show up at 4 loops. However two recent papers, one by Howe and collaborators, and one by Kallosh, argue that R^5 cannot be made supersymmetric. This result can also be cross-checked by looking for R^5 in the expansion of the 5-graviton amplitude in string theory. Stieberger has found that it does not appear (and I recently checked his result). Accordingly I don't expect to see any divergence in 5-graviton scattering in four dimensions until at least the 5 loop order.

- *Korthals-Altes:*

You use dimensional regularization. Which variant did you use? You would like to respect supersymmetry without violating other sacrosanct features.

- *Dixon:*

There are two closely-related variants of dimensional regularization that preserve supersymmetry by keeping the number of bosonic and fermionic degrees of freedom the same. 1) Dimensional reduction (DRED), valid for $D<4$; 2) Four-dimensional helicity scheme (FHD), valid for $D>4$. We use the second version because we like to use the helicity formalism, and if $D<4$ we don't have two transverse dimensions with which to define helicity states. Also there are certain algebraic relations (involving the $D=4$ Levi-Civita tensor, for example), which are needed to maintain supersymmetry at higher loops that hold if $D>=4$, but are hard to maintain in $D<4$.

- *Cerchiai:*

Can you elaborate more on how the full E7(7) symmetry of N=8 D=4 SUGRA and the SU(8) R-symmetry could be used for your amplitudes?

- *Dixon:*

One possible application would be to try to use this symmetry to directly forbid counterterms, without even having to compute any amplitudes. (This is work in progress with J.

Broedel). The U-duality group for N=8 D=4 SUGRA is E7(7)[R] (over the real numbers). It gets broken to a discrete subgroup E7(7)[Z] (over the integers) when going to full string theory, or even just when including the masses of black hole states. That is to say, the masses of the heavy states depend on where you are in the coset space E7(7)[R]/SU(8)[R] in which the 70 scalars live. The first relevant term when going beyond N=8 D=4 SUGRA is proportional to R^4. If R^4 were not invariant under E7(7)[R], and if the E7(7)[R] is not spoiled by anomalies, then this would be a simple argument why the R^4 counterterm does not appear at 3 loops. A paper by N. Marcus in 1985 shows that E7(7)[R] (or at least the SU(8)[R] subgroup) is not spoiled by 1-loop anomalies. However, our current indication is that the R^4 term is actually invariant under E7(7)[R], so the symmetry argument would not forbid it. A second way to try to use E7(7)[R] would be in the construction of the scattering amplitudes. Maybe it would lead to a re-organization of the contributing loop integrals in such a way that the good ultraviolet behaviour is even more manifest. But this idea is currently only an ambitious, ill-defined, and yet-to-be realized project.

- *Cifarelli:*

In your conclusions you ask the question about the finiteness of your theory at higher loops and mention some "partial evidence" that it will stay indeed finite. Could you better illustrate this point?

- *Dixon:*

The evidence I was referring to there, is based on the inspection of multi-loop graphs, looking for lower-loop subgraphs that can be surrounded by (generalized unitarity) cuts. For example, we know from the 1-loop "No Triangle Theorem" that we can rule out contributions with 1-loop triangle subgraphs. From our knowledge of the 2-loop 4-graviton amplitude, we can further constrain the form of 2-loop subgraphs, and so on. These constraints tell us something about the ultraviolet behaviour as some subset of the loop momenta get large, but not all of them. That's why I referred to it as partial evidence.

- *Marienfeld:*

If gravitational waves were detected, would this give any input to supergravity?

- *Dixon:*

Unfortunately it probably would not give too much input. It is hard to imagine that quantum corrections to graviton scattering could have any role in observable gravitational wave phenomena, which are at very long wavelengths and low energies, where the corrections should be incredibly tiny.

- *Kovetz:*

Are planar diagrams easier to calculate and can they be used in order to simplify the non-planar calculation?

- *Dixon:*

Yes, the planar contributions are much easer to calculate, due to dual-(super)conformal invariance. This symmetry, roughly speaking, requires the integrals to be invariant under inverting the momenta. Originally it was a semi-empirical rule (Drummond et al., hep-th/0607160), but now it is understood better. In any case, it limits the number of integrals appearing in the full answer and tells us exactly what the numerator is for each integral, up to an overall constant which can be determined using generalized unitarity. Using this symmetry, the planar 4-gluon amplitude in N=4 super-Yang-Mills theory was pushed up to 5 loops by Bern et al. (0705.1864), while the non-planar contributions are still stuck at 4 loops. Moreover, the planar information can be exported to the non-planar diagrams through a kind of kinematic Jacobi identity fulfilled by the tree diagrams, called the twist identity. Using this identity, some of the non-planar graphs can be written as the difference of two planar ones.

Is the best superstring model NP complete?

Spaces of quantum field theories

Michael R. Douglas

Simons Center for Geometry and Physics *and* IHES

June 29, 2010

Abstract

We discuss the concept of a space of quantum field theories. This idea arose out of study of the renormalization group in the 1970's, but has never been fully developed. It has much in common with the mathematical concept of a space of Riemannian manifolds, which has been developed very considerably in recent years, leading to spectacular results such as Perelman's proof of the Poincaré conjecture. We discuss the analogies, beginning with Friedan's computation of the beta function for the two-dimensional nonlinear sigma model, and moving on to the linear sigma model, which finds analogies with recent works on the Ricci flow by Tian and other mathematicians.

1 Introduction

String/M theory, our best candidate for a fundamental theory unifying all the forces, is formulated in ten (and eleven) space-time dimensions. To describe the real world, we must follow Kaluza and Klein, and postulate that six (or seven) of the spatial dimensions form a small compact manifold. The first quasi-realistic models of this type were proposed in 1985 by Candelas, Horowitz, Strominger and Witten. They started with the then just-developed $E_8 \times E_8$ heterotic string, and proposed that the extra six dimensions formed a Calabi-Yau manifold. These were essentially the only class of manifolds then known to carry nontrivial Ricci flat metrics, and they had the bonus that low energy supersymmetry would be a natural outcome. At the time, only a few Calabi-Yau manifolds were known, and it was not unreasonable to hope that consistency combined with fitting the current data would select out a few candidates, which would make new and testable predictions.

In the years after this was proposed, it was found that there were thousands of topologically different Calabi-Yau manifolds, and the original idea that superstring theory might allow only one or a few favored compactifications started to seem like wishful thinking. The discovery of superstring duality in the mid-90's and the invention of many more classes of compactification, involving branes, G_2 manifolds, and many other new structures, further strengthened this impression.

Meanwhile, strong evidence was accumulating from observational cosmology that the expansion of the universe was accelerating in a way which could only be fit by assuming the existence of "dark energy," either a cosmological constant or something similar. The question of whether this was consistent with string theory was studied intensively starting around 2000. Today, there are fairly convincing arguments (though not yet proof) that it is consistent, but only in the context of a "landscape" of vacua (long-lived metastable solutions), realizing a wide variety of different values for the cosmological constant, out of which the observed small value is chosen by anthropic selection. According to this idea, we live in a vacuum with a small cosmological constant because in the others, the structures we see around us such as stars and galaxies, clearly necessary for our existence, would not have been able to form in the early universe. Whatever one's opinion about this type of argument, it is certainly true that our universe does not have a large cosmological constant, and thus just from the point of view of fitting the data we are instructed to choose solutions with small cosmological constant. Since these exist, then following the usual rules of science, by which a theory is ruled out only if it cannot fit the data, string theory remains a viable candidate theory.

Much of the discussion of the landscape has focused on the question of whether testable predictions can still be derived from string theory, perhaps by showing that some seemingly consistent possibilities cannot arise from string compactification, or perhaps by better understanding early (Planck scale) cosmology. One popular approach is to try to derive the probabilities for different vacua to be created within quantum cosmology (the so-called "measure factor"). While this would clearly be a sensible thing to do in other contexts, many find the claim that there are many universes which each have some probability to be observed by us very hard to accept.

There are other less obvious but equally striking consequences of the landscape ideas. In my lectures here in 2006, I emphasized one of them, that the structure of the landscape is such that we may never be able to identify the particular vacua with a small cosmological constant and thus which fit our data. This is because, even in simplified toy models such as the Bousso-Polchinski model of flux vacua, the problem of finding vacua with small cosmological constant can be shown to be in the class which computer scientists call **NP complete**. Without reviewing this in detail, the general expectation is that for such a problem, there is no way to find solutions (here vacua with small cosmological constant) which is much faster than simply trying the candidates one by one. Even in the best case with the most supersymmetric cancellations, this would mean trying at least 10^{60} candidates to find one which works, a task which is beyond any computer we could ever build. While this does not mean that we cannot make predictions from string/M theory, it is certainly strange to think that we might never be able to identify the solution which describes our universe, even if we had convincing experimental evidence that string theory was right.

The Bousso-Polchinski model, while tractable, is highly oversimplified compared to the entire landscape. In particular, the aspect of the problem which involves choosing a particular six or seven dimensional compactification manifold is completely left out. This is mostly because there is no concrete picture of the "space of possible solutions of string theory." And the answers to the question of how hard it is to find vacua – is the problem NP complete, or easier, or harder? – depend very much on what this space looks like. In the worst case, it might be infinite, and the question might even be undecidable. We must have some picture of this space of possibilities, to think about this question.

In these talks, I will describe some ideas which could help in making such a picture. We will take as our starting point the "world-sheet" (or perturbative) string theory, in which a solution is a two-dimensional conformal field theory (CFT) satisfying a few simple constraints, such as that the Virasoro central charge is the same as that for six dimensional flat space. However the general ideas could be applied to think about other spaces of quantum field theories, and thus could have many other applications, as searching through the set of possible quantum theories for the models which fit our data is one of the most basic problems of modern theoretical physics.

2 Spaces of quantum field theories

The basic idea of what we are after is some sort of mathematical structure (a manifold, a family of algebras, or something else) which contains **all** of the quantum field theories satisfying certain conditions, chosen so that it would be possible in principle to vary parameters (say, coupling constants) to move continuously between the different theories.

Among the conditions, we would certainly want to fix the dimension of space-time, we might want to require a certain amount of supersymmetry, and we might need to fix other quantities. We might also place conditions to cut down on the size of the space. A naive example of this would be to insist that our theories can be defined using no more than N

fields. This is a bit naive because the "number of fields" is not gauge invariant and not invariant under dualities. At strong coupling, it may not be obvious how many fields we need to define a particular QFT.

Although one could talk about a space of any type of QFT, here we focus on 2d QFT and CFT, to address the question raised in the introduction. There are many ways to define these theories, such as

- Spin systems (Ising model, Potts model, ...)
- Interacting fermi models (Luttinger liquid, Gross-Neveu model)
- Nonlinear sigma models (the $O(N)$ model, Calabi-Yau models)
- Landau-Ginzburg models and linear sigma models.

So part of the problem is to understand how these different definitions are related, and whether any single type of definition subsumes the others.

In addition, we don't want a purely abstract or mathematical framework; we would like to somehow describe the many physical phenomena which have been discovered, such as

- T-duality
- Mirror symmetry
- Change of topology and change of dimension
- Asymptotic freedom and mass generation

A good theory of the "space of QFTs" would incorporate all of these examples and phenomena.

2.1 Renormalization group

Perhaps the most useful general knowledge we have about QFT is the framework of the **renormalization group** (RG). Let us review it. We start with a "basepoint" QFT T_0 for example one defined by a functional integral with an action S_0. We then consider the set of theories obtained by perturbing T_0, in other words adding to the action the integrals of local operators O_i with couplings g^i,

$$Z[t^i] = \int [D\Phi] \ \exp -S_0 + \sum_i g^i \int O_i. \tag{1}$$

Since products of operators at coincident points diverge, this requires renormalization, which introduces a scale μ. Varying μ, and varying the couplings according to the RG,

$$\mu \frac{d}{d\mu} g^i = \beta^i(g),$$

leaves the QFT invariant. By generalizing this to position-dependent couplings, one can see that the RG also controls the scale dependence of correlation functions.

It is useful to think of the couplings g^i as coordinates on a "space of QFTs." The beta function β^i is then a vector field on this space, whose integral lines are RG flows. If we start a flow from two nearby points g and $g + \delta g$, the two RG flows can converge (an irrelevant perturbation), diverge (relevant) or stay nearby (marginal). At leading order, this distinction just follows from the scaling dimension of the coupling constants, which can be read off from the dimensions of the operators in Eq. (1) – couplings which scale like positive powers of mass are relevant, and negative powers of mass are irrelevant. Dimensionless couplings can be "marginally relevant" or "marginally irrelevant" because of logarithmic divergences; these correspond to "asymptotically free" theories in the UV and IR respectively.

In QFTs defined by adding interactions to free bosonic and fermionic field theory, the beta function can be computed in perturbation theory, as a Taylor series in the coupling constants g^i. This is also possible for perturbations around other solvable theories, for which we have explicit correlation functions. One always finds that the number of relevant and marginal directions is **finite**. Following a flow to the IR, it asymptotes to a fixed point with $\beta = 0$. This fixed point is a CFT. Similarly, one can add a relevant operator to a CFT, to get a flow to another CFT.

As we move too far away from a basepoint QFT, since the coupling of a relevant operator grows under RG flow, eventually perturbation theory will break down. On the other hand, it might be that the flow is approaching a different QFT or CFT which can itself be used as a basepoint, perhaps by perturbation theory, perhaps in some other way. We can think of the RG and other observables as defined on "local patches" around the two basepoints, and try to find a region in which the two descriptions are both valid. The map from the couplings g_1^i describing perturbations around one basepoint, to the couplings g_2^i around another, can then be thought of as a change of coordinates on the same "space of QFTs." Thus, just as we describe manifolds by introducing coordinate patches, so too we can build up a description of the "space of QFTs."

If we could find a large enough set of patches so that **all** of the flows which start within the set stay within the set, we would have a **complete** space of QFTs, or "theory space." The problem of combining overlapping patches is not so easy, because the natural bases for the operators O_i in two patches are usually different. The intuition of a "space of QFTs" suggests that this is a **geometric** problem, and that the relation between two sets of couplings g_1^i and g_2^i is a diffeomorphism.

Making this precise requires choosing a renormalization scheme, and the dependence on this scheme can presumably be thought about geometrically as well. But, since the most interesting cases are overlaps where couplings get strong, describing these in a precise way is generally beyond present technique.

2.2 General results

Especially in $d = 2$, there are some general results, many due to Sasha Zamolodchikov:

- The two-point function provides a natural **metric** on theory space,

$$G_{ij,Zam}(g) = \langle O_i(1)\, O_j(0) \rangle .$$

In string theory, each coupling g^i corresponds to a space-time field, and this metric determines the kinetic terms

$$\int d^4x G_{ij,Zam}(g) \frac{\partial g^i}{\partial x^\mu} \frac{\partial g^j}{\partial x^\mu} .$$

One can go on to define the connection, curvature, and so on.

- The c-theorem. One can extend the conformal central charge c to a function defined on all of (unitary) theory space, which decreases under RG flow. It is an outstanding open question whether such a c-function can be defined in $D > 2$ QFT.

See [3] for references and a recent discussion, including a new proof of the c theorem.

So we have some ingredients of a general picture, but not the actual picture. What would such a general picture look like?

The first point to realize and accept is that, in almost all cases, a space of 2d QFTs is going to be very complicated, too complicated to get a complete classification or "map." The basic argument for this, which we will develop shortly, is that the problem of classifying QFTs with a given central charge c, is similar to (and includes as a limit) the problem of classifying manifolds with dimension dim $= c$. But for dim ≥ 4, no complete classification of manifolds can ever be made, for a variety of reasons. This point, and an excellent introduction to Riemannian geometry as mathematicians understand it now, can be found in [1].

Despite such difficulties, there are important questions which cannot be answered without some concept of theory space, such as

- Is theory space connected (by paths)? Given two QFTs X and Y, are they in the same component? If so, is the path between them "finite length" or "infinite length" ?
- How "big" is theory space? Can we show that the number of QFTs satisfying some conditions is finite, and estimate the number?
- How many parameters (or measurements) do we need to characterize a QFT? At weak coupling, we can answer this, but more generally?

These are analogous to mathematical questions about manifolds which **can** be answered. Indeed, there has been a lot of progress in this type of mathematics, namely the theory of spaces of Riemannian manifolds, and Ricci flow.

2.3 Sigma models

The best way to motivate the relevance of the theory of manifolds for QFT is to recall Dan Friedan's 1980 work on 2d nonlinear sigma models (or NLSMs) [2].

Given a manifold M, we can think of the fields of a sigma model as a map Φ from a base space Σ (say a Riemann surface) to M. The action is then

$$S = \int_\Sigma G_{ij}(\Phi)\partial\Phi^i \cdot \partial\Phi^j$$

and is determined by the choice of a Riemannian metric G_{ij} on M. Friedan's main result was that the renormalization theory is **generally covariant**, so that observables transform sensibly under changes of coordinate $\Phi \to \Phi'$, with the usual transformation rule for the metric $G_{ij}(\Phi)$. Any observable which "carries indices" will thus transform as a tensor. For example, the beta function for the metric is a symmetric tensor, which at the n'th perturbation theory is a local functional of the couplings involving $2n$ derivatives,

$$\beta[g_{ij}] = -R_{ij} + \mathcal{O}(R^2).$$

If we look at the RG flow Eq. (??), keeping only the leading R_{ij} term, and define $t = -\log\mu$, this leads to

$$\frac{d}{dt}g_{ij} = \beta[g_{ij}] = -R_{ij}. \tag{2}$$

In mathematics, this is known as **Ricci flow**.

At leading order, the fixed points $\beta = 0$ corresponding to CFTs are Ricci flat manifolds, which solve the vacuum Einstein equations. However, we would need to better understand the corrections, which in string theory are proportional to α' (the inverse string tension), to know if these are really CFTs.

In any case, a large family of 2d QFTs can be obtained as nonlinear sigma models. We should include all of the relevant and marginal couplngs, and we might choose to study the supersymmetric case. This leads to the space of QFTs familiar in superstring theory: a QFT is a manifold M carrying the following data:

- The metric G_{ij},
- An antisymmetric tensor B_{ij},
- A function T on M (the "tachyon field"). This is the potential in the world-sheet action, and
- Another function D on M (the dilaton), which couples to the world-sheet curvature. In mathematics, a similar function is often introduced, but it is usually thought of as defining a second volume form $e^{-nD}\sqrt{G}$ (so-called "metric-measure spaces").

At leading order, the more general beta functions are a gradient of an action:

$$\beta^I = \mathcal{G}^{IJ}\frac{\partial}{\partial\phi_J}\int_M \sqrt{G}e^{-2D}\left(R + 4||\partial D||_G^2 + \frac{1}{12}H^2 + ||\partial T||^2 + 4T^2\right)$$

The c function at leading order is

$$c = \dim M - \frac{\alpha'}{4}\left(R + 4\Delta D - (\partial D)^2\right) + \ldots.$$

For a long time, it was not completely clear whether this could be extended to all orders, or beyond perturbation theory. This was recently shown in [3].

Although we know there are others, the weakly coupled NLSMs are certainly a large and interesting set of QFTs. What would it mean to talk about a "space" of these QFTs?

We can certainly follow the definition we gave above, where the patches are defined as perturbations around a set of reference metrics G^α. But doing this does not seem too illuminating. What can anyone say about this space?

2.4 Space of manifolds with metric

Let us cite some general mathematical results about the "space of all Riemannian manifolds," before discussing how this is defined.

It is useful to mark out the possibilities along three general axes – all of these have QFT analogs.

- Dimension (In 2d QFT, this is central charge c)
- Diameter = the maximum distance between a pair of points on M. (In QFT, this is the inverse of the gap in the radial Hamiltonian $H = L_0 + \bar{L}_0$, not the mass gap).
- Topological complexity $\sim$ sum of Betti numbers. (In QFT, the number of relevant, marginal or low dimension operators).

Another important characterization is by curvature. Consider $D = 2$, the Riemann surfaces. In $D = 2$ the total (or average) curvature is directly tied to topology by the Gauss-Bonnet formula,

$$\chi = 2 - 2g = \frac{1}{2\pi}\int_M \sqrt{g}R,$$

where g is the genus (number of handles). Since the topology of a Riemann surface is completely determined by its genus, we can exhibit the entire "space of 2d (closed) manifolds."

- Positive curvature – sphere, genus 0.

- Zero curvature – torus, genus 1.

- Negative curvature – genus ≥ 2.

Note that there are an infinite number of negatively curved manifolds, all with complicated topology. On the other hand, there are very few possibilities with positive or zero curvature.

Locally, the Ricci flow $\dot{g}_{ij} = -R_{ij}$ causes curvature to diffuse and average, much like the flow of heat (mathematically, both are governed by parabolic PDE's). Thus, starting from any distribution of curvature, in the IR the metric will tend towards constant curvature. The global behavior is then governed by the average sign:

- For zero curvature (torus), one reaches a fixed point (the flat torus). This is a CFT which can be used for string compactification.
- With negative curvature, the manifold grows indefinitely.
- With positive curvature, the manifold shrinks towards zero size. Eventually, the flow hits a singularity.

These mathematical results can tell us something about the early cosmology of string theory. Suppose we start with two compact dimensions, which form a Riemann surface. One can show that, in some situations relevant for cosmology (*e.g.* a linear dilaton), the time evolution is reasonably approximated by Ricci flow; let us grant this.

Then, although there are an infinite number of Riemann surfaces, all of them with $g \geq 2$ will decompactify. We are left with a finite number of possibilities, $g = 1$ and $g = 0$.

To use $g = 0$, we would need to deal with the singularity of RG flow. The simplest way to do this is to turn on flux. This produces a positive energy proportional to an inverse power of the size of M, and stabilizes the manifold. The well known $AdS_p \times S^q$ solutions are of this form.

The same general picture holds with more extra dimensions – they should have positive or zero curvature. Negative curvature is difficult to get:

- For sigma models without flux, the beta function runs the wrong way (IR free).
- Flux makes a positive contribution to the stress tensor, which is balanced by positive curvature. Negative energy objects (like orientifold planes) are highly constrained.
- More generally, the KK reduction of the Einstein term $\int_M \sqrt{g}R$ leads to a potential which is the opposite sign of the curvature – so negative curvature tends to roll to infinite volume.

Perhaps string theory also contains some *dynamical* explanation for why the extra dimensions should have a maximal diameter, or should not be topologically complicated?

As I showed in joint work with Bobby Acharya, a short set of conditions,

- Fixed dimension D.

- Curvature bounded below – it can be positive but it can't be very negative.
- Diameter bounded above. As the manifold gets larger, there is more room for topological structure,

suffice to guarantee that there are only a **finite** number of distinct possible topologies for the extra dimensions.

2.5 Ricci flow

Let us turn to better understanding RG flow. Although physical RG flow is not precisely Ricci flow, perhaps the mathematical work on Ricci flow can give us useful inspiration.

In 1982, Hamilton proposed to use Ricci flow to study the topology of (real) 3-manifolds. This was motivated by Thurston's conjectured classification of 3-manifolds – they are all supposed to be built from "prime pieces" which each carry one of 8 canonical homogeneous metrics. In particular, the constant positive curvature case is realized only by S^3 and its finite quotients (one cannot attach "pieces"), thus this conjecture implies the Poincaré conjecture in 3 dimensions.

On some level, Hamilton showed that Ricci flow starting from a more general metric leads to one of the 8 canonical types. However, in higher dimensions, Ricci flows generally lead to singularities. The simplest example is to consider two manifolds connected by a "neck" which looks like $S^2 \times L$ (a line, say $[0,1]$). The positive curvature of S^2 makes it shrink, while the zero curvature of the other two-planes means that the line remains roughly the same. Thus one gets a singularity in which the neck shrinks down to a line.

In fact, this singularity is useful in realizing Thurston's conjecture: by cutting M along the neck, we obtain his conjectured decomposition into two pieces. However, the appearance of singularities means that we need to go beyond Ricci flow, at least as defined by smooth metrics and curvatures. We need a way to deal with singularities.

Perelman's great advance was to find ways to control and fix up (do surgery on) the singularities. He showed that, very generally, the approach to a singularity looks like a "neck" of the form we discussed. After cutting the neck in two, one can continue the flow, and repeat the procedure. This process will eventually terminate, thereby realizing Thurston's decomposition.

In the physics of sigma models, one should not have to postulate the continuation of the RG through a singularity. Rather, once we take into account strong coupling effects, such as α' corrections, instantons, or other new light fields, we should find a non-singular physical continuation of the flow.

Now the sigma model beta functions have α' corrections,

$$\dot{g} = -R + \alpha' R^2 + \dots$$

in the bosonic models, and $-R + (\alpha')^3 R^4 + \dots$ in the supersymmetric models. One might think that understanding these corrections is the key to understanding the singularities.

But, this does not seem to be right – there is a lot of evidence that α' corrections are not qualitatively important, and other effects govern the physics of these singularities. The most general argument to this effect uses the linear sigma model – we will explain this at length shortly.

In the case of the sigma models with $(2,2)$ supersymmetry, one can show directly that α' corrections, though they exist, do not fix up the singularities.

As for $N = 1$ supersymmetry in $d = 4$, in these models, M must be a complex Kähler manifold, so its metric can be characterized by a complex structure and a Kähler form. As an example, a flat metric on a rectangular two-torus is characterized by two "radii" (or periods) R_1 and R_2. The complex structure is $\tau = iR_2/R_1$, while the Kähler form is $\omega = R_1R_2d^2z$.

One can show that in a $(2,2)$ sigma model, under RG flow, the complex structure of M is fixed, while the Kähler form ω flows as

$$\beta(\omega_{ij}) = -R_{ij} + \alpha'\partial(\text{total derivative terms}).$$

Furthermore, one can show that

$$\beta(R_{ij}) = \partial(\text{total derivative terms}).$$

Now, the integral of ω over a two-cycle C, is the volume of the cycle,

$$\mathrm{Vol}\,(C) = \int_C \omega.$$

It evolves under the RG simply as

$$\mu\frac{d}{d\mu}\mathrm{Vol}\,(C) = -\int_C R_{ij}dz^i d\bar{z}^j = \text{constant}$$

$$\mu\frac{d}{d\mu}\mathrm{Vol}\,(C) = -\int_C R_{ij}dz^i d\bar{z}^j = \text{constant}$$

(More mathematically, this integral of R_{ij} is the first Chern class $c_1 \cdot C$, and is a topological invariant.)

So, if $c = \int_C R > 0$ to start, it will stay that way, and an RG flow starting at $\mathrm{Vol}\,(C) = V$ (in string units) will become singular in at most time V/c. Any α' corrections could only speed up singularity formation.

We need to find some other way to continue through the singularities of the Ricci flow.

3 Linear sigma models

Physics fixes up these singularities, in a variety of other ways. The original example of this was the case $M = S^{N-1}$, in other words the $O(N)$ model. Here the volume shrinking to zero is the flow to strong coupling.

As was found in the 70's [4], the best way to understand the resulting physics is to use the **linear sigma model**. Instead of $N-1$ coordinates on S^{N-1}, we we use N coordinates ϕ^i on $\mathbb{R}^N$, and define the sphere by a constraint $|\phi|^2 = R^2$. This can be implemented by a Lagrange multiplier λ, and the action

$$S = \int d^2\sigma \sum_{i=1}^{N} [(\partial\phi^i)^2 + \lambda(\phi^i)^2] - \lambda R^2.$$

We can understand the physics by computing an effective potential for λ. Since the ϕ fields appear quadratically, the only quantum correction is at one loop:

$$V_{eff} = -\lambda R^2 + \lambda \int \frac{d^2k}{k^2+\lambda} = -\lambda R^2 + \pi\lambda \log\frac{\mu^2}{\lambda}.$$

This is minimized at

$$R^2 = \pi \log \frac{\mu^2}{\lambda}$$

so

$$\lambda = \mu^2 C \exp -R^2/\pi.$$

Because of the coupling $\lambda|\phi|^2$, this is also the mass for the ϕ fields. Thus the "zero size sphere" is transmuted into a massive theory. There is no sphere in the IR.

This picture was generalized to the $(2,2)$ supersymmetric sigma models by Witten (1993). We can realize a family of Kähler manifolds depending on a Kähler class ζ, by introducing an auxiliary $U(1)$ gauge field, and taking ζ as the Fayet-Iliopoulos term,

$$\sum_{i=1}^{n+1} Q_i|\phi^i|^2 = \zeta.$$

The case of all $U(1)$ charges $Q_i = +1$ defines the $\mathbb{CP}^n$ model, which is much like the $O(N)$ model. One can also have mixed sign charges. For example, the case $n=3$, $Q_1 = Q_2 = +1$ and $Q_3 = Q_4 = -1$, is the "resolved conifold" geometry in three complex dimensions. (These are "affine toric manifolds.")

As we will show below, there is a beta function for ζ,

$$\beta\zeta = -\sum_i Q_i,$$

One can show that the target space (solution of D-term constraint mod $U(1)$ gauge symmetry) always has a two-cycle C with $\mathrm{Vol}\,(C) = \zeta$, and $\int_C R = c_1(C) = \sum_i Q_i$, so this is exactly the same as our previous equation $\beta\mathrm{Vol}\,(C) = -\int_C R$.

In addition, there is a field which plays the role of λ (a scalar in the $U(1)$ gauge multiplet usually called σ), and all the other ingredients we used in the discussion of the $O(N)$ theory.

Now, there are two cases:

- If only positive (or only negative) Q_i appear, the physics works the same way as the $O(N)$ model. The σ component of the vector multiplet gets a vev, which stops the RG at $\zeta_{eff} = 0$. This leads to a mass gap, and this part of the geometry disappears.

- If both positive and negative Q_i appear, the solutions to the D-term constraint will change as ζ changes sign. For the conifold, this leads to a "flop," while more generally it leads to a *topologically different* manifold.

  Thus, the RG continues through the singular geometry, and produces a different geometry on the other side. On the other side, the curvature is negative, so the new two-sphere generally continues to grow without bound.

Independently, mathematicians (Tian, Yau, Donaldson, Chen, Cao, Song...) have developed a picture of the space of Kähler-Ricci flows, determined by c_1 of the manifolds, and explaining how to fix up (do surgery on) the singularities. In general one can find a $d+1$-dimensional manifold with a $U(1)$ acting by isometries, such that the quotient is the sequence of metrics obtained by Ricci flow, where the flow time $\log t = \mu = \zeta$.

This leads to a mathematical picture which agrees with the physics:

- If only positive (or only negative) Q_i appear, the continuation is done by replacing the shrinking cycle with a point.

- If both positive and negative Q_i appear, continuing to negative ζ does the flop or flip. If one considers the "normalized Ricci flow"

$$\frac{d}{dt}\omega = -c_1(M) + \omega$$

  which can have a fixed point for positive c_1, this will only have a fixed point for the "minimal model," a particular distinguished metric among all those which can be reached by birational transformations.

3.1 Why is the linear sigma model good?

There is a philosophical objection to basing our definitions on the linear sigma model. Ever since Einstein invented general relativity, we have been told that although it can be very convenient to picture space-time as embedded in a higher dimensional space, this is only a fiction – the geometry of general relativity is *intrinsic.*

On the other hand, the successes of the LSM, in particular that all of the fields (the extra coordinates and Lagrange multipliers) seem to play important physical roles, suggests that, in some sense, we really should think of (at least some) curved manifolds as naturally embedded in higher dimensional spaces. We need to study both intrinsic and *extrinsic* geometry.

Ultimately, the reason why the LSM is natural, is that

- The RG is an averaging operation, and

- Averaging only makes sense in a linear (or affine) space.

We can think of the renormalization group as defined by a process of averaging. In spin systems, one takes a "block" of spins, and defines an average over the spins in the block. One then integrates over all configurations with a given fixed value for the average spins, to define the Boltzmann weight of a configuration of block spins. Finally one rescales to match the original conventions.

The same definition of the RG can be made in field theory. Thus, the crucial step is to define the average value of the field in some region. In the sigma model, this means we must be able to average a set of points $x_1, x_2, \ldots x_n \in M$, or functions which take values in M.

However, in a general manifold, there is no way to add points, or to average them. One must postulate a linear (or affine) structure in order to do this.

On a Riemannian manifold, one can use geodesic flow to define a map from T_pM, the tangent space to a point, to its neighborhood. A tangent vector v at p, maps into the point obtained by starting at p with velocity v, and moving along a geodesic for time 1. The components of v are *normal coordinates.* They are good for some subset in T_pM but become multivalued beyond this.

In normal coordinates, the coefficients of the Taylor expansion of the metric are local tensors, made from the Riemann tensor and its derivatives. Friedan used this to define covariant renormalization of sigma models.

However, given that there is no preferred point p, this definition is not completely natural either. And, at strong coupling (small M, small cycles, large curvature), the fields fluctuate over regions of M larger than the validity of the normal coordinates.

Since we need a linear structure to define the RG, it may be more sensible to postulate one on all of M. This brings us directly to the linear sigma model. The fundamental idea of the LSM is that, by choosing an appropriate embedding, we can simplify the RG, and make the dynamical variables which emerge at strong coupling manifest.

As a final note in passing, recent work in statistics of high dimensional data sets has the same problem, to average data points which live in non-Euclidean (or even unknown) manifolds M. The same device of embedding is found useful.

References

[1] *A panoramic view of Riemannian geometry*, M. Berger, Springer 2007.

[2] D. H. Friedan, "Nonlinear Models In Two + Epsilon Dimensions," Annals Phys. **163**, 318 (1985).

[3] D. Friedan and A. Konechny, "Gradient formula for the beta-function of 2d quantum field theory," arXiv:0910.3109 [hep-th].

[4] *Gauge Fields and Strings*, A. Polyakov, Harwood 1987.

CHAIRMAN: M. R. DOUGLAS

Scientific Secretaries: O. Gamayun, Y. Klopot

DISCUSSION I

- *Smirnov:*

I have a question about sigma models defined on Riemann surfaces. When you described the renormalization group flow for these models, as I understood, you said that for genus zero the structure of the flow is that the sphere just collapses to a point. And for genus 2 and more the surface grows to infinity. Genus 1 is the only case when we have a fixed point of the flow. The question is: what does this mean from the conformal field theory point of view? Does it mean that this is the only case when we have a conformal field theory limit of these models?

- *Douglas:*

Yes, basically the answer is yes. If you only use the metric, no other couplings, you are only going to get a fixed point with genus 1. If you turn on other couplings, you can do other things. For example, with target space S^3 you can turn on Neveu-Schwarz flux to get a fixed point, the WZW model for the SU(2) group. You can also get more target spaces as flows out of UV fixed points. For example, you can get the growing higher genus surfaces in the IR. In a sense it is a higher dimensional model; I'll explain that next time.

- *Gamayun:*

In your lecture you mentioned compactification and renormalization group flow. I would like to ask whether there is any connection between these things, because it seems that shrinking of the sphere under RG flow is the natural way to shrink the extra dimensions.

- *Douglas:*

The basic connection which people understand is that you can make solutions of string theory in which the time evolution looks like the Ricci flow. So one example: if you turn on a dilaton field with a large time-like derivative, the second order supergravity equations approximately reduce to the first order Ricci flow equations. In this context, our RG discussions would imply that positively curved spaces spontaneously compactify, and negatively curved spaces spontaneously decompactify. But it is an open research question to understand this more generally.

- *Gamayun:*

So, then one more question: is it true that we should compactify on something with genus zero, just because it shrinks?

- *Douglas:*

Not necessarily. Even in this simplest case of two dimensions, something else could come out. For example, the two-torus can also shrink, because besides the “alpha prime” renormalization group which comes from world-sheet dynamics, there can be additional quantum dynamics: there could be quantum corrections to the potential coming from the Casimir energy, and from instantons. Also, there could be cosmological dynamics. In space-time, the size of the torus corresponds to a massless field, and it could be that the initial conditions give this field a velocity, so the torus will grow or shrink linearly with time. So there are many possibilities; while the sphere shrinks faster, there could be other effects which favor the torus.

- *Aschieri:*

I liked very much the picture of the different phases of quantum field theories. But where are the transition functions relating to these theories? They are not just given by coordinate transformation but more general ones, e.g. in the case of theories related by T-duality. In some sense non-commutative geometry emerges. Is there a new geometry related with these different patches of QFT's? What is your general view on the subject, rather than specific examples?

- *Douglas:*

Yes, part of the project is to develop more interesting geometry. People do a lot of different things in this direction. There are some phenomena that we know are described by non-commutative geometry, mostly involving D-branes and open strings. There is a generalized geometry that people like Chris Hull have worked on, when you try to treat the winding modes of the string just like the momentum modes, leading to a geometry in twice as many dimensions. There is another type of generalized geometry which is useful for adding Neveu-Schwarz flux, and other massless supergravity fields. So there are many open directions that people are exploring. And geometry is a language. People have different attitudes about general relativity. On the one hand there was Einstein -- he got to a certain point using very clever physical arguments and he realized that the way to go further might have already been worked out by the mathematicians, and indeed there was Riemannian geometry. And there was the opposite attitude of people like Feynman and Weinberg. They said that since you can show that the theory of gravity is uniquely determined by consistency conditions, you don't even need to know about general covariance. You can read such a discussion in Weinberg's textbook on gravitation. So, geometry is a language. It's something that I think will help us to understand the discussion much better than without it. But it does not tell you what to say, we have to come up with the physical arguments first and then find the language and, yes, part of the language will be this new geometry.

CHAIRMAN: M. R. DOUGLAS

Scientific Secretaries: E. D. Kovetz, I. Yeletskikh

DISCUSSION II

- *Burda:*

I have a question about Witten's generalization of this *O(N)* σ-model story. There, we obtain an equation with U(1) charges as coefficients, and consider different cases, depending on signs of this charges. So, is it possible to ask about the case when one or several of these charges are equal to zero. What kind of reduction do we obtain in this case, and what happens with topology?

- *Douglas:*

If some coordinates are uncharged under a U(1), they do not participate in the transition associated to that U(1) and it's Fayet-Iliopoulos terms. In other words, we can think of the target as a direct product of a part which flows and another part which to a good approximation does not. Of course, those coordinates might participate in some later flow or transition.

- *Gamayun:*

There is a 'regularization' of Ricci flow in σ-models by means of the QFT methods, which lead to the dimensional transmutation. And there are purely mathematical regularizations (adding ω). What is the connection between these two 'regularizations'?

- *Douglas:*

The short answer is that the mathematicians simply postulate a 'regularization' to solve their problems. It is ad hoc and this is an advantage of physics ideas. However, in some cases, like the flop we discussed, the two ideas do agree. It is not completely clear why, perhaps this is just on grounds of simplicity, perhaps a deeper explanation can be found.

Erice Lecture on Microscopic Gravity

Gia Dvali

Arnold Sommerfeld Center for Theoretical Physics
Department für Physik, Ludwig-Maximilians-Universität München
Theresienstr. 37, 80333 München, Germany
Max-Planck-Institut für Physik
Föhringer Ring 6, 80805 München, Germany
CERN, Theory Division
1211 Geneva 23, Switzerland
CCPP, Department of Physics, New York University
4 Washington Place, New York, NY 10003, USA

Abstract

In this lecture we review a fundamental connection between the number of particle species and the nature of gravitational interaction at the shortest length-scales. The key role in establishing this connection is played by the well-understood properties of macroscopic black holes, combined with effective field-theoretic approach to gravity. One universal outcome is, that the existence of particle species in low energy theory inevitably replaces the Planck length by a longer fundamental scale. Application of this phenomenon for the solution to the Hierarchy Problem fixes the number of species at 10^{32}.

1 The Hierarchy Problem

The main theoretical motivation for the searches of beyond the standard model physics at LHC is the Hierarchy Problem. There exist also less direct motivations, such as the dark matter and unification of the gauge couplings, but these also become connected to new physics at TeV energies only in the context of the concrete solutions to the hierarchy problem, such as for example a low energy supersymmetry. I therefore first elaborate on the concept of the hierarchy problem.

The hierarchy problem is often presented as a puzzle of an unnaturally small number, the ratio of the weak interaction scale ($M_W \sim 10^2$GeV) to the Planck scale ($M_P \sim 10^{19}$ GeV), or equivalently the product of the Newtons constant $G_N \equiv M_P^{-2}$ and the square of the W-boson mass,

$$\frac{M_W^2}{M_P^2} \sim 10^{-34} \,. \tag{1}$$

The key point however is not the smallness of the number *per se*, but its UV-sensitivity. The value of the weak scale in the Standard Model is set by the vacuum expectation value of the Higgs field, which itself is determined by the mass term. The Higgs mass term is quadratically sensitive to the UV cutoff of the theory.

Notice, that in the absence of gravitational interaction, one could have argued that the problem is not so severe, since one could hide divergences into the renormalization of the physical quantities. One could also argue that the cutoff sensitivity is an artifact of regularization, which would be absent if one would use regularization schemes in which cutoff-dependence is not explicit.

However, in the presence of gravity the above arguments no longer work, since gravity provides an ultimate cutoff to the shortest possible length-scales, and thus cutoff becomes physical. This is due to the fact, that gravity is a theory with minimal length.

2 The Role of Gravity

In Einstein's gravity, the Planck length is *the shortest* length-scale of nature, and any attempt of probing the shorter distances will instead probe the larger length-scales.

Let us start with the pure Einsteinian theory of gravity in four-dimensions, in which the only propagating degree of freedom is a massless spin-2 graviton, $h_{\mu\nu}$. No other propagating species are assumed at this point. All the sources, other than the ones composed out of gravitons, will be considered as external sources. This requirement uniquely fixes the action in form of the Einstein-Hilbert action,

$$S_{EH} = \int d^4x \, M_P^2 \sqrt{-g}\, \mathcal{R} \,. \tag{2}$$

For definiteness, we have set the cosmological constant to zero. In this way, we shall always consider observers on asymptotically-flat spaces.

In our discussion the central role is played by the Planck mass $M_P \sim 10^{19}$GeV and by the corresponding Planck length, $L_p \equiv M_P^{-1} \sim 10^{-33}$ cm. The overall numerical factors of order one will be ignored throughout the lecture.

Field theoretically, Planck scale defines the coupling of graviton to energy momentum sources *universally*,

$$h_{\mu\nu}\frac{T^{\mu\nu}}{M_P}, \tag{3}$$

where $T_{\mu\nu}$ is an arbitrary energy-momentum tensor. The important fact is, that the above universality property is true to all orders in non-linear interactions of graviton. That is, one can either think of the above coupling as the coupling to an external source, or as a self-coupling of graviton to its own energy-momentum tensor in a given non-linear order. Due to the above crucial property, Einstein's gravity viewed as a quantum field theory possesses an universal strong coupling scale, M_P. This fact shall play the key role in what follows.

In Einstein's gravity, L_P represents the absolute lower bound on any distance that can ever be resolved. Distances $L \ll L_P$, cannot be probed, *in principle*!

The fundamental reason for such an obstacle is the existence of black holes (BH). BH-formation interferes with any attempt of extracting information from beyond the Planck length and produces an insuperable barrier. In fact, harder we try to go beyond L_P, with a larger and more classical BH we shall end up. Physics of such a BH has nothing to do with short distances and is entirely governed by the infra-red (IR) gravity.

The above reasoning can only be violated if the theory possesses an agent that could violate energy positivity condition. Then, using such an agent, one could encode (and extract) information at arbitrarily short distances, without paying the energy price.

Due the existence of the minimal length, in theory with Einstein gravity an elementary particle with a mass m much larger that the Planck mass makes no sense. Instead any point-like object with such mass becomes a classical black hole of the size $R = mL_P^2$, since the gravitational radius of such a particle exceeds its Compton wave-length.

This fact makes the hierarchy problem real. Indeed, in the presence of gravity, no question arises why the Higgs is not heavier than the Planck mass, since it it were heavier it would no longer be an elementary particle, but a classical black hole. Thus, gravity provides an universal regulator in form of the Planck scale, and the stability of the Higgs mass versus this scale requires an explanation.

It turns out [4–6], that the value of the regulator scale is profoundly defined by the nature and number of particle species that are present in low energy theory. We shall now discuss this relation.

3 Generalized Planck Length

Before turning to arbitrary particle species, we wish first to extend our analysis to the theories that include new gravitational degrees of freedom on top of the Einsteinian massless graviton, and see how the notion of Planck length generalizes to such theories.

Consider a theory that includes additional gravitational degrees of freedom. We shall define such degrees of freedom as the ones that are sourced by the conserved energy momentum tensor $T_{\mu\nu}$ and which in the classical limit and for localized sources amount to the local deformations of an asymptotically-flat metric $g_{\mu\nu} = \eta_{\mu\nu} + \bar{h}_{\mu\nu}$. By conservation of the source, to the linear order these include only spin-0 and spin-2 fields. Thus, a small metric perturbation around the flat background ($\bar{h}_{\mu\nu}$) in such theories will contain not only the massless spin-2 state, $h_{\mu\nu}$, but also an arbitrary number of massive spin-2 ($h^i_{\mu\nu}$) and spin-0 ($\phi^j_{\mu\nu}$) fields:

$$\bar{h}_{\mu\nu} = \frac{1}{M_P}\left(h_{\mu\nu} + \sum_i c_2(i)\, h^i_{\mu\nu} + \sum_j c_0(j)\, \phi^j_{\mu\nu}\right). \tag{4}$$

Possible spin-1 fields cannot couple to the conserved sources to the linear order, and therefore do not appear in the above decomposition. However, they can participate in non-linear interactions.

The one-graviton exchange amplitude among the two sources $T_{\mu\nu}$ and $t_{\mu\nu}$ is highly restrictive and takes the following form,

$$T^{\mu\nu}\langle \bar{h}_{\mu\nu}\bar{h}_{\alpha\beta}\rangle t^{\alpha\beta} = \frac{1}{M_P^2}\left(\frac{T_{\mu\nu}t^{\mu\nu} - \frac{1}{2}T^{\mu}_{\mu}t^{\nu}_{\nu}}{p^2} + \sum_i \rho_2(i)\frac{T_{\mu\nu}t^{\mu\nu} - \frac{1}{3}T^{\mu}_{\mu}t^{\nu}_{\nu}}{p^2 + m_i^2} + \sum_j \rho_0(j)\,\frac{T^{\mu}_{\mu}t^{\nu}_{\nu}}{p^2 + m_j^2}\right) \tag{5}$$

where we have separated the contributions from massless spin-2, massive spin-2 and spin-0 poles. We have normalized the relative strengths to the one of a zero mode graviton. In case of continuum, the discrete sum has to be replaced by the integral. The crucial point is, that all the spectral densities $\rho_2(i) \equiv |c_2(i)|^2$ and $\rho_0(j) \equiv |c_0(j)|^2$ must be semi-positive,

$$\rho_2(i) > 0, \quad \rho_0(j) > 0\,, \tag{6}$$

in order for the theory to be ghost-free. The above structure uniquely defines the linearized metric produced by an arbitrary source. For instance, the gravitational potential produced by a localized non-relativistic mass $T_{\mu\nu} = \delta_\mu^0 \delta_\nu^0 M \, \delta(r)$ is give by,

$$\bar{h}_{00}(r) = \frac{M}{M_P^2 r} \left(\frac{1}{2} + \frac{2}{3} \sum_i \rho_2(m_i) e^{-m_i r} + \sum_j \rho_0(m_j) \, e^{-m_j r} \right) . \tag{7}$$

What would be an analog of L_P as of the shortest observable distance in such a theory? This scale can be found in terms of the BH properties in the following way. [1]

Since we are interested in theories that in deep-IR flow to Einsteinian gravity with the only propagating degree of freedom being a massless spin-2, we shall assume the existence of a mass gap $M_c \equiv 1/R_c$ corresponding to a first massive excitation in expansions (4) and (5). In such a theory, at distances $r \gg R_c$, the only propagating degree of freedom is a massless graviton and gravity is pure Einsteinian, with all the usual properties. In particular, in such a theory there must exist Schwarzschild BHs of radius $R \gg R_c$. The mass-to-radius dependence for such a BH is given by an usual Einsteinian relation, $R(M) = ML_P^2$. Now, since $R \gg L_P$, we have $R \gg M^{-1}$. In the other words, by taking M large, the Schwarzschild radius of such a BH can be made arbitrarily larger than its Compton wavelength.

Now, let us start decreasing M parametrically. Of course, for $R(M) > R_c$ we are in Einstein's gravity and $R(M)$ decreases linearly with M. Once $R(M)$ drops below R_c, we are no longer in Einsteinian regime and dependence of $R(M)$ on M will change, in general.

If $R(M)$ and M^{-1} meet for some $M = M_*$, then the (first) meeting point,

$$R(M_*) = M_*^{-1}, \tag{8}$$

marks the analog of the Planck scale M_*. The corresponding length scale $L_* \equiv M_*^{-1}$ plays the same role as the Planck length plays in Einstein gravity. In particular semi-classical black holes exist with masses above M_* (sizes larger than L_*). Notice that because of semi-positivity of the spectral representations, there is a strict inequality $L_* \geqslant L_P$.

[1] Our analysis automatically puts the lower bound on the length scale down to which the semi-classical BHs can exist. In concrete models (although we are not aware of any fully consistent theory of this sort) the BHs may cease to exist even at larger scales, but this is not in conflict with our findings, since we derive the lower bound on the semi-classicality length.

4 Gravity and Species

In the above discussion we saw that introduction of the additional gravitational messengers in the theory replaces M_P by a smaller scale M_*, or equivalently, L_P is replaced by a longer length L_*. This is due to the fact that by requirement of the absence of ghosts gravity is attractive, and the existence of more messengers make it stronger, not weaker. In other words, with more messengers gravity hits the strong coupling scale at lower energies.

It turns out, that the fact of increase of the effective Planck length $L_P \to L_*$ remains true even if the species in question are not gravitational. That is, introduction of arbitrary particle species in the low energy theory *by consistency* results into replacement of L_P by a longer length, the lower bound on which is set by the number of species [4–6]. We shall now discuss this phenomenon.

Consider Einstein gravity in D space-time dimensions, in which gravitational interaction is mediated by a D-dimensional massless particle of spin-2, the graviton $h_{\mu\nu}$. The corresponding D-dimensional Planck mass we shall denote by M_D. As we have discussed, the shortest observable distance in such a theory is given by D-dimensional Planck length $L_D \equiv M_D^{-1}$.

Let us now try to couple this theory to N particle species. For simplicity we shall take the species to be light. As we shall see, this seemingly-innocent deformation of the theory dramatically affects the gravitational dynamics. In fact, in the presence of light species, it is no longer consistent to assume that gravity is mediated by a single massless graviton, but rather the new gravitational degrees of freedom *must be introduced necessarily* [4–6]. These degrees of freedom are necessary in order to UV-complete theory at the new fundamental scale, L_N, which is *larger* than the D-dimensional Planck length L_D,

$$L_N \equiv N^{\frac{1}{D-2}} L_D \,. \tag{9}$$

We shall refer to L_N as the *species scale*.

Effect of species on gravity was also considered in some perturbative [7] and cosmological [8] contexts.

In order to prove the above, following [4, 5], we shall reproduce the two ways of reasoning.

4.1 Bound on the UV-Scale From Black Hole Evaporation

The existence of scale L_N can be immediately deduced from the properties of the semi-classical BHs. Our focus will be on classically-static, neutral, non-rotating BHs on asymptotically-flat D-dimensional space. In theory of Einsteinian gravity we can always design such BH to be neutral under all possible long-range gauge fields (such as electric field), so that the only characteristic quantum number is its mass M, which determines its Schwarzschild radius $R(M)$ in the following way,

$$R(M)^{D-3} = M L_D^{D-2} . \tag{10}$$

The crucial property is, that such a semi-classical BH radiates thermally with the Hawking temperature given by the inverse size $T = R^{-1}$. The evaporation of a semi-classical black hole obeys the following law of the temperature-change,

$$\frac{1}{T}\frac{dT}{T} = \Gamma_{total} , \tag{11}$$

where Γ_{total} is the total rate of the emission of all the particle species. The evaporation rates into the heavy species of mass m are suppressed by the Boltzmann factors $e^{-\frac{m}{T}}$, whereas all the species with masses $\ll T$ are produced *democratically*. Democratic evaporation of semi-classical BHs follows from the no-hair [2] properties of the classical one.

Thus, for seeing how the number of species N contributes into the evaporation rate, it is most useful to normalize the latter to that of a massless graviton. The rate of graviton emission, by general covariance is given by

$$\Gamma_{gr} = T \, (T \, L_D)^{D-2} . \tag{12}$$

So by democracy of semi-classical BH evaporation, in the presence of N species we have ,

$$\frac{1}{T}\frac{dT}{T} = \Gamma_{gr} N . \tag{13}$$

Using the explicit form of (12) for Γ_{gr} we can rewrite (13) as,

$$\xi(T) \equiv \frac{1}{T^2}\frac{dT}{T} = (T \, L_D)^{D-2} N . \tag{14}$$

Where we have defined a BH semi-classicality parameter ξ. The crcial point is, that for any semi-classical BH ξ must satisfy,

$$\xi(T) \ll 1 . \tag{15}$$

Breakdown of the above conditionn inevitably signals violation of the semi-classicality approximation, and thus departure from classical Einsteinian gravity at distances corresponding to the Schwarschild radius at which the condition (14) is violated. It is easy to see, that this violation takes place exactly for the temperature $T = 1/L_N$. Thus, in the presence of species, gravity cannot remail in a semiclassical Einsteinian regime, and must be UV-completed at the scale L_N. This observation inevitably implies the existence of new gravitational degrees of freedom, that must modify semi-classical BH dynamics at distances larger or of order L_N.

4.2 Species Scale from the Black Hole Barrier of Information-Storage

An alternative proof of the existence of the species scale [5] is based on impossibility of resolving species identities at the length scales shorter than L_N. This is a fundamental obstacle created by gravity.

This is in full accordance with principles of holography [1] according to which the maximal information stored by a BH is given by the area in Planck units. Species identities represent a particular form of information.

For a physicists, existence of N distinct particle species means that he/she can label them, at least in principle, and further distinguish these labels by physical measurements. Let Φ_j be the particle species and $j = 1, 2, ...N$ be their labels. Let us now show, that resolving these labels at distances beyond L_N is fundamentally impossible. The reasoning is similar to why one cannot resolve distances beyond L_P in theory without species, with the difference that the existence of species makes this limiting distance to grow.

Indeed, any process of decoding the label of an unknown particle localized within the space-time box of size L, involves comparing the unknown particle with all N different sample species. Thus, any such measurement must involve localization of N various species (or of the equivalent information) within the same box.

This fact automatically limits the size of the box from below. Indeed, localization of each sample particle within the size L costs energy $E = 1/L$, which implies that the total energy localized within the region L is at least $E_{Total} = N/L$. The corresponding Schwarzschild radius then is,

$$R(E_{Total})^{D-3} = \frac{N}{L} L_D^{D-2}. \tag{16}$$

The key point is, that species measurement scale L cannot be decreased arbitrarily, since eventually its Schwarzschild radius will exceed its own size, and localized species can no longer be resolved. Any further attempt to decrease L will result into the creation of an even bigger BH. The critical size, beyond which the resolution of species is no longer possible is readily derived by equating the Schwarzschild radius to the the localization size L,

$$L^{D-3} = R^{D-3} = \frac{N}{L}L_P^{D-2}, \tag{17}$$

which gives the bound on L set by L_N. This constraint is very powerful. It tells us that gravity makes it impossible to resolve the species beyond the scale L_N, which means that the new gravitational degrees of freedom must enter into the game.

Note that our bound is completely insensitive to the possible loss of information by BHs (for the review see [3]), since irrespective of such violations, the information can at best be retrieved on the time scales of BH evaporation, which for semi-classical BHs is much longer than their size.

So far, we did not specify the nature of species. Now it is time to discuss this issue.

4.3 Kaluza-Klein Species

The simplest case is when species have just right quantum numbers to fill the KK tower of new d compact flat dimensions of radius R_c. In such a case everything falls nicely into the places and relation between L_N and L_D exactly reproduces the usual geometric relation between the $D+d$-dimensional (L_{D+d}) and D-dimensional Planck lengths,

$$L_N = L_{D+d} = L_D\,(R_c/L_{D+d})^d\,. \tag{18}$$

The reason why L_N came out equal to the L_{D+d}, becomes immediately clear if we notice that $N = (R_c/L_{D+d})^d$ is simply the number of KK species. It is obvious, that the fundamental length scale of the theory is L_{D+d}, and this is precisely what species counting tells us.

The field theoretic reason of why KK species decrease the fundamental gravity scale is due to the fact that they provide more messengers of gravitational interaction, which make gravity stronger at short distances. Consider for example a 5-dimensional theory compactified on a circle of radius R_c. As it is well-known, from the 4-dimensional point of view this is a theory of the tower of massive spin-2 states. A one-graviton exchange amplitude between the two conserved energy-momentum sources $T_{\mu\nu}$ and $\tau_{\mu\nu}$ is such a case takes the following form,

$$T^{\mu\nu} \langle \bar{h}_{\mu\nu} \bar{h}_{\alpha\beta} \rangle t^{\alpha\beta} = \frac{1}{M_P^2} \sum_n \frac{T_{\mu\nu} t^{\mu\nu} - \frac{1}{3} T_\mu^\mu t_\nu^\nu}{p^2 + n^2/R_c^2} . \tag{19}$$

Although each KK graviton couples by $1/M_P$-suppressed interaction, because of multiplicity the strong coupling *universally* happens at the scale $M_*^3 = M_P^2/R_c$, which is simply a 5-dimensional Planck mass. The same scale sets the crossing point for the BH Schwarzschild radius $R(M)^2 = M\, R_c/M_P^2$ and its inverse mass. Theory contains no other strong coupling scale, and because of this, distances shorter than $L_5 \equiv (R_c/M_P^2)^{-1/3}$ cannot be probed, in principle. This result is not surprising since in deep-UV the KK theory is just a five dimensional pure-Einstein gravity.

4.4 Non-Gravitational Species and Hierarchy Problem

Since the existence of the scale L_N follows from the general semi-classicality properties of the large black holes and is insensitive to the precise nature of species, we are lead to the conclusion that in any theory with N particle species, the gravitational cutoff is lowered to $M_* = 1/L_N$. This universally fixes the number of particle species in any theory that solves the hierarchy problem by bringing gravitational cutoff to TeV energies to be $N \sim 10^{32}$. From this perspective the large extra dimensional solution to the hierarchy problem [9] is a particular case of many-species approach, in which 10^{32} species are the KK gravitons. What our findings tell us is, that this effect is universal:

Consistent theories with many particle species must have a low gravitational cutoff.

Acknowledgments

It is pleasure to thank Prof. Zichichi for invitation and hospitality at the Erice summer school.

References

[1] G. 't Hooft, "Dimensional reduction in quantum gravity", gr-qc/9310026.

L. Susskind, "The World As A Hologram", J. Math. Phys. **36**, 6377 (1995), hep-th/9409089.

L. Susskind and E. Witten, "The Holographic bound in anti-de Sitter space", arXiv:9805114 [hep-th].

[2] W. Israel, *Phys. Rev.* **164** (1967) 1776; *Commun. Math. Phys.* **8**, (1968) 245;

B. Carter, *Phys. Rev. Lett.* **26** (1971) 331.

J. Hartle, *Phys. Rev.* **D 3** (1971) 2938.

J. Bekenstein, *Phys. Rev.* **D 5**, 1239 (1972); *Phys. Rev.* **D 5**, (1972) 2403; *Phys. Rev. Lett.* **28** (1972) 452.

C. Teitelboim, *Phys. Rev.* **D 5** (1972) 294.

[3] For a review see, J. Preskill, "Do Black Holes Destroy Information?", arXiv:hep-th/9209058

[4] G. Dvali, "Black Holes and Large N Species Solution to the Hierarchy Problem," arXiv:0706.2050 [hep-th];

G. Dvali and M. Redi, "Black Hole Bound on the Number of Species and Quantum Gravity at LHC," Phys. Rev. **D77** (2008) 045027, arXiv:0710.4344 [hep-th];

G. Dvali and D. Lust, "Power of Black Hole Physics: Seeing through the Vacuum Landscape," JHEP, **06** (2008) 47, ArXiv:0801.1287 [hep-th];

G. Dvali, "Nature of Microscopic Black Holes and Gravity in Theories with Particle Species", arXiv:0806.3801 [hep-th];'

[5] G. Dvali and C Gomez, "Quantum Information and Gravity Cutoff in Theories with Species", Phys. Lett. **B674** (2009) 303, arXiv:0812.1940 [hep-th] .

G. Dvali and C. Gomez, "Strong Coupling Holography", arXiv:0907.3237 [hep-th]

[6] R. Brustein, G. Dvali and G. Veneziano, " A Bound on Effective Gravitational Coupling from Semiclassical Black Holes", JHEP 0910:085,2009, arXiv:0907.5516 [hep-th]

[7] G. Dvali and G. Gabadadze, "Gravity on a brane in infinite-volume extra space," *Phys. Rev.* **D 63**, 065007 (2001) [hep-th/0008054].

G. Veneziano, "Large-N bounds on, and compositeness limit of, gauge and gravitational interactions," JHEP **0206**, 051 (2002) [arXiv:hep-th/0110129].

[8] R. Brustein, D. Eichler, S. Foffa and D. H. Oaknin, "The shortest scale of quantum field theory," Phys. Rev. D **65**, 105013 (2002) [arXiv:hep-th/0009063].

J. D. Bekenstein, "Is The Cosmological Singularity Thermodynamically Possible?," Int. J. Theor. Phys. **28**, 967 (1989).

R. Brustein, D. Eichler and S. Foffa, "A braneworld puzzle about entropy bounds and a maximal temperature," Phys. Rev. D **71**, 124015 (2005) [arXiv:hep-th/0404230].

R. Brustein, "The generalized second law of thermodynamics in cosmology," Phys. Rev. Lett. **84**, 2072 (2000) [arXiv:gr-qc/9904061].

R. Brustein and A.J.M. Medved, "Bounds on Black Hole Entropy in Unitary Theories of Gravity", arXiv:1003.2850 [hep-th]

[9] N. Arkani-Hamed, S. Dimopoulos and G. R. Dvali, "The hierarchy problem and new dimensions at a millimeter," Phys. Lett. B **429**, 263 (1998) [arXiv:hep-ph/9803315]; I. Antoniadis, N. Arkani-Hamed, S. Dimopoulos and G. R. Dvali, "New dimensions at a millimeter to a Fermi and superstrings at a TeV," Phys. Lett. B **436**, 257 (1998) [arXiv:hep-ph/9804398].

CHAIRMAN: G. DVALI

Scientific Secretaries: F. Giannuzzi, Y. Peters

DISCUSSION I

- *Preghenella:*

If black-hole production sets in at LHC energies, could it somehow destroy the discovery potential of the LHC?

- *Comment by Zichichi*:

It would be great if at LHC we would discover at least 6 leptons with the same charge totally unpredicted.

- *Dvali:*

It would be really fantastic if we would discover black holes at the LHC. This would mean that we can study the regime of quantum gravity experimentally. If we are above the strong gravity cut-off in the collision center of mass energies, we can produce black holes. We can make this prediction thanks to very restrictive nature of gravity. We know that many things in gravity are unique. Let me give an example: The higher you go in spin of a messenger of a given interaction, the more restrictive becomes the theory. For instance, if we start with spin-zero or spin one half, and go to spin one and higher, these higher spins always come with gauge invariance. There is no consistent theory of an elementary particle with the spin starting from one and higher, which does not exhibit the gauge invariance. The theory of spin-one is therefore much more restrictive than a theory of spin zero. It turns out that spin-two is even more restrictive. For example, you can have a bunch of massless spin-one particles and there is no inconsistency in that. But, you cannot have this situation with spin-two. You can only have an unique spin-two particle which is massless, and this is Einstein's general relativity. Let's start again with gravity. So, yes, if we start producing black holes at the LHC, we suck up all the energy of the high energy collision into producing black holes. In other words, black holes are the most probable end-point of the highest energy collisions. Sooner or later we start forming them at sufficiently high energy and starting from there they will dominate the collision process. At higher energies we will get bigger and more and more classical objects, which will then decay softly and more and more democratically into the all possible stuff.

- *Parsamyan:*

How should a typical black-hole event look like in Atlas, for instance? Will it be easily distinguishable?

- *Dvali:*

The above-mentioned softness and mass-dependent democracy are model-independent properties of micro black-holes. On top of this, in any given theory (for example, depending on what manifold you compactify, what branes you have there, etc) black holes may exhibit specific properties which differ from model to model. A model-independent qualitative property of the black hole production is that, at threshold center of mass energies, it is virtually impossible to distinguish a micro black hole from a particle. So, you produce a state which evaporates so quickly, that you cannot even call it an evaporation process in a well-defined thermal sense. It is rather a quantum decay of a particle. So, it just decays as any other new unstable particle would do. So, for instance, at the threshold energies of their production, it will be virtually impossible to distinguish a black hole from say a supersymmetric particle with the same gauge charges.

However, the distinctive feature of a black hole is that once you go above the threshold energy of their production, the two specific properties set in. Let us compare the following two situations. Let's say we perform two collision experiments in parallel and produce a new resonance, e.g. Z' on one side, and a black hole on the other side. How do these two events compare? Above the threshold of Z' production, we get harder and harder products coming out of the collision. For the black hole this is different. Once you go well above the threshold, instead of producing a harder stuff, you produce a bigger black hole, which is softer. It's like producing a more fuzzy classical object. It recycles the collision energy and decays into softer quanta. So, with higher collision energy, the resulting black hole is bigger, the corresponding Hawking temperature is lower and it delivers softer and softer stuff. There is a correlation between the center of mass energy, the mass of the object that you produce and the softness of the decay products that you observe. The picture is as if we produce a heavy particle that almost continuously looses its mass, because it evaporates. On top of these model-independent properties, you have a bunch of exotica which are model dependent. And also there is one model independent aspect that should be there, that the black holes should become more and more flavor democratic with the higher mass. So, maybe LHCb might be the right place to look for the flavor properties, because somehow for heavy enough black holes, you should see all the Standard Model flavors produced democratically in the final state.

- *Gamayun:*

How does Planck mass appears in the renormalizable SM?

- *Dvali:*

Planck mass does not appear in the SM. The Planck mass is the parameter in Einstein's gravity. Without gravity, the hierarchy problem in the SM would be much less severe. So would be much less reason to worry. What gravity brings is that point-like particles (including the virtual ones) heavier than the Planck mass, are no longer particles, but rather the classical black holes. Thus, they are big classical objects. Such virtual states don't have any significant quantum effect, they are too heavy to contribute in the quantum loops, just as other classical objects. This is why, once we take into account gravity, the Planck mass appears to be a natural regulator, and the hierarchy problem becomes real.

- *Bettini:*

Prof. Zichichi recalled this morning the "gap problem", namely the difference by about two orders of magnitude between grand unification scale and Planck scale. The argument assumes SUSY particles exist. On the other hand, if you add the number of helicity states of SUSY particles to the 140 of the SM, you have a total number that might move the Planck scale down and close or almost close the "gap".

- *Dvali:*

Indeed, the existence of many particles species in grand unified theories explains the consistency of the gap. If we try to unify couplings at the Planck mass, it is inconsistent precisely because of this result. If we take a typical GUT (e.g., SO(10)) the number of helicities that are available there are of the order of a thousand. This tells us that the GUT scale has to be below the Planck scale by consistency with quantum gravity scale. Whenever we have a new physics that comes with the point-like particle species, this new physics cannot happen at the Planck scale, since particle species make gravity strong before reaching the Planck mass.

- *Zichichi:*

Supposed there is no difference from Fermi to the Planck scale, at each TeV you have hundreds of particles.

- *Dvali:*

Therefore this would restrict those kind of theories, because these particles have to obey this bound. For instance, if we consider the compactifications Michael Douglas was talking about, this counting is interesting, since the point is that if you talk about topologies of the manifold, these topologies are subject to index theorems: Non-trivial topologies usually imply that you have some zero modes in the manifold. This is precisely the great thing about this complementary approach, because it tells you that somehow we can restrict the possible topologies by simple counting of modes. If you have a complicated topology, which gives you too many states, it has to obey the bound on species, or something highly non-trivial must be going on in this theory. This is a tool which would restrict any new physics which comes with many degrees of freedom.

- *Tannenbaum:*

We have increased the energy of accelerators by a factor of almost $\sim 10^6$ (10MeV to 10TeV) in less than a century. If we continue at this rate, it is conceivable to be able to accelerate protons to the Planck energy in less than 200 years, no big deal in "Erice time". If you collide two protons, each with energy greater than E_{Planck}, could you then make a conventional gravity black hole which would then accrete and destroy earth?

- *Dvali:*

Let's suppose we collide protons at the Planck energy, and assume that there is Einstein's gravity valid all the way up to these energies. At that point you start producing strongly interacting objects, which you could call quantum black holes. The black holes that you produce at M. Planck would not be democratic and would decay like very short-lived particles within the Planck time. These black holes would pose no danger for the earth. So, you have to go way above the Planck mass in energy in order to produce a classical black hole that would be a problem for the earth. Even in such a case, it still would not be a problem for the universe, of course. To pose such a problem, you would need to produce a black hole which is as large as the universe (that is, to over-close the Universe), but for this, there is simply not enough energy within the observable part of the Universe itself. We know this cosmologically, since the Universe is extremely flat, meaning that the energy density is very close to the critical, but not larger. Coming back to the earth, say, if you make a black hole of radius of order 1 cm, which is the same as the mass of the earth, it will grow pretty quickly. But, to make a black hole of 1 cm radius, you need much more than the Planck energy. Then yes, once you have enough energy localized within the Schwartzschild radius, you start producing such black holes. But, with the modern-like technology, you would probably destroy your apparatus before you even form such a big black hole, because you will have to localize the enormous energy (of the order of the earth's mass) in that tiny (1 cm) radius. So, in general, I think it important to educate the general public who is worried about this kind of stuff. What they have to understand is how much we really know, and that the scientific facts are not just somebody's opinion, these are conclusions that are based on excellent arguments.

- *Baldini:*

Can you repeat your answer to the question why black holes are not produced abundantly by very high energy cosmic rays that eat the earth?

- *Dvali:*

Assuming that strong gravity scale is TeV, they are being produced, but extremely rarely. Unfortunately, the high energy cosmic rays are not (yet) a good probe for new physics, such as quantum gravity, since we need collisions with high momentum transfers, i.e. hard collisions with momentum exchange of the order of a TeV. These kind of detectable events in high energy cosmic rays are extremely rare. When they go through the atmosphere, the collisions are typically much softer. The high momentum transfer events are rare, but of course they must be happening. So, that's why we think that the black holes should be produced in these events. However, the current experimental statistics is still not good enough for putting the strong restrictions. But, this is a good direction for the future experiments.

- *Badziak:*

If we could have acceleration with energies much above the threshold, we would produce quasi-classical black-holes. Does it mean that the lifetime of these black holes could be of the order of a second? Can it accrete matter and become dangerous for the earth?

- *Dvali:*

Of course, it depends on your energy in the collision, whether you can go well enough above the threshold. There are some exotic possibilities in which you can prolong the lifetime of a black hole even for moderate (around TeV) energies. For instance, if you have a negative curvature in the extra dimensional bulk (with one dimension transverse to the brane) . In such a case, the gravity of the brane is known to be repulsive. So the otherwise-unbounded particles are pushed into the extra dimension by brane gravity. In such a case if you produce a particle on "our" brane, it can accelerate away from the brane, and then collide with its orbifold image at the opposite end of the extra dimension and form a black hole. Because of warping of the extra space, this black hole can look pretty classical from the point of view of the local geometry in the bulk, and thus, be pretty long-lived. Thus, if you will produce this state, it will sit there for a long time, and for us it will look like a stable long-lived particle. Hypothetically, such a black hole could grow if we could feed it with energy. However, in practice, it is impossible to reach a dangerous growth, since the accretion rate is not high enough. We would need a much higher accretion rate in order for this black hole to grow and become dangerous. So it will not pose any danger, but such a black hole could have a very spectacular signature if it carries an electric charge. In such a case, the black hole gets attached to the brane via an electric flux tube. So it will sit there and gradually loose its mass and then be pulled back by the electric flux tube and explode into the SM particles. None of these exotica can be dangerous for us, because the accretion rate is negligible compared to their decay rate.

- *Aschieri:*

I would like to understand your picture of a fundamental scale of gravity, dependent on the number of particle species in the context of inflation, where you just have one field.

- *Dvali:*

One of the uses of this bound is to restrict the inflationary vacua of the theory. If you have many particles then your scale of inflation is also restricted because you cannot make inflation with a Hubble parameter exceeding the scale of quantum gravity. At least you can no longer treat your theory in the usual inflationary, classical, way. This is a very powerful restriction on the number of particles and the quantum gravity scale in the theory. So we can restrict the possible inflationary vacua by taking into account the relation between the particles that are available in a given inflationary vacuum and their inflationary coupling parameters.

- *Kurkov:*

In your consideration you need lower order of perturbation theory for quantum gravity. It is well known that quantum gravity is non-renormalizable theory. Why do you think it is sufficient to use such incorrect theory for description of fundamental problem, such as "Problem of Hierarchy"?

- *Dvali:*

In my lecture I did not use any quantum-perturbative description of gravity or anything like this. In deriving the bound I entirely relied on the black hole argument, which is entirely non-perturbative. So in deriving this bound, and in particular the part that gravity has to become strong earlier than m_{Planck}, I didn't use any perturbation theory. Once we discovered that gravity is becoming strong earlier than in Einstein's theory, below the strong gravity scale we can use the graviton propagator, which is not a quantum gravitational concept. Let's consider something entirely classical, that you can understand completely in terms of the Newtonian laws. For example, consider a planet orbiting around the sun. You can equivalently understand the same process by the exchange of a virtual graviton. This obviously is not quantum gravity. To describe the perihelion precession of mercury, you have to take into the account non-linearity. In terms of the Feynman diagram, it is just an exchange, involving a three-graviton vertex, between the sun and the Earth, and this gives exactly the mercury perihelion precession as computed by Einstein. Thus, in General Relativity, this process can be described in two equivalent ways. One is the geometric way of computing the metric by solving the Einstein's equation by expanding perturbatively in the small parameter which is the ratio of the gravitational radius of the sun and the distance from the sun's center. The equivalent way is to compute tree-level Feynman diagrams with virtual graviton exchanges.

- *Arnowitt:*

The question being asked is that if you had a "Lamb shift" correction and the theory is non-renormalizable, the corrections may be enormous, whether or not the one graviton exchange agrees with experiment. That needs a "threshold theorem" that says one graviton is correct for large distances (low energy).

- *Dvali:*

The point is, that none of these diagrams contain loops. If there were loops, there would be contributions which would simply renormalize the low energy parameters, e.g., gravitational coupling, or the masses of sources and probes. To compute Keplerian orbits you don't have to know quantum gravity, otherwise the nature would be non-understandable.

- *Douglas:*

One can better answer Kurkov's question by bringing in the concept of effective theory. We know empirically that the long distance theory of gravity is general relativity. Dvali's arguments work from long distances, where the effective theory is good, and tries to say at what short distance it will break down.

- *Dvali:*

Suppose you have some ultraviolet theory of quantum gravity, which is most likely finite. The way string theory accomplishes this we know, because it gives you interference from infinite number of higher spin states and this works in a beautiful way. Let's say there is a string theory of quantum gravity at short distances, and suppose we do all the proper renormalization, and include these corrections in some parameters which could for instance run with energy. Then, I would try to write down some effective theory for this interaction at low energies. This effective action will be expandable in powers of some inverse mass scale – the scale where gravity gets strong and quantum. There will then be the following terms in the action. We would start from Einstein's action and then there will be higher order invariant terms which are suppressed by appropriate powers of the quantum gravity scale. For any physical problem for which the curvatures are smaller than this fundamental scale, you can ignore these corrections and you are in Einsteinian gravity. This scale is not the Planck scale if you have N particles in your theory, this scale is much lower. The corrections from quantum theory of gravity will start much earlier. Once you are below this strong gravity scale you can use tree-level Feynman diagrams for the classical gravity description.

- *Vikman:*

Comment: We could invert the logic in the bound $M_* = M_{Planck}/\sqrt{N}$. If we measure B mode for CMB and confirm the simplest inflationary scenario with $M \sim 10^{13}$ GeV we could infer that $N_{max} \sim 10^{12}$ different fields with masses less than 10^{13} GeV.

- *Dvali:*

Yes, that's an excellent comment. You can invert the argument and get information about the number of fields if you know what is the scale of inflation.

CHAIRMAN: G. DVALI

Scientific Secretaries: P. Burda, A. Smirnov

DISCUSSION II

- *Cifarelli:*

Can you tell more about how you derived flavor hierarchy in masses, which you have shown just at the end of the lectures this morning?

- *Dvali:*

Yes, unfortunately I was a bit telegraphic in my discussion about flavor masses. There are several points which come from two different directions. First, this idea can already be formulated within four-dimensional field theories. We can say that perhaps the smallness of mixings between different generations of quarks and leptons and their mass hierarchy has to do with them having a notion of a ``nearest neighbor'' in the generation space. So, let us say, that the third generation is the heaviest, because it is the closest to the Higgs and thus has the strongest coupling to it. And then, the other generations somehow get masses from the mixing with the third generation, and this mixing is hierarchically due to the fact that some of the generations are closer to each other. This idea was suggested by Fritzsch within four-dimensions. This is the concept of nearest neighbor families. Of course, in a four dimensional theory there is no precise sense of what this concept of being a neighbor means, because there is no space in which the distance between families of quarks and leptons can be defined. Now, suppose we have a theory in which we have N particle species, and the larger is N so much the better for our argument. When we have these N particle species, what gravity tells us is that immediately this particle species develop sense of locality in the space of species. In other words, a new space dimension emerges. Of course, whenever we have particle species, which we can label by some index j which runs from 1 to N, then, I can always say, that this index j is now a new coordinate. In the other words, I can think of some imaginary space of species in which the distance is measured by j. This is not what we are talking about. Without gravity this labelling is just a mathematical triviality, because, of course, whenever I have N objects, I can always label them, and then declare that my label is a coordinate. Now, with gravity, what is highly non-trivial is that this index "j" really acquires a meaning of a real dimension, in the sense that it does affect gravity. And, moreover, as we have seen from the considerations of unitarity in the production of micro black holes, there is a sense of true locality in this space. So let's think of this "j" as a dimension and let's think of the localization sites in this dimension corresponding to different values of "j". So, consider two localization sites, say j= 1 and j=K, in this dimension. Now, suppose I do a collision experiment at the first site (j=1) in this imaginary space of species. That is, I am colliding particles with label j=1. So, as I argued, by unitarity, the smallest black holes predominantly decay back to the same flavor j=1 from which they were created. When we increase the energy in the collision, the resulting black hole becomes heavier and becomes more democratic. It can decay not only in j=1 but into the other species, say j=K, also. So the picture literally is, as if the species are separated in a true dimension. And the existence of these species changed gravity classically at short distances. So whenever we introduce many species, they are forced to reassemble themselves in a real dimension. And of course, this argument works better and better when N is large. If we extrapolate this result to the standard model flavors, we conclude that they also are separated in a new dimension. This indicates that since flavors live in a real microscopic dimension, the hierarchy of the fermion masses and mixing angles could be the result of them being neighbors in real dimension at different distances from the site where Higgs lives in this dimension.

- *Kovetz:*

Regarding that image you had, about co-dimension one brane and black hole near, and the brane tube that is formed between them, could you explain how this brane tube is formed?

- *Dvali:*

Yes, this story with a gauge tube attached to the brane is a very interesting one on its own. The point goes as follows. Suppose, you want to localize a gauge theory on a surface (on a lower dimensional submanifold) in such a way that from the point of view of an observer living on this submanifold the gauge theory is in a Coulomb phase. So, in other words, if I take point-like charges, let's say plus(+) and minus(-), localized on the surface, they interact with a Coulomb law corresponding to the dimension of the surface and not to the dimension of the whole space in which this surface is embedded. Then, you can prove that in order for this to happen, the bulk of the theory should have a property of a dual superconductor. That is, the charges in the bulk must be in the confining phase under the same gauge interaction. Why? Because, if this is not the case, then the electric flux lines created by charges on the surface will end on the outer medium, and will create images, and charges on the surface would be screened. This would mean that your effective low dimensional theory on the surface, will not be in the Coulomb phase, but rather it will be in the Yukawa (Higgs) phase. So, to localize the massless gauge field on a surface, the bulk should have property of repelling the flux lines. And this is the property of the dual superconductor. Whenever this is the case the gauge field flux lines can only enter the bulk in form of the flux tubes, that are either closed (closed strings) or terminate on the lower dimensional surface where the massless gauge field is localized (open strings).

- *Coccia:*

May you comment more on possible experimental tests for supporting some of your conclusions?

- *Dvali:*

If we are motivated by the solution of the hierarchy problem, which means that the strong gravity scale is about a few TeV, then, there are several predictions. First, there is a good indication that in such theories there must be a new degrees of freedom which open up around the sub millimeter scale. In the case of large extra dimensions, if the compactification radius is around hundred microns, one experimental signal would be simply change of the gravity law at that scale. This change is within the experimental reach in table-top studies. There is also a model-independent degree of freedom, which corresponds to the local fluctuations of the compactification volume. In other words, if I take some four-dimensional point given by coordinate X, let us say X = 0 corresponding to the location of this room, and if I squeeze the volume of the extra dimension at this four-dimensional location and then let it go, there will be a wave propagating in our three space dimensions. This is exactly analogous to the following situation. Consider for instance, an elastic tube. Let the coordinate along the tube describe our 3 dimensional space and the circle be a compact extra dimension. If I locally squeeze the tube (change the compactification radius) and then release it, there will be a wave propagating along the tube. From our (found dimensional) point of view, this wave looks like a scalar particle. There are arguments indicating that this scalar particle is typically light and would correct the Newton's force. This is another reason for the experimental tests of the Newton's gravitational law at short distances, and this test is extremely important, because it is very important to understand what is the validity domain of pure Einstein's gravity, and when the new degrees of freedom open up. Moreover, there also exists a whole spectrum of experiments which try to perform precision measurements of gravity at different length scales. For instance, Adelberger's group performed so far the most accurate measurement at the shortest length scales.There also are other ideas to use cantilevers – cryogenic oscillators to test gravity at short distances. Moreover, people are trying to improve precision at different length scales. In this respect, there is a very interesting experiment, which is actually on-going, the so-called Lunar Ranging –experiment.

Basically what they do is that they shoot laser beams to the moon, which get's reflected from lunar surface. There are mirrors on the moon's surface which were brought there by Apollo-11 mission in 1969. The statistics accumulated by now is so impressive, that basically today we know the lunar orbit by an accuracy of approximately half cm. There are new generation planned experiments which will improve that. So if quantum gravity scale is low, then there is a general qualitative expectation that we should look for new gravitational degrees of freedom and therefore we should look for deviations from Newton Law at any possible distance.

Other experimental signals of TeV gravity include strong gravity effects at LCH, such as productions of black holes, which I have partially covered in my lectures. See lectures by Lust for string productions at LHC.

- *Giannuzzi:*

About the figure with one black hole and one brane, and a string between them. The string breaks in order to reach minimal energy and the brane enters the black hole. Does this happen for any black hole or does it depend on the mass of the black hole?

- *Dvali:*

It turns out that this is always true. In other words, it is true whenever we want to stabilize the black hole at the finite distance from the brane. The configuration is the following: even if originally this black hole was attached to the brane by a tube of the electric flux, the balance between the different parameters (such as, the black hole mass, warping of the space and the brane tension) is such that it is energetically favorable for the string to shrink and pull out the brane tube. In other words the flux tube gets replaced by the brane tube.

- *Giannuzzi:*

Yes, but does it depend on the mass of the black hole?

- *Dvali:*

No. That is the point, that the effect is independent on the mass, as long as the black hole is stabilized at a finite distance from the brane. Actually what we have learned, for me was a surprise. Because we could have said that 4-dimensional black hole no-hair theorem should not apply here. Since we have a black hole whose radius is smaller than the size of extra dimension, then from the point of view of 4-dimensional observer, this is not a 4-dimensional Schwarzschild black hole. So why should this black hole obey a 4-dimensional no-hair theorem? And a priori there would be no crime if these micro black holes would not obey it. But amazingly they do, no matter how small are they. Even if they are black holes of some underlying high-dimensional theory and from the point of view of 4-dimensions they are not really Einsteinan black holes, they still obey the 4D no-hair theorem. This is amazing and there must be some fundamental reason for it.

- *Monin:*

As far as I understood the cross section for black hole creation is purely geometrical meaning that it grows with the energy. What about unitarity in this case? And second question: in other paper there was a discussion of these processes, it was concluded that first, the production of a classical black hole is exponentially suppressed and, second, that the production of arbitrary number of smaller black hole is more probably (due to the splitting of the initial energy).

- *Dvali:*

So, the first question is whether the cross section at high energies violates unitarity. I do not see why it should violate unitarity. With energy growth, you just form a bigger object and then this bigger object is very long lived. For instance, let us take an extreme case. Suppose we collide say e-plus(+) and e-minus(-) or two photons with the impact parameter less than 3 km and with the energy of the solar mass. We will form some macroscopic black hole with the solar mass, with Schwarzschild radius of about 3 km, and then this object evaporates very slowly. A priori, there is no unitarity violation in this process. So, the second question is actually a very interesting one. What you are referring to is that to produce a non-perturbative objects, for instance, a soliton, in two-particle collision is an exponentially suppressed process. The reason why there is this exponential suppression in the production of solitons is precisely because they are solitons. For instance, let us take a topological soliton. In order to form a soliton you need many particles in a coherent state, and usually there is an exponentially suppressed probability to get the required coherence in the collision of two particles. So in the weakly coupled field theory the production of solitons, for instance, of magnetic monopoles, are expected to be exponentially suppressed. So may be if you have many particles, for example, very heavy ions colliding, and can create a very many particles environment then this is a different story. But in a two particle collision the solitons production must be exponentially suppressed. I think this was the intuition, when people argued that may be even production of black holes should be exponentially suppressed, because the black holes are also classical objects. So if it is exponentially suppressed to produce monopoles, why it is not the same for the black holes? In this respect, the black holes a crucially different. The reason is, that you can think about the black holes as the Noether charges, rather than solitons. The point is like this, think about electric charge, if I have an elementary electric charge, then at large distances it does produce Coulomb electric field. In a sense the electric charge does imply at large distances the classical configuration of the field, but to produce a charge in two particle collision, is not exponentially suppressed, because charges is Noether charge. The same is true about the black holes. Once you bring enough energy within the small volume, it has no choice other than to form a black hole. So the black hole does not require the coherence that is essential for soliton production.

- *Takaki:*

In your lecture this morning, you always talked about small black holes, big black holes... But you never give us some kind of units. For me it would be more easy to see how big they really are.

- *Dvali:*

Of course, as I already said in my lecture, the numbers are model dependent. First of all, we don't know where the scale M* is. But in order to be interesting for LHC, this scale has to be around few TeV and not higher. So, then you can ask the following question – suppose this scale is 1 TeV, how displaced could be the standard model flavors in extra dimensions? That is a valid question. And then, there is a limit to how much you can displace these flavors. Why? Because you cannot displace an electron and the muon arbitrary far in extra dimension, because they have to share a photon and other gauge fields. In the other words, by gauge invariance, we know that we cannot displace electron and a muon, because they both have to couple to a photon, to the same massless photon. And the fact, that we have not seen Kaluza-Klein excitations of photon or of a Z-boson, or other gauge fields, puts the bound on how much you can displace standard model particles within the extra dimension. And again, these bounds are about inverse few hundred GeV. So, typically the distance between the standard model particles in the extra dimension cannot be larger than inverse few hundred GeV. This is a pretty rigorous bound, because it comes from the fact, that different standard model particles share the same gauge interactions. Of course, if we have some hypothetical hidden sector, they might be displaced arbitrarily in the extra dimension, sine the only thing we share is gravity.

Supergravity: Foundations and Applications

Anna Ceresole[a,b] and Sergio Ferrara[b,c]

[a] *INFN, Sezione di Torino, Via Pietro Giuria 1, 10125 Torino, Italy*

[b] *Physics Department, Theory Unit, CERN,*
CH -1211, Geneva 23, Switzerland

[c] *INFN - Laboratori Nazionali di Frascati,*
Via Enrico Fermi 40, I-00044 Frascati, Italy

ABSTRACT

We briefly discuss the foundations of Supergravity as a gauge theory of supersymmetry, its mathematical structure and the basic lagrangians for N-extended supersymmetry. As applications, we describe the basic elements of the supersymmetry breaking mechanism and various aspects of the physics of extremal black holes.

Contribution to the Proceedings of the 47th Erice International School of Subnuclear Physics "The most unexpected at LHC and the status of high energy frontier"
Erice, Sicily: 29 August - 7 September 2009
Lecture delivered by Prof. Sergio Ferrara

Contents

1 Foundations

Supergravity, the gauge theory of supersymmetry, was discovered in 1976 [1, 2] and since then it has been in the mainstream of a broad field of research focused on the quest of a unified theory for all fundamental quantum interactions. Its many facets range from models for particle physics phenomenology, Kaluza-Klein supergravity and extra dimensions of spacetime, understanding the geometric nature of fundamental interactions in extended supergravities, and black hole physics [3, 4, 5, 6, 7]. Nowadays, supergravity is also broadly recognized as the core of string theory, where it arises as an effective field theory limit.

The algebraic structure of four dimensional supersymmetry was brought to the attention of theoretical particle physicists in the second half of 1973 by Wess and Zumino [8]. They had been inspired by a similar structure, found by Gervais and Sakita in 1971 [9], already present in two dimensions in the dual-spinor models constructed in 1971 by Neveu and Schwarz [10] and by Ramond [11]. Supersymmetry had first appeared in mathematics in 1971 [12], although with a misinterpretation of Majorana spinors, and in early 1973 by Volkov and Akulov[13] when spontaneously broken supersymmetry was brought up to explain the masslessness of the neutrino as a Goldstone fermion. The relevance of supersymmetry for quantum field theory, especially in view of its remarkable ultraviolet properties and its marriage with Yang-Mills gauge invariance, was soon established in early 1974[14], and it brought to a series of dramatic events that culminated, after two intense years, in the formulation of supergravity by Freedman, van Nieuwenhuizen and S.F., at the Ecole Normale in Paris and at SUNY Stony Brook. Soon after, a simplified version (first-order formulation) was presented by Deser and Zumino [2]. While in the second formalism, the spin-3/2 four-fermion interaction has the meaning of a contact term (similar to seagull terms in scalar electrodynamics or non-Abelian gauge theories) required by fermionic gauge invariance, in the first-order formalism it has the meaning of a torsion contribution to the spin connection from spin-3/2 matter. This discovery also implied that any supersymmetric system coupled to gravity should manifest local supersymmetry [3, 15, 17]

From a philosophical point of view, supergravity is an extension of Einstein's theory of General Relativity based on an enlargement of the spacetime symmetry: the principle of General Covariance is extended to contemplate diffeomorphisms in a new entity, called "Superspace"[18] . The mathematical background of supersymmetry [19] and supergravity relies on an extension of ordinary Lie Algebras that includes the supersymmetry generators, which require a graded commutator. Surprisingly enough, unlike General Covariance, local supersymmetry can be spontaneously broken via a mechanism that is analogous to the Higgs phenomenon of non abelian gauge theories. This gives rise to a supergravity potential that allows to make contact with phenomenological applications [15, 16]. Among other very interesting applications of supergravity, there is the outfall of the electric-magnetic duality symmetry of the theory coupled to scalar matter[21], which plays a fundamental role in the study of black hole solutions: for extremal black holes, the effective potential of the theory leads to the computation of the universal Bekenstein-Hawking entropy formula [7].

1.1 Mathematical Structure

Supersymmetry and supergravity are based on an algebraic structure called "Graded Lie Algebra" (GLA) [19, 22]. GLA's are extensions of ordinary Lie Algebras. The commutation relation among a set of elements A_m,

$$[A_m, A_n] = f_{mn}{}^{l} A_l \tag{1.1}$$

of a D-dimensional Lie Algebra, are extended to a d-dimensional grading representation by introducing a new set of generators $\{Q_\alpha\}$, such that

$$[A_m, Q_\alpha] = S_{m\alpha}{}^{\beta} Q_\beta \tag{1.2}$$

$$\{Q_\alpha, Q_\beta\} = F_{\alpha\beta}{}^{m} A_m \tag{1.3}$$

satisfying graded Jacobi identities

$$[X_A, X_B\}, X_C\} + \text{cyclic}(ABC) = 0 \tag{1.4}$$

The graded Poisson (commutator) bracket is given by

$$[X_A, X_B\} = X_A X_B - (-)^{AB} X_B X_A \,, \tag{1.5}$$

where $A = 0, 1$ for even or odd generators respectively: in this way, it becomes an anticommutator for two odd generators. In the superalgebra of supersymmetry, Q_α are spinorial generators (representations of Clifford Algebras), while the A_m bosonic generators are associated with the Poincarè (or Conformal) algebra plus internal symmetries: more specifically one can have

- Poincarè : $ISO(D-1,1)$
- Conformal: $SO(D,2)$
- Anti de Sitter: $SO(D-1,2)$

The automorphism group of the superalgebra is the R-symmetry of the theory.

The theoretical motivations behind this supersymmetric extension is the enlargement of space-time with a number d of "fermionic dimensions" . The translation generators P_μ associated to the spacetime coordinates x^μ (with $\mu = 1, \ldots, D$) are extended by supercharges Q_α, with $\alpha = 1, \ldots, 2^{D/2}$ (or $2^{D/2-1}$ for chiral spinors) for even D, while $\alpha = 1, \ldots, 2^{\frac{D-1}{2}}$ for D odd. P_μ and Q_α act as supertranslations in an abstract space made of bosonic coordinates x^μ and fermionic coordinates θ^α, such that

$$\theta^\alpha \theta^\beta = -\theta^\beta \theta^\alpha \tag{1.6}$$

that is, the fermionic coordinates are Grassman numbers. The space $M_{(D,d)}$ with points (x^μ, θ^α) is called "Superspace" by physicists [18, 23] and "supermanifold" by mathematicians [24, 25] and it is labelled by the dimensions of the bosonic and fermionic coordinates respectively. For example, in $N = 1$, $D = 4$ supersymmetry, one has $M_{(4,4)}$, while M-theory [26]is defined on a $D = 11$ dimensional spacetime and it is based on the supermanifold $M_{(11,32)}$ [27, 25, 24].

Another essential tool in supergravity is given by properties of **Majorana Spinors in D-dimensions**. Reality properties of spinors depend on the signature of space-time $\rho = s - t$ ($|s-t|$) and moreover spinors enjoy Bott periodicity (mod 8) [24, 28, 22]. Symmetry properties of spinors depend on the space-time dimension $D = s + t$ rather than on ρ. Using the above reasonings, one obtains that spinors are real whenever

1) D even:

- $\rho = 0$ (real, chiral)
- $\rho = 2, 6$ (real, not chiral)
- $\rho = 4$ (pseudoreal , chiral)

2) D odd:

- $\rho = 1, 7$ (real)
- $\rho = 3, 5$ (pseudoreal).

Note that in Physics one calls chiral spinors what in Mathematics are called semi-spinors, and physicists call Majorana (or pseudo-Majorana) what mathematicians would call real or pseudoreal (quaternionic) spinors[22, 24].

Massless particles in D-dimensions are defined by

$$P_\mu P^\mu = 0 \tag{1.7}$$

and they form irreducible representation of the little group $SO(D-2)$ of the light-like momentum. Graviton states correspond to a symmetric traceless tensor with dimensions $\frac{(D-2)(D-1)}{2} - 1$. Rank k antisymmetric tensors, correspond to $\frac{(D-2)(D-3)\cdots(D-k))}{k!}$ massless states. Spinors correspond to irreducible representations of spin $D-2$. For spin 1/2 , if D is even there are $2^{\frac{D-2}{2}}$ or $2^{\frac{D-4}{2}}$ states for chiral spinors, and $2^{\frac{D-3}{2}}$ states for D odd. For the spin 3/2 gravitino, which is a gamma-traceless vector-spinor, one has $(D-3)2^{\frac{(D-2)}{2}}$ and $(D-3)2^{\frac{D-4}{2}}$ states for chiral spinors in D even, $(D-3)2^{\frac{(D-3)}{2}}$ states for D odd.

Supersymmetry algebras in D dimensions are associated to a supersymmetry charge Q_α that, in D even, is either a Majorana spinor, with $2^{D/2}$ states, or chiral Majorana spinor, with $2^{D/2-1}$ states. More precisely, in ($D = 2\,mod\,8$) the supercharge is chiral Majorana, in ($D = 6\,mod\,8$) it is chiral pseudo-Majorana and in ($D = 4\,mod\,8$) is Majorana (not chiral). In fact, in the latter case, chiral spinors are complex, and a Majorana spinor is made by a chiral spinor plus its (antichiral) complex conjugate. For D odd, the dimension of Q_α is $2^{\frac{D-1}{2}}$ and it is Majorana for $D = 1, 3$ mod 8, and pseudo-Majorana for $D = 5, 7$ mod 8.

The R-symmetries are: $SO(N)$ for $D = 2, 3, 9, 10, 11$, $USp(2N)$ for $D = 5, 6, 7$ and $U(N)$ for $D = 4, 8$. A physical requirement for supergravity theories with a graviton being the top elicity state is that $dimQ_\alpha \leq 32$. This imposes a restriction on the spacetime dimension, $D \leq 11$ and the number of supersymmetry charges is $N \leq 8$. at $D = 4$ [29, 30].

The **SuperPoincaré Algebra with p-brane charges** [31] corresponds to a central extension of the supersymmetry algebra given by

$$\{Q^A_\alpha, Q^B_\beta\} = \ldots \gamma^{\mu_1\ldots\mu_p}_{\alpha\beta} Z^{AB}_{\mu_1\ldots\mu_p} + \ldots \tag{1.8}$$

Electric and magnetic charges q_p and p_{D-p-4} obey a Dirac-Schwinger-Zwanziger quantization condition, $q_p p_{D-p-4} \sim 2\pi n$. p-branes can be BPS saturated . For maximal supersymmetry the horizon geometry of black extremal p-branes is universal [32] and is given by $AdS_{p+2} \times S_{D-p-2}$: this is the extremal p-brane horizon geometry, which is widely used in the AdS/CFT correspondence [33]. Dyonic branes can exist for $p = D - 4 - p$, where they satisfy the quantization rule $q_\Lambda p'^\Lambda \mp q'_\Lambda p^\Lambda = 2\pi n$ (minus for p even, plus for p odd). This is a reflection of electric-magnetic duality, given by $Sp(2n, R)$ for p even and $SO(l, m)$ for p odd [20, 21].

The basic example is M theory, which is 11-D supergravity [34]. The theory contains a graviton $g_{\mu\nu}$ with 44 states, a three index antisymmetric tensor $A_{\mu\nu\rho}$ with 84 states and one gravitino $\psi_{\mu\alpha}$ with 128 states: this leads to 128 bose and 128 fermi states, for a total of 256 states.

The $N=1$ susy algebra reads

$$\{Q_\alpha, Q_\beta\} = (\gamma_\mu)_{\alpha\beta}P^\mu + (\gamma_{\mu\nu})_{\alpha\beta}Z^{\mu\nu} + (\gamma_{\mu_1\cdots\mu_5})_{\alpha\beta}Z^{\mu_1\cdots\mu_5} \tag{1.9}$$

which contains no scalar central charge, but it has only 2-brane (electric) and 5 brane (magnetic) charges.

Conformal supersymmetry in D-1 dimensions is related to supergravity in AdS(D) space, where $AdS(D) = SO(D-1,2)/SO(D-1,1)$. AdS(D) supergravity exists for $D \leq 7$, which implies that superconformal field theories have the highest dimension D=6 [29]. The bound is saturated by M-theory compactified on $AdS(7) \times S_4$. The relevant simple superconformal algebras are some real forms of the so called unitary $U(N,M)$ and orthosymplectic series $OSp(N,2M)$. For $D=4$ we have $U(N|M=(2,2))$ with $N_{max}=4$. For $D=3,6$ we have $OSp(N|2M)$ with $N_{max}=8$ and $M_{max}=2$. For $D=5$ we have the exceptional superalgebra F_4 [19].

The use of graded Lie Algebras in lagrangian field theories has allowed physicists to evade the famous **Coleman Mandula theorem** [35] by replacing it with its supersymmetric version, elaborated by Haag, Lopuszansky and Sohnious[36]. Briefly, the HLS theorem states that supersymmetry is the most general relativistic symmetry of an S matrix which enjoys Poincare' symmetry and describes the scattering on a finite number of massive particles. The bosonic symmetry of the S-matrix still remains a direct product of its relativistic symmetries times an internal symmetry.

Let us remember that elementary particles in four dimensions are classified by irreducible representations of the Poincare' group. For finite mass M and spin J they have $2J+1$ states, for zero mass and helicity $\lambda \neq 0$ they have two states of opposite helicity ± 1. Supersymmetry assembles these states in supermultiplets having the same mass but different spin content. The HLS supersymmetrry algebra in four dimensions is

$$\begin{array}{ll} \{Q^A_\alpha, Q_{\dot\alpha B}\} = 2(\sigma_\mu)_{\alpha\dot\alpha}P^\mu\delta^A_B & \\ \{Q^A_\alpha, Q^B_\beta\} = \epsilon_{\alpha\beta}Z^{AB} & Z^{AB} = -Z^{BA} \\ [Q^A_\alpha, P^\mu] = 0 & [Q^A_\alpha, Z^{BC}] = 0 \\ [Z^{AB}, Z^{CD}] = 0 & [Z^{AB}, P_\mu] = 0 \end{array} \tag{1.10}$$

where Q^A_α and $Q^A_{\dot\alpha}$ are related by hermitian conjugation, $Q^A_{\dot\alpha} = (Q^A_\alpha)^*$ as we have been using here Weyl and Van der Veerden spinors according to Bagger and Wess[37]. Central charges appear at least in two important physical applications: in $N=2$ and in $N=4$ spontaneously broken supersymmetric gauge theories, they are associated to the massive spectrum of both perturbative massive vectors and non perturbative monopole-dyons massive states[38] . In supergravity they play a crucial role in the description of extremal black holes, their BPS properties, and the attractor mechanism[40]. In all these theories, massive short multiplets correspond to a representation of the HLS superalgebra with non-vanishing Z^{AB}. Through an R-symmetry rotation, Z^{AB} can always be skew diagonalized and brought to "normal form" [49]

$$Z^{AB} = \begin{pmatrix} z_1\epsilon & & 0 \\ & \ddots & \\ 0 & & z_{N/2}\epsilon \end{pmatrix} \tag{1.11}$$

where N even has been assumed, the skew eigenvalues are $z_1, \ldots, z_{N/2}$ and $\epsilon = \left(\begin{smallmatrix} 0 & 1 \\ -1 & 0 \end{smallmatrix}\right)$. The BPS bound demands that $M \geq |z_i|$. Saturation of the BPS bound gives the notion of BPS state. The minimal saturation $M = |z_1| > |z_i|$, $i \neq 1$, corresponds to a $1/N$ BPS state, while maximal saturation takes place for $M = |z_1| = \cdots = |z_{N/2}|$ for 1/2 BPS states [48]. The highest spin of the shortest massive multiplet in these two cases is $(N-1)/2$ and $N/4$ respectively.

For $N = 8$, supersymmetry and E_7 invariance imply that there are three kinds of BPS states, corresponding to 1/8, 1/4 and 1/2 preserved supersymmetry [48, 41, 42]. The 3/8 shortening is not realized.

Let us now give in some detail the structure of 4D multiplets in terms of Poincare multiplets as previously discussed. $N = 1$ massive multiplets, denoted by (M, J) contain $4(2J+1)$ states, corresponding to 4 Poincaré multiplets $(m; J-1/2, J^-, J^+, J+1/2)$. For massless multiplets, we get four states with helicities $\pm|\lambda|, \pm|\lambda+1/2|$. For $N \geq 1$, massive long multiplets correspond to $2^{2N}(2J+1)$ states with highest spin $J + N/2$. Conversely, the massless multiplets contain helicity states ranging form $-\lambda$ to $-\lambda + N/2$, the smallest multiplet having $\lambda_{max} = N/4$ for even N. PCT conjugation reverses the helicity, so that PCT self conjugate massless multiplets contain 2^{N+1} states unless $\lambda = N/4$, in which case the multiplet is self conjugate and contains 2^N states. This holds for $N = 4n$ (multiple of 4). The two standard examples of this latter situation are N=4 Yang Mills theory and N=8 Supergravity. In these cases the R symmetry lacks the $U(1)$ factor and becomes $SU(N)$, that is $SU(4)$ and $SU(8)$ respectively. A consequence of multiplet shortening and BPS saturation is the fact that, in spontaneously broken N=4 Yang Mills theories, all massive vector multiplets must be 1/2 BPS. Analogously, all massive Kaluza-Klein gravitons in toroidal compactifications must be 1/2 BPS.

1.2 Supergravity lagrangians and susy rules in N-extended supergravities

The basic geometric fields of N-extended supergravities in $D = 4$ are spin 2,1,0 bosons and spin 3/2 and 1/2 fermions:

$$V_\mu^a(x), A_\mu^\Lambda(x), \phi^i(x), \psi_{\mu\alpha\, A}(x), \lambda_I(x) \tag{1.12}$$

where Λ is the total number of vectors (graviphotons and matter photons), I is the number of spin 1/2 fermions (both from gravity and matter multiplets), and $A = 1, \ldots, N$.

$\phi^i(x)$, the scalar fields, are coordinates of a non-compact manifold $\mathcal{M}$, with positive definite metric $g_{ij}(\phi)$, whose holonomy group depends on the number N of supersymmetries. The theory is largely restricted by local symmetries such as spacetime diffeomorphism invariance, gauge invariance of spin 1 gauge fields and diffeomorphisms of the manifold spanned by the scalar field sigma model. N extended supersymmetry fixes to a large extent the lagrangian, in particular it restricts the nature of the scalar fields manifold. $N = 1$ chiral multiplet scalars are coordinates of a Hodge-Kahler manifold. N=2 vector multiplet and hypermultiplet scalars are coordinates of a Special-Kahler and a quaternionic manifold respectively, while for $N > 2$ the scalar manifold is a symmetric space uniquely fixed by supersymmetry. Theories with $N \leq 4$ allow in principle for an arbitrary number of matter multiplets. In non-linear σ-models based on G/H cosets, the vector field strengths and their dual are in a symplectic representation of G, while the fermions are in a representation of H.

The general form of a supergravity lagrangian is

$$L = L_B + L_{2F} + L_{4F} \tag{1.13}$$

where the bosonic lagrangian L_B is given by

$$\sqrt{g}^{-1}\mathcal{L}_B = -\frac{R}{2} + \frac{1}{2}g_{ij}(\phi)\partial_\mu\phi^i\partial^\mu\phi^j + Im\mathcal{N}_{\Lambda\Sigma}F^\Lambda \wedge^* F^\Sigma + Re\mathcal{N}_{\Lambda\Sigma}F^\Lambda \wedge F^\Sigma - V(\phi) \tag{1.14}$$

where $\mathcal{N}_{\Lambda\Sigma}$ is a complex symmetric matrix with $Im\mathcal{N} < 0$ and $\Lambda = 1, \ldots n$ for n vector fields. This implies that $\mathcal{N} \subset Sp(2n, \mathbb{R})/U(n)$. L_{2F} is the part of the lagrangian bilinear in the fermions, which in particular contains the fermionic mass deformations of the theory

$$L_{F2} = L_{kin} + S_{AB}\bar{\psi}^A_\mu\gamma^{\mu\nu}\psi^B_\nu + iN^A_I\bar{\lambda}^I\gamma^\mu\psi_{\mu A} + M^{IJ}\bar{\lambda}_I\lambda_J + hc \tag{1.15}$$

Here S_{AB}, N^A_I and M^{IJ} are the fermionic mass matrices which are related to the fermionic supergravity transformation laws as follows

$$\delta\psi_{\mu A} = \nabla_\mu\epsilon_A + \cdots + 2S_{AB}\epsilon^B \tag{1.16}$$

$$\delta\lambda_I = \cdots + N^B_I\epsilon_B\,. \tag{1.17}$$

Invariance under supersymmetry relates L_{F2} to $V(\phi)$ and in particular determines that

$$\delta^A_B V(\phi) = -3\bar{S}^{AC}S_{CB} + N^A_I N^I_B \tag{1.18}$$

with $(N^A_I)^* = N^I_A$ and $S_{AB} = S_{BA} = (\bar{S}^{AB})^*$.

For Lorentz invariant configurations, the supergravity potential $V(\phi)$ computed at the critical point $V_0 = V|_{\partial V=0}$ is the vacuum energy of the theory. If $N^A_I = 0$, meaning that supersymmetry is unbroken , then $V(\phi)$ is non positive and it only vanish if $S_{AB} = 0$ at its critical points $\partial_\phi V = 0$ (Poincare supersymmetry). If S_{AB} is different from zero, we have anti de Sitter supersymmetry. Note that broken supersymmetry may occur with vanishing vacuum energy, provided that at the critical point $3\bar{S}^{AC}S_{CA} = N_I^A N_A^I$ (with no sum over A). Also note that partial supersymmetry breaking may occur if N_A^I vanishes for some values of A.

So far we have identified in the local supersymmetry variations those terms which are relevant to discuss mass deformations in Lorentz invariant backgrounds. These deformations correspond to "gauged supergravities". Other relevant backgrounds are black hole solutions where the relevant deformations are induced by the gauge field strength fluxes, i.e. the electric and magnetic charges of the black holes. In order to write the relevant terms in the susy variations in this case, it is useful to split the I index of the spin 1/2 fermion as follows: $\lambda_I = (\Lambda_{MA}, \lambda_{ABC})$ where λ_{MA} are the matter fermions and λ_{ABC} are the fermions of the gravity multiplet. Then the relevant transformations are

$$\delta\psi_{\mu A} = \nabla_\mu \epsilon_A + T_{ABC\mu\nu}\gamma^\nu \epsilon^B + \cdots \tag{1.19}$$

$$\delta\lambda_{MA} = \cdots + T_{M\mu\nu}\gamma^{\mu\nu}\epsilon_A + \cdots \tag{1.20}$$

$$\delta\lambda_{ABC} = \cdots + T_{[AB\mu\nu}\gamma^{\mu\nu}\epsilon_{C]} + \cdots \tag{1.21}$$

in terms of the antiselfdual parts of the dressed graviphotons and matter photons T,

$$T^-_{M,\mu\nu} = h_{\Lambda,M}F^{-\Lambda}_{\mu\nu} - f^\Lambda_M G^-_{\Lambda,\mu\nu} \tag{1.22}$$

$$T^-_{AB,\mu\nu} = h_{\Lambda,AB}F^{-\Lambda}_{\mu\nu} - f^\Lambda_{AB} G^-_{\Lambda,\mu\nu}\,, \tag{1.23}$$

and the selfdual part T^+ of T being zero. Denoting by $n_v = N(N-1)/2 + m$ the total number of vector fields, the (h, f) matrices form two complex $n_v \times n_v$ scalar field dependent matrices with the properties $\mathcal{N} = hf^{-1}$

$$f = \frac{1}{\sqrt{2}}(A - iB)\,, \qquad h = \frac{1}{\sqrt{2}}(C - iD) \tag{1.24}$$

where the real matrix $S = \binom{AB}{CD}$ is an element of $Sp(2n_v, R)$.

The extremal black hole potential, which controls the black hole properties, in particular the Bekenstein-Hawking entropy formula, is given in terms of the matrix $\mathcal{N}$ and the charges p^Λ, q_Λ by

$$V_{BH} = -\frac{1}{2}(q_\Lambda - \mathcal{N}_{\Lambda\Sigma}p^\Lambda)Im\mathcal{N}^{-1\Lambda\Sigma}(q_\Delta - \bar{\mathcal{N}}_{\Delta\Pi}p^\Pi)\,. \tag{1.25}$$

For extremal regular black holes the entropy is given by [40]

$$S_{BH} = \frac{1}{4}A_H = \pi V_{BH}|_{crit} \qquad , \qquad V_{BH}|_{crit} = V(\phi = \phi_{crit})\,, \tag{1.26}$$

with $\phi_{crit} = \phi_H$ and ϕ_H is the value of the scalar field at the horizon.

2 Applications

2.1 Spontaneous Supersymmetry Breaking

Surprisingly enough, unlike general covariance, it is possible to "spontaneously break" local supersymmetry as it happens in spontaneously broken non abelian gauge theories as the Higgs effect. The massless spin 3/2 gravitino $\psi_{\mu\alpha}$ eats the spin 1/2 fermion (the goldstino) and becomes massive. This is the supersymmetric version of the Higgs mechanism and it is called the "SuperHiggs" [16, 3, 15, 17]. The gravitino mass is given by

$$m_{3/2} \sim \frac{1}{\sqrt{3}} \frac{1}{M_P} < F_u > \tag{2.1}$$

where $< F_u >= E_S^2$, the order parameter of supersymmetry breaking, is the vacuum expectation value of a dimension 2 (auxiliary) field, such that

$$J_{\mu\alpha} \sim < F_u > \gamma_\mu \lambda \,. \tag{2.2}$$

If $< F_u >= 0$ supersymmetry is unbroken.

In effective supersymmetry breaking, the "goldstino", the goldstone fermion of spontaneously broken supersymmetry, enters in the supercurrent as

$$J_{\mu\alpha} = E_S^2 \gamma_\mu \lambda_g = m_{3/2} M_P \gamma_\mu \lambda_g \,. \tag{2.3}$$

From the (Yukawa) coupling between the goldstino and ordinary matter (ϕ_M, ψ_M), we have the effective interaction

$$g_M \bar{\lambda}_g \psi_M \phi_M \tag{2.4}$$

Combining the previous two formulae, supercurrent conservation implies

$$E_{SM}^2 = \Delta m_M^2 = g_M m_{3/2} M_P = g_M E_S^2 \,. \tag{2.5}$$

The last formula indicates that two distinct scenarios of supersymmetry breaking can occur, depending on whether $g_M << 1$ or $g_M \sim O(1)$. The first situation is encountered when $m_{3/2} \sim E_{SM}$, so that $g_M \sim E_{SM}/M_P << 1$. This is the gravity mediation mechanism where supersymmetry behaves as a softly broken theory at low energy. The second situation corresponds to $E_S \sim E_{SM}$ so that $m_{3/2} \sim E_S^2/M_P << E_{SM}$. This is the situation where the gravitino is almost massless, and spontaneous breaking with a non decoupled goldstino survives at low energy.

2.2 Extremal Black Holes

Black holes, one of the most stunning consequences of General Relativity, enjoy thermodynamical properties in a generalized phase space where the quantum mechanical attributes are their (ADM) mass, charge, spin and scalar charges (see for instance [39]) .

Supergravity, in virtue of its invariance under local Poincare' transformations, naturally includes General Relativity, as it describes the (supersymmetric) coupling between gravitation and other fields. For this reason, supergravities and particularly Maxwell-Einstein (super)-gravity theories with scalar fields, arising as effective theories from Superstrings or M-theory, are nowadays extensively used to reach a better understanding of black holes, which appear among their classical solutions[40, 7].

Unlike Schwarzschild black holes, charged (Reissner-Nordstrom) and/or spin (Kerr-Newman) black holes can be *extremal*, *i.e* with vanishing temperature, in which case their event and Cauchy horizons coincide. In formulae, the extremality parameter is given by

$$c = 2ST = \frac{1}{2}(r_+ - r_-) \to 0 \tag{2.6}$$

where c measures the surface gravity and $S = log\ \mathcal{N}$ is the black hole entropy which counts the number $\mathcal{N}$ of microstates. In Supergravity, it is given by the famous Bekenstein Hawking area formula

$$S = \frac{1}{2}A_H = \pi R_H^2 \,. \tag{2.7}$$

where R_H is the effective radius for extremal charged black holes. Like all the relevant quantities in this context, R_H must be an electric-magnetic duality invariant function of the charges $Q = (p^\Lambda, q_\Lambda)$, $(\Lambda = 1, \ldots, n_v)$ [21] .

Various examples of entropy formulae in terms of electric-magnetic charges and duality symmetries are

Reissner-Nordstrom: $S \sim p^2 + q^2$, with $U(1)$ symmetry

Axion-Dilaton: $S \sim |p_1 q_2 - p_2 q_1|$, with $SL(2) \times SO(2)$ symmetry

$N = 4$ **supergravity:** $S \sim |p^2 q^2 - (p.q)^2|^{1/2}$, with $SL(2) \times SO(6, n)$ symmetry

$N = 8$ **supergravity:** $S \sim \sqrt{|I_4|}$ with $E_{7(7)}$ symmetry.

One of the main reasons why extremal, static, spherically symmetric black hole solutions have been extensively analyzed for supergravities in different space-time dimensions and with various degrees of supersymmetry, is that they enjoy a remarkable property: the scalar field trajectories $\phi^i(\tau)$, in terms of the radial evolution parameter $\tau = -1/\rho$ from asymptotic infinity ($\tau \to 0$) to the horizon ($\tau \to -\infty$), behave as dynamical systems, losing any dependence on their initial values $\phi^i_0 = \phi^i(\tau)|_{\tau\to 0}$. The scalars evolve to a common value at the horizon, $\phi^i_H = \phi^i_H(Q)$, which they reach with zero velocity ($\dot{\phi}^i \to 0$), where they entirely depend on the electric-magnetic charge vector Q of the asymptotic configuration. This attractive feature is called *Attractor Mechanism* [40] in analogy with dynamical systems. The attractive trajectories have a universal geometry of spacetime near the horizon which is given by the $AdS_2 \times S_2$ Bertotti-Robinson metric. In this way, extremal black holes behave as solitons interpolating between maximally symmetric geometries of (super)spacetime: Minkowski for $\tau \to 0$ and the conformally flat metric for $\tau \to -\infty$ [5, 32].

Consider the 4D lagrangian

$$\mathcal{L} = -\frac{R}{2} + \frac{1}{2} g_{ij}(\phi)\partial_\mu \phi^i \partial^\mu \phi^j + I_{\Lambda\Sigma} F^\Lambda \wedge^* F^\Sigma + R_{\Lambda\Sigma} F^\Lambda \wedge F^\Sigma \tag{2.8}$$

where $I = Im\,\mathcal{N}_{\Lambda\Sigma}, R = Re\,\mathcal{N}_{\Lambda\Sigma}$, the vector kinetic matrix $\mathcal{N}_{\Lambda\Sigma}$ depends on the scalar fields, and the vector field strength F^Λ, $(\Lambda = 1, \ldots, n_v)$ and its dual $G_{\Lambda\mu\nu} \sim \frac{^*\delta\mathcal{L}}{\delta F^\Lambda_{\mu\nu}}$ form a symplectic vector whose fluxes give rise to the electric and magnetic charges

$$q_\Lambda = \frac{1}{4\pi}\int_{S2} G_\Lambda\,, \qquad p^\Lambda = \frac{1}{4\pi}\int_{S_2} F_\Lambda\,. \tag{2.9}$$

For static, spherically symmetric extremal black holes, the metric ansatz (in the extremal case, $c = 2ST = 0$)[40] is

$$ds^2 = -e^{2U}dt^2 + e^{-2U}\frac{d\tau^2}{\tau^4} + \frac{1}{\tau^2}(d\theta^2 + \sin\theta^2 d\phi^2)\,. \tag{2.10}$$

The main object in these investigations is the effective potential [40]

$$V_{BH} = -\frac{1}{2}Q^T \mathcal{M} Q = \frac{1}{2}Z_{AB}\overline{Z}^{AB} + Z_M Z^M\,, \tag{2.11}$$

given as a quadratic form in the electric-magnetic charge vector Q=(p,q), with $\mathcal{M}(\phi^i)$ a $2n_v \times 2n_v$ real symmetric $Sp(2n_v, R)$ matrix

$$\mathcal{M}(\phi^i) = \begin{pmatrix} I + RI^{-1}R & -RI^{-1} \\ -I^{-1}R & I^{-1} \end{pmatrix}. \tag{2.12}$$

For N extended theories, it can also be written as above in terms of the squares of the central charges and possibly matter charges of the N extended supersymmetry algebra (A,B are SU(N) indices, while M is the fundamental representation of the matter group, if present) [7].

The extrema of this potential as a function of the scalar fields, $\partial_i V_{BH} = 0$, give the attractor solutions, while the values of V_{BH} at the attractor points yield the entropy.

This potential arises upon reducing the general 4D (or higher D) lagrangian to the one-dimensional almost geodesic action describing the radial evolution of the $n+1$ scalar fields $(U(\tau), \phi^i(\tau))$.

$$S = \int \mathcal{L} d\tau = \int (\dot{U} + g_{i\bar{j}}\dot{\phi}^i \dot{\phi}^{\bar{j}} + e^{2U} V_{BH}(\phi(\tau), p, q) \tag{2.13}$$

complemented with the hamiltonian constraint (in the extremal case)

$$\dot{U}^2 + g_{ij}\dot{\phi}^i\dot{\phi}^j - e^{2U} V_{BH}(\phi(\tau), p, q) = 0\,. \tag{2.14}$$

It follows that for $\tau \to 0$ (radial infinity)

$$M^2_{ADM} = V_{BH}(\phi_\infty, p, q) - \frac{1}{2} g_{ij}\Sigma^i\Sigma^j \tag{2.15}$$

while for $\tau \to -\infty$ (horizon)

$$S_{BH} = \frac{1}{4}A_H = \pi V_{BH}(\phi_H, p, q) = V_{BH}|_{\phi_{crit}} \tag{2.16}$$

showing *attractor behaviour* [40]. For BPS states

$$S_{BH} = \pi |Z_{highest}|^2_{crit} \qquad M^2_{ADM} = |Z_{highest}|^2_\infty . \tag{2.17}$$

Black holes are then obtained as solutions of the equations of motion

$$\frac{d^2U}{d\tau^2} = e^{2U} V_{BH} \tag{2.18}$$

$$\frac{d^2\phi^i}{d\tau^2} = g^{i\bar{\jmath}} \frac{\partial V_{BH}}{\partial \phi_{\bar{\jmath}}} e^{2U} . \tag{2.19}$$

Black hole attractors were first uncovered within $d = 4$, $N = 2$ supergravity, where the underlying geometric structure of moduli space is given by special geometry[44] and the duality is encoded in $Sp(2n+2)$ symplectic transformations. Duality invariant quantities are those that remain unchanged under the simultaneous action of the duality group on the charge vector $Q = (p^\Lambda, q_\Lambda)$ and on the scalar fields (expressed through the symplectic sections of special geometry (X^Λ, F_Λ), with $\Lambda = (0, i) = 0, \dots, n$).

An important feature is that the extremal black hole potential (2.11) can be written in terms of a real "superpotential" scalar function $W(\phi)$ by

$$V_{BH} = W^2 + 2g^{ij}\partial_i W \partial_j W . \tag{2.20}$$

W also governs the first order flow (BPS) equations for the radial evolution of the scalar fields ϕ^i and the warp factor U from asymptotic infinity towards the black hole horizon :

$$U' = -e^U W , \qquad \phi'^i = -2e^U g^{ij}\partial_j W . \tag{2.21}$$

This superpotential W also encodes many basic properties of the extremal black hole, such as the entropy at the horizon $S_{BH} = \pi W^2_H$, the ADM mass $M_{ADM} = W_\infty$, and the scalar charges at infinity, $\Sigma_i = \partial_i W_\infty$. By using the asymptotic behaviour of the warp factor at spacial infinity $\tau \to 0$, one has $U \to -M_{ADM}\tau$ and at the horizon $\tau \to \infty$

$$e^{-2U} \to \frac{1}{4\pi} A_H \tau^2 \tag{2.22}$$

One gets model independent expressions for ADM mass, scalar charges and entropy as a function of W

$$M^2_{ADM} = W^2_{\tau\to 0} = W^2_\infty \tag{2.23}$$

$$\Sigma^i = -2g^{ij}\frac{\partial W}{\partial \phi^j}|_{\tau\to 0} = -2g^{ij}\frac{\partial W}{\partial \phi^j}|_\infty \tag{2.24}$$

$$S = \frac{1}{4}A_H = \pi W^2|_{\tau\to -\infty} = \pi W^2|_H \tag{2.25}$$

Moreover, the horizon value of the scalar fields ϕ^i_H is a critical point for W in the moduli space of the theory, $\partial W/\partial\phi^r|_{\phi^i=\phi^i_H} = 0$. For supersymmetric configurations, it has long been known that W is given by the modulus of the central charge $|Z(z,\bar{z};q,p)|$ for the specific case of the N=2 supersymmetry algebra[40]. The flow equations are nothing but the BPS equations for the preserved supersymmetry. The BPS flows end at the critical points of the central charge $D_iZ = (\partial_i + \frac{1}{2}\partial_i K)Z = 0$, which are also critical points of the full potential (2.20). However, it was uncovered [45] that the same set up also works for the non BPS branch[53], provided one identifies a real function $W(z,\bar{z})$ of the scalar fields z^i and of the electric-magnetic charges (q,p) which plays the role of a *fake* superpotential. The extrema of W now describe the non supersymmetric attractors, which can be stable or display runaway behaviour. For singular black hole solutions, W displays a runaway behaviour $\phi_H \to \infty, W \to 0$. The important point is that, using this fake superpotential, the task of constructing full solutions is simplified in that one has to solve first order BPS-like equations rather than second order equations of motion for the scalar fields, as in the supersymmetric case. The "fake superpotential" was introduced in [45] where some features of domain walls in "fake supergravity theories"[52] (where, however, the fake superpotential is an $SU(2)$ valued matrix) were tailored to the context of black holes.

For $N \geq 2$, in the above equations $|Z|$ is substituted by the highest eigenvalue $|Z_{highest}|$ of the appropriate central charge matrix.

For all $N > 2$ supergravities and for $N = 2$ theories based on symmetric spaces, a first order flow also exists for all non-BPS configurations in terms of a *fake superpotential* W. One can prove the validity of the BPS bound along the flow

$$M_{ADM} > |Z_i|\,. \tag{2.26}$$

At the horizon

$$\dot{\phi}^i|_{\tau\to-\infty} = 0 \to \partial_i W_H = 0 \quad (W_H \leftrightarrow W_{crit})\,. \tag{2.27}$$

Example: Black Holes in N=8 supergravity

As already noticed above, U-duality of the underlying extended supergravity is known to dictate many important features [43]. For a generic (continuous) U-duality group G, it was clarified in [48] that different supersymmetry features of a given solution are characterized by G-invariant conditions on the central charges. This is perfectly illustrated by the example of maximal $N = 8$ supergravity. Here there are 70 scalar fields parametrizing the manifold $E_{7(7)}/SU(8)$, which is a symmetric coset space. There are 28 vectors (electric) and 28 dual vectors (magnetic) covering the 56 dimensional fundamental representation of $E_{7(7)}$, so that the charge vector is $Q = (p^\Lambda, q_\Lambda)$ with $\Lambda = 1,\ldots,28$. The 56 spin $\frac{1}{2}$ form the completely antisymmetric representation Ξ_{ABC} of $SU(8)$.

The susy algebra contains an antisymmetric central charge matrix $Z^{AB}(\phi, Q)$,

$$\{Q^A_\alpha, Q^B_\beta\} = \epsilon_{\alpha\beta} Z^{AB} \qquad , \qquad Z^{AB} = -Z^{BA} . \tag{2.28}$$

It turns out that the area of the horizon for both regular 1/8-BPS and non-BPS attractors, is proportional to the square of the absolute Cartan quartic invariant of the 56 dimensional representation of $E_{7(7)}$.

$$I_4 = Tr(Z\overline{Z})^2 - \frac{1}{4}(Tr\, Z\overline{Z})^2 + 4(Pf\, Z + Pf\, \overline{Z}) . \tag{2.29}$$

This expression, involving the traces of the matrix $Z_{AB}\bar{Z}^{BC}$ and of its square as well as the real part of the Pfaffian of $Z_{AB}(\phi, Q)$,

$$PfZ = \frac{1}{2^4 4!} \epsilon^{abcdefgh} Z_{AB} Z_{CD} Z_{EF} Z_{GH} \tag{2.30}$$

actually depends only on the electric and magnetic charges $Q = (p, q)$ and not on the scalar fields. This leads to a classification of the attractors in terms of orbits of the charge vector Q [50, 51]. A positive or negative sign in $\sqrt{\pm I_4}$ tells apart a BPS from a non-BPS attractor [47].

By an $SU(8)$ transformation, one can reach the canonical bases where the antisymmetric central charge matrix is (skew)diagonalized and takes the form

$$Z_{AB} = \begin{pmatrix} \rho_0 & & & \\ & \rho_1 & & \\ & & \rho_2 & \\ & & & \rho_3 \end{pmatrix} \otimes \begin{pmatrix} 0 & 1 \\ -1 & 0 \end{pmatrix} e^{i\varphi/4} \tag{2.31}$$

in terms of its skew eigenvalues $z_i = \rho_i e^{i\varphi/4}$, $i = 0, 1, 2, 3$ which amount to 5 independent parameters. The attractor equation becomes

$$\partial_i V = 0 \to z_i z_j + z_k^* z_l^* = 0 , (i \neq j \neq k \neq l) \tag{2.32}$$

which yields two solutions

a) $z_h \neq 0$, $z_{i \neq h} = 0$

b) $|z_i| = \rho$, $arg Pf Z = \pi$

Considering the quartic invariant in this basis,

$$I_4 = [(\rho_0 + \rho_1)^2 - (\rho_2 + \rho_3)^2][(\rho_0 - \rho_1)^2 \tag{2.33}$$
$$- (\rho_2 - \rho_3)^2] + 8\rho_0\rho_1\rho_2\rho_3(\cos\phi - 1) \tag{2.34}$$
$$\rho_0 \geq \rho_1 \geq \rho_2 \geq \rho_3 \tag{2.35}$$

one finds that there are two disjoint regular orbits:

i) 1/8 BPS: with $I_4 > 0$, for $\rho_0 \neq 0$ and $\rho_1 = \rho_2 = \rho_3 = 0$. The duality orbit is $E_{7(7)}/E_{6(2)}$ and the moduli space is $E_{6(2)}/SU(2) \times SU(6)$.
ii) non BPS: with $I_4 < 0$, for $|z_i| = \rho$ and $\arg Pf\, Z = \pi$. The duality orbit is $E_{7(7)}/E_{6(6)}$ and the moduli space: $E_{6(6)}/USp(8)$.
There are three additional "small" orbits, preserving respectively 1/8, 1/4 and 1/2 supersymmetry, which are singular, having $I_4 = 0$ and thus zero horizon area.

The entropy of the two regular branches is given by

$$S_{1/8BPS} = \pi V_{BH}|_{crit} = \pi \rho_0^2 \,, \tag{2.36}$$

$$S_{nonBPS} = \pi V_{BH}|_{crit} = \pi (2\rho_0)^2 \,. \tag{2.37}$$

Also in the $N = 2$ case, the basic question is the identification of a complete set of duality invariant quantities to be used as a basis for W. The answer was provided for the $N = 2$ case in [56]. It turns out that the superpotential is not simply given by a linear combination, but rather by a non-polynomial expression (containing radicals) of the basic duality invariant quantities.

By now, quite some work has been done to find the fake superpotential W in many cases [45, 54, 59, 60, 61], until an universal procedure for its construction in $N = 2$ special geometries has been established in [62], based on a generalization of the results obtained for the *stu* model. This universal procedure has a natural interpretation also beyond the t^3 and the *stu* model. It actually states that one can obtain W for a generic symmetric space geometry in $N = 2$ theories by looking at the *stu* model and substituting the invariants with the generic invariants of the desired model. This procedure can be further generalized to the maximally extended $N = 8$ theory using the fact that the *stu* model is both a subsector of $N = 8$ and a model in $N = 2$ where we know how to describe W in terms of invariants. The results of [62] agree with the outcome of studies of black hole evolution with the method of time-like reduction to 3 dimensions in [63].

To conclude, let us note that recent results [64] point to an interesting role played by black holes with singular geometry in relation to String Theory. It has been recently found that $N = 8$ black holes with vanishing area of the horizon ($I_4 = 0$), can be interpreted as dynamically reduced black branes with regular geometry. For example, 1/8 BPS black holes with $I_4 = 0$ correspond to reduced (wrapped) black holes (black strings) at D=5 (with three charges q_1, q_2, q_3, $J_{MAX} = 7/2$). Conversely, 1/4 BPS black holes with $\partial I_4 = 0$ correspond to reduced and wrapped black strings at $D = 6$ (two charges q_1 and q_2, $J_{MAX} = 3$. Finally, 1/2 BPS black holes with $\partial^2 I_4 = 0$ correspond to KK states i.e. massless particles in higher dimensions $J_{MAX} = 2$. If only regular black holes, with $AdS_2 \times S_2$ horizon geometries are to be retained in a consistent theory of gravity, one may not need to open up extra dimensions[65]: weather this feature could hint to a possible finite theory of D=4, $N = 8$ supergravity [66] without the need of string theory, is a question that is presently under close investigation.

Acknowledgement: We would like to thank the Ettore Majorana Foundation and Centre for Scientific Culture for the stimulating scientific environment and the kind hospitality during the School. This work is supported in part by the ERC Advanced Grant no. 226455, *"Supersymmetry, Quantum Gravity and Gauge Fields"* (*SUPERFIELDS*) and by MIUR-PRIN contract 20075ATT78.

References

[1] S. Ferrara, D. Z. Freedman and P. Van Nieuwenhuizen, "Progress Toward a Theory of Supergravity", Phys. Rev. **D13** (1976) 3214.

[2] S. Deser and B. Zumino, "Consistent Supergravity", Phys. Lett. **62B** (1976) 335.

[3] P. Van Nieuwenhuizen, "Supergravity", Phys. Rept. **68** (1981) 189.

[4] M. J. Duff, B.E. W. Nilsson and C.N. Pope, "Kaluza-Klein Supergravity", Phys. Rept. **130**, 1 (1986).

[5] M. J. Duff, R. R. Khuri and J. X. Lu, "String solitons", Phys. Rept. **259**, 213 (1995) [arXiv:hep-th/9412184].

[6] O. Aharony, S. S. Gubser, J. M. Maldacena, H. Ooguri and Y. Oz, "Large N field theories, string theory and gravity", Phys. Rept. **323**, 183 (2000) [arXiv:hep-th/9905111].

[7] L. Andrianopoli, R. D'Auria, S. Ferrara and M. Trigiante, "Extremal Black Holes in Supergravity", Lect.Notes Phys.737:661 (2008), [arXiv:hep-th/0611345]

[8] J. Wess and B. Zumino, "Supergauge Transformations in Four-Dimensions", Nucl. Phys. **B70** (1974) 39; B. Zumino, in "Renormalisation and invariance in Quantum Filed Theory", ed. E. Caianello (Plenum Press, New York, 1974), p. 367

[9] J.-L. Jervais and B. Sakita, "Field Theory Interpretation Of Supergauges In Dual Models", Nucl. Phys. **B34** (1971) 633.

[10] A. Neveu and J.H.Schwarz,"Factorizable dual model of pions", Nucl. Phys. **B31** (1971) 86.

[11] P. Ramond, "Dual Theory for Free Fermions", Phys. Rev. Lett **D3** (1971) 2415

[12] Yu. A. Gol'fand and E. P. Likhtman , "Extension of the Algebra of Poincare Group Generators and Violation of p Invariance", JETP Lett, **13** (1971) 323

[13] D. V. Volkov and V. P. Akulov, "Is the Neutrino a Goldstone Particle?", Phys. Lett. **46B** (1973) 109

[14] S. Ferrara and B. Zumino, "Transformation Properties of the Supercurrent", Nucl. Phys. **B79** (1974) 413; A. Salam and J. Strathdee, "Supersymmetry and Nonabelian Gauges", Phys. Lett. **51B** (1974) 353

[15] S. Ferrara in "Supersymmetry", vol 2, World Scientific, Singapore and North Holland, Amsterdam.

[16] P. Fayet and S. Ferrara, "Supersymmetry," Phys. Rept. **32** (1977) 249.

[17] S. Weinberg, "The Quantum Theory of Fields", Vol. 3, Cambridge University Press (May 9, 2005)

[18] A. Salam and J. Strathdee, "Supergauge Transformations", Nucl. Phys. **B76** (1974) 477

[19] A. Van Proeyen, "Tools for supersymmetry", hep-th/9910030.

[20] M. K. Gaillard and B. Zumino, "Duality Rotations For Interacting Fields", Nucl. Phys. B **193** (1981) 221.

[21] P. Aschieri, S. Ferrara and B. Zumino, "Duality Rotations in Nonlinear Electrodynamics and in Extended Supergravity", Riv. Nuovo Cim. **31**, 625 (2009) [arXiv:0807.4039 [hep-th]].

[22] R. D'Auria, S. Ferrara, M. A. Lledo and V. S. Varadarajan, "Spinor algebras", J. Geom. Phys. **40** (2001) 101 [arXiv:hep-th/0010124].

[23] S. Ferrara, J. Wess and B. Zumino, "Supergauge Multiplets and Superfields", Phys. Lett **51B** (1974) 239

[24] P. Deligne, "Notes on Spinors", in Quantum Fields and Strings: A Course for Mathematicians. American Mathematical Society, (1999).

[25] B. Kostant, "Graded Manifolds, Graded Lie Theory, And Prequantization", *In *Bonn 1975, Proceedings, Differential Geometrical Methods In Mathematical Physics*, Berlin 1977, 177-306*

[26] E. Witten, "String theory dynamics in various dimensions", Nucl. Phys. B **443** (1995) 85 [arXiv:hep-th/9503124].

[27] V. G. Kac, "A Sketch of Lie Superalgebra Theory", Commun. Math. Phys. **53** (1977) 31; V. G. Kac, "Lie Superalgebras", Adv. Math. **26** (1977) 8; V. G. Kac, J. Math. Phys. **21** (1980) 689.

[28] R. Coquereaux, "Modulo 8 periodicity of Real clifford Algebras and Particle Physics", Phys. Lett. **115B** 389 (1982).

[29] W. Nahm, "Supersymmetries and their representations", Nucl. Phys. **B135** (1978) 149.

[30] M. Gell-Mann, P. Ramond and R. Slansky, "Color Embeddings, Charge Assignments, And Proton Stability In Unified Gauge Theories", Rev. Mod. Phys. **50**, 721 (1978).

[31] P. K. Townsend, "P-brane democracy", arXiv:hep-th/9507048.

[32] G. Gibbons and P. K. Townsend, "Vacuum interpolation in supergravity via super p-branes", Phys.Rev.Lett.**71**, (1993) 3754 .

[33] J. M. Maldacena, "Lectures on AdS/CFT" *Prepared for Theoretical Advanced Study Institute in Elementary Particle Physics (TASI 2002): Particle Physics and Cosmology: The Quest for Physics Beyond the Standard Model(s), Boulder, Colorado, 2-28 Jun 2002*

[34] E. Cremmer, B. Julia and J. Scherk, "Supergravity theory in 11 dimensions," Phys. Lett. B **76**, 409 (1978).

[35] S. Coleman, J. Mandula, "All Possible Symmetries of the S Matrix", Phys. Rev. **159** 1251 (1967).

[36] R. Haag, J. Lopuszanski and M. Sohnius, All Possible Generators of Supersymmetries of the S Matrix, Nucl. Phys. B88 257 (1975).

[37] J. Wess and J. Bagger, "Supersymmetry and supergravity," *Princeton, USA: Univ. Pr. (1992) 259 p*

[38] N. Seiberg and E. Witten, "Monopole Condensation, And Confinement In N=2 Supersymmetric Yang-Mills Theory," Nucl. Phys. B **426** (1994) 19 [Erratum-ibid. B **430** (1994) 485] [arXiv:hep-th/9407087].

[39] S. Ferrara, K. Hayakawa and A. Marrani, "Lectures on Attractors and Black Holes," Fortsch. Phys. **56**, 993 (2008) [arXiv:0805.2498 [hep-th]].

[40] S. Ferrara, R. Kallosh and A. Strominger, *"N=2 extremal black holes,"* Phys. Rev. D **52** (1995) 5412; A. Strominger, "Macroscopic entropy of $\mathcal{N}= 2$ *extremal black holes*, Phys. Lett. **B383**, 39 (1996); S. Ferrara and R. Kallosh, "Supersymmetry and attractors", Phys. Rev. D **54**, 1514 (1996); S. Ferrara, R. Kallosh, "Universality of supersymmetric attractors", Phys. Rev. D**54**, 1525 (1996); S. Ferrara, G. W. Gibbons and R. Kallosh, "Black holes and critical points in moduli space", Nucl. Phys. B **500**, 75 (1997).

[41] S. Ferrara and M. Gunaydin, "Orbits of exceptional groups, duality and BPS states in string theory", Int. J. Mod. Phys. A13 (1998) 2075;

[42] S. Ferrara and R. Kallosh, "On N = 8 attractors," Phys. Rev. D **73**, 125005 (2006) [arXiv:hep-th/0603247].

[43] C. M. Hull and P. K. Townsend, Nucl. Phys. B**438** 109 (1995); M. Cvetic and C. M. Hull, "Black holes and U-duality" Nucl. Phys. B **480** (1996) 296 [arXiv:hep-th/9606193].

[44] A. Strominger, "Special Geometry," Commun. Math. Phys. **133** (1990) 163.

[45] A. Ceresole and G. Dall'Agata, "Flow Equations for Non-BPS Extremal Black Holes," JHEP **0703** (2007) 110 [arXiv:hep-th/0702088].

[46] J. Strathdee, "Extended Poincare Supersymmetry", Int. J. Mod. Phys.**A2** (1) 273 (1987).

[47] R. Kallosh and B. Kol, "E(7) Symmetric Area of the Black Hole Horizon," Phys. Rev. D **53**, 5344 (1996) [arXiv:hep-th/9602014].

[48] S. Ferrara and J. M. Maldacena, "Branes, central charges and U-duality invariant BPS conditions," Class. Quant. Grav. **15**, 749 (1998) [arXiv:hep-th/9706097].

[49] S. Ferrara, C. A. Savoy and B. Zumino, "General Massive Multiplets In Extended Supersymmetry," Phys. Lett. B **100** (1981) 393.

[50] S. Bellucci, S. Ferrara, M. Gunaydin and A. Marrani, "Charge orbits of symmetric special geometries and attractors," Int. J. Mod. Phys. A **21** (2006) 5043 [arXiv:hep-th/0606209].

[51] L. Andrianopoli, R. D'Auria and S. Ferrara, "U-duality and central charges in various dimensions revisited," Int. J. Mod. Phys. A **13**, 431 (1998) [arXiv:hep-th/9612105]; L. Andrianopoli, R. D'Auria and S. Ferrara, "Five dimensional U-duality, black-hole entropy and topological invariants," Phys. Lett. B **411**, 39 (1997) [arXiv:hep-th/9705024].

[52] D. Z. Freedman, C. Nunez, M. Schnabl and K. Skenderis, "Fake Supergravity and Domain Wall Stability," Phys. Rev. D **69** (2004) 104027 [arXiv:hep-th/0312055]. A. Celi, A. Ceresole, G. Dall'Agata, A. Van Proeyen and M. Zagermann, "On the fakeness of fake supergravity," Phys. Rev. D **71**, 045009 (2005) [arXiv:hep-th/0410126].

[53] E. G. Gimon, F. Larsen and J. Simon, "Black Holes in Supergravity: the non-BPS Branch," JHEP **0801** (2008) 040 [arXiv:hep-th/0710.4967].

[54] L. Andrianopoli, R. D'Auria, E. Orazi and M. Trigiante, "First Order Description of Black Holes in Moduli Space," JHEP **0711** (2007) 032 [arXiv:hep-th/0706.0712].

[55] L. Andrianopoli, R. D'Auria, E. Orazi and M. Trigiante, "First Order Description of D=4 static Black Holes and the Hamilton-Jacobi equation," [arXiv:hep-th/0905.3938].

[56] B. L. Cerchiai, S. Ferrara, A. Marrani and B. Zumino, "Duality, Entropy and ADM Mass in Supergravity," Phys. Rev. D **79** (2009) 125010 [arXiv:0902.3973].

[57] A. Ceresole, G. Dall'Agata, S. Ferrara and A. Yeranyan, "First order flows for N=2 extremal black holes and duality invariants," Nucl.Phys.B824:239-253,2010 [arXiv:0908.1110] .

[58] M. J. Duff, J. T. Liu and J. Rahmfeld, "Four-Dimensional String-String-String Triality," Nucl. Phys. B **459**, 125 (1996) [arXiv:hep-th/9508094]; K. Behrndt, R. Kallosh, J. Rahmfeld, M. Shmakova and W. K. Wong, "STU black holes and string triality," Phys. Rev. D **54** (1996) 6293 [arXiv:hep-th/9608059].

[59] S. Bellucci, S. Ferrara, A. Marrani and A. Yeranyan, "stu Black Holes Unveiled," [arXiv:hep-th/0807.3503].

[60] G. Lopes Cardoso, A. Ceresole, G. Dall'Agata, J. M. Oberreuter and J. Perz, "First-order flow equations for extremal black holes in very special geometry," JHEP **0710** (2007) 063 [arXiv:hep-th/0706.3373].

[61] S. Ferrara, A. Gnecchi and A. Marrani, "d=4 Attractors, Effective Horizon Radius and Fake Supergravity," Phys. Rev. D **78** (2008) 065003 [arXiv:hep-th/0806.3196]; J. Perz, P. Smyth, T. Van Riet and B. Vercnocke, "First-order flow equations for extremal and non-extremal black holes," JHEP **0903** (2009) 150 [arXiv:hep-th/0810.1528]; K. Goldstein and S. Katmadas, "Almost BPS black holes," JHEP **0905** (2009) 058 [arXiv:hep-th/0812.4183]; I. Bena, G. Dall'Agata, S. Giusto, C. Ruef and N. P. Warner, "Non-BPS Black Rings and Black Holes in Taub-NUT," JHEP **0906** (2009) 015 [arXiv:hep-th/0902.4526]; P. Galli and J. Perz, "Non-supersymmetric extremal multicenter black holes with superpotentials," [arXiv:0909.5185].

[62] A. Ceresole, G. Dall'Agata, S. Ferrara and A. Yeranyan, "Universality of the superpotential for d=4 extremal black holes", Nucl. Phys. **B** (2010), in press [arXiv:0910.2697].

[63] G. Bossard, Y. Michel and B. Pioline, "Extremal black holes, nilpotent orbits and the true fake superpotential" [arXiv:0908.1742].

[64] M. Bianchi, S. Ferrara and R. Kallosh, "Observations on Arithmetic Invariants and U-Duality Orbits in N =8 Supergravity," JHEP **1003**, 081 (2010) [arXiv:0912.0057 [hep-th]]; "Perturbative and Non-perturbative N =8 Supergravity," arXiv:0910.3674 [hep-th].

[65] M. B. Green, H. Ooguri and J. H. Schwarz, "Decoupling Supergravity from the Superstring," Phys. Rev. Lett. **99**, 041601 (2007) [arXiv:0704.0777 [hep-th]].

[66] Z. Bern, L. J. Dixon and R. Roiban, "Is N = 8 Supergravity Ultraviolet Finite?," Phys. Lett. B **644**, 265 (2007) [arXiv:hep-th/0611086]; Z. Bern, J. J. Carrasco, L. J. Dixon, H. Johansson and R. Roiban, "The Ultraviolet Behavior of N=8 Supergravity at Four Loops," Phys. Rev. Lett. **103**, 081301 (2009) [arXiv:0905.2326 [hep-th]].

CHAIRMAN: S. FERRARA

Scientific Secretaries: B.L.Cerchiai, A.Shcherbakov

DISCUSSION I

- *Davey:*

What precisely is the relationship between the mass of black holes and the central charge of the supersymmetry algebra?

- *Ferrara:*

The central charges appear in the supersymmetry algebra, namely in the anticommutator of two spinors:

$$\{Q_\alpha^A, Q_\beta^B\} = Z^{AB} = Z\,\varepsilon^{AB}$$

Now, let's consider a charged black hole, like the Reissner–Nordström black hole, and the supersymmetry algebra in the background of its metric. Since the central charge commutes with all the elements in the supersymmetry algebra, it takes a constant value on a particle state and it is described by a number. So, for massless particles it is not present, but for massive particles you can have a non vanishing central charge and you can prove that the mass of the particle is always bigger or equal to the eigenvalues of the central charge:

$$M \geq |Z|$$

This is the so called BPS bound and black holes which saturate it are called extremal:

$$M = |Z|$$

For such a black hole you find that in Planck units the mass is equal to the charge. This is what is called a BPS black hole or a black hole which preserves supersymmetry. So, the central charge plays a role in black hole physics, in the sense that extremal black holes have a non vanishing central charge and supersymmetric black holes have a value of the central charge which is equal to their mass. There is a similar statement for p-branes in higher dimensional supergravity.

- *Davey:*

How much supersymmetry are you talking about in this case? For instance, what happens for $\mathcal{N}=1$?

- *Ferrara:*

Since the central charge is antisymmetric in the number of supersymmetry, in order to have this phenomenon, you need at least $\mathcal{N}=2$ in 4D. Then the central charge is described by just a number, because it is an antisymmetric 2 x 2 matrix, and a black hole which saturates the central charge preserves half of the original supersymmetry. So, you start with a theory which has 8 supersymmetries, as long as you have only the massless fundamental object, and then when you add the black hole, which is a solitonic object, it breaks 4 of the supersymmetries and preserves only 4 out of the 8. This is called a ½ BPS black hole. If the mass doesn't saturate this bound, then the black hole does not preserve any supersymmetry.

- *Davey:*

What happens in the N=4 case?

- *Ferrara:*

The simplest example is N=2 supergravity, where you have the Reissner–Nordström black hole, which breaks 4 out of the 8 original supersymmetries. But you have a similar phenomenon in N=4 and for higher N, where there is a variety of black holes, which can preserve different fractions of the original supersymmetry.

- *Smirnov:*

As you said in your morning lecture, supersymmetry is just an extension of the symmetry algebra of space-time by some additional generators, i.e. the spinor charges. Could you please explain why supersymmetry is the only possible extension, or at least why it is so natural in the mathematical sense?

- *Ferrara:*

The question is whether you are able to construct a symmetry which generalizes space-time symmetry and which at the same time can be implemented in a Lagrangian field theory and has a realization in standard theories, like gauge theories or scalar field theories or gravity. So, the question is not just a mathematical problem of enlarging a symmetry, but there is also the physical question of whether this mathematical structure can be implemented in a conventional Lagrangian field theory. And in this sense supersymmetry seems to be unique, simply because there are no other examples.

The striking fact about supersymmetry is that it can be implemented in a Lagrangian field theory, and in the case of gauge theories, it just reduces the coupling constant of the theory. This yields a particular relation between the coupling in the Higgs scalar potential and the gauge coupling, which is one of the sources of the application of supersymmetry to the standard model, where you get a prediction for the Higgs mass. It is a fascinating fact that this symmetry, which actually was discovered 35 years ago in a completely different context in the framework of string theory, can be implemented in 4D, as was shown by Wess and Zumino, who developed the first renormalizable supersymmetric model, the Wess-Zumino model, which is the simplest example of a theory containing only scalars and spin ½ Majorana particles and which was then extended to gauge theories, both Abelian and non-Abelian and then eventually to gravity. It is completely non trivial that this kind of symmetry can be merged with gauge invariance and with general covariance. It is a quite unique structure with these properties, there are no other known examples of symmetries which enlarge space-time and can be implemented in conventional Lagrangian theory.

- *Mastrolia:*

The anomaly diagrams in a non-Abelian gauge theory are 1-loop 3-point graphs. Which Feynman diagrams enter in the Green-Schwarz anomaly?

- *Ferrara:*

They are actually hexagon diagrams, i.e. six-point. The reason is that in the 1-loop anomaly the number of points in the diagram is in some sense dimension dependent. It is 3-point if you are in 2D, which gives you the famous Schwinger mass of 2D electrodynamics, then it is 3-point in 4D and in general there is a relation between the anomaly diagram and the $n={}^{D}/_{2}+1$ point polygon amplitudes, which is connected to 1-loop anomalies. So, in the case of 10D it is a hexagon diagram. In these diagrams only fermions must circulate: in the loops there are the gauginos of the Yang-Mills multiplet and then there are the gravitino and the spin ½ particle of the gravity multiplet. The main point is that the anomalies can be there because all these fields are chiral:

they are Majorana chiral fermions in ten dimensions. And such Majorana fermions can induce these gravitational anomaly and mixed gauge-gravitational anomalies. There is a paper by Witten and Alvarez-Gaumé where they performed a general analysis of these anomalies in D dimensions and their cancellation. It is a highly non trivial result that these anomalies are cancelled in 10D $\mathcal{N}$=1 SUGRA for two particular choices of gauge groups.

- *Burda:*

As I understand, the quantized SUGRA theory can in principle be considered as a full consistent theory of quantum gravity. But then in a quantum gravity theory our classical space-time should emerge. Instead, we usually start from some classical space-time, Minkowski or AdS, and then just consider a theory of linear fluctuations of the metric over the classical expression. So, the question is: do we actually understand how this classical space-time emerges in SUGRA or string theories?

- *Ferrara:*

This is a very complicated question. In string theory, as in supergravity, gravity is regarded mostly in a geometrical way, as a conventional Lagrangian field theory. This is also how 't Hooft, Veltmann and others look at the divergences of gravity in loops. You consider the graviton as an ordinary gauge field of spin 2 and then you expand this field around Minkowski space and you look at scattering amplitudes with the usual Feynman rules for a spin-2 particle together with lower spin particles. This is e.g. the way Dixon and others look at $\mathcal{N}$=8 supergravity in 4D. For quantum gravity here we mean a theory which is renormalizable or finite in the sense that you get computable scattering amplitudes in terms of only a small number of parameters, and that you don't have new divergences when you increase the number of loops or of legs in the amplitudes. So, this is what is called in supergravity a possible quantum theory of gravity, in the same way as string theory is considered to resolve the problem of quantum gravity. In string theory you look at the perturbative expansion in the string coupling and the problem of divergences is resolved by the fact that there is a natural ultraviolet cut-off determined by the string length. But gravity is still treated as a conventional field theory, in which you make a perturbative expansion with respect to some background space-time.

- *Burda:*

But this does not answer the question of how our classical Minkowski space-time emerges from the theory.

- *Ferrara:*

Eventually, the question should be answered in a theory of quantum gravity, but nobody knows how, because gravity can have various different backgrounds: Minkowski space-time is one, but there are other interesting backgrounds, like anti de Sitter space, and in cosmology de Sitter space. It therefore really depends on which background you choose, and it depends on which are the epochs you are exploring in this theory of quantum gravity. Quantum gravity should explain the Big Bang, the evolution of the universe, particle physics, so it is quite a large and ambitious program.

- *Burda:*

In your morning lecture you made several remarks on applications of the SUGRA construction to AdS-like space-times, because of the existence of the AdS/CFT conjecture. But what about more realistic dS-like backgrounds? Is it possible to make all these steps for dS-like space-times?

- *Ferrara:*

In principle in a supergavity theory you can have a de Sitter background as well as an anti de Sitter one. The main difference is that if you want a background that preserves supersymmetry, then you have only the possibility of having a Minkowski or and AdS space-time. A dS background is incompatible with the preservation of supersymmetry. You can have a de Sitter solution in supergravity, but this solution will be a solution with no residual supersymmetry.

- *Cifarelli:*

This morning you mentioned something about spontaneously broken Higgs mechanism. Do you have a favourite guess about $m_{3/2}$ mass?

- *Ferrara:*

This is a subject of my tomorrow lecture. In fact, different scenarios of supersymmetry yield very different range of the gravitino mass. I will give only two examples to show the difference between these two cases, which were actually envisaged at the very beginning of supersymmetry, when Pierre Fayet from the Ecole Normale explored possible scenarios of supersymmetry in connection with applications to the Standard Model. But I will talk about this tomorrow.

- *Aschieri:*

You recalled that D=11 is the maximal dimension for supersymmetric theories due to the problems of higher spin theories. How robust is this finding, in the sense that one could consider theories with an infinite tower of spins? Is there a relation to a kind of supergravity in 12 dimensions?

- *Ferrara:*

There's been a lot of work going on about higher spin theories, e.g. by Vassiliev, Sagnotti and others. To my recollection, a first problem in these higher spin theories is that in order to have a consistent coupling of the higher spins with the lower spins which preserves gauge invariance, you have to go from a Minkowski background to an AdS one. The second point is that as far as I understand these higher spin theories need to be formulated as an underlying theory of some new algebraic structure related to these higher spin fields. However, this is not my subject. But I think that higher spin theories arise when you consider extended objects, like in string theory. Here you have higher spin theories, but all the higher spin fields are massive. On the other hand, in AdS you can have higher spin theories where all higher spin fields are massless, and there is the expectation that there should be an unbroken phase of string theory in which all the particles should become massless and that somehow string theory is related to some broken phase of a theory which has a huge symmetry. There's been work done on this subject, but as far as I know there's not been enough progress which could point towards a distinct application of higher spin theories to ordinary gravity or supergravity.

As far as 12 dimensions is concerned, it would be interesting, but as far as I know there is no generalization of conventional supergravity to 12 dimensional space-time.

CHAIRMAN: S. FERRARA

Scientific Secretaries: D. Tapia Takaki, D. Epifanov

DISCUSSION II

- *Burda:*

Could you explain more in detail the issue of black hole entropy?

- *Ferrara:*

In General Relativity, the black hole entropy is given by the Bekenstein-Hawking entropy-area formula, which essentially relates geometrical quantities to thermodynamical quantities. The Bekenstein-Hawking formula simply states that the entropy of a black hole is one-quarter of the area of its event horizon. In presence of vector fields charged with respect some gauge group, and for instance in supersymmetric theories of gravity, the Bekenstein-Hawking formula contains the contribution of the electric and magnetic charges of the black hole, which are generally conserved quantities due to the overall gauge-invariance of the physical system under consideration. For example, in N=8 supergravity in four space-time dimensions, there are 28 magnetic and 28 electric charges, and the black hole entropy will be a function of them, suitably constrained by the duality-invariances of the theory itself. The issue of scalar fields coupled to the black hole background is crucial.

For simplicity, let us consider a static (*i.e.* independent on time), spherically symmetric black hole, coupled to a unique scalar field. Such a scalar field undergoes an evolution dynamics, determining a *flow* from asymptotical spatial infinity to the event horizon of the back hole itself. Under some very general assumptions, the dynamics of the evolution flow of the scalar field is governed by a second-order differential equation, which remarkably becomes first-order for those flows which preserve some amount of the supersymmetries pertaining to the asymptotical background.

Here a surprising phenomenon takes place: regardless of supersymmetry-preserving features of the flow, the value taken by the scalar field at the horizon of the black hole depends purely on the conserved (electric and magnetic) charges, and not on the asymptotical value of the scalar, which is the initial condition of its evolution. In other words, the dynamics of radial evolution of the scalar field, despite being deterministic, is such that it *"loses"* memory of its initial conditions.

This is what in Statistical Mechanics is called an *"attractor"*, and thus the phenomenon sketchily described above has been dubbed *Attractor Mechanism*, as it was discovered in mid 90's by Andy Strominger, Renata Kallosh, and myself.

The *Attractor Mechanism* prescribes the "velocity" of the scalar field to vanish when the horizon of the black hole is reached, namely:

$$\dot{\phi}(r)\xrightarrow[r\to r_H]{}0$$

This provides us with a way to compute the black hole entropy. Indeed, as I proved in a paper with Kallosh and Gibbons written more than years ago, the *Attractor Mechanism* admits a *"critical implementation"*, in the sense that the (purely charge-dependent) values of the scalars at the horizon of the black hole are nothing but the critical points of a suitably defined *"effective back hole potential"* V.

It should be stressed that *Attractor Mechanism* strictly holds only for the so-called "extremal" black holes, which are stable gravitational objects with vanishing temperature (and thus no Hawking radiation), but non-vanishing area of the event horizon (which turns out to be unique, as well).

As I have mentioned, the black hole entropy is a function of the electric and magnetic charges. Its precise functional form is completely determined by the electric-magnetic duality properties of the theory under consideration. Considering once again N=8 supergravity, the black hole entropy is invariant under finite transformations of the exceptional group E_7(7), acting on the charges themselves. Remarkably, there exists only one independent invariant of E_7(7) in the representation in which the charges actually sit, namely the 56-dimensional fundamental representation of the group itself.

- *Scherbakov:*

Could you comment on the universality of Bekenstein-Hawking formula? Could it be applied "small" black holes?

- *Ferrara:*

Black holes whose classical entropy (*i.e.* the entropy computed with the usual two-derivative Einstein-Hilbert action) vanishes, are usually named "small" black holes. For such a broad class of gravitational objects, the Bekenstein-Hawking formula needs to be generalized, including higher–derivative terms in the gravitational sector of the theory.

In general, besides the Einstein-Hilbert term (containing the Ricci intrinsic scalar curvature R) there appear many other terms, with increasingly higher powers of R itself, and in general with derivative terms of order higher than two, namely:

$$L = \sqrt{-g}\,(R + cR^{2} + \ldots)$$

This is implemented by the so-called Wald formula, which yields a non-vanishing result also for the entropy of those black hole which are usually dubbed "small", in order to distinguish them from the "large" black holes, whose entropy (computed with the Bekenstein-Hawking formula) does not vanish.

$A=0 \rightarrow$ *"small"* black hole
$A_H \neq 0 \rightarrow$ *"large"* black hole

- *'t Hooft:* :

In a previous discussion, Professor Zichichi raised the important question of the actual existence of extremal black holes, which have vanishing temperature. He said: " ... *then they do not exist !*". I actually agree with this statement. Even if they do not emit Hawking radiation, such extremal black holes may decay by tunneling effects, thus neutralizing their excessive charges. Consequently, one should conceive an extremal black holes as the limit result of a dynamical process involving more realistic black holes, with finite temperature. Do you agree with this point of view?

- *Ferrara:*

Yes, I do.

- *Aschieri:*

In the study of extremal black that you presented, there are two sets of symmetries entering the game: the duality rotations, and supersymmetry. How many results are still valid when only duality rotational symmetry is present?

- *Ferrara:*

The Attractor Mechanism does not rely on supersymmetry at all. In general, it describes the near-horizon dynamics of scalar fields coupled to a theory of gravity with vector fields. In order to obtain well-defined values of the scalar fields at the horizon, one has demand that the "scalar velocity" vanishes, thus fixing the entropy in terms of electric and magnetic charges only. However, as I said, supersymmetry modifies the form of the differential equations describing the radial evolution dynamics of the scalar fields. Indeed, within a supersymmetric theory all attractor scalar flows are governed by first-order differential equations. And this holds true not only for the scalar configurations preserving some supersymmetry, but also for those ones which break all supersymmetries, and are thus non-supersymmetric (but within a supersymmetric theory).

This will generally not be the case for a non-supersymmetric theories, in which radial flows are usually of second-order.

- *Korthals-Altes:*

In yesterday's lecture you discussed the Dirac consistency condition on electric and magnetic charges, and also the Zwanziger-Schwinger condition on several dyonic objects, where the usual minus sign could be replaced by a plus sign. Could you spend some more words about it?

- *Ferrara:*

Yes, I discussed Zwanziger-Schwinger conditions in higher space-time dimensions, and with an arbitrary number of U(1)'s. Thus, the duality group changes from Sp(2n,R) in four space-time dimensions to SO(n,n), and the minus sign can be changed to plus.

- *Klopot:*

Could you please give more details on the relation between central charges and the effective black hole potential?

- *Ferrara:*

Yes, of course. In general, there are two different formulae for the effective black hole potential V, governing the attractor behavior of scalars coupled to extremal black holes. The most general formula expresses V in terms of the kinetic vector matrix, *i.e.* in terms of the scalar-dependent matrix which determines the structure of the kinetic term for the vector field strengths coupled to the considered black hole. This formula holds for any theory of gravity with a well defined kinetic vector term, and it is independent on supersymmetry. In a supersymmetric theory of gravity, this formula can be recast in another, suggestive form, involving only the so-called central charges, which are the objects which centrally-extend the supersymmetry algebra, by entering in the anti-commutators of two supercharges. Considering the case of N=8 supergravity (in which there are no matter multiplets), this second formula reads

$$V_{BH} = \sum_{A<B=1}^{8} Z_{AB}\,\overline{Z}^{AB}$$

The central charges previously mentioned are organized into an 8x8 skew-symmetric complex central charge matrix.

$$Z_{AB}(\phi^i, p, q)$$

Such a matrix is a function of the scalars, as well as of the electric and magnetic charges, of the theory. By computing the critical value of V, one determines the black hole entropy through the aforementioned Bekenstein-Hawking entropy-area formula, namely:

$$S = \frac{1}{4} A = \pi V_{BH} \Big|_{\phi=\Phi_H}$$

Orientifold string vacua and strings at the LHC*

Dieter Lüst[1]

Arnold Sommerfeld Center for Theoretical Physics
Department für Physik, Ludwig-Maximilians-Universität München
Theresienstr. 37, 80333 München, Germany

Max-Planck-Institut für Physik
Föhringer Ring 6, 80805 München, Germany

Abstract

We consider extensions of the standard model based on open strings ending on D-branes, with gauge bosons due to strings attached to stacks of D-branes and chiral matter due to strings stretching between intersecting D-branes. Assuming that the fundamental string mass scale is in the TeV range and the theory is weakly coupled, we discuss possible signals of string physics at the Large Hadron Collider (LHC).

---

*Proceedings of the Erice 2009 International School of Subnuclear Physics
[1]dieter.luest@lmu.de, luest@mppmu.mpg.de

1 Introduction

String theory, being the theory of one-dimensional extended objects, contains an infinite number of elementary particles, when seen from the field theory point of view. Specifically, the excitation spectrum of a relativistic string contains a tower of infinitely many higher spin excitations, whose number grows exponentially with the masses of the excited string modes. In fact, these states follow the well-known Regge excitation pattern with masses

$$M_n^2 = nM_s^2\,, \tag{1}$$

where M_s is the intrinsic string scale ($M_s^{-2} = \alpha'$), and n counts the n$^{th.}$ excitation level.

In string compactifications from ten to four space-time dimensions there are several additional generic mass scales and particles, which are related to the internal geometry of the compact space. These are Kaluza-Klein particles, whose masses scale inversely with some internal radii, possible winding states, as well as moduli fields that determine the parameters of the underlying geometry. In particular, in intersecting D-brane compactifications with extra dimensions, there are several distinct mass scales and moduli fields, which could be also of experimental relevance. Basically, each of these mass scales are related to some generic particles, which arise during the compactification process.

A particularly interesting scenario is given by D-brane compactifications with a string scale M_s being near the scale of the Standard Model (SM), i.e. $M_s = \mathcal{O}(\text{TeV})$ [1, 2], which implies that some of the extra dimensions must be much larger than M_s^{-1}. Moreover, since M_s sets the fundamental scale of gravity in the higher dimensional space, gravitational effects will start to play an important role already at the low string scale. In fact, in this scenario there is an explosion of string states at the TeV scale, and spectacular, model independent stringy signatures are expected at the LHC in case the string scale is low [3, 4, 5, 6]. D-brane compactifications with a low string scale are indeed possible, because the open string gauge interactions are confined to the lower dimensional world volumes of the D-branes, but on the other hand the gravitational force is enormously weakened by the fact, that the closed string gravitons can leak out into the large transversal bulk space. Of course, one has to investigate, if low strings scale D-brane compactifications can be dynamically realized by a certain moduli potential, a question, which was addressed in ref. [7, 8].

2 Length scales and string excitations in D-brane orientifold compactifications

2.1 Various choices for the string scale

In the following we will consider type II orientifold compactifications with D-branes, with gauge bosons due to open strings attached to stacks of D-branes and chiral matter due to open strings stretching between intersecting D-branes (for reviews see [9, 10]). Let us recall that there are three basic, dimensional parameters in type II string compactifications with wrapped and/or intersecting D-branes.

Of course, first there is the fundamental string scale, given in terms of the slope parameter α' as:

$$(1): \quad M_s = \frac{1}{\sqrt{\alpha'}} \,. \tag{2}$$

The string scale is related to the fundamental 10-dimensional Planck mass M_{10} through a dimensionless string coupling constant g_s in the following way,

$$M_{10}^8 = \frac{1}{g_s^2} M_s^8 \,. \tag{3}$$

Second, we have to compactify from ten to four dimensions on an internal 6-dimensional space $\mathcal{M}_6$; so as second basic string mass scale we introduce the volume V_6 of the internal space:

$$(2): \quad m_6 \equiv \frac{1}{V_6^{1/6}} \,. \tag{4}$$

When measuring masses in terms of Planck units, it is useful to introduce a dimensionless overall volume, given in string (or 10-dimensional Planck) mass units, as

$$V_6' = V_6 M_s^6 = \frac{M_{\text{Planck}}^2}{M_s^2} \,. \tag{5}$$

We are considering scenarios, where the SM is located at some stacks of D(3+p)-branes, which are wrapped around some internal p-cycles inside $\mathcal{M}_6$: $\Sigma_p \subset \mathcal{M}_6$. So, the third basic string mass scale is related to the volume $V_p^{\parallel}$ of the sub-space Σ_p, around which the D-branes are wrapped:

$$(3): \quad m_p^{\parallel} = \frac{1}{(V_p^{\parallel})^{1/p}} \,. \tag{6}$$

Alternatively we can introduce the volume $V_{6-p}^{\perp}$ of the space transversal to the D-branes,

$$(3'): \quad m_{6-p}^{\perp} = \frac{1}{(V_{6-p}^{\perp})^{1/(6-p)}} \,, \tag{7}$$

where the following relation holds:

$$V_6 = V_p^{\|} V_{6-p}^{\perp} \,. \tag{8}$$

These three fundamental dimensional parameters of D-brane models are linked to two 4D physical observables in the following way:

(A) the strength of gravitational interactions, as determined by the 4D Planck mass:

$$(A): \quad M_4^2 = 8\, e^{-2\phi_{10}}\, M_s^8\, \frac{V_6}{(2\pi)^6} \,. \tag{9}$$

Here $g_s = e^{\phi_{10}}$ is the string coupling constant.

(B) the strength of the gauge interactions, as determined by the 4D gauge coupling:

$$(B): \quad g_{Dp}^{-2} = (2\pi)^{-1}\, M_s{}^p\, e^{-\phi_{10}}\, V_p^{\|} \,. \tag{10}$$

Hence, for fixed string coupling $g_s = e^{\phi_{10}}$, knowing M_{Planck} and g_{Dp}^{-2} leaves one of the three dimensionful string parameters undetermined (unlike for the heterotic string). Since the 4D gauge coupling must not be too small, we take it of order unity. This fixes $(V_p^{\|})^{-1/p}$ to be of order M_s, i.e. $(V_p^{\|})^{-1/p} \simeq M_s$. This leaves only M_s (or V_6) as free parameter. Then, the transversal volume $V_{6-p}^{\perp}$ is of the oder

$$V_{6-p}^{\perp} = V_6 M_s^p \,. \tag{11}$$

Let us recall that there are several natural choices for the string scale M_s. The perhaps still most natural choice is to identify M_s with $M_4 \equiv M_{\text{Planck}}$. These are the high string scale models (like heterotic string compactifications):

$$M_s \equiv M_{\text{Planck}} \simeq 10^{19}\ \text{GeV} \,. \tag{12}$$

In this case, V_6 is of order $V_6^{1/6} \simeq M_{\text{Planck}}^{-1}$.

Alternatively, one often assumes the existence of a GUT gauge group and makes the identification:

$$M_s \equiv M_{GUT} \simeq 10^{16}\ \text{GeV} \,. \tag{13}$$

In this scenario, V_6 is of order

$$V_6^{1/6} \simeq 10^{-15}\ \text{GeV}^{-1} \,. \tag{14}$$

Lower string scale compactifications identify M_s with the scale of spontaneous supersymmetry breaking:

$$M_s \equiv M_{SUSY} \simeq 10^{11}\ \text{GeV} \,. \tag{15}$$

Table 1: The different mass scales in D-brane models

	M_s (GeV)	L_s (m)	$M_6 = V_6^{-1/6}$ (GeV)	$V_6^{1/6}$ (m)	$M_2^\perp = (V_2^\perp)^{-1/2}$ (GeV)	$(V_2^\perp)^{1/2}$ (m)
(o)	10^{19}	10^{-35}	10^{19}	10^{-35}	10^{19}	10^{-35}
(i)	10^{16}	10^{-32}	10^{15}	10^{-31}	10^{13}	10^{-29}
(ii)	10^{11}	10^{-27}	10^{6-7}	$10^{-(22-23)}$	10^{3}	10^{-19}
(iii)	10^{3}	10^{-19}	$10^{-14/6}$	10^{-14}	10^{-13}	10^{-3}

This scenario implies that V_6 is of order

$$V_6^{1/6} \simeq 10^{-(6-7)}\ \mathrm{GeV}^{-1}\,. \tag{16}$$

The most radical, but perhaps also most natural choice is to identify M_s right away with the scale of the SM, i.e. with the TeV scale:

$$M_s \equiv M_{SM} \simeq 10^3\ \mathrm{GeV}\,. \tag{17}$$

Therefore in this scenario, V_6 is very large, it is of order

$$V_6^{1/6} \simeq 10^{14/6}\ \mathrm{GeV}^{-1}\,, \tag{18}$$

and hence this scenario is also called *very large extra dimension scenario.*

Let us summarize these four possibilities, and the corresponding energy and length scales in the table 1 (we use that 1 GeV $\sim$ $(10^{-16}$ m$)^{-1}$, we have set $g_s \simeq 1$, $p = 4$, and we neglect factors of 2π). V_6' takes the following values:

$$V_6' = 1, 10^6, 10^{16}, 10^{32}\,. \tag{19}$$

2.2 Several generic particles in brane compactifications and their associated length scales

Now let us recall several generic particles that are related to the different mass scales in D-brane compactifications, as well as their masses, measured in the Einstein frame, and expressed in terms of Planck units.

(i) String Regge excitations:

The most generic and model independent stringy excitations are the so-called Regge modes. Their masses are proportional to the string scale M_s, and in terms of Planck units given by:

$$M_{\mathrm{Regge}} = \sqrt{n} M_s = \frac{\sqrt{n} M_4}{\sqrt{V_6'}}\,. \tag{20}$$

Here n denotes the oscillator number of these states. In case they are open strings, the stringy Regge excitations can carry the SM quantum numbers, i.e.

they can be produced by pp-collisions via gauge interaction in the TeV scale string scenario, as we will discuss in the next chapter.

(ii) Transversal direction KK particles:

For the masses of the KK particles related to the large volume cycles one obtains:

$$M_{KK}^{\perp} = \frac{m}{(V_{6-p}^{\perp})^{1/(6-p)}} = m\ M_4 (V_6')^{\frac{8-p}{2p-12}}, \tag{21}$$

where m counts the internal closed string KK momentum of each state. In the four scenarios discussed above $M_{KK}^{\perp}$ takes the following values (here for $m = 1$, $p = 4$):

$$M_{KK}^{\perp} = 10^{19}, 10^{13}, 10^{3}, 10^{-13}\ \text{GeV}. \tag{22}$$

These KK particles are closed strings, and they are neutral under the SM gauge group, and hence interact only gravitationally with SM particles.

(iii) Small (longitudinal) cycle KK particles:

For the masses of the KK particles related to the small volume cycles one obtains:

$$M_{KK}^{\|} = \frac{m'}{(V_p^{\|})^{1/p}} \simeq m'\ M_s = \frac{m'\ M_4}{(V_6')^{1/2}}. \tag{23}$$

Here m' is the open string KK momentum along the wrapped $(3+p)$-branes. In the four scenarios discussed above $M_{KK}^{\|}$ takes the following values ($m' = 1$):

$$M_{KK}^{\|} = 10^{19}, 10^{16}, 10^{11}, 10^{3}\ \text{GeV}. \tag{24}$$

These KK particles are open strings, and hence they carry SM quantum numbers, i.e. they are the KK excitations of the SM fields. Hence they can be directly produced at the LHC, as will be discussed in the following.

(iv) Overall volume modulus:

Related to the overall six-dimensional volume V_6, there is a generic closed string modulus field T in any compactification. In the low string scale compactifications, the tree-level mass of this modulus may be unacceptably small, (e.g., its typical tree level mass was calculated using the moduli potential calculated in [8]: $M_T = \frac{M_{\text{Planck}}}{(V_6')^{3/2}}$).

However, the point [11] we would like to stress is, that there is a model-independent lower bound on the typical mass that the volume modulus should get from the quantum correction, which is given by

$$\Delta M_T^2 \simeq \frac{\langle T_\mu^\mu T_\mu^\mu \rangle}{M_4^2} \gtrsim N_{SM} \frac{\Lambda_{SM}^4}{M_4^2}, \tag{25}$$

where Λ_{SM} is the scale that cuts-off the contribution to the $\langle T^{\mu}_{\mu} T^{\mu}_{\mu} \rangle$ correlator from the loops of the standard model particles, and N_{SM} is the number of particles contributing. Lowest experimental bound on Λ_{SM} is around TeV, which (ignoring the accidental cancellations among different Standard Model species) translates as the expected lowest value of the volume modulus mass around

$$\Delta M_T \simeq 10^{-13}\ \text{GeV}\,, \tag{26}$$

which corresponds to a length scale in sub-millimeter range. This is the range where the existing table-top fifth force experiments start to penetrate.

(v) Small (longitudinal) cycle volume moduli:

Related to the small cycles of longitudinal volume $V_p^{\|}$, there are a moduli fields t with tree level masses (here for $p = 4$)

$$M_t = \frac{M_4}{V_6'} \tag{27}$$

In the four scenarios discussed above M_t takes the following values:

$$M_t = 10^{19}, 10^{13}, 10^{3}, 10^{-13}\ \text{GeV}\,. \tag{28}$$

Again, these t-moduli are neutral and interact only gravitationally with SM fields.

(vi) Mini black holes:

Let us also briefly comment on the possible creation of (mini) black holes in string scattering processes. In contrast to the string Regge excitations and the KK (winding) modes, the black hole states are of non-perturbative nature, and their masses are basically determined by the scale of gravity in the higher dimensional bulk theory. More precisely, the energy threshold for black hole production is given by:

$$E_{\text{b.h.}} \simeq \frac{M_{\text{string}}}{g^2_{\text{string}}}\,. \tag{29}$$

Comparing this with the masses of the perturbative string excitations in eq.(20), one recognizes that one first produces $n^* \sim 1/g^4_{\text{string}}$ string states before one reaches $E_{\text{b.h}}$. In case of weak string coupling, n^* is a quite large number, e.g. for the string coupling constant being of the same order as the gauge coupling, $g_{\text{string}} \approx 0.2$, $n^* \approx 625$. Hence black hole production will not be of any relevance for our further discussion.

3 Production of heavy string Regge excitations and KK/winding states at the LHC

The production of string Regge excitations will lead to new contributions to standard model scattering processes, like QCD jets or scattering of quarks into leptons or gauge bosons, which can be measurable at LHC in case the string scale is low [12, 13, 3, 4, 5, 6]. Second there are the KK and winding excitations along the small internal dimensions, i.e. KK and winding excitations of the SM fields. Their masses depend on the internal volumes, and they should be also near the string scale M_{string}.

For those amplitudes involving four gauge bosons or two gauge bosons and two matter fermions, the amplitudes do not depend on the geometry of the underlying Calabi-Yau spaces [3]. This model independence still also holds for the four–fermion matter amplitudes, but only w.r.t. their dependence on the four-dimensional kinematical variables s, t, u. On the other hand, the four–fermion amplitudes do depend on the internal Calabi-Yau geometry and topology.

The general structure of a four point amplitude of four open string states is as follows. Let Φ^i, $i = 1, 2, 3, 4$, represent gauge bosons, quarks of leptons of the standard model realized on three or more stacks of intersecting D-branes. The corresponding string vertex operators V_{Φ^i} are constructed from the fields of the underlying superconformal field theory (SCFT) and contain explicit (group-theoretical) Chan-Paton factors. In order to obtain the scattering amplitudes, the vertices are inserted at the boundary of a disk world-sheet, and the following SCFT correlation function is evaluated:

$$\mathcal{A}(\Phi^1, \Phi^2, \Phi^3, \Phi^4) = \sum_{\pi \in S_4/Z_2} V_{CKG}^{-1} \int_{\mathcal{I}_\pi} \left(\prod_{k=1}^{4} dz_k \right) \langle V_{\Phi^1}(z_1)\; V_{\Phi^2}(z_2)\; V_{\Phi^3}(z_3)\; V_{\Phi^4}(z_4) \rangle \,. \tag{30}$$

Due to the extended nature of strings, the world–sheet string amplitudes are generically non–trivial functions in α' in addition to the usual dependence on the kinematic invariants and degrees of freedom of the external states. In the effective field theory description this α'–dependence gives rise to a series of infinite many resonance channels due to Regge excitations and/or new contact interactions. Generically, tree–level string amplitudes involving four gluons or amplitudes with two gluons and two fermions are described by the Euler Beta function depending on the kinematic invariants $s = (k_1 + k_2)^2$, $t = (k_1 - k_3)^2$, $u = (k_1 - k_4)^2$, with $s + t + u = 0$ and k_i the four external momenta. The whole amplitudes $A(k_1, k_2, k_3, k_4; \alpha')$ may be understood as an infinite sum over s–channel poles with intermediate string states $|k; n\rangle$ exchanged, as it can be seen in the figure 1. After neglecting kinematical factors the string amplitude $A(k_1, k_2, k_3, k_4; \alpha')$ assumes the form

$$A(k_1, k_2, k_3, k_4; \alpha') \sim -\frac{\Gamma(-\alpha' s)\, \Gamma(1 - \alpha' u)}{\Gamma(-\alpha' s - \alpha' u)} = \sum_{n=0}^{\infty} \frac{\gamma(n)}{s - M_n^2} \tag{31}$$

as an infinite sum over s–channel poles at the masses

$$M_n^2 = M_s^2\, n \tag{32}$$

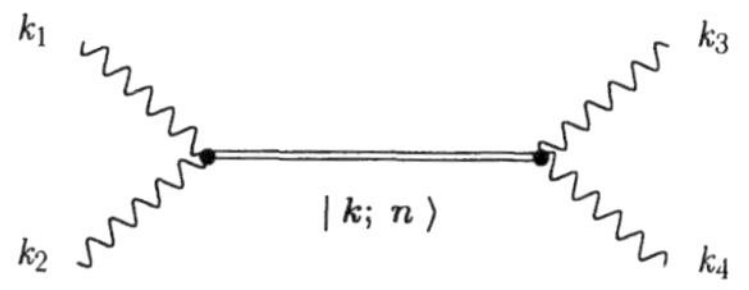

Figure 1: Exchange of an infinite tower of Regge excitations in open string scattering processes.

of the string Regge excitations. In eq.(31) the residues $\gamma(n)$ are determined by the three–point coupling of the intermediate states $|k;n\rangle$ to the external particles and given by

$$\gamma(n) = \frac{t}{n!}\frac{\Gamma(-u\alpha'+n)}{\Gamma(-u\alpha')} = \frac{t}{n!}\prod_{j=1}^{n}[-u\alpha'-1+j] \sim \frac{(-\alpha'\,u)^n}{n!} \quad , \tag{33}$$

with $n+1$ being the highest possible spin of the state $|k;n\rangle$.

3.1 Two gluon, two quark scattering amplitudes

For example, we now consider the following correlation function between two gauge bosons and two matter fermions [3]:

$$\langle V^{(0)}_{A^x}(z_1,\xi_1,k_1)\, V^{(-1)}_{A^y}(z_2,\xi_2,k_2)\, V^{(-1/2)}_{\psi^{\alpha_3}_{\beta_3}}(z_3,u_3,k_3)\, V^{(-1/2)}_{\bar\psi^{\beta_4}_{\alpha_4}}(z_4,\bar u_4,k_4)\rangle\ . \tag{34}$$

The fermion vertex operators are boundary changing operators, being inserted at the intersection of brane stack a and b. Specifically, the chiral fermion vertex operators of the quarks and leptons are:

$$\begin{aligned} V^{(-1/2)}_{\psi^{\alpha}_{\beta}}(z,u,k) &= g_\psi [T^{\alpha}_{\beta}]^{\beta_1}_{\alpha_1} e^{-\phi(z)/2}\, u^\lambda S_\lambda(z)\, \Xi^{a\cap b}(z)\, e^{ik_\rho X^\rho(z)}\ , \\ V^{(-1/2)}_{\bar\psi^{\beta}_{\alpha}}(z,\bar u,k) &= g_\psi [T^{\beta}_{\alpha}]^{\alpha_1}_{\beta_1} e^{-\phi(z)/2}\, \bar u_{\dot\lambda} S^{\dot\lambda}(z)\, \overline{\Xi}^{a\cap b}(z)\, e^{ik_\rho X^\rho(z)}\ . \end{aligned} \tag{35}$$

These vertices connect two segments of disk boundary, associated to stacks a and b, with the indices α_1 and β_1 representing the string ends on the respective stacks. The internal field $\Xi^{a\cap b}$ of conformal dimension 3/8 is the fermionic boundary changing operator. In the intersecting D-brane models, the intersections are characterized by angles θ_{ba}. Then $\Xi^{a\cap b}$ can be expressed in terms of bosonic and fermionic twist fields σ and s:

$$\Xi^{a\cap b} = \prod_{j=1}^{3} \sigma_{\theta^j_{ba}}\ s_{\theta^j_{ba}} \quad , \quad \overline{\Xi}^{a\cap b} = \prod_{j=1}^{3} \sigma_{-\theta^j_{ba}}\ s_{-\theta^j_{ba}}\ . \tag{36}$$

The fact that fermions originate from the same pair of stacks, say a and b is forced upon us by the conservation of twist charges, in a similar way as

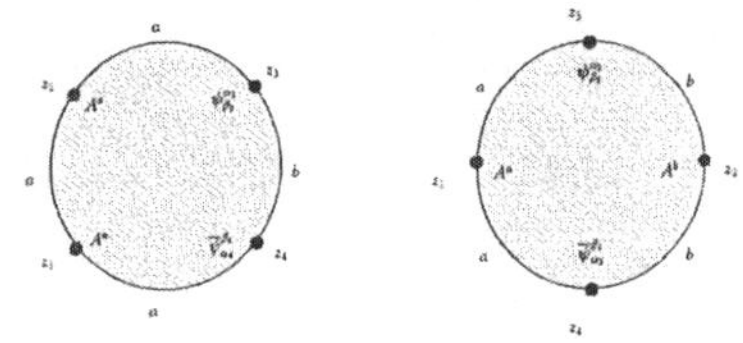

Figure 2: The two gauge boson - two fermion disk diagrams.

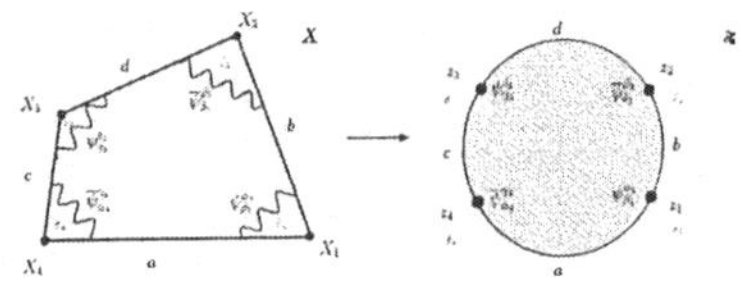

Figure 3: The SM by four stacks of intersecting branes and the corresponding four fermion disk diagram.

their opposite helicities are forced by the internal charge conservation. It follows that both gauge bosons must be associated either to one of these stacks, say $(x, y) = (a_1, a_2)$, or one of them is associated to a while the other to b, say $(x, y) = (a, b)$. The corresponding disk diagrams are shown in the figure 2. Using these informations one obtains the following tree level (squared) amplitudes for two gauge boson, two fermion scattering processes [3],

$$|\mathcal{M}(gg \to q\bar{q})|^2 \ = \frac{t^2+u^2}{s^2}\Big[\frac{1}{6}\frac{1}{ut}(tV_t+uV_u)^2-\frac{3}{8}V_tV_u\Big] \tag{37}$$

and

$$|\mathcal{M}(gq \to gq)|^2 \ = \frac{s^2+u^2}{t^2}\Big[V_sV_u-\frac{4}{9}\frac{1}{su}(sV_s+uV_u)^2\Big] \tag{38}$$

with

$$V_s = \frac{s\,u}{tM_s^2}B(-s/M_s^2, -u/M_s^2) = \frac{\Gamma(1-s/M_s^2)\,\Gamma(1-u/M_s^2)}{\Gamma(1+t/M_s^2)}, \tag{39}$$

and with similar expressions for V_t and V_u. Note that there can be only the exchange of heavy Regge states and no KK-states. Hence also these two amplitudes are completely independent from the internal geometry.

3.2 Four quark scattering amplitudes

Next consider the scattering amplitude of four Standard Model fermions [3]. In the most general case, all fermions are at different intersections, as seen in figure 3. Then, without going into further details the corresponding tree level four-fermion scattering amplitudes take the following form:

$$
\begin{aligned}
|\mathcal{M}(qq \to qq)|^2 &= \frac{2}{9}\frac{1}{t^2}\left[(sF^{bb}_{tu})^2 + (sF^{cc}_{tu})^2 + (uG^{bc}_{ts})^2 + (uG^{cb}_{ts})^2\right] \\
&+ \frac{2}{9}\frac{1}{u^2}\left[(sF^{bb}_{ut})^2 + (sF^{cc}_{ut})^2 + (tG^{bc}_{us})^2 + (tG^{cb}_{us})^2\right] \\
&- \frac{4}{27}\frac{s^2}{tu}(F^{bb}_{tu}F^{bb}_{ut} + F^{cc}_{tu}F^{cc}_{ut}) \qquad (40)
\end{aligned}
$$

Here the functions F and G depend on α and now also on the masses of the internal KK (and winding) states. Therefore these amplitudes depend on the internal geometry and are not anymore model independent. Furthermore, due to the quantum numbers of the fermions there are no s-channel poles, but in the t,u-channels there is the exchange of Regge, KK and winding modes, as it can be seen by appropriate expansions of the functions F and G.

3.3 Signals for lowest mass strings at the LHC

In this section we will determine the contribution from the exchange of excited, heavy Regge states to dijet processes at the LHC [4]. The first Regge excitations of the gluon (g) and quarks (q) will be denoted by g^*, q^*, respectively. The first excitation of the C will be denoted by C^*. In the following we isolate the contribution to the partonic cross section from the first resonant state. Note that far below the string threshold, at partonic center of mass energies $\sqrt{s} \ll M_s$, the form factor $V(s,t,u) \approx 1 - \frac{\pi^2}{6}su/M_s^4$ and therefore the contributions of Regge excitations are strongly suppressed. The s-channel pole terms of the average square amplitudes contributing to dijet production at the LHC can be obtained from the general formulae given in in the previous subsection. However, for phenomenological purposes, the poles need to be softened to a Breit-Wigner form by obtaining and utilizing the correct *total* widths of the resonances [14]. After this is done, the contributions of the various channels are collected in figure 4. In this figure we show a representative plot of the invariant mass spectrum, for $M_s = 2$ TeV, detailing the contribution of each subprocess. The QCD background has been calculated at the partonic level from the same processes as designated for the signal, with the addition of $qq \to qq$ and $q\bar{q} \to q\bar{q}$. Finally we estimate (at the parton level) the LHC discovery reach, namely one may calculate a signal-to-noise ratio, with the signal rate estimated in the invariant mass window $[M_s - 2\Gamma,\, M_s + 2\Gamma]$. The top two and bottom curves in figure 5 show the behavior of the signal-to-noise ratio as a function of the string scale for three integrated luminosities (100 fb^{-1}, 30 fb^{-1} and 100 pb^{-1}) at the LHC. It is remarkable that within 1-2 years of data collection, *string scales as large as 6.8 TeV are open to discovery at the* $\geq 5\sigma$ *level.* For 30 fb^{-1}, the presence of a resonant state with mass as large as 5.7 TeV can provide a signal of convincing significance ($S/N \geq 13$). The bottom curve in figure 5, corresponding to data collected in

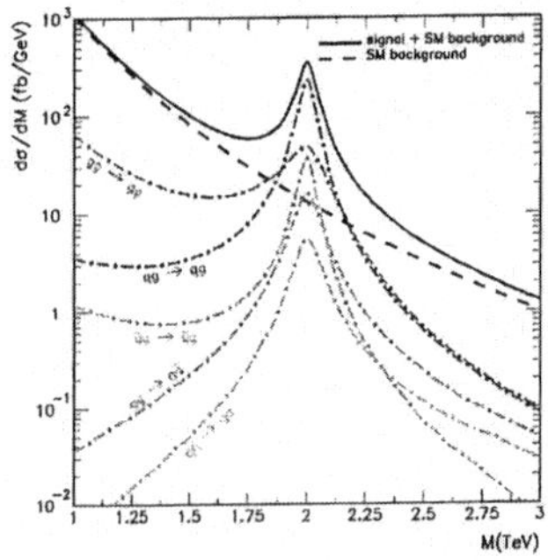

Figure 4: Dijet cross sections.

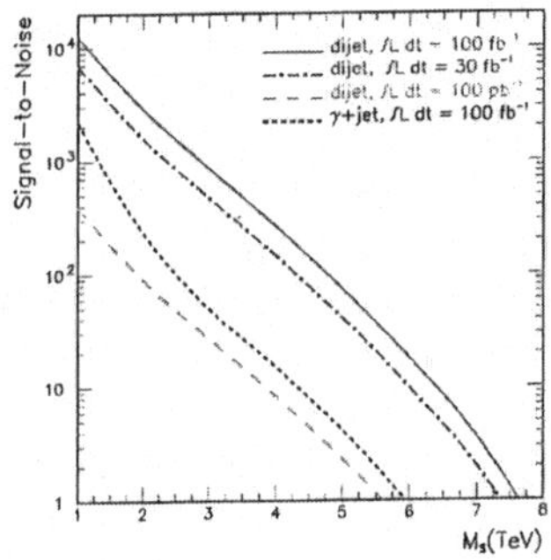

Figure 5: Signal to noise ratio.

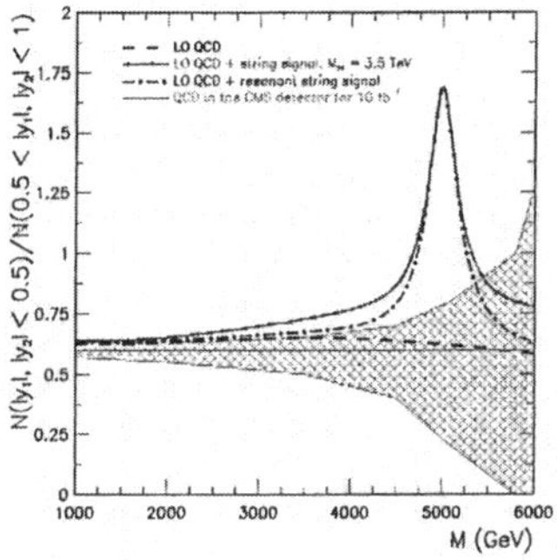

Figure 6: For integrated luminosities of 10 fb^{-1} the expected value (solid line) and statistical error (shaded region) of the dijet ratio of QCD in the CMS detector is compared with LO QCD (dashed line), LO QCD + lowest massive string excitations (dot-dashed), and LO QCD + lowest massive string excitations + non-resonant string scattering in $qq \to qq$ (dot-solid). We have taken for the KK mass $M_{KK} = 3.5$ TeV and $M_s = 5$ TeV.

a very early run of 100 pb^{-1}, shows that a resonant mass as large as 4.0 TeV can be observed with 10σ significance! Once more, we stress that these results contain no unknown parameters. They depend only on the D-brane construct for the standard model, and *are independent of compactification details.*

Next, using the result for the four quark scattering amplitude (40), one can also compute the contribution of the KK particles to the hadronic cross sections at LHC. Although there are no s-channel resonances in $qq \to qq$ and $qq' \to qq'$ scattering, KK modes in the t and u channels generate calculable effective 4-fermion contact terms. These in turn are manifest in an enhancement in the continuum below the string scale of the R ratio for dijet events. For $M_{\text{KK}} \leq 3$ TeV, this contribution is small, but can be detected at the LHC with 6σ significance above SM background. In combination with the simultaneous observation in dijet events of a string resonance at $M_s > M_{\text{KK}}$, this would consolidate the stringy interpretation of these anomalies. In figure 6 we show the contribution of the lightest KK mode for $M_{KK} = 3.5$ TeV.

4 Conclusions

The fundamental string mass scale is an arbitrary parameter. If it is in the range of few TeVs, vibrating string modes, i.e. Regge excitations predicted by critical superstring theory, will be observed at the LHC. One can also extend the discussion of the scattering processes with the best potential for revealing the effects of Regge excitations, to the case of five external particles [6]. The corresponding amplitudes describe processes underlying three jet production, photon (or electroweak bosons) plus dijets, jets associated to Drell-Yan lepton pairs and many other processes that can be used not only to discover Regge excitations of quarks and gluons but also to determine their masses, spins, decay widths, branching ratios and other properties as predicted by superstring theory.

Without any doubt, the most interesting property of full-fledged string disk amplitudes is their universality, i.e. their model-independence. The amplitudes involving arbitrary number of gluons and at most two fermions do not depend on the compactification details. One can exhibite a larger universality, between the amplitudes involving one quark-antiquark pair plus a number of gluons and their purely gluonic counterparts obtained by replacing the quark-antiquark pair by two gluons. In fact one can expect the relation between purely gluonic amplitudes and those involving one quark-antiquark pair to remain valid for N-point amplitudes [6].

If fundamental strings are discovered at the LHC, the landscape problem will be nullified, however a whole class of new, difficult problems will come into focus. With the present understanding of superstring theory, it is not clear what to expect beyond the string threshold. The name used for describing this energy domain is "trans-Planckian," dating back to the times when the string mass was automatically assumed to be near the Planck mass. It is possible that we will be exploring such trans-Planckian physics very soon.

Acknowledgements

We like to thank L. Anchordoqui, H. Goldberg, S. Nawata, O. Schlotterer, S. Stieberger and T. Taylor for the very pleasant and fruitful collaboration on that subject.

References

[1] N. Arkani-Hamed, S. Dimopoulos and G. R. Dvali, "The hierarchy problem and new dimensions at a millimeter," Phys. Lett. B **429**, 263 (1998) [arXiv:hep-ph/9803315].

[2] I. Antoniadis, N. Arkani-Hamed, S. Dimopoulos and G. R. Dvali, "New dimensions at a millimeter to a Fermi and superstrings at a TeV," Phys. Lett. B **436**, 257 (1998) [arXiv:hep-ph/9804398].

[3] D. Lüst, S. Stieberger and T. R. Taylor, "The LHC String Hunter's Companion," Nucl. Phys. B **808**, 1 (2009) [arXiv:0807.3333 [hep-th]].

[4] L. A. Anchordoqui, H. Goldberg, D. Lüst, S. Nawata, S. Stieberger and T. R. Taylor, "Dijet signals for low mass strings at the LHC," Phys. Rev. Lett. **101**, 241803 (2008) [arXiv:0808.0497 [hep-ph]].

[5] L. A. Anchordoqui, H. Goldberg, D. Lüst, S. Nawata, S. Stieberger and T. R. Taylor, "LHC Phenomenology for String Hunters," Nucl. Phys. B **821**, 181 (2009) [arXiv:0904.3547 [hep-ph]].

[6] D. Lüst, O. Schlotterer, S. Stieberger and T. R. Taylor, "The LHC String Hunter's Companion (II): Five-Particle Amplitudes and Universal Properties," arXiv:0908.0409 [hep-th].

[7] V. Balasubramanian, P. Berglund, J. P. Conlon and F. Quevedo, "Systematics of Moduli Stabilisation in Calabi-Yau Flux Compactifications," JHEP **0503**, 007 (2005) [arXiv:hep-th/0502058].

[8] J. P. Conlon, F. Quevedo and K. Suruliz, "Large-volume flux compactifications: Moduli spectrum and D3/D7 soft supersymmetry breaking," JHEP **0508**, 007 (2005) [arXiv:hep-th/0505076].

[9] D. Lüst, "Intersecting brane worlds: A path to the standard model?," Class. Quant. Grav. **21**, S1399 (2004) [arXiv:hep-th/0401156].

[10] R. Blumenhagen, B. Körs, D. Lüst and S. Stieberger, "Four-dimensional String Compactifications with D-Branes, Orientifolds and Fluxes," Phys. Rept. **445**, 1 (2007) [arXiv:hep-th/0610327].

[11] G. Dvali and D. Lüst, to appear.

[12] S. Cullen, M. Perelstein and M. E. Peskin, Phys. Rev. D **62**, 055012 (2000) [arXiv:hep-ph/0001166].

[13] D. Chialva, R. Iengo and J.G. Russo, "Cross sections for production of closed superstrings at high energy colliders in brane world models," Phys. Rev. D **71**, 106009 (2005) [arXiv:hep-ph/0503125].

[14] L. A. Anchordoqui, H. Goldberg and T. R. Taylor, "Decay widths of lowest massive Regge excitations of open strings," Phys. Lett. B **668**, 373 (2008) [arXiv:0806.3420 [hep-ph]].

CHAIRMAN: D. LUEST

Scientific Secretaries: M. Badziak, T. Jelinski

DISCUSSION I

- *Davey:*

How are the D6 branes oriented in 10D space in the IIA scenario and how are their mutual relations related to 4D physics?

- *Luest:*

In type IIA compactifications D6-branes are wrapped on 3D submanifolds (strictly speaking 3-cycles) of 6D internal space. In the generic case those 3-cycles inside Calabi-Yau (CY) 6-fold intersect at point in six dimensions. In other word, it means that gauge bosons effectively live in 6+1 dimensions. So non-Abelian world-volume field theory of gauge bosons is seven-dimensional, whereas fermions basically live only in 3+1 dimensions. So this gives you also nice hierarchy of dimensions. On the other hand in type IIB one has D7-branes which wrap 4D submanifolds (4-cycles) of CY. Here world-volume theory of gauge bosons is 7+1 dimensional. Generically two 4-cycles in 6D space intersect along 2D space (complex curves) such that chiral fermions effectively live in 6D space-time. One can also look at compactifications from the M-theory point of view. Here one compactifies space-time from eleven dimensions to four dimensions on a 7D manifold. Gauge bosons live on 3D singularities inside 7D internal space. These singularities are basically nothing else than world-volumes locus of D6-branes in type IIA theory and chiral fermions live on a 0D conical singularities. These points precisely correspond to the above mentioned intersections of D6-branes. There is also another one point of view. Namely non-perturbative uplift of IIB teory – so called F-theory. Now gauge bosons live on 4D singularities, so called ADE singularities, and chiral fermions live on 2D singularities which are intersection curves of D7-branes.

- *Marienfeld:*

Can you comment on possible candidate groups for GUT models: SU(5) , SO(10) ?

- *Luest:*

Particularly SU(5) is the most simple example of the GUT theory. SU(5) can be quite easily engineered via string compactifications with D-branes so SU(5) with the gauge bosons in the 24 dimensional representation with the matter fields in 5 and \bar{5} representation for example. To engineer SU(5) you start with two stacks of branes. The first stack consists of five branes (D6 branes in type IIA or D7 branes in type IIB). Another stack is given by the extra brane which intersects the first stack. You can look on the open string sector. There are open strings which are stretched between one stack and its orientifold image (orientifold mirror). This gives you 10D representation of U(5) or SU(5). Engineering SO(10) GUT theory is much more complicated because of some reasons which I did not explain. Can you have open strings with the endpoints on the branes such that these endpoints give rise to spinorial representation of orthogonal groups? If you analyze it you will see that this is not possible. You can add only fundamental representations or what is in the direct product of 2 fundamental and anti-fundamental representaions which is provided by the orientifold. But no spinor representations are possible in this perturbative construction. In SO(10) GUTs the matter transforms according to 16D representation of SO(10) so basic spinor representation is 16D representation. They transform as 10+\bar{5}+1 representation if you decompose SO(10) into SU(5) multiplets. In 10 you have doublet of lefthanded u and d quarks and righthanded u quark and the electron. In \bar{5} you have a doublet of lefthanded neutrino and electron and righthanded d quark. In SO(10) you have also righthanded neutrino. So this is the field content. It turns out that this decomposition 10+\bar{5}+1 is not possible in the perturbative open strings theory. It is indeed possible when one can go to the other classes of compactifications. People obtained SO(10) GUTs by heterotic string compactifications . In type II orientifold it is not possible unless you go to SU groups. If you look at this from the non-perturbative point of view you get extra degrees of freedom. These are not fundamental strings which are stretched between D-branes but the so called D1 strings which are non-perturbative from the strings point of view. These extra non-perturbative degrees of freedom also give rise to spinorial representation. They give also rise to exceptional gauge groups which are again possible in the heterotic string. To answer your question once again: when you look at the simple SU(5) GUT scenario, it can be done in this context.

- *Tapia Takaki:*

Can you describe what are the expected properties for an unbroken SUSY?

- *Luest:*

In field theory supersymmetry is motivated as a solution to the hierarchy problem. The hierarchy problem is twofold. One is that there is a large difference in scales in particle physics. The scale of the Standard model is much much lower than the scale of the GUT theory, for example, or the Planck mass. This is the first hierarchy problem but there is also the second, technical aspect. Namely, if you consider radiative corrections to the Higgs mass you would see that the Higgs mass is pushed up to the highest possible scale so the Standard Model is not stable against the radiative corrections. In the string theory there is also similar phenomena. In the string theory without supersymmetry you also deal with several kinds of instabilities. They are called Neveu-Schwarz (NS) tadpoles. There are similar divergences that you also encounter here, not quite but similar in spirit. To avoid these instabilities you also need supersymmetry. Of course, there is a big unsolved problem in string compactifications. You have to break supersymmetry eventually. To do it in a controlled way and do not encounter again instabilities at higher loop is a problem which is not yet completely solved. People are, of course, considering many scenarios of supersymmetry breaking. I mentioned the scenario in which supersymmetry is broken in the hidden sector by some effects: by fluxes and/or non-perturbative effects like gaugino condensation and then by some matter fields it is transmitted to the observable sector and give rise to soft supersymmetry breaking coupling there but the breaking of supersymmetry is discussed in the framework of effective field theory. However, to be really honest one should go to fully non-supersymmetric string compactification with largely separated scales which is completely stable. I think it is fair to say that nobody really succeded to do it.

- *Tapia Takaki:*

Do you expect one of your SM to be stable enough to live long enough to observe some evidence of it?

- *Luest:*

It could be in principle possible but nobody have complete survey of all the possibilities. You can do case by case study. The statistical methods that I have shown you in my lecture was under the question how many brane embeddings can give rise to the spectrum of the Standard Model. Not much more.

- *Cerchiai:*

Can you tell us more about your opinion on the anthropic principle?

- *Luest:*

String theory gives room to that kind of anthropic principles or anthropic speculations. My attitude is tended to go down to Earth. I want to know a constructive path whether we can realize something which we know to exist. In a way to give kind of existence proof. Of course, I would be then interested in kind of statistical questions as likelihoods. And I would go even more down to Earth. Namely, are there any features of the models that we can discover in nearby experiments.

- *Shcherbakov:*

How does the modification of the Standard Model described by Prof. Shaposhnikov affect the number of string vacua admitting Standard-Model-like spectrum?

- *Luest:*

The number of possible string vacua, if you are a purist, is a God given number. We don't know it but perhaps at the very end of the day some clever physicist will find the number of possible string vacua and also clasify them. Whether Prof. Shaposhnikov's model is contained in one of them, I don't know because I haven't studied this problem so I have to leave your question unanswered.

- *Kovetz:*

Taking you back to the SO(10)/SU(5) GUT discussion, since SO(10) is preferable to SU(5), don't you consider this as a weakness of this family of models?

- *Luest:*

You can consider this as a weakness but this is technical problem. There are also non perturbative uplifts of these model and they may include SO(10) GUTs.

CHAIRMAN: D. LUEST

Scientific Secretaries: D.Nesterov, L.Parisi, G.Corcella

DISCUSSION II

- *Davey:*

Let's suppose at the moment that String theory correctly does all, at least describes high energy physics in some sense and that in some sense one can get the Standard Model at least at low energy. What proportion of these good models have any kind of experiment law, a kind of input into theory?

- *Luest:*

That's a very good question. If you remember the scenario which I was discussing today with the LHC implications, which means that the string scale is low, you will constraint very much further what kind of six dimensional compact spaces you can use. This question was analyzed and the answer basically is that the topology and geometry of the six-dimensional space must be rather peculiar. These are the so-called Swiss-cheese manifolds which have a very big volume on one hand but they must allow for small holes inside. The reason why there must be small holes inside is that we need large enough gauge coupling constants. If also the cycles around which you have D-branes should be big, then the gauge coupling constants would be much too small following the relation which I have shown you that the gauge coupling constant is proportional to the volumes of the wrapped cycles. You can only use very peculiar Calabi-Yau spaces which satisfy strong conditions. One of the conditions is that they must have negative Euler number which of course is not very rare. But there must be other conditions which I do not discuss here. So, basically your question is how many of the total Calabi-Yau spaces which lead to reasonable physics satisfy these criteria. It is the interesting question to which I do not have any good statistic answer at the moment.

- *Davey:*

So we just don't know? Don't we?

- *Luest:*

No, we don't know. But I expect that the fraction will be still quite big in view of big number of 10^{500} or 10^{200} which Michael Douglas was quoting in the discussion. I don't have a figure but I think it's big enough and deserves to be studied.

- *Inguglia:*

Do you know how to detect the chargeless string with very low mass?

- *Luest:*

If you refer to a particular overall volume modulus then the only way to detect it is by very sensitive gravitational experiments, which are sensitive to the very short distances where gravitational force is modified. It is similar to tests where one wants to measure deviations from the Newton law. Basically you have to set up very precise measurements of the gravitational constant to see effects of these "sterile" or "neutral" states. Indeed several experiments were done and the limits are such that the masses of those states must not be much below this 10^{-13} GeV. So this kind of predictions is already at the border of this kind of experiments.

- *Inguglia:*

You also have talked about the fifth force in this case due to the very light mass. Which kind of effects does the fifth force produce?

- *Luest:*

Fifth force experiments are like modifying gravity at short distances. If you have other carriers which have a very long range and indeed if the mass is low then the range of the force is very long – it is just like a modification of gravity so you can do one of the same experiments to get the deviations from ordinary gravitational interactions and also to discover the fifth forces - these additional forces.

- *Kovetz:*

Are there viable models based on IIA constructions?

- *Luest:*

Not so many. This is the question of getting large radii and having spaces with holes – Swiss cheese manifolds. These are analyzed in the context of type IIB with wrapped D7 branes and the IIA case is not really worked out yet. I think it is also an interesting problem one should have a look at what is a mirror description of those Calabi-Yau manifolds which allow for large volume in type IIB. I think one can do it using kind of standard techniques. So looking also on some toy examples for which type IIB description is known and then you should map these on type IIA side where you have 3 cycles and D6 branes. I think explicitly it was not done yet.

- *Ferrara:*

In the scenarios where there are different possible string scales can you comment on what the natural scale of cosmological constant is and how the cosmological problem is considered and resolved?

- *Luest:*

I think that the problem of cosmological constant is not really a priori related to the choice of the scales you make because in any case all these scales as you know a priori are much larger than the cosmological constant and a priori you would expect that the cosmological constant is at least as big as the string scale chosen. In the TeV string scale scenario a priori you again would expect that the cosmological constant is also in the TeV range. This is so since supersymmetry is broken at this scale. So I think basically you are facing a similar problem as in models where the string scale is at the Plank mass. You can solve the problem by having some miraculous cancellations which nobody knows yet or statistically scaling all the landscape and looking on the fraction of those compactifications for which the cosmological constant by accident is indeed very small. So either you follow the conventional approach (but I think nobody really succeeded to show models for which the cosmological is small at every order in perturbation theory) or you prefer more speculative, sophistical or even anthropic point of view.

- *Ferrara:*

May be it is not realistic but there is a class of models with no-scale supergravity where the tree-level cosmological constant is very small. Can you get such kind of models in these scenarios?

- *Luest:*

No, not really, because we do not assume that tree-level supersymmetry is unbroken. Just on the contrary I assume that basically supersymmetry is broken at the string scale. I think these no-scale models are not really a good starting point for these considerations. But I might be wrong.

- *Ferrara:*
But in no-scale model the supersymmetry is broken.

- *Luest:*
Yes, the supersymmetry is broken in the controlled way. But the scalar potential due to these factors of three has a property that the cosmological constant is at the tree-level vacuum. Indeed you break supersymmetry but the cosmological constant is still zero at the minimum where the supersymmetry is broken.

- *Aschieri:*
If the string scale is of the order of TeV, should we have already observed supersymmetry at current experiments? More generally, is there any theoretical reason why the supersymmetry breaking and string scales are to be very close?

- *Luest:*
A priori you do not need any supersymmetry if the string scale is low. According to the ADD scenario, the hierarchy problem is not solved by the presence of the supersymmetry but rather just putting the scale of fundamental physics in the TeV range. You do not face hierarchy problem at all. In string theory, the situation is a bit more complicated. Suppose you could completely decompactify the theory, if you would start in 10 dimensions, you know that the 10-dimensional superstring is supersymmetric. Therefore there will be always some relicts of the supersymmetry, however I do not really assume any particular breaking mechanism. So supersymmetry can also be broken by some orbifold kind of projections or by some peculiar projections where you project out some of the supersymmetric states. Your question is hard to answer unless you go to the specific model. Here I want to discuss only model-independent generic quantities. There might be superpartners of the same order of mass at the string scale, but it might also happen that you may have no superpartners for the Standard Model particles in the conventional sense. However the states I discussed are completely generic, they are true for all compactifications which I have discussed. The question of the supersymmetry on the other hand is more model-dependent.

-*'t Hooft:*
If string theory could become visible already at LHC, how can you avoid the occurrence of gravitational degrees of freedom?

- *Luest:*
Gravity is weakly coupled due to one important relation: the Planck mass squared is given in terms of the 8th power of the string scale multiplied by the volume of the volume of the 6-dimensional space. Even if you lower the fundamental scale of gravity which is M_{string} to the TeV region, the effective Planck mass of the effective scale of gravity is still as higher as 10^19 Gev, i.e. the Plank mass does not change it's values. So the gravitational interactions still have the same size as in .conventional gravity theories.

- *'t Hooft:*
You have an effective string theory at the QCD scale, with open strings describing quarks. How can you avoid the presence of closed strings with there gravitational degrees of freedom?

- *Luest:*

Those closed strings effectively coupled only with the effective Planck mass which is 10^{19} GeV. Of course, you can have a new type of contact interactions between the open strings, which might lead to problems, such as flavor changing neutral currents. But all the forces which are carried by closed strings which do not have Standard Model gauge quantum numbers still have couplings of the order of the known Planck mass. You can put it also like G.Dvali was describing it, namely at TeV scale we have of the order of 10^{32} particles, which dilute the effective gravitational coupling constant, so that the gravitational interaction stays small. Such 10^{32} particles are also related to the Kaluza-Klein excitations of the internal space: it is a framework pretty similar to the one described in G.Dvali's lectures.

- *Badziak:*

People working in string theory usually follow two different pictures: a global picture, with the issue of moduli stabilization, and a local one, where one constructs a more realistic models like MSSM with no exotics, etc. Would you think that separating these two approaches is justified?

- *Luest:*

I think it's justified. Indeed, many of the features depend only on the local properties of the Standard Model. Often there is no need to know how D-brane quiver of the Standard Model is globally embedded into any compact space. It is much simpler just discussing a local picture: in this way, one avoids all the complications to find the compact spaces, which realize the embedding of the local geometry. The global picture is much more ambitious, as it aims at satisfying all consistency relations like tadpoles cancellations: assuming the local setup, there exist global manifolds where even the local picture can be consistently embedded. The local picture is anyway very useful, as many questions can be answered just relying on the local geometry.

Status of dark matter and neutrino physics

A. Bettini
Dipartimento di Fisica G. Galilei dell'Università and INFN, Padova
Laboratorio Subterráneo de Canfranc, Spain

Abstract

Tremendous progress both in the quantity and in the quality of the observational cosmological data in the last fifteen years, associated with theoretical work in their understanding has lead to a "standard cosmological model". This model is phenomenological and is consistent with data within an accuracy of a few per cent. The energy-budget of the Universe shows that the matter we know is only a small fraction of the total. The largest fraction is invisible or, as it is commonly called, dark. This lecture will summarise the status of the direct experimental searches for dark matter.
The phenomena of neutrino oscillation and neutrino flavour conversion in matter, not foreseen by the Standard Model, have been firmly established. The neutrino mass spectrum and their mixing angles have been partially measured. I shall review the status of the experiments presently in data-taking, construction or in project.

1. Status of the ΛCDM cosmological model

The value of the average mass density in the Universe has been determined from a number of observational data. It is important to notice that these data are relative to very different scales (dimensions) and to different redshifts (epochs) and that they have been obtained with different techniques, corresponding to different underlying physics. Consistently and independently, they provide the same value within the uncertainties. At the last scattering epoch, 372 000 years after the initial explosion (at redshift z=1089), dark matter had shaped the primordial density fluctuations through its gravitational potential. The effects are visible in the spectrum of the anisotropies of the Cosmic Microwave Background (CMB). In particular the height of the first peak in the spectrum gives the total matter density. The matter density is sensed at later epochs through its gravitational effects, ranging form the kinetic of the clusters and super-clusters of galaxies to the large-scale galaxy correlation function, to the gravitational lensing. Given as a fraction of the critical density, the matter density Ω_m is $\Omega_m h^2=0.1358\pm0.0037$[1], where $h=H_0/100$ is the adimensional ratio between the Hubble constant $H_0=70.5\pm1.3$ km/(s Mpc)[1] and the arbitrary value of 100 km/(s Mpc). In a round number, $h^2\approx0.5$. Notice the 3% and 2% accuracies of these two fundamental quantities. Remember that 1 Mpc=3.09×10^{22} m, or 30.9 Zm. The strange units for the Hubble constant are due to historical reasons, corresponding, in round numbers, to $H_0=2.3\times10^{-18}$ s^{-1}, which is the reciprocal of a time, 13Gyr, of the order of the age of the universe.

The contribution of normal matter, called "baryonic" in cosmology, $\Omega_b=0.0456\pm0.0015$[1] is determined mainly by two independent data sets: primordial nucleosynthesis, and CMB anisotropy spectrum. These data show that the largest fraction of matter is "dark"; its constituents are not baryons and do not have strong or electromagnetic interactions. We can try to detect them through weak interactions, if they have any. These particles are generically called WIMPs (Weakly Interacting Massive Particles). Evidence for local existence of dark matter comes from the rotation curves of the galaxies, including our own. Dark matter is here, around us, now.

Neutrinos a priori are candidates for dark matter. Not in practice. Neutrinos affected the Large-Scale Structures (LSS) formation at scales typically of a few tens Mpc (at z=0.1-0.2). Neutrinos, at those epochs, moved with speeds larger than the escape velocities from the smaller structures; freely streaming from higher to lower density regions they tended to suppress the formation of the smaller structures. Data on the LSS show no effect, implying that the fraction of dark matter due to neutrinos is less than a few percent.

There are no dark matter candidates in the Standard Model, but several have been proposed by theorists. The most credited is in SUSY, the neutralino, a mixture of the superpartners of the weak bosons and higgs, which is stable if *R*-parity is conserved. Clearly two complementary searches are in order: the search of the cold dark matter (CDM) in the Universe, to understand the largest fraction of its mass and its artificial production, which will be in the LHC range, for a precise study of their properties.

2. The present status of neutrino physics

Contrarily to the assumptions of the Standard Model, neutrinos are massive, their stationary states, namely the mass eigenstates, are not the states of definite flavour and flavour lepton numbers are not conserved.

Neutrino oscillations and flavour conversion in matter are discoveries of the underground laboratories, looking at natural sources (sun and cosmic rays). Confirmation and improvement in precision were achieved in reactor and accelerator experiments. Vacuum oscillation has been discovered in ν_μ produced by cosmic rays in the atmosphere by the SuperKamiokande experiment lead by Y. Totzuka. It is a kinematic phenomenon and, at the first order, does not depend on the sign of the square mass difference, Δm^2, neither on the substitution $\theta_{ij}\Leftrightarrow\pi/2-\theta_{ij}$, where θ_{ij} is the relevant mixing angle, θ_{23} in this case. Flavour conversion in matter is a dynamical phenomenon, due to the $\nu_e e$ interaction potential and has been discovered in the solar ν_es. It depends on the sign of the square mass difference, δm^2 and on the substitution $\theta_{ij}\Leftrightarrow\pi/2-\theta_{ij}$, where the mixing angle is θ_{12} in this case.

The neutrino flavour states are linear combinations of the mass eigenstates ν_1, ν_2 and ν_3, with mass m_1, m_2 and m_3 respectively. The mixing matrix can be written as the product of three rotations including a phase factor, as in the case of quarks, and a fourth diagonal matrix with two more phases. The latter, called Majorana phases, can be absorbed in the wave functions only if neutrinos and antineutrinos are different (Dirac) particles. With $c_{ij}=\cos\theta_{ij}$, and $s_{ij}=\sin\theta_{ij}$ we have

$$\begin{pmatrix}\nu_e\\ \nu_\mu\\ \nu_\tau\end{pmatrix}=\begin{pmatrix}1&0&0\\0&c_{23}&s_{23}\\0&-s_{23}&c_{23}\end{pmatrix}\begin{pmatrix}c_{13}&0&s_{13}e^{i\delta}\\0&1&0\\-s_{13}e^{-i\delta}&0&c_{13}\end{pmatrix}\begin{pmatrix}c_{12}&-s_{12}&0\\s_{12}&c_{12}&0\\0&0&1\end{pmatrix}\begin{pmatrix}1&0&0\\0&e^{i\alpha}&0\\0&0&e^{i\beta}\end{pmatrix}\begin{pmatrix}\nu_1\\ \nu_2\\ \nu_3\end{pmatrix}\qquad(1)$$

In total there are nine quantities to measure, three masses, three mixing angles and the three phases. The latter, if $\neq 0$ and $\neq\pi$, give *CP* violation effects in the lepton sector. Majorana phases are irrelevant in oscillation and matter conversion phenomena; they appear only (for any practical purpose) in double beta decay.

Different groups have performed global fits including all the available data, providing values for the three mixing angles (only an upper limit for θ_{13}) and two squared mass differences (only the absolute value of the larger one). Results can be summarised as follows[2] (2 σ uncertainties):

$$\left|\Delta m^2\right|\equiv\left|m_3^2-\left(m_2^2+m_1^2\right)/2\right|=2400^{+500}_{-620}\ \mathrm{meV}^2\qquad \delta m^2\equiv m_2^2-m_1^2=79.2\pm7.1\ \mathrm{meV}^2$$
$$\sin^2\theta_{12}=0.312^{+0.040}_{-0.034}\qquad \sin^2\theta_{23}=0.466^{+0.136}_{-0.100}\qquad \sin^2\theta_{13}<0.036\ .\qquad(2)$$

Recently it has been observed[3] that a "tension" present between the solar neutrino data on one side and the reactor neutrino KAMLAND experiment on the other can be interpreted to point to a non-zero value of θ_{13}, which however is only a 1.5 σ effect. Another independent 1.5 σ effect has been reported by the MINOS collaboration[4]. MINOS is a neutrino detector located in the Soudan Underground Laboratory (SUL) illuminated by a the ν_μ neutrino beam (NUMI) from Fermilab at 730 km distance. Designed as a ν_μ disappearance experiment, MINOS has a limited sensitivity to ν_e appearance. 37 ν_e like selected events have been observed to be compared to an expected background of $27\pm5_{\mathrm{stat}}\pm2_{\mathrm{syst}}$. Clearly, no conclusion can be drawn on both these statistically non significant effects.

There are two basic methods to search for θ_{13}: ν_e disappearance, in an accelerator experiment and ν_e appearance in a reactor experiment. The corresponding oscillation maximum amplitudes are

$$A(\nu_e \to \nu_x) = \sin^2(2\theta_{13}) \approx 4\theta_{13}^2, \qquad A(\nu_\mu \to \nu_e) = \sin^2(\theta_{23})\sin^2(2\theta_{13}) \approx 2\theta_{13}^2 \qquad (3)$$

The presently running accelerator experiments, the above mentioned MINOS[5] and OPERA[6] on the CNGS beam from CERN at LNGS[7], may provide a limited improvement in the present limit. In a near future, the T2K[8] experiment, which will start in 2010, on a neutrino beam generated at the JPARC facility at Tokai shooting on the SuperKAMIOKANDE detector at 300 km distance has a sensitivity $\theta_{13}^2 \le 10^{-3}$, if no event is observed. On a longer time scale, we have the FNAL NOvA[9] project, which might start in several years from now on a neutrino beam from FNAL to a new location, Ash River, at 820 km distance. While the sensitivity in θ_{13} will be similar to that of T2K, the longer path underground will make the experiment sensitive to the sign of Δm^2, provided θ_{13} is not too small.

Reactor experiments search for $\bar{\nu}_e$ disappearance. It is immediate to see that, given the neutrino energy spectrum and the value of Δm^2, the optimum flight length is around 2 km. These experiments measure the antineutrino energy spectrum at the source (near detector) and at a certain distance (far detector). The expected difference due to the disappearance being extremely small, the detectors must be as identical as possible and the sensitivity is completely dominated by systematic effects. The DoubleCHOOZ[10] experiment is expected to start data taking in 2011, with one source and two detectors at 1 km distance. Its sensitivity, in case of no event, will be $\theta_{13}^2 \le 10^{-2}$. The much more complex Daya Bay[11] experiment in China, expected to start in 2012, will have three sources (nuclear power stations) and three detectors at different distances. The sensitivity is expected to be $\theta_{13}^2 \le 2 \times 10^{-3}$.

The "atmospheric" oscillation, namely the one at Δm^2, is expected to be dominantly $\nu_\mu \to \nu_\tau$ with the amplitude $P(\nu_\mu \to \nu_\tau) = \sin^2(2\theta_{23})(1.27\Delta m^2(\text{eV}^2)L(\text{km})/E(\text{GeV}))$. Two experiments are presently running MINOS and OPERA with similar flight lengths L=730 km. To have a high oscillation probability, with a maximum value around 75%, MINOS employs a relatively low energy beam $\langle E_\nu \rangle \approx 3$ GeV. Its main contribution is increasing the accuracy in $|\Delta m|^2$. OPERA is designed to investigate whether the oscillation is really $\nu_\mu \to \nu_\tau$ or not. To do that the τ lepton produced in the target by the ν_τ must be produced and observed. Consequently the neutrino energy should be high enough, $\langle E_\nu \rangle \approx 17$ GeV. The maximum oscillation probability is consequently small, around 15%, and the number of expected events is small too. The emulsion technique used by OPERA will however allow to identify the tau events unambiguously and with high efficiency.

Notice that disappearance and appearance experiments are complementary due to the interplay between the contributions of $\theta_{13}^2 \neq 0$ and those of the CP violations that are different in the two cases.

An important aspect of the solar neutrino physics is being investigated by the BOREXINO experiment at LNGS. Electron neutrinos are produced in the central high density regions of the Sun. In matter the mass eigenstates are not ν_1, ν_2 and ν_3. Indeed, it can be shown, that the mass eigenstate corresponding to ν_2 in the Sun core is essentially ν_e. Neutrinos moving towards the surface go through regions of decreasing density, down to the vacuum. As shown in Fig. 1a , the level crossing condition is $1/\beta = 2\cos\theta_{12}$, where β is the ratio between the kinetic and the dynamic term in the Hamiltonian

$$\beta = \frac{\Delta m^2 / (2E_\nu)}{\sqrt{2}G_F n_e} \tag{4}$$

where n_e is the numeric electron density. This is the MSW[12] effect. At the Sun densities the resonance is met only if the energy is high enough, larger than a few MeV. At the lowest energies the phenomenon is the oscillation, as in vacuum (Fig. 1b).

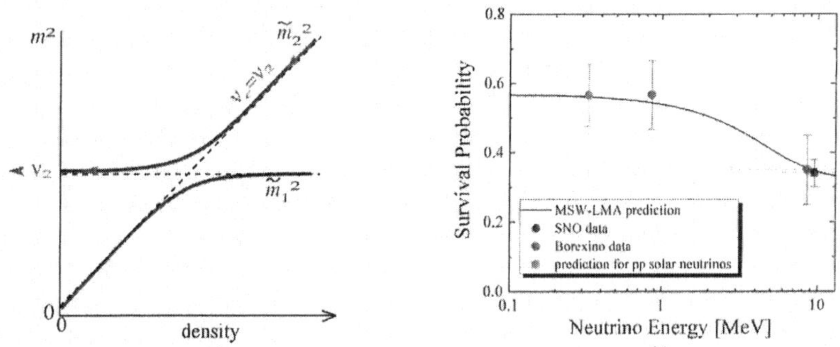

a) b)

Fig. 1 a) The MSW effect, level crossing of the mass eigenvalues in variable density matter; b) Electron neutrino survival probability in the sun as a function of energy as observed by BOREXINO

BOREXINO[13] detects the electrons hit be neutrinos in the process $\nu_e + e \rightarrow \nu_e + e$ in a volume of liquid scintillator that acts both as the target and as the detector medium. The light emitted by the electron track is detected by an array of photomultipliers surrounding the medium. Being the signal simply a small flash of light happening very rarely, the experiment had to achieve background levels lower than the status of art by several orders of magnitude. It was onbtained with an intense and demanding R&D programme over several years.

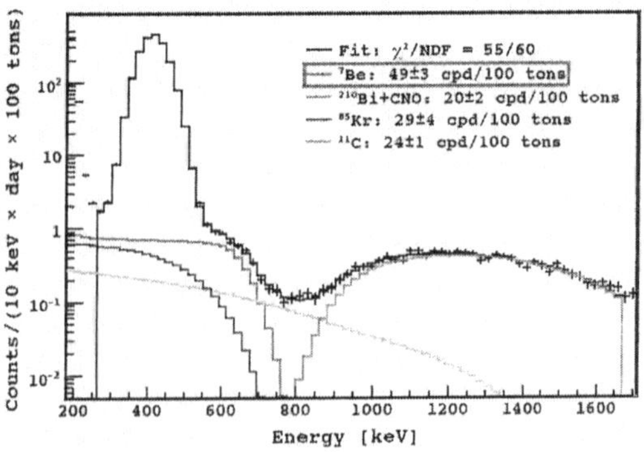

Fig. 2. Borexino energy spectrum (192 days of live time) and fit. A similar fit after statistical subtraction of ^{210}Po by exploiting the α–β separation capability of the detector gives consistent results

The structure of the experiment is onion-like: the innermost physical volume is a nylon sphere containing 300 t of hyper-pure liquid scintillator, doped pseudocumene (PC). However, only the innermost 100 t are used as fiducial volume. The "fiducialisation" is performed off-line by reconstructing the event location from the distribution of light on the PMs and controlled with calibration sources data. The nylon vessel is surrounded by a further volume of PC inside a large stainless steel sphere (1350 m^3) hosting 2212 PMs on its surface. Finally, the S/S sphere is contained in 2100 m^3 of ultra-pure water in a tank equipped with 208 PMs. This acts as a passive shield for neutrons and gammas and as an active veto of the cosmic muons.

Monochromatic neutrinos from the Be line in the solar spectrum, produce a characteristic Compton-edge like feature in the electron energy spectrum at about 700 keV. This spectrum as measured in real time by BOREXINO[14] is shown in Fig. 2, together with its interpretation in terms of signal and backgrounds contributions. The fit gives a value for the Be neutrino flux Φ (^{7}Be)=(5.12±0.51)×10^{11} $m^{-2}s^{-1}$, in good agreement with the theoretical expectations of the solar model (5.08±0.5) ×10^{11} $m^{-2}s^{-1}$. BOREXINO, not only observed the Be neutrinos (in real time) for the first time, but also measured, in the same experiment, the high energy (Bo) neutrinos, exploring for the first time the MSW-Vacuum oscillation transition. Results are shown in Fig.1b.

3. Neutrino nature and masses

Information on the absolute value of neutrino masses comes from: the measurement of the end point energy and shape of the beta decay spectrum, the search for neutrino-less double beta decay, and the measurement of large structures spectrum in the Universe. Each of them has provided only upper limits till now.

As already mentioned in §1, cosmology gives a stringent limit on the neutrino mass density, Ω_ν, a quantity that is closely related to the sum of neutrino masses by the simple expression

$$\Omega_\nu \approx \frac{\sum_{i=1}^{3} m_i}{94h^2} \approx \frac{\sum_{i=1}^{3} m_i(\text{eV})}{47} \tag{5}$$

A rather conservative limit is $\sum m_i < 600$ meV, which corresponds, for each eigenvalue, to m_i <200 meV. This, albeit indirect, is the lowest limit on neutrino masses. In the next few years the sensitivity of the Planck mission data associated with gravitational lensing may bring the sensitivity as low as $\Sigma m_i \approx 150$ meV[15], which is close to the lower limits from the oscillations $\Sigma m_i > 50$ meV for $\Delta m^2>0$ and $\Sigma\, m_i > 100$ meV for $\Delta m^2<0$.

The measurement of the electron spectrum in the beta decay of a nucleus near its end-point gives a direct determination of the quantity

$$m_e^2 = \sum_{i=1}^{3} \left|U_{ei}\right|^2 m_i^2 = c_{13}^2 c_{12}^2 m_1^2 + c_{13}^2 s_{12}^2 m_2^2 + s_{13}^2 m_3^2 \tag{6}$$

The present lower limit is m_e > 2.3 eV from the Mainz[16] and Troitz[17] experiments on ^{3}H decay. Tritium is used due to its extremely low endpoint energy (18.6 keV). The next step is the KATRIN[18] experiment under construction at Karlsruhe aiming to increase by an order of magnitude the sensitivity, down to 0.2 eV (detection limit at 5σ=0.35 eV).

Nuclides that are energetically stable against beta decay, such as the even-even nuclei, can decay via the second order weak decay $(A,Z) \rightarrow (A,Z+2) + 2e^- + 2\bar{\nu}_e$. This is a very slow, but standard, process, the two-neutrino double-beta decay ($2\nu 2\beta$). Its diagram at quark level is shown in Fig. 3a). If neutrinos are Majorana particles, namely their own antiparticle, the quark level process shown in Fig. 3b) is possible, the neutrino-less double beta ($0\nu 2\beta$) decay

$$(Z,A) \rightarrow 2e^- + (Z+2,A) \tag{7}$$

The process violates the lepton number by two units. At the quark level the relevant Feynman diagrams are given in Fig. 3.

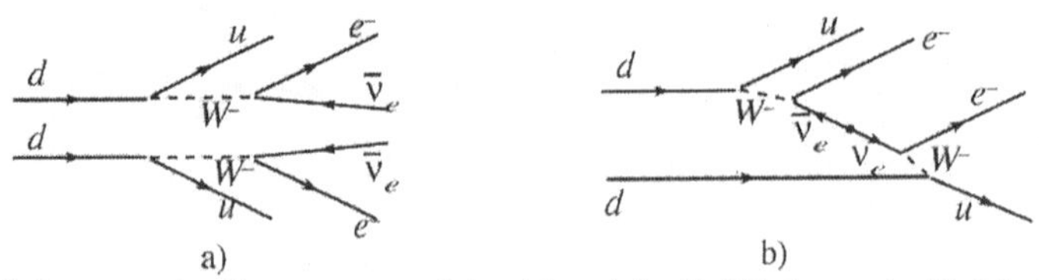

Fig. 3. Lowest order diagrams at quark level for: a) the ($2\nu2\beta$) decay; the ($0\nu2\beta$) decay

The inverse lifetime is given by

$$T_{1/2}^{-1} = G_Z\left(Q_{\beta\beta}\right)\left|M^{0\nu}\right|^2 M_{ee}^2 \tag{8}$$

where G_Z is the phase space volume, depending on the nucleus, which is easily calculated and $M^{0\nu}$ is the "nuclear matrix element", which takes into account that the decay happens in a nucleus, and M_{ee} is the relevant mass matrix element, the "Majorana mass"

$$M_{ee} = \left|\sum_{i=1}^{3} U_{ei}^{\ 2} m_i\right| = c_{13}^2 c_{12}^2 m_1 + c_{13}^2 s_{12}^2 m_2 e^{i\alpha} + s_{13}^2 m_3 e^{i\beta} \tag{9}$$

Notice that the phase factors can induce cancellations, as we will discuss soon.

Experimentally one should detect the two electrons of (7) and measure their energies, better the sum of their energies, which gives a monochromatic spectrum, with a line at the Q-value of the decay, called $Q_{\beta\beta}$. This quantity, typically a few MeV, is known, for much nuclides, within uncertainties of the order of 10^{-4}. Consequently, we know where to search and the energy resolution is a fundamental request for such experiments.

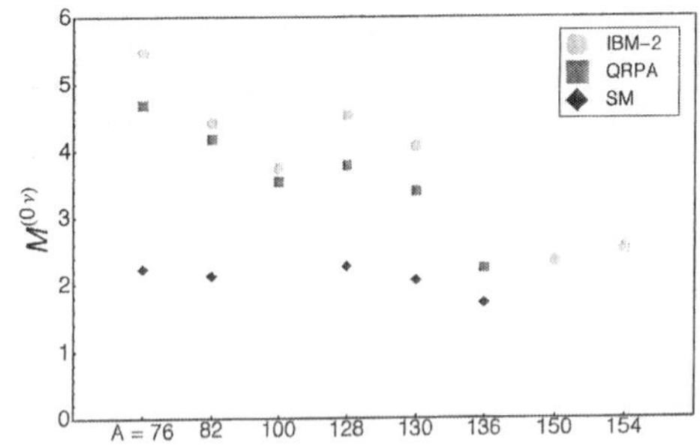

Fig. 4. from [19]. Neutrinoless double beta decay matrix elements in the formulation of Shimkovic et al.[20] for IBM-2[19], QRPA[21] and ShM[22]

Experiments obviously measure, or limit, the half life $T_{1/2}$. To extract M_{ee} from that, we need to know the nuclear matrix element $M^{0\nu}$. These are very difficult calculations, because the "double-beta active" nuclei have complicated structures. Approximations and truncations are necessary, cancellations between large terms take place in the calculation, etc. However, enormous progress has been achieved by the theorists in the recent years. Three basic theoretical techniques have been developed: the "Quasi Particle Random Phase Approximation" (QRPA), the "Shell Model" (ShM) calculations and the "Interacting Boson Model" (IBM-2). Fig. 4 shows examples of the results. One can notice that QRPA and IBM-2 agree within the expected uncertainties (of the order of 30%) for all the nuclides, while ShM agrees with them at the larger values of A, but not at the smaller ones. The reasons are not known, but are being investigated.

Fig. 5 shows the Majorana mass, calculated with the known values of the mixing angles, as a function of the lightest neutrino mass. As anticipated, M_{ee} may be very small. Indeed, it may even be zero. Notice that if neutrinos are Dirac particles, obviously M_{ee}=0, but the opposite is not true: it may be M_{ee}=0 even if they are Majorana.

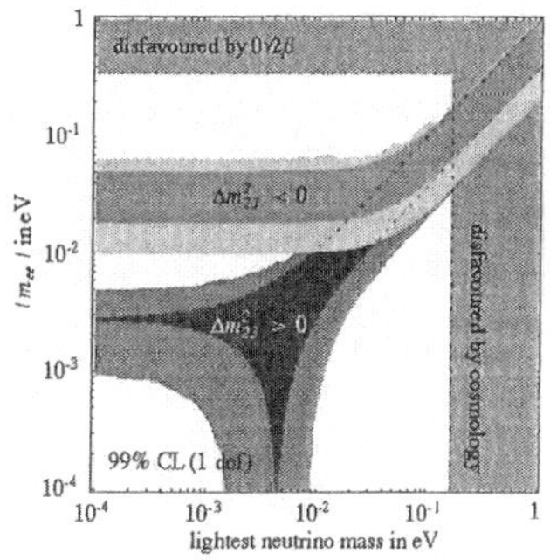

Fig. 5. Majorana neutrino mass vs. lightest neutrino mass [27]

To have an idea of the orders of magnitudes let us evaluate the expected half lives for three of the isotopes used in the present most sensitive experiments for M_{ee}=100 meV, using the QRPA[21] matrix elements. The half-lives are: for ^{56}Ge: 2.2×10^{26} yr, for ^{130}Te: 0.55×10^{26} yr, for ^{136}Xe: 2.15×10^{26} yr. How many events are expected in one tone year exposure? For ^{76}Ge enriched at 86% (the nuclei of the GERDA[23] experiment and MAJORANA proposal) 14 events, in natural TeO_2 (the salt used by CUORE[24]) 10 events [or 36 events in 1 t yr of enriched ^{130}Te] and in ^{136}Xe enriched at 85% (the isotope used by EXO200[25] and NEXT[26]) 14 events. According to eq. (7) the count rate scales with the square of M_{ee}. Consequently, the experiment being built today, with a sensitive mass of between 50 kg and 200 kg, may hope to reach about 100 meV, provided the backgrounds may be reduced to be close to zero. Let us look more closely to the effects of backgrounds.

The sensitivity of an experiment to the lifetime τ in presence of a background rate b is proportional to the factor

$$F_\tau \propto \left(\frac{MT}{b\Delta E}\right)^{1/2} \tag{10}$$

where M is the sensitive mass, T the exposure time and ΔE the energy resolution. We see that, say, a factor two in energy resolution is equivalent to a factor 2 in sensitive mass, and similarly for the background rate. More precisely, the sensitivity to M_{ee}, which is inversely proportional to the square root of the lifetime, scales as the fourth root of the sensitive mass in presence of constant background. This implies that the background must be reduced substantially to zero in the region of interest, namely an energy interval of the order of the energy resolution (FWHM) around the known position of the signal. In these conditions the sensitivity to M_{ee} scales as the square root of the exposure, $F_M \propto \sqrt{MT}$.

Even after eliminating all the backgrounds external and internal to the detector (remembering that surfaces are always sites of background sources), the irreducible background due to the $2\nu2\beta$ decay remains. Near the end-point of the two-electron kinetic energy K_{ee} the $2\nu2\beta$ decay spectrum is proportional to $\left(Q_{\beta\beta} - K_{ee}\right)^5$. One can then evaluate that the signal to background for half-lives $T^{2\nu}_{1/2}$ and $T^{0\nu}_{1/2}$ is

$$\frac{S}{B} \simeq \frac{m_e Q^5_{\beta\beta}}{7(\Delta E)^6}\frac{T^{2\nu}_{1/2}}{T^{0\nu}_{1/2}} \tag{11}$$

The sixth power dependence on ΔE shows that no compromise on energy resolution is tolerable. The best resolution is achieved by the Ge diodes with $\Delta E \approx 4$ keV FWHM at Q_{M}, followed by the TeO_2 bolometers of CUORICINO with $\Delta E=8$ keV, a figure which is being substantially improved for CUORE. EXO200, which is a liquid TPC, aims to achieve $\Delta E \approx 100$ keV FWHM, while NEXT with a pressure (10 bar) gas TPC aims at $\Delta E \approx 25$ keV FWHM.

4. Direct search for dark matter

The direct searches for dark matter look for the WIMPs elastic scattering on nuclei, detecting the energy deposited by the recoiling nucleus.

If dark matter particles are neutralinos (χ), their interaction cross sections can be calculated, in principle, on the basis of SUSY models. However, the unknown parameters of the theory are so many that the results span over many orders of magnitudes in uncertainty. Two basic possible couplings can be distinguished: a) "spin-dependent" (SD), an axial vector coupling to the nucleons spin in which only unpaired nucleons couple; b) "spin-independent" (SI), a scalar interaction with the mass of the nucleons, proportional to A^2. This makes the search in the SI sector more sensitive and we will concentrate on that. The WIMPs flux is uncertain too, depending on the assumed model of the Galactic halo model.

The results of the calculations and of the experiments are usually represented in the plane of the cross-section, reduced to a single nucleon, σ_p vs. the WIMP mass m_χ. This corresponds to projecting on a plane the multidimensional parameter space (local χ density, SI and SD cross section ratio, ratio of SD coupling to neutrons and to protons, local velocity, etc.).

There is no firm theoretical upper limit on m_χ, but one can take reasonably 1000 GeV; the lower limit is about 6 GeV from cosmology. The theoretical expectations for the cross sections have even more uncertainties, ranging from 10^{-5} to 10^{-12} pb. The corresponding rates depend on target nucleus and on energy threshold. Typically, they may range from 1 to 10^{-7} counts/(kg keV yr). The "ultimate" detector should have a sensitive mass of the order of 10 t with a background rate smaller than a few counts per year in a 10 keV window at, say, 30 keV. Today experiments (not looking for modulation) have exposures of the order of 10 – 100 kg d, thresholds at several keV and few events background.

Experimentally, the target coincides with the detector, in which one tries to detect the nucleus elastically hit by the WIMP. The energy is deposited in three forms: as ionisation charge, scintillation light or phonons.

The experimental challenges are the extremely low signal rate, small energy deposits (several keV) and a signal spectrum decreasing with increasing energy, similarly to backgrounds. The detectors must be operated in an underground laboratory, but this is not enough. Specific R&D is needed for achieving the highest radio-purity and for the systematic screening of all the set-up components, including the shields and the detector itself.

We can distinguish three basic backgrounds: 1. electromagnetic, due to γ's, which is the dominant one; 2. radioactive contaminants of the surfaces and their surroundings, which release only part of the energy in the sensitive volume, simulating WIMP signal; 3. neutrons, which give nuclear recoils indistinguishable from WIMPs.

Backgrounds can be modelled, but the accuracy and the reliability of the models is limited. Hence, evidence for dark matter cannot be claimed solely by observing a discrepancy with the background model. Consequently, the experiments develop, in a blind analysis, selection criteria to define a "background free" region in the experimental parameter space. Then within the assumed model, for each mass m_χ, the maximum possible signal rate allowed by the data is calculated. The result is reported as an "exclusion plot" in the σ_p vs. m_χ plane.

A second method, used so far only by the DAMA collaboration, searches for a specific positive signature, namely the annual modulation of the signal rate expected for the following reason. The solar system moves in the Galaxy with a speed of about 230 km/s. The velocity of our detector in the halo is modulated, with one year period, by the speed of the Erath around the Sun of about 30 km/s. Correspondingly, the signal rate is expected to be maximum in June, when the two velocities are in the same direction, minimum in December when they are opposite. Indeed the DAMA (100 kg) and the DAMA/LIBRA (250 kg) experiments, with an overall exposure of 300 555 kg d over 7 years of a set of hyperpure NaI detectors (analysis threshold = 2 keV) have observed the annual modulation with all the expected characteristics[28]. Notice that DAMA is the only experiment that does not apply any pulse shape analysis to exclude the electromagnetic "backgrounds". It is consequently the sole experiment sensitive to a possible electromagnetic signal.

The DAMA observation is model independent. However, the extraction from the data of the values of the WIMP mass and cross section depends on the assumed model and so does any comparison with experiments not looking for modulation and employing different target nuclei. Consequently, an independent experiment aiming to confirm or dispute the DAMA result must be able to search for the annual modulation with low enough background rate and sufficient sensitivity both to nuclear and electron recoils. Presently, the KIMS[29] experiment in the Y2L underground laboratory in Korea is running with more than 100 kg of CsI and might be able to reach these conditions. The ANAIS[30] experiment in the LSC underground laboratory in Spain aims to 100-150 kg of NaI detector. It is presently in the R&D phase to reach a K contamination in the detector crystals below 1 ppb.

The largest fraction of the experiments not looking for modulation can be divided in two classes: those using bolometers and those using noble fluids.

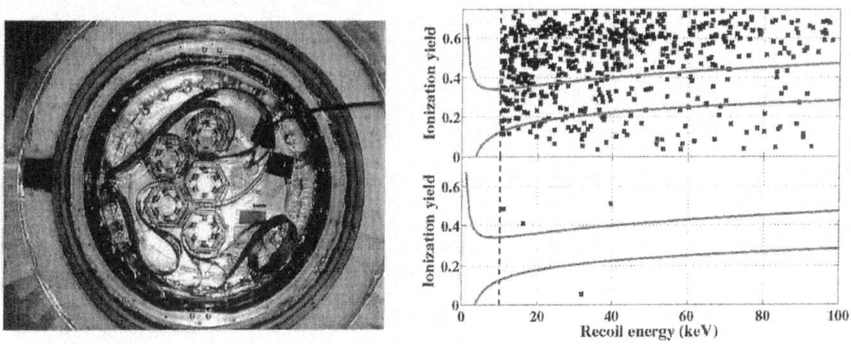

a) b)

Fig.6a) The ZIP detectors in the cryostat; and b) Ionization yield versus timing parameter for calibration data in one of the CDMS Ge detectors

In a bolometer, the energy of the recoiling nucleus or electron is released in three different forms: as heat, i. e. as phonons, as light, i. e. as photons and as ionisation, i. e. as charges. The fractions of energy in each form is different for the heavy, non relativistic recoiling nuclei and for the light, relativistic recoiling electrons (the main background). One then measures two of these quantities to discriminate the background from the signal.

The presently running experiments are EDELWEISS II in the LSM laboratory in France, using 10 kg of Ge bolometers, and measuring ionisation and phonons, CRESST2 at LNGS in Italy, using 5 kg of $CdWO_4$ and measuring light and heat and CDMS in the SUL laboratory in the USA. We will discuss CDMS that has produced the most sensitive limits, while EDELWEISS II and CRESST2 are struggling to reach the background free condition.

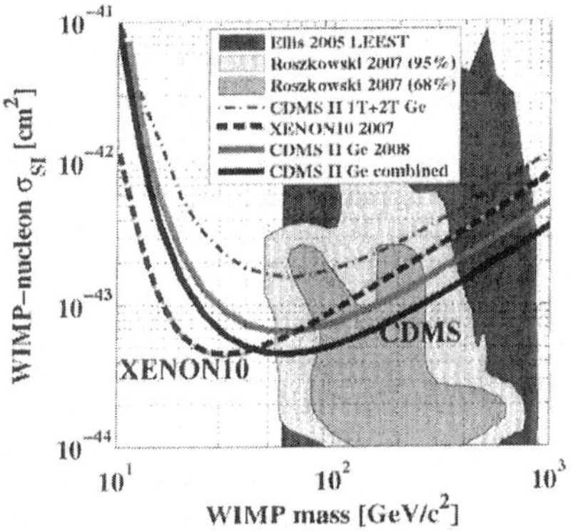

Fig.7 Limits on the SI cross section from CDMS and XENON10 compared with theoretical expectations

The basic detector element, the bolometer, of CDMS is called ZIP. ZIP detectors are assembled in towers of 6 units. CDMS operates with 5 towers of overall 30 ZIPs, 11 of which are made of Si (1.1 kg in total) and 19 of Ge (4.75 kg), as shown in Fig. 6a. The former are meant to control the neutron background, because, due to its small atomic number, Si is not expected to be sensitive to WIMPs (SI coupling). Each of the ZIPs is read out by four phonon sensors and by two charge collectors. The former give the recoil energy and, by detecting the phonons before thermalisation, allow to discriminate, by looking at the pulse shape, between bulk and surface events. In particular, photons interacting in the electrodes, which are located on the surface, give small ionisation and fast pulses. A reduction of this background of a factor 200 has been achieved with this technique, resulting in zero-background conditions, as shown in Fig. 6b.

In a 120 kg d WIMP exposure, no event in the signal region after cuts have been found, to be compared with an expected background rate of 0.6±0.5[31]. The corresponding exclusion curve is shown in Fig. 7 CDMS is working toward larger sensitive masses. It will gain linearly with the exposure, as long as it remains background free.

Noble fluids, Ar, Xe and Ne, look very promising for WIMPs search for the following attractive characteristics: they can be easily assembled in large masses, up to hundreds tons, at reasonable costs; techniques to purify large quantities of such liquids at extreme radiopurty levels are available, as those developed by BOREXINO. With liquids, self-shielding structures can be built, with the central fiducial mass shielded by the external one without any surface between them (avoiding possible backgrounds); the external mass can be instrumented and used as an active veto (μ, anti-Compton, etc.) as shown schematically in Fig. 8a. In this configuration, the photomultipliers are potentially an important source of gamma ray background that can be easily at the level of a Bq for each PM. Intense research in collaboration with industry, notably Hamamatsu, is going on with very important results. The QUPID "all quartz" 3'' photomultiplier has a radiopurity as low as 20 μBq.

The physical process of energy transfer to the medium from the recoil electrons on one side and nuclei on the other gives two independent means can be exploited to discriminate the dominant electromagnetic background. They are based on the detection of the ionisation and of the scintillation light, as follows. Part of the energy deposited is released as scintillation light emitted by excited dimers ($Xe_2^* \to 2Xe$ + photons; $Ar_2^* \to 2Ar$ + photons). The de-excitation curves have two components: one fast from the singlet $^1\Sigma$ states, which are populated mainly by the heavy recoils, and one slow from the triplet $^3\Sigma$ states, populated mainly by the light recoils. This behaviour has been discovered and studied several years ago by T. Doke and collaborators[32].

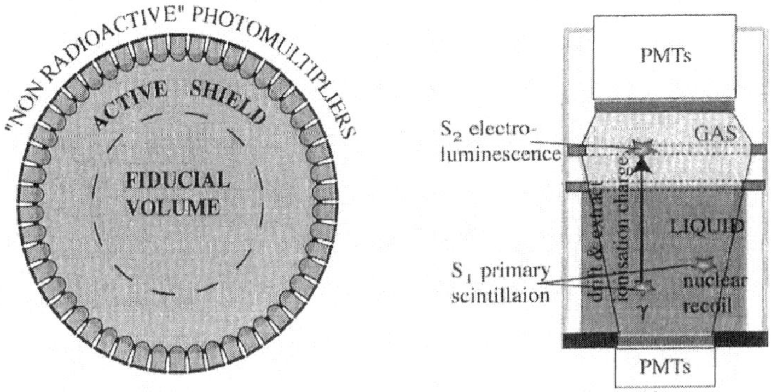

a) b)

Fig. 8. a) Principle of a self-screening noble liquid detector, b) two phase TPC

Moreover, the ratio of ionisation to scintillation is much smaller for the slow-moving nuclear recoils than for electrons. The ionisation charges can be detected by the proportional scintillation produced in the very strong electric field near to a thin anode wire. As we proposed in 1993[33], with the same photo-sensors one can then measure two light pulses, the first, S_1 due to the primary scintillation, the second, S_2 due to the proportional scintillation. The ratio S_2/S_1 is large for minimum ionising particles (γ background) and small for nuclear recoils, or α particles. The evolution of the method lead to the dual-phase detectors[34]. As shown schematically in Fig. 8b, a uniform electric field is applied to drift the electrons, produced in the liquid, vertically up to the surface and extract them into the gas phase (with almost 100% efficiency); here they are amplified using appropriate grid structures and the proportional scintillation light is detected. A few considerations are in order. The ionisation electrons are very few and must be collected with high efficiency; in particular, the electric field must be accurately designed, in order, for example, not to "lose" electron in the non-uniformities close to the race-track. Photons must be efficiently collected too, because they are so few (e.g. ≈200 at 16 keV in Xe). Another important aspect is the following. Clearly, the liquid in the innermost regions of the detector are shielded against incoming background radiation (typically gammas) by the outer parts of the same liquid. One can then define a central "fiducial volume" that may be background free, or almost so. It is then important to be able to measure the position of the events with enough resolution for an accurate "fiducialisation". This can be achieved basically in two ways: building the detector as an image chamber or using arrays of photomultipliers.

The XENON project foresees a series of experiments of increasing mass and decreasing background rate. The basic design is a 3D position sensitive TPC, providing full vertex reconstruction. The outer LXe is employed as an active veto. A 10 kg prototype, XENON10 has successfully completed its data taking in a 136 kg d exposure at LNGS. Fig. 8a) shows S_2/S_1 vs. the recoil energy. The signal region, defined through a blind analysis, contains 10 events, which are interpreted as due to background, mainly statistical leakage from the electron recoil band (at higher values of S_2/S_1) and "anomalous" events due to non active Xe regions. The corresponding exclusion curve[35] is reported in Fig. 7b.

The main background sources in XENON10 have been identified in a number of set-up elements located inside the shields. The configuration has been redefined in XENON100, which is now running at LNGS. If it will be successful the next step will be XENON1000.

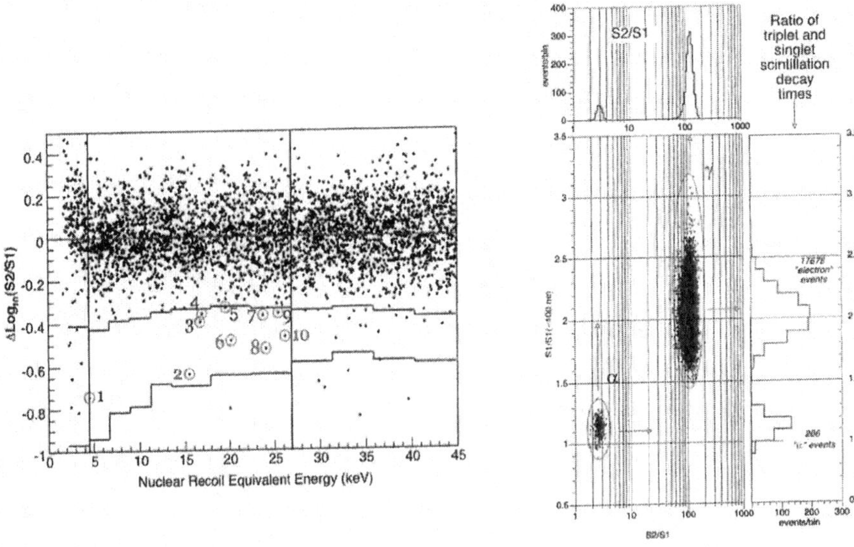

a) b)

Fig. 9. a) XENON10 S_2/S_1 vs. nuclear recoil energy; b) slow/fast vs. S_2/S_1 in LAr from WARP

Two projects are based on dual phase Ar TPC: ArDM, with 850 kg mass presently under test at CERN and proposed for underground future running at the LSC laboratory, and WARP already in an advanced stage of construction at LNGS. The main advantage of LAr is the huge background reduction that can be achieved with the fast/slow discrimination, due to the above mentioned very large difference between the two decay times. Simulations and laboratory data of the WARP collaboration show that discrimination power up to 10^8 and larger can be achieved. Fig. 9b) shows the separation between the signals from alfa particles and gammas obtained with calibration data by WARP.

However, in Argon, one has to fight against the enormous background due to the beta decay of the ^{39}Ar, which amounts to 1 Bq/kg. This isotope has been produced by the nuclear explosions in the atmosphere. It limits, in particular, to about one ton the size of the liquid Ar TPC, because in presence of multiple events within a maximum drift time the association of S_1 and S_2 to each of them becomes ambiguous. The research of fossil Ar sources and on isotope separation is being pursued.

5. Conclusions

In the next years we might assist to extremely important discoveries in neutrino physics and dark matter search. The smallest mixing angle θ_{13} may be found to be different from zero and measured, provided its value is close to the present limit, opening the route to the search for CP violation in the lepton sector. The absolute mass of neutrinos may be determined if it is not smaller of, say, 100 meV. The nature of neutrinos, Majorana or Dirac, might be determined, under the same conditions. SUSY and neutralinos might be discovered at LHC, if they exist. The WIMP component of dark matter might be discovered all the way down to cross section values of 10^{-47} cm^2 in underground laboratories.

References

1. J. Dunkley et al., Astrophys. J.Suppl.**180** (2009) 306-329; arXiv:0803.0586

2. Fogli et al., hep-ph/0805.2517v2
3. A. B. Balantekin and D. Yilmaz, J. Phys. **G 35** (2008) 075007; G. L. Fogli, E. Lisi, A. Marrone, A. Palazzo and A. M. Rotunno, Phys. Rev. Lett. **101** (2008) 141801; M. Maltoni and T. Schwetz, arXiv:hep-ph/0812.3161
4. M. V. Diwan, "Recent Results from the MINOS experiment", proceedings of the 13th International Workshop on Neutrino Telescopes", Venice, March, 2009, arXiv:hep/ex/0904.3706
5. P. Adamson et al., Phys. Rev. Lett. **101** (2008) 131802
6. M. Guler et al., LNGS-P25-00, 2000, CERN-SPSC-2000-028, CERN-SPSC-P-318; R. Acquafredda et al., New J. Phys. **8** (2006) 303; R. Acquafredda et al., JINST **4** (2009) P04018
7. G. Acquistapace et al., CERN-YELLOW-98-02 (1998); R. Baldy et al., INFN-AE-99-5 (1999)
8. Y. Itow et al. arXiv:hep-ex/0106019
9. D. S. Ayres et al., arXiv:hep-ex/0503053; http://www-nova.fnal.gov
10. F. Ardellier et al., arXiv:hep-ex/0606025
11. X. Guo et al., arXiv:hep-ex/0701029
12. L. Wolfenstein, Phys. Rev. **D 17** (1978) 2369; S. P. Mikheev and A. Y. Smirnov, Sov. J. Nucl. Phys. **42** (1985) 913 [Yad. Fiz. **42** (1985) 1441]
13. G. Alimonti et al., Nucl. Instrum. Meth. **A 600** (2009) 568
14. G. Bellini et al., arXiv: asrto-ph/ 0910.3367 and references therein
15. Planck collaboration. ESA-SCI(2005) 1
16. C. Kraus et al., *Eur. Phys. J.* **C40** (2005) 447
17. V. M. Lobashev et al., *Nucl. Phys. B Proc. Suppl.* **91** (2001) 280
18. KATRIN Collaboration, Design Report 2004. NPI ASCR Rez EXP-01/2005MS-KP-0501
19. J. Barea and F. Iachello, Phys. Rev. **C79** (2009) 044301
20. F. Shimkovic, G. Pantis, J.D. Vergados and A. Faessler, Phys. Rev. **C 60**, 055502 (1999)
21. F. Shimkovic, A Faessler, V. Rodin, P. Vogel and J. Engel, Phys. Rev. **C 77**, 045503 (2008)
22. E. Caurier, J. Menéndez, F. Nowacki and A. Poves, Phys. Rev. Lett. **100**, 052503 (2008)
23. I. Abt et al. LNGS P38/04; hep-ex /04040390 and http//www.mpi–hd.mpg.de/GERDA/proposal.pdf
24. R. Ardito et al. "CUORE: a cryogenic underground observatory for rare events" hep-ex/0501010/ and references therein
25. R. Neilson *et al*, *J. Phys.: Conf. Ser.* **136** (2008) 042052 and references therein
26. F. Grañena et al. "NEXT, a HPGXe TPC for neutrinoless double beta decay searches", Letter of Intent (EXP-05-LoI-2009) at Canfranc National Laboratory (3 April 2009). See also F. Nova et al. (2008) EOI-08 LSC and D. Nygren: "0ν double beta decay, WIMPs and Xenon. Will the search converge?" talk at NEUTRINO 2008
27. F. Feruglio, A. Strumia, F. Vissani, *Nucl. Phys.* **B637** (2002) 345
28. R. Bernabei et al., Europ., Phys. Journ. C **56** (2008) 333
29. H. S. Lee et al., Phys. Rev. Lett. **99** (2007) 091301
30. J. Amaré et al., "ANAIS, Annual modulation with NaI scintillator" Letter of Intent (EXP-01-LoI-2009) at Canfranc National Laboratory (September 2009)
31. Z. Ahmed et al., Phys. Rev. Lett. **102** (2009) 011301
32. A. Hitachi et al., Phys. Rev. **B 27** (1983) 5279
33. P. Benetti et al., Nucl. Instr. Meth.; **A327** (1993) 203
34. D. Cline et al., Astrop. Phys. **12** (2000) 377; also: M. Yamashita et al. Astrop. Phys. **20** (2003) 79.
35. J. Angle et al., Phys. Rev. Lett. **100**, 021303 (2008)

CHAIRMAN: A. BETTINI

Scientific Secretaries: R. Preghenella, A. Alici

DISCUSSION

- *Inguglia:*

Do you think that there are some connections between the fact that the energy of the inner part of a star is not enough to allow thermonuclear fusion and the electron neutrino deficit?

- *Bettini:*

No, the fusion reactions in the Sun and in the other stars do happen, at an energy far below the Coulomb barrier. That implies that the speed of the reaction is very small, but not zero. Fortunately it is not too fast. If it were, the star would explode as an H-bomb. Instead, it takes few billions of years to burn. That has nothing to do with neutrino oscillations. The phenomenon was clarified by Bethe seventy years ago.

- *Inguglia:*

What do you think about the possibility of correlation between quarks and leptons by considering the value of θ_{12} for leptons and θ_C for quarks.

- *Bettini:*

I have no idea; nobody knows why the mixing angles, both for neutrinos and quarks, have the values they have.

- *Cifarelli:*

Could you illustrate some of the experimental techniques used to search for dark matter? And what is in your opinion the most significant and reliable limit they can provide?

- *Bettini:*

Probably the most promising of the present techniques are based on the two-phase TPC with Xenon, Argon or Neon. The sensitive volume is liquid. In general there are basically three types of background: the electromagnetic background due to gamma rays scattering on electrons, the background due to neutrons scattering on nuclei, which recoil similarly to those hit by a WIMP, and the background due to contaminations of the surfaces. The last one is reduced, in the TPC, by defining a fiducial volume and by reconstructing the position of the event, without any physical surface. The mechanisms of energy release to the medium by recoiling electrons (relativistic particles) and slowly recoiling nuclei are different. We can then discriminate between the two cases. The fractions of energy released as charge and light are different. The initial light is detected by the photomultipliers, while the ionisation charges are drifted out of the liquid volume and amplified in the gas above it; the scintillation light detected by the photomultipliers gives a second signal. So one has two pulses separated by the electrons drift time, giving the position along the field direction. In the noble liquids an ion produced by the ionisation process forms quickly a dimmer with another atom, in an excited state. This may be a triplet or a singlet. Heavy and light recoils tend to populate differently the triplet and the singlet states. Moreover, the two have different decay times. In the case of Xenon the excitation decay time is 4 ns for the singlet, 22 ns for the triplet. In Argon is 7 ns against 1.6 μs. So, this handle is more powerful in Ar. However, Ar has also negative sides, in particular the presence of the ^{39}Ar radioactive isotope.
Even if WIMPs are the neutralinos, SUSY contains so many unknown parameters that the theoretical expectations on the WIMP-nucleus cross-section span over some ten orders of

magnitudes as shown in figure 1. The mass is also unknown. If LHC will discover neutralinos and measure a number of SUSY parameters, the parameter space will certainly shrink dramatically and the results of the direct search experiments will become more reliable. In order to have a feeling, assume the muon g–2 "anomaly" to be real and to be due to SUSY. The search area will be reduced to the darker (green in colour) area in figure 1. You can easily imagine that with the help of LHC we will precisely know where to look for dark (neutralino) matter.

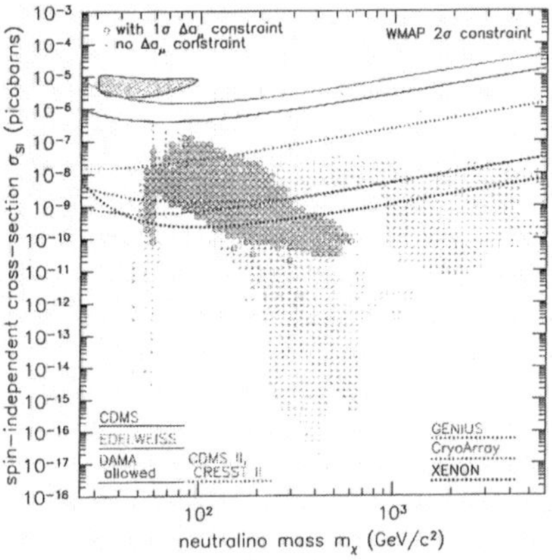

Figure 1

- *Klopot:*

What is the current experimental status of the sterile neutrino problem? One can consider a model which contains in addition to tree flavour states at least one additional sterile state.

- *Bettini:*

There is no experimental evidence for sterile neutrinos. However, you can always assume they mix with a mixing angle small enough to avoid the present, and future, experimental limits.

- *Klopot:*

Are there any specific experiments which can review this kind of neutrinos if they exist?

- *Bettini:*

The LEP experiments measured with high accuracy the width of the Z^0, finding no contribution beyond that of the three known neutrinos. Moreover, the global fit to the oscillation experiments, in particular the solar ones and Kamland, do not need any sterile neutrino; rather the fit becomes worst if you include them.

- *Baldini:*

Why some people claim that positron/electron excess observed in the cosmos could be evidence for dark matter?

- *Bettini:*

There is only a solid evidence, the excess of positrons observed by Pamela. They observed that the ratio between positrons and electrons increases with energy, while the "normal" sources should give a decreasing behaviour. Positron and antiprotons are rather rare, so they can give signals for dark matter. And somebody interpreted the Pamela discovery as an evidence for that.

However, Pamela does not observe any anomaly in the antiproton to proton ratio. One is then forced to assume a very peculiar behaviour of dark matter. Moreover, if it were dark matter its density should be much higher and its cross section much larger than expected. On the other hand, there are much more mundane explanations. The contribution of the few known nearby pulsars can well explain the observed spectrum.

- *Baldini:*

Why the new experiments looking for dark matter cannot check the annual modulation?

- *Bettini:*

DAMA had a sensitive mass of 100 kilos, now the LIBRA of 250 kilos; both took data for several years running continuously with a very low background level. Up to now no other experiment has such a mass. XENON10, the most massive one, was 10 kilos. XENON-100, which is running now, is about 100 kilos. To me, even if some other experiment will find a positive evidence, the final confirmation will need to observe an annual modulation. However, the modulation is only a few percent of the total, so a large mass and very low background are needed.

- *Calivà:*

Is the $0\nu\beta\beta$ decay the only way to investigate the nature of the neutrino or are there other ways?

- *Bettini:*

Up to now there are none. One can think for example to send an electron beam into a target and look for muons coming out, i.e. electron to muon transitions. But calculations show that this process is so rare that it is far from the present techniques by several orders of magnitude. The main problem is that all the effects that distinguish between Majorana and Dirac neutrinos are proportional to the square of the ratio between neutrino mass and energy. There is a new idea by M. Yoshimura. Consider an atomic transition forbidden by parity. The energy is of the order of eV, not much higher than neutrino mass. The transition may happen by weak interactions with the emission of 2 neutrinos, a neutrino and antineutrino if they are Dirac, two neutrinos if they are Majorana. Clearly the probability of such a spontaneous transition is too small by many orders of magnitude to be observable. However, one can think to force the transition by a tuned LASER beam of very high intensity. The present technology does not seem to be too far from making a real experiment possible.

- *Cifarelli:*

In case of Supernova explosion, can you say something about neutrino mass?

- *Bettini:*

No, you cannot say anything significant. Suppose, for example, that electron neutrinos are detected. They did not necessarily have this flavour when produced at the supernova, due to MSW transitions and oscillations on route. One might, however, extract some average neutrino mass from the dispersion of the signal, which is at the source some 10 s long. For a Supernova at 10–100 kpc the sensitivity is of the order of about 10 eV only. So it is worse than the already known limits.

- *Dixon:*

Can you explain why the slight tension between Kamland and solar neutrino experiments can be relieved by adding a small θ_{13}?

- *Bettini:*

The effect seen by Kamland is neutrino oscillations. Flavour change in solar neutrinos is dominated by (vacuum) oscillations at low energy, matter effects at high energies. Moreover, increasing or decreasing θ_{13} on one side and increasing or decreasing θ_{12} on the other, have different effects on the survival probability in oscillation and in matter effects. These facts give some sensitivity to θ_{13}. One needs a value $\theta_{13}\neq 0$ to fit both types of data. But the effect is statistically very weak.

Experimental evidence for Pointlike Baryons at $q^2 = 4M_{\mathcal{B}}^2$

Rinaldo Baldini Ferroli[a,b], Simone Pacetti[a,b] Adriano Zallo[b], and Antonino Zichichi[a,c,d]

[a]Museo Storico della Fisica e Centro Studi e Ricerche "E. Fermi", Rome, Italy
[b]INFN, Laboratori Nazionali di Frascati, Frascati, Italy
[c]INFN and Department of Physics, University of Bologna, Bologna, Italy
[d]CERN, Geneva, Switzerland

1 Introduction

The significance of baryon time-like form factors (FF) as crucial tools to understand subnuclear physics was pointed out long time ago by Conversi, Massam, Muller and Zichichi [1]. They looked at the annihilation $p\overline{p} \to e^+e^-$ to investigate proton FF's in the time-like region and to obtain first experimental signals of a structured $p\overline{p}\gamma$ vertex. To get rid of the hadronic background, the preshower technique was exploited achieving a π/e rejection factor of the order of 10^6.

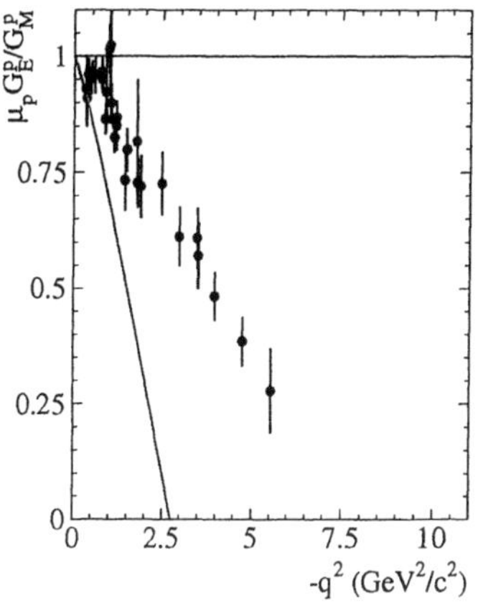

Figure 1: The solid circles indicate the polarization data for the ratio $\mu_p G_E^p/G_M^p$, the blue line represents the "scaling dogma", while the red curve is the prediction given in Ref. [2].

In the space-like region the exploration of proton FF's received a strong boost from the latest polarization data on the ratio G_E^p/G_M^p, shown in fig. 1. These data, collected at the Jefferson Laboratory [3] and M.I.T. [4], show a very unexpected behavior in contrast with the results obtained with the usual Rosenbluth separation technique. Indeed according to these previous data sets, all obtained by means of Rosenbluth technique, the space-like "scaling dogma" was established, which provides the identity $G_E^p = G_M^p/\mu_p$, the blue line in fig. 1. Conversely the new data not only do not agree at all with the scaling dogma, but they show an unexpected decreasing behavior that approaches to zero, a linear extrapolation gives $G_E^p/G_M^p = 0$ around $q^2 = -10.$ GeV^2/c^2. Such a peculiar space-like behavior was predicted in a farsighted and actually with no free parameters description of the ratio G_E^p/G_M^p(red curve in fig. 1), based on a soliton interpretation of the nucleon (the soliton radius,the only free parameter, was just the expected one) given by Massam and Zichichi [2] more than 30 years before the polarization data were collected.

2 Baryon cross section and form factors

The Feynman diagram shown in fig. 2 represents in Born approximation two crossed processes: the $e\mathcal{B} \to e\mathcal{B}$ elastic scattering, in the space-like direction, and the annihilations $e^+e^- \leftrightarrow \mathcal{B}\overline{\mathcal{B}}$ in the time-like direction ($\mathcal{B}$ stands for baryon).

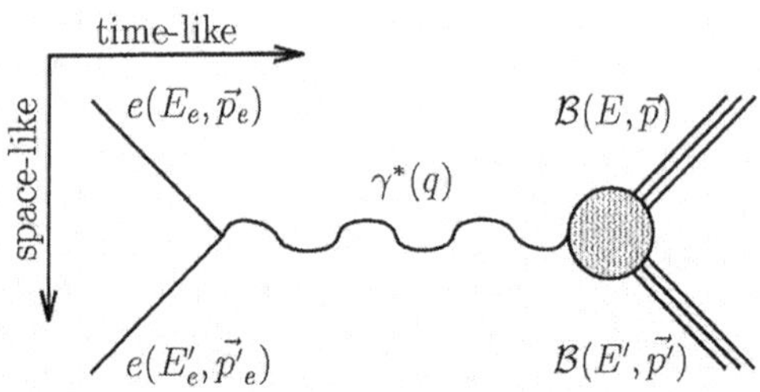

Figure 2: Feynman diagram for the scattering $e\mathcal{B} \to e\mathcal{B}$ and annihilations $e^+e^- \leftrightarrow \mathcal{B}\overline{\mathcal{B}}$.

The "blob" on the baryon vertex indicates that the electromagnetic coupling does not have an elementary structure. This vertex is described by Γ^μ which, following gauge and Lorentz invariance, is written in terms of two FF's as

$$\Gamma^\mu = F_1(q^2)\gamma^\mu + i\frac{\sigma^{\mu\nu}q_\nu}{2M_\mathcal{B}}F_2(q^2)\,, \tag{1}$$

where F_1 and F_2, the so-called Dirac and Pauli FF's, are Lorentz scalar functions of q^2 (q, as indicated in fig. 2, is the 4-momentum of the photon). Besides the Dirac and Pauli FF's we can equivalently use other pairs of structure functions, as for instance the Sachs electric and magnetic FF's, G_E and G_M, defined as

$$\begin{aligned} G_E &= F_1 + \frac{q^2}{4M^2}F_2 \\ G_M &= F_1 + F_2 \end{aligned} \,, \tag{2}$$

where M is the baryon mass.

In general the FF's are analytic functions defined in the whole q^2 complex plane with a cut along the positive real axis, from the first inelastic theoretical threshold $s_{\rm th} = 4M_\pi^2$ up to infinity. Such a cut is due to the opening of infinite allowed intermediate states I_n ($n = 1, 2, \ldots$), it is just through those states that the production reaction proceeds, i.e.:

$$e^+e^- \to \mathcal{B}\overline{\mathcal{B}} = \sum_n e^+e^- \to I_n \to \mathcal{B}\overline{\mathcal{B}}\,. \tag{3}$$

Since the I_n are physical states, the total amplitude takes a non-vanishing imaginary part that can be obtained through the usual spectral decomposition. It follows that: in the whole space-like region ($q^2 < 0$) and in the portion of the time-like region below the theoretical threshold ($0 \leq q^2 \leq s_{\rm th}$) the FF's are real, while over the cut, i.e. for $q^2 > s_{\rm th}$ they are complex.

Using eq. (1), in terms of the Sachs FF's given in eq. (2), we can compute the differential cross section formulae. For the elastic scattering in the laboratory frame we have the famous Rosenbluth formula [5]

$$\frac{d\sigma_{\rm el}}{d\Omega} = \frac{\alpha^2 E'_e \cos^2\frac{\theta}{2}}{4E_e^3 \sin^4\frac{\theta}{2}} \left[G_E^2 - \tau\left(1 + 2(1-\tau)\tan^2\frac{\theta}{2}\right) G_M^2\right] \frac{1}{1-\tau}, \qquad \tau = \frac{q^2}{4M^2}, \quad (4)$$

while, the annihilation ($e^+e^- \to \mathcal{B}\overline{\mathcal{B}}$) cross section, in the center of mass frame (cms), reads

$$\frac{d\sigma_{\rm annihi}}{d\Omega} = \frac{\alpha^2 \beta C}{4q^2} \left[(1+\cos^2\theta)|G_M|^2 + \frac{1}{\tau}\sin^2\theta |G_E|^2\right], \quad (5)$$

where θ is the scattering angle and the other symbols refer to the labeling of fig. 2. This formula, that has been obtained for the first time by Berman, Cabibbo, Gatto and Zichichi [6], depends on the moduli of the FF's. For a pointlike fermion pair it is $G_E = G_M = 1$.

Particular attention has to be paid to the factor C which multiplies the whole cross section of eq. (5). This is the final state interaction effect, the Coulomb enhancement factor, not included in the Born approximation, that accounts for the final-state electromagnetic interaction between the produced baryons. It corresponds to the squared value of the Coulomb scattering wave function at the origin, assumed as a good approximation in the case of a long range interaction added to a short range one, the Sommerfeld-Schwinger-Sakharov rescattering formula [7]

$$C(q^2) = \begin{cases} 1 & \text{for neutral baryons} \\ \dfrac{\pi\alpha/\beta}{1 - e^{-\pi\alpha/\beta}} & \text{for charged baryons} \end{cases}, \qquad \beta = \sqrt{1 - \frac{4M^2}{q^2}}. \quad (6)$$

This factor, having a weak dependence on the fermion pair total spin, is assumed to be the same for G_E and G_M and hence can be factorized.

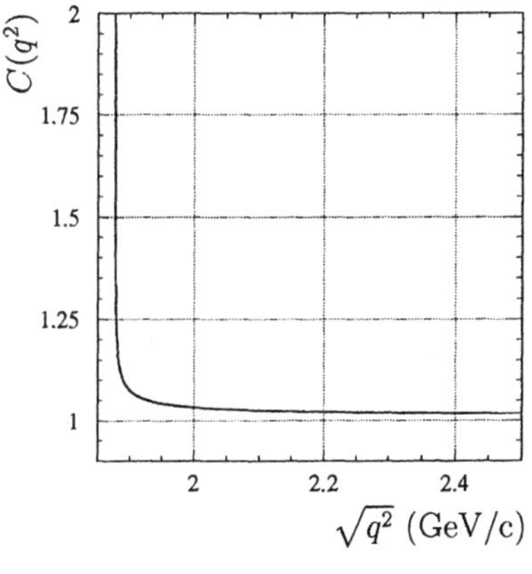

Figure 3: Coulomb enhancement factor as a function of the cms momentum from eq. (6).

Very near threshold, in case of charged baryons, the Coulomb factor behaves as

$$C(q^2 \to 4M^2, \beta \to 0) \sim \frac{\pi\alpha}{\beta}, \tag{7}$$

so that the phase-space factor β, see eq. (5), is cancelled and the cross section is expected to be finite and not vanishing even exactly at threshold. However, as it is shown in Fig. 3, as soon as the fermion relative velocity is no more vanishing, actually few MeV above the threshold, it is $C \sim 1$ and Coulomb effects can be neglected.

3 $\sigma(e^+e^- \to p\overline{p})$ at threshold

The recent measurement of the $e^+e^- \to p\overline{p}$ cross section performed by *BABAR* by means of the initial state radiation technique (ISR) [8] has highlighted unexpected features. Indeed, exploiting the advantages of the ISR technique, for the first time, the near-threshold region became accessible to experiments and it is just there that *BABAR* found an amazing feature of the proton. At threshold the ISR technique has important advantages: non zero detection efficiency and very good invariant mass resolution (about 1 MeV).

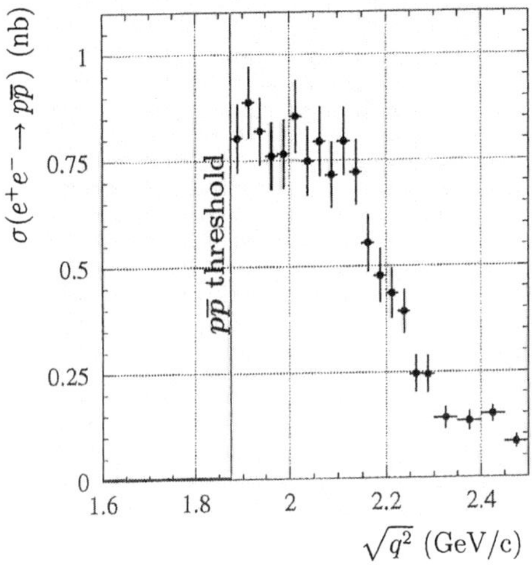

Figure 4: $e^+e^- \to p\overline{p}$ total cross section, as measured by *BABAR* [8]. The gray vertical line indicates the production threshold.

Figure 4 shows the near-threshold data on the total $p\overline{p}$ total cross section, as measured by *BABAR*. The formula for the total cross section, in case of the proton, is

$$\sigma(e^+e^- \to p\overline{p}) = \frac{4\pi\alpha^2\beta C}{3q^2}\left[|G_M^p|^2 + \frac{2M_p^2}{q^2}|G_E^p|^2\right]. \tag{8}$$

As already stated, taking the limit towards the threshold, the Coulomb correction cancels the closure of the phase-space by making the cross section finite. Using the previous formula and eq. (7) we can compute this finite value in terms of well known constants and FF's as

$$\sigma(e^+e^- \to p\overline{p}) \underset{\sqrt{q^2}\to 2M_p}{\longrightarrow} \frac{\pi\alpha^2\beta C}{2M_p^2}|G^p(4M_p^2)|^2 = \frac{\pi^2\alpha^3}{2M_p^2}\cdot|G^p(4M_p^2)|^2 \simeq 0.85\cdot|G^p(4M_p^2)|^2 \text{ nb},$$

where G^p is the common value that, according to analyticity as well as S wave dominance, G_E^p and G_M^p take at $q^2 = 4M_p^2$. The numerical factor, 0.85 nb, is strikingly similar to the measured one, shown in fig. 4. Therefore the Coulomb interaction dominates the energy region near threshold and it is found [9]

$$\boxed{|G^p(4M_p^2)| = \mathbf{0.99 \pm 0.04}(\text{stat})\pm\mathbf{0.03}(\text{syst})}.$$

That is, a $p\overline{p}$ pair produced at threshold, $\sqrt{q^2} = 2M_p$, in spite of the very complicated quark structure of the proton, **behaves like a pointlike fermion pair** $(G_E = G_M = 1)$. In the following this feature is suggested to be a general one for baryons. S-wave is peculiar of fermion pairs at threshold. In the case of meson pair production, the total angular momentum conservation requires a P-wave, that vanishes at the origin as well as the quoted Coulomb enhancement factor and for the cross section a β^3 behavior near threshold is expected.

4 Other charged baryon form factors at threshold

4.1 The $\Lambda_c^+\overline{\Lambda}_c^-$ channel

Recently Belle has measured for the first time the $e^+e^- \to \Lambda_c^+\overline{\Lambda}_c^-$ cross section [10]. The anomalously high $\Lambda_c^+\overline{\Lambda}_c$ production rate, close to the $p\overline{p}$ one, suggests a similar FF behavior. Still the cross section is not zero at threshold.

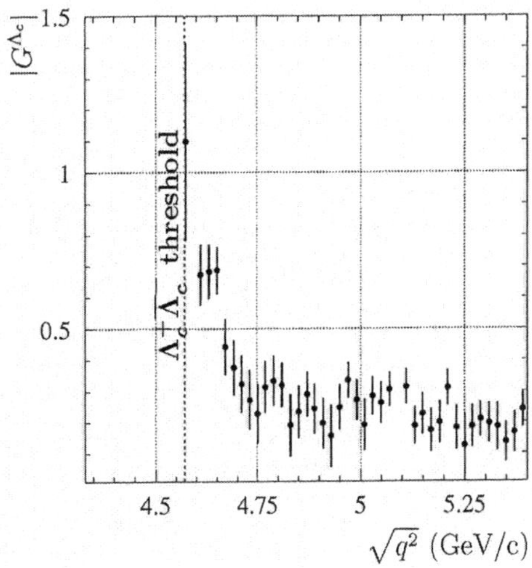

Figure 5: The Λ_c effective FF, extracted from the $\sigma(e^+e^- \to \Lambda_c^+\overline{\Lambda}_c^-)$ cross section measured by Belle [10].

In fig. 5 the modulus of the G^{Λ_c} effective FF is reported, obtained under the hypothesis that $|G_E^{\Lambda_c}| = |G_M^{\Lambda_c}|$. That is expected only at threshold, but this is the best one can do lacking angular distributions. On the $\Lambda_c^+\overline{\Lambda}_c$ cross section given in Ref. [10] a recipe to get the threshold behavior must be used because, due to the finite energy-resolution, the events at threshold are spread in an energy bin that extends itself below $\sqrt{q^2} = 2M_{\Lambda_c}$. Hence, a systematic uncertainty of ~30% is estimated for this procedure, to be added in quadrature to the other systematic errors. Taking into account the Coulomb correction, the statistical errors and the overall systematic quoted errors added in quadrature, mostly due to the uncertainties in the Λ_c branching ratios, it is found:

$$|G^{\Lambda_c}(4M^2_{\Lambda_c})| = 1.1 \pm 0.3(\text{stat}) \pm 0.4(\text{syst}).$$

The central value, even though with large errors, is close, in a tantalizing way, to the aforementioned suggested feature: again baryon pairs, produced at threshold, behave as pointlike fermions. The hypothesis it is a general feature for baryons is strengthened by the gap between $4M^2_{\Lambda_c}$ and $4M^2_p$ (always keeping in mind the uncertainties of the Λ_c FF data). Extrapolating to $e^+e^- \to \Lambda_b\overline{\Lambda}_b$ it should be at threshold $\sigma(e^+e^- \to \Lambda_b\overline{\Lambda}_b) \simeq 23\,\text{pb}$, which means one of the most important hadronic channels at those energies.

4.2 The $p\overline{N}(1440)$+c.c. channel

Among the background processes studied by *BABAR* to perform the $e^+e^- \to p\overline{p}$ cross section measurement there is also the reaction $e^+e^- \to p\overline{p}\pi^0$ [8].

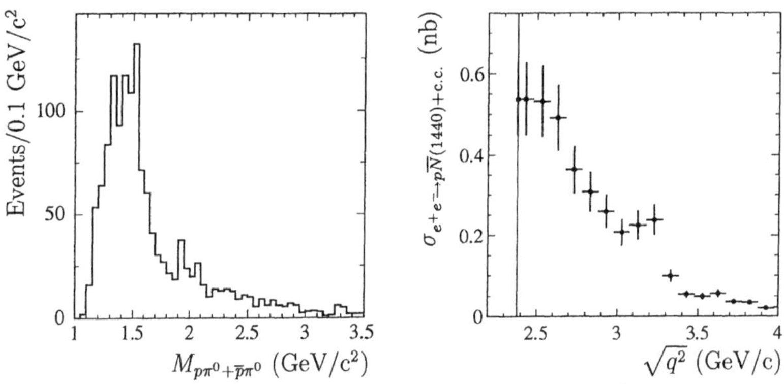

Figure 6: Left: $p\pi^0 + \overline{p}\pi^0$ invariant mass distribution. Right: total cross section for $e^+e^- \to p\overline{N}(1440)$+c.c., the vertical red line indicates the $p\overline{N}(1440)$ production threshold.

Since the $p\pi^0 + \overline{p}\pi^0$ invariant mass distribution, shown in the left panel of fig. 6, is dominated by the broad $N(1440)$ resonance [11], an orbital excited state of the proton, we can easily extract the cross section for the process $e^+e^- \to p\overline{N}(1440) + \overline{p}N(1440)$, introducing a first order correction for the N(1440) width. Such a cross section is reported on the right panel of fig. 6, where the red line indicates the production threshold. Also for this process the threshold value is non-zero.

To extract the corresponding effective FF we have to use a more general function suitable for baryon-antibaryon pairs with different masses. Using also in this case the Coulomb correction formula of eq. (6) and taking the threshold limit for $q^2 \to (M_{N(1440)} + M_p)^2$, we get

$$\sigma_{e^+e^-\to p\overline{N}(1440)+\text{c.c.}} = \frac{16\pi^2\alpha^3 M_p^{3/2} M_{N(1440)}^{3/2}}{(M_p + M_{N(1440)})^5} \left|G^{pN(1440)}\right|^2 = 0.49 \cdot \left|G^{pN(1440)}\right|^2 \text{nb}, \qquad (9)$$

from which, comparing with the data;

$$|G^{pN(1440)}| = 1.04 \pm 0.09\,! \tag{10}$$

Surprisingly, the pointlike behavior at threshold holds also for **transition FF's**.

5 The neutral baryons puzzle

The peculiar features of $e^+e^- \to p\overline{p}$ and other charged baryons cross sections have been observed by *BABAR* [12] also in the case of $e^+e^- \to \Lambda\overline{\Lambda}$, $\Sigma^0\overline{\Sigma^0}$, $\Lambda\overline{\Sigma^0}$, even if within bigger experimental errors. In particular the cross section $\sigma(e^+e^- \to \Lambda\overline{\Lambda})$, shown in fig. 7a, is different from zero at threshold, being 0.20 ± 0.05 nb. The cross sections $\sigma(e^+e^- \to \Sigma^0\overline{\Sigma^0})$ and $\sigma(e^+e^- \to \Lambda\overline{\Sigma^0})$, reported in fig. 7a and b, have been measured by *BABAR* for the first time.
At threshold, assuming a smooth extrapolation from the first data point, it is $\sigma(e^+e^- \to \Sigma^0\overline{\Sigma^0}) = 0.03 \pm 0.01$ nb and $\sigma(e^+e^- \to \Lambda\overline{\Sigma^0}) = 0.047 \pm 0.023$ nb. In all these cases final state Coulomb effects **should not be taken into account** and a finite cross section at threshold is completely unexpected.

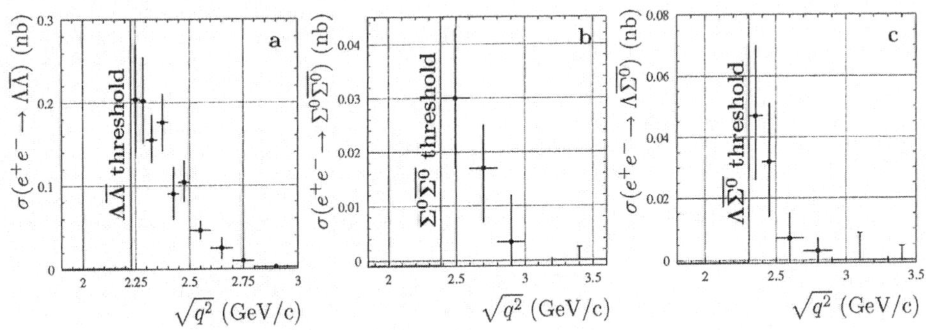

Figure 7: $e^+e^- \to \Lambda\overline{\Lambda}$ (a), $e^+e^- \to \Sigma^0\overline{\Sigma^0}$ (b), and $e^+e^- \to \Lambda\overline{\Sigma^0}$ (c) total cross sections, as measured by *BABAR* [12]. The gray vertical lines indicate the production thresholds.

Another interesting feature of these neutral cross sections is their substantially different absolute value, hundreds of pb for $\Lambda\overline{\Lambda}$, only few tens of pb for $\Sigma^0\overline{\Sigma^0}$ and $\Lambda\overline{\Sigma^0}$. Nevertheless, they appear well compatible if one invokes flavor SU(3) symmetries. In particular, according to the U-spin invariance [13]:

$$G_{\Sigma^0} - G_\Lambda + \frac{2}{\sqrt{3}} G_{\Lambda\overline{\Sigma^0}} = 0\,, \tag{11}$$

This relationship is in fair agreement with the *BABAR* data, assuming real same-sign FF's at threshold:

$$M_{\Sigma^0}\sqrt{\sigma_{\Sigma^0\overline{\Sigma^0}}} - M_\Lambda\sqrt{\sigma_{\Lambda\overline{\Lambda}}} + \frac{2}{\sqrt{3}}\overline{M_{\Lambda\Sigma^0}}\sqrt{\sigma_{\Lambda\overline{\Sigma^0}}} = (-0.06 \pm 6.0) \times 10^{-4}\,.$$

6 Conclusions

The $e^+e^- \to \mathcal{B}\overline{\mathcal{B}}$ cross sections, as measured by *BABAR*, do not vanish at threshold. In the case of $e^+e^- \to p\overline{p}$ this behavior is explained by the $p\overline{p}$ Coulomb enhancement factor and it comes out that $|G^p(4M_p^2)| = 1$, that is **proton pairs behave as pointlike fermions**. This cross section is remarkably flat near threshold: it turns out that S- and D-wave have opposite trends, producing this peculiar behavior and the S-wave contribution has a steep drop above threshold, consistent with Coulomb dominance.
A pointlike threshold behavior is suggested also in the cases of the Λ_c FF, recently achieved by Belle, as well as for the pair $p\overline{N}(1440)$. The latter appears as a really awkward feature because, dealing with a transition FF, the presence and the manifestation of a structure should be mandatory to explain the "transition" itself.
A consistent framework of neutral strange baryon FF's is obtained requiring flavor U-spin symmetry.Neutral strange baryons show a non-vanishing cross section at threshold too. A phenomenological interpretation might be given in terms of a "remnant" of quark pair Coulomb interaction before hadronization.

References

[1] M. Conversi, T. Massam, A. Zichichi, and T. Muller, Nuovo Cim. **40**, 690 (1965).

[2] T. Massam and A. Zichichi Nuovo Cim. A**40**, 1137 (1966).

[3] M.K. Jones *et al.*, Phys. Rev. Lett. **84**, 1398 (2000);
O. Gayou *et al.*, Phys. Rev. C **64**, 038202 (2001);
O. Gayou *et al.*, Phys. Rev. Lett. **88**, 092301 (2001).

[4] B.D. Milbrath *et al.*, Phys. Rev. Lett. **80**, 452 (1998); Phys. Rev. Lett. **82**, 2221(E) (1999).

[5] M.N. Rosenbluth, Phys. Rev. **79**, (1950) 615.

[6] A. Zichichi, S. M. Berman, N. Cabibbo and R. Gatto, Nuovo Cim. **24**, 170 (1962).

[7] A. D. Sakharov, Zh. Eksp. Teor. Fiz. **18**, 631 (1948) [Sov. Phys. Usp. **34**, 375 (1991)];
A. Sommerfeld, *Atombau und Spektralliniem* (Vieweg, Braunschweig, 1944), Vol. 2, p.130;
J. Schwinger, *Particles, Sources, and Fields*, Vol. III, p. 80.

[8] B. Aubert *et al.* [*BABAR* Collaboration], Phys. Rev. D **73**, 012005 (2006) [arXiv:hep-ex/0512023].

[9] R. Baldini, S. Pacetti, A. Zallo and A. Zichichi, Eur. Phys. J. A **39**, 315 (2009) [arXiv:0711.1725 [hep-ph]]; R. B. Ferroli, S. Pacetti and A. Zallo, arXiv:1008.0542 [hep-ph].

[10] G. Pakhlova *et al.* [Belle Collaboration], Phys. Rev. Lett. **101**, 172001 (2008) [arXiv:0807.4458 [hep-ex]].

[11] C. Amsler et al., Phys. Lett. B **667**, 1 (2008).

[12] B. Aubert *et al.* [*BABAR* Collaboration], Phys. Rev. D **76**, 092006 (2007) [arXiv:0709.1988 [hep-ex]].

[13] D. Park, *Introduction to strong interactions* (W. A. Benjamin, Inc., New York, 1966).

CHAIRMAN: S. PACETTI

Scientific Secretaries: P. Aschieri, B. Parsamyan

DISCUSSION

- *Dixon:*

What happens to the dispersion relation for $R(q^2)$ if the denominator $G_M(q^2)$ in the ratio has a zero for some complex q^2, i. e. a pole for $R(q^2)$?

- *Pacetti:*

In our analysis we assumed no zeros for $G_M(q^2)$. Such an assumption is based on a previous work [Eur. Phys. J. C11 (1999) 709], where we analyzed the single proton magnetic form factor and we found that the hypothesis of no zeros was well verified.

- *Dixon:*

Why didn't you include the SLAC elastic *ep* data, with Rosenbluth separation, which go out to much larger space-like q^2 than the data you showed?

- *Pacetti:*

The SLAC data on electron-proton elastic scattering have been extracted using the Rosenbluth method. This method is more sensitive to radiative corrections or two-photon contribution with respect to the transfer polarization technique, used at JLab. In particular, it is quite difficult to extract the value of the electric form factor because, at high transfer momentum, it represents only a tiny fraction of the cross section. Then, even though the SLAC measurements of the elastic scattering cross section are reliable, the extraction of single form factors, as well as of their ratio, is affected by uncontrolled sources of errors. This is the reason why we considered only polarization data.

- *Tannebaum:*

At JLab the ratio $\mu G_E^2/ G_M$ has been also measured using the standard Rosenbluth technique [Phys. Rev. Lett. 94 (2005) 142301]. The explanation of differences between Rosenbluth and polarization methods is the two-photon exchange contribution, which was well known at SLAC. It is crucial that the measurements be redone with both electron and positron scattering to resolve this issue.

- *Pacetti:*

There are two schools of thought regarding the reconciliation of polarization and Rosenbluth data. In one case the difference is attributed to the role played by the two-photon contribution amplitude, which is underestimated in the Rosenbluth separation method with respect to the polarization technique. In the other case, instead, the cause of the discrepancy is identified with an unsuitable implementation of radiative corrections. Of course, as you said, redoing the experiments with electron and positron should give a final answer to this problem.

- *Wolf:*

The phase of the ratio G_E/ G_M shows a rapid variation as function of the energy in the center of mass, such a variation could be interpreted in terms of resonant contributions?

- *Pacetti:*

In general, the region where we expect the main contributions, namely vector meson resonances, is the so-called "unphysical region", where, by definition, there are no data. In our procedure, to avoid the problem of introducing a model to describe these resonances, we have chosen to use the ratio between the electric and magnetic form factors. In such a way, similar resonant contributions appear in both numerator and denominator and hence cancel out. Indeed, we have obtained a phase which shows onlysmall variations below the physical thereshold. The main aim of this analysis was to construct an analytic parametrization for the ratio to verify the connection between space-like and time-like data and to make predictions in high-energy regions not yet experimentally investigated, while no reliable information on low-energy resonances can be extracted.

- *Haidt:*

What happens, if you consider the $\pi^+ \pi^-$ production instead of the proton-antiproton?

- *Pacetti:*

The baryon-anti-baryon production represents a special channel for this kind of analysis, indeed, only in this case, the phase space vanishing is exactly cancelled out by the Coulomb correction. Such a cancellation does not occur in meson-anti-meson production processes. Indeed, in these cases the cross section at the threshold must be zero, because the phase space vanishes as the velocity to the third power, while the Coulomb correction gives always the same factor, proportional to the inverse of the velocity, as in the proton-antiproton case. As a consequence, it is very difficult to extract any information by the threshold data.

- *Mastrolia:*

You showed a tow-loop diagram for the form factors F_1 and F_2. Up to which order in perturbation theory are they known? Can the analytic calculation of the two-loop QED corrections for the electron form factors be re-cycled to improve the analysis of proton form factors?

- *Pacetti:*

The Coulomb factor is an exact soft-photon re-summation to all orders. Since the standard Feynman method, in the limit towards the threshold, fails, such a correction factor is obtained solving a non-relativistic Schrödinger equation for two electric charges. This result, exact for point-like particles in the limit of vanishing velocity, is just the correction we used.

- *Epifanov:*

What about a similar analysis for neutrons?

- *Pacetti:*

Unfortunately, for time-like neutron form factors there are not enough data to attempt any investigation.

- *Cifarelli:*

Can you tell me about further data from new experiments that could confirm your findings?

- *Pacetti:*

There are at least two experiments that could give important answers in the near future. VEPP2000, at Novosibirsk, is an electron-positron collider, which hosts two detectors, with a maximum energy in the center-of-mass of 2 GeV. This means that there is a window of less than 200 MeV to observe nucleon-anti-nucleon production. Nevertheless it should be possible to obtain some data close to the threshold, eventhough the absence of a boost makes it difficult to reach the exact threshold energy. Another experiment , already running, is BESIII at the BEPCII electron-positron collider in Beijing. In this case the machine can scan a large energy range: from about 2 GeV up to 5 GeV. This allows not only to measure directly the nucleon form factors at these energies, but also to investigate lower-energy regions using the Initial State Radiation technique, successfully exploited at flavor machines.

Neutrino masses, dark matter, baryon asymmetry and inflation can be explained at once

Mikhail Shaposhnikov
Institut de Théorie des Phénomènes Physiques, EPFL, CH-1015 Lausanne, Switzerland

Abstract

I will discuss a proposal for a unified solution of the problems of neutrino masses, dark matter, baryon asymmetry of the Universe and inflation, which *does not require* introduction of any new energy scale besides already known, namely the electroweak and the Planck scales. This point of view, supplemented by a requirement of simplicity, has a number of experimental predictions which can be tested, at least partially, with the use of existing accelerators such as CERN SPS and the LHC, with current and future X-ray telescopes, and with the Planck mission.

1 Introduction

The Standard Model (SM) of elementary particles has successfully predicted a number of particles and their properties. However, there is no doubt that the SM is not a final theory. Indeed, over the last several decades it became increasingly clear that it fails to explain a number of *observed* phenomena in particle physics, astrophysics and cosmology. These phenomena *beyond the SM* (BSM) are:
(i) *Neutrino oscillations* (transition between neutrinos of different flavours).
(ii) *Baryon asymmetry* (excess of matter over anti-matter in the Universe).
(iii) *Dark matter* (some 80% of all matter in the Universe consists of unknown particles).
(iv) *Inflation* (a period of the rapid accelerated expansion in the early Universe).
(v) *Dark energy* (late time accelerated expansion of the Universe).
This list of *well-established observational* drawbacks of the SM is complete at

present time. All the other BSM problems are those of theoretical fine-tuning: the "gauge hierarchy problem", strong CP-problem, etc. There are few other problems in astrophysics and cosmology, such as the origin of high energy cosmic rays, the existence of 0.511 MeV annihilation line in the direction of the Galaxy center, Pamela anomaly, etc, which may signal that the SM is not complete. However, these phenomena do not provide a "smoking gun" signature for BSM physics as several standard physics explanations were proposed to deal with them. There are also several anomalies in particle physics experiments, such as discrepancy between experiment and theory prediction of anomalous magnetic moment of muon, LSND anomaly, evidence of the neutrinoless double decay presented by a part of the Heidelberg group, etc. However, none of these anomalies has been confirmed by other experiments.

Once the SM is not a fundamental theory, one has to ask oneself: "At what energies the SM should be superseded by some other, more fundamental theory?"

The existence of gravity with the coupling related to the Planck scale $M_{Pl} = G_N^{-1/2} = 1.2 \times 10^{19}$ GeV (G_N is the Newtonian gravitational constant) implies that the cut-off is *at least* below the Planck scale. If the cutoff is identified with M_{Pl}, the low energy Lagrangian can contain all sorts of higher-dimensional SU(3)×SU(2)×U(1) invariant operators, suppressed by the Planck scale:

$$\mathcal{L} = \mathcal{L}_{\rm SM} + \sum_{n=5}^{\infty} \frac{\mathcal{O}_n}{M_{Pl}^{n-4}} \; . \tag{1}$$

Here $\mathcal{L}_{SM}$ is the Lagrangian of the SM. These operators lead to a number of physical effects that cannot be described by the SM, such as neutrino masses and mixings, proton decay, etc.

Alternatively, one can put a cut-off $\Lambda \ll M_{Pl}$ in (1). This would imply that new physics (and new particles) appear much below the Planck scale at energies $E \sim \Lambda$. If $\Lambda \gg M_W$, where M_W is the mass of the weak W-boson, the resulting theory suffers from the so-called *gauge hierarchy problem*, i.e. the problem of quantum stability of the mass of the Higgs boson against quantum corrections from heavy particles.

These lectures are devoted to the description of the scenario for BSM physics and its consequences for astrophysics and cosmology, in which no any new energy scale (besides the *electroweak* and the *Planck* scales) is introduced. In such an approach the hierarchy problem gets shifted to the Planck scale and one has no firm reasons to believe that the field theoretical logic is still applicable to it. A review of these ideas can be also found in [1, 2].

2 Construction of the Minimal Model

The assumption of the absence of a new energy scale between the Fermi and Planck scales is very powerful, as it allows to conclude that new light particles must be added to the Standard Model. Indeed, the observed pattern of neutrino oscillations cannot be explained by the action (1) with the Planck scale cutoff. The lowest order five-dimensional operator

$$\mathcal{O}_5 = A_{\alpha\beta}\left(\bar{L}_\alpha\tilde{\phi}\right)\left(\phi^\dagger L^c_\beta\right) \tag{2}$$

leads to the Majorana neutrino masses of the order $m_\nu \sim v^2/M_{Pl} \simeq 10^{-6}$ eV (here L_α left-handed leptonic doublets, the index $\alpha = e, \mu, \tau$ labels generations, ϕ is a Higgs doublet with $\tilde{\phi}_j = i(\tau_2)^k_j\phi^*_k$, c is the sign of charge conjugation, and $v = 174$ GeV is the vacuum expectation value of the Higgs field). At the same time, the atmospheric neutrino mass difference sets the scale $m_{\rm atm} \simeq 0.05$ eV, where $m_{\rm atm} = \sqrt{|\Delta m^2_{\rm atm}|}$, $|\Delta m^2_{\rm atm}| = 2.40^{+0.12}_{-0.11} \times 10^{-3}$ eV^2.

A way to extend the SM suggests itself after a look at the fermionic content of the SM, see Fig. 1, left panel. All left fermions of the SM, except neutrinos, have their right counter-parts. Let us complete this table by adding three right handed neutrinos N_I, see Fig. 1, right panel (other possible names for the new particles are singlet fermions, massive neutral leptons, sterile neutrinos, or Majorana fermions). We will call the resulting theory the νMSM – the Neutrino Minimal Standard Model and use the notations $I = 1, 2, 3$ or $I = e, \mu, \tau$ interchangeably. It is defined by a renormalisable Lagrangian,

$$\mathcal{L} = \mathcal{L}_{SM} + i\bar{N}_I\partial_\mu\gamma^\mu N_I - \left(F_{\alpha I}\,\bar{L}_\alpha N_I\tilde{\phi} - \frac{M_I}{2}\,\bar{N}^c_I N_I + h.c.\right), \tag{3}$$

where $F_{\alpha I}$ are new Yukawa couplings. The Majorana masses M_I are consistent with the gauge symmetries of the SM. Without loss of generality, the Majorana mass matrix can be chosen in diagonal form; we keep the number of N_I $\mathcal{N}$ arbitrary for the moment, though only the case with $\mathcal{N} = 3$ will be studied in some detail. Interestingly, this theory provides the electric charge

SM fermions

quarks

left	u	d	c	s	t	b
right	u	d	c	s	t	b
left	ν_e	e	ν_μ	μ	ν_τ	τ
right		e		μ		τ

leptons

νMSM fermions

quarks

left	u	d	c	s	t	b
right	u	d	c	s	t	b
left	ν_e	e	ν_μ	μ	ν_τ	τ
right	N_e	e	N_μ	μ	N_τ	τ

leptons

Figure 1: Femions of the SM (left) and of the νMSM (right).

quantization without Grand Unification, coming from the requirement of cancellation of gauge and gravitational anomalies [3].

To find the best strategy for the study of this model, note that till now none of the 19 parameters of the SM has been found theoretically. They are all taken from different experiments. Since we even cannot explain the known parameters of the SM, there are little chances (at least at present) that we can find νMSM parameters theoretically. Let us see therefore whether νMSM agrees with particle physics experiments, can explain neutrino masses, dark matter, baryon asymmetry of the Universe, and provide cosmological inflation, making the Universe flat, homogeneous and isotropic, and producing the necessary spectrum of perturbations for some choice of parameters. The parameters, found in this way can be used then to fix the experimental plans to search for new physics.

3 Neutrino masses

If the Dirac masses $M_D = F_{\alpha I}\langle\phi\rangle$ are much smaller than Majorana masses M_I, *the type I see-saw formula* holds [4, 5, 6, 7]:

$$(m_\nu)_{\alpha\beta} = -\sum_{I=1}^{\mathcal{N}} (M_D)_{\alpha I} \frac{1}{M_I} (M_D^T)_{I\beta}, \tag{4}$$

where m_ν is a 3×3 matrix of active neutrino masses, mixings and (possible) CP-violating phases. An elementary analysis of (4) shows that the number of right-handed singlet fermions $\mathcal{N}$ must be at least two to fit the data of neutrino oscillations. If there were only one sterile neutrino, then two active neutrinos would be massless. With two singlet fermions only one of the active neutrinos is massless, which does not contradict to experiment. Moreover, in this case the number of new parameters in the Lagrangian (3) is 11 (they can be counted as follows: 2 Majorana masses, 2 Dirac masses, 4 mixing angles and 3 CP-violating phases) and is larger than the number of parameters (7) describing the mass matrix of active neutrinos with one zero eigenvalue. In other words, already for $\mathcal{N}=2$ the Lagrangian (3) can describe the pattern of neutrino masses and mixings observed experimentally. Of course, the situation gets even more relaxed for $\mathcal{N}=3$, when the family structure of the νMSM coincides with that of the SM. With this choice, the νMSM contains 18 new parameters: 3 Majorana masses of new neutral fermions N_i, and 15 new Yukawa couplings in the leptonic sector (corresponding to 3 Dirac neutrino masses, 6 mixing angles and 6 CP-violating phases).

The see-saw formula (4) leaves the mass of singlet neutrinos be a free parameter: multiplying M_D by any number x and M_I by x^2 does not change the right hand side of formula (4). Therefore, the choice of M_I is a matter of theoretical prejudice, which cannot be fixed by experiments with active neutrinos only.

A naturalness argument of [8], which we reproduce below, indicates that the masses M_I should be smaller than 10^7 GeV. Indeed, the stability of the Higgs mass against radiative corrections due to N_I-exchange in the loops requires

$$\frac{1}{16\pi^2}\sum_{\alpha I}|F_{\alpha I}|^2 M_I^2 \lesssim M_H^2\ , \tag{5}$$

where M_H is the Higgs mass. Combining this with the see-saw formula (4) leads to

$$M_I \lesssim \left(\frac{16\pi^2 M_H^2 v^2}{\sqrt{|\Delta m^2_{\rm atm}|}}\right)^{\frac{1}{3}} \simeq 10^7 \mathrm{GeV}\ . \tag{6}$$

Any M_I mass below this value is "natural". Moreover, the choice $M_I = 0$ increases the symmetry of the Lagrangian, leading to lepton number conservation. This means that small M_I are stable against radiative corrections and thus are "natural" in the t'Hooft sense [9]. In the νMSM approach we choose M_I below the electroweak scale, considering the new particles on the same footing as known quarks and leptons, all having the masses roughly in GeV region (except of the t-quark).

To estimate what this implies for the Yukawa couplings, we take the larger of two mass splittings $|\Delta m^2_{\rm atm}|$ and find from the see-saw relations (4) that

$$\frac{|F|^2}{M_I\,[\text{GeV}]} \sim \frac{\sqrt{|\Delta m^2_{\rm atm}|}}{v^2} \sim 10^{-12}. \tag{7}$$

The condition $M_I \lesssim 10^2$ GeV implies that $|F| \lesssim 10^{-7}$.

In fact, the full see-saw formula (4) leaves some freedom in the choice of Yukawa couplings. As it was mentioned above, the observed existence of two mass splittings (solar and atmospheric, [10]) requires at least two sterile neutrinos. Therefore, for $\mathcal{N} = 3$ one of the singlet fermions may be decoupled from active fermions without spoiling the explanation of neutrino mixing. This point is important for discussion of the Dark Matter, see below.

Another possible choice of parameters of Lagrangian (3) (suffering from the hierarchy problem of [8]) relates the smallness of neutrino masses to the largeness of the Majorana mass terms M_I. Indeed, taking the Yukawa couplings $F_{\alpha I} \sim 1$ and the Majorana masses in the range $M_I \sim 10^{10} - 10^{15}$ GeV one can get the neutrino masses, required by experimental observations. An attractive feature of this scenario is that the new scale may be associated with the Grand Unification scale. The model (3) with this choice of parameters can also give rise to the baryon asymmetry of the Universe through leptogenesis [11] and anomalous electroweak number non-conservation at high temperatures [12]. We will not discuss this possibility in these lectures. For reviews of GUT-scale see-saw see e.g. [13]. Yet another option is to assume that M_I can be as small as eV. It is discussed in [14, 15] and will not be considered here.

4 Dark matter

Though the νMSM does not have any extra stable particle in comparison with the SM, the lightest singlet fermion, N_e, may have a life-time τ_{N_e} greatly exceeding the age of the Universe and thus play the role of a dark matter particle [16, 17, 18, 19]. The main decay mode is $N_e \to 2\nu\bar{\nu}, 2\bar{\nu}\nu$ and

$$\tau_{N_e} = 5 \times 10^{26}\,\text{sec} \left(\frac{1\ \text{keV}}{M_1}\right)^5 \left(\frac{10^{-8}}{\theta^2}\right), \tag{8}$$

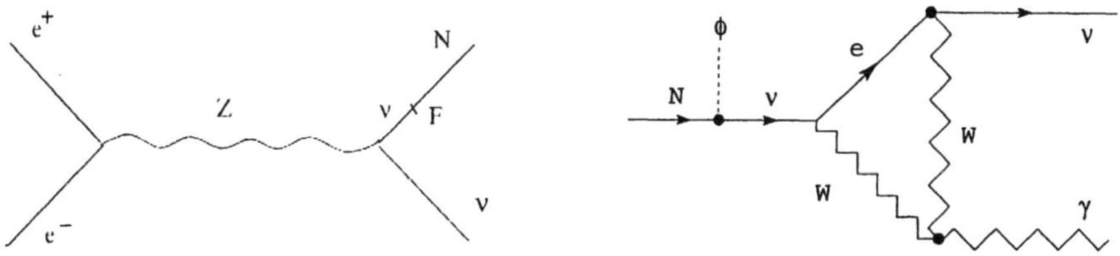

Figure 2: **Left.** The diagram for production of N_e in the early Universe. **Right.** Radiative decay of N_e.

where the mixing angle θ_1 is the ratio of Dirac and Majorana masses,

$$\theta_1^2 = \sum_{\alpha=e,\mu,\tau} \frac{v^2|F_{\alpha 1}|^2}{M_1^2} . \tag{9}$$

For example, choosing M_1 in keV region and $\theta^2 \sim 10^{-8}$ leads to a life-time exceeding the age of the Universe by 10 orders of magnitude.

For the lightest singlet fermion N_e to be a legitimate DM candidate, the following additional conditions should be satisfied.
(i) *Cosmological production.* N_e are created in the early Universe in reactions $l\bar{l} \to \nu N_e$, $q\bar{q} \to \nu N_e$, etc (see Fig. 2). We should get the correct DM abundance.
(ii) *X-rays.* N_e decays radiatively, $N_e \to \gamma\nu$, producing a narrow line which can be detected (see Fig. 2). This line can be searched for with the use of X-ray satellites such as Chandra or XMM-Newton. It has not been seen yet. This condition happens to be stronger than requirement of the stability of DM sterile neutrino on cosmological time scale.
(iii) *Structure formation.* If N_e is too light it may have considerable free streaming length and erase fluctuations on small scales. This can be checked by the study of Lyman-α forest spectra of distant quasars and by other means. In particular, the analysis of the smallest DM-dominated objects (dwarf spheroidal galaxies of the Milky Way, dSphs) gives the most conservative constraint $M_{\text{DM}} \geq 400$ eV [20, 21].

Let us discuss cosmological production of sterile neutrinos. In the region of the parameter space admitted by X-ray observations (point (ii) above) sterile neutrinos were never in thermal equilibrium in the early Universe due to νMSM interactions. Our assumption of the absence of new physics all the way from the Fermi to Planck scale fixes their concentration after inflation to zero [22]. Then the DM abundance can be computed as a function of the parameters of the νMSM.

Qualitatively, the rate of DM *sterile* neutrino production Γ_N at temperatures below the electroweak scale in charge symmetric plasma can be written as [23]

$$\Gamma_N \sim \Gamma_\nu \theta_M(T)^2 \,, \tag{10}$$

where $\Gamma_\nu \sim G_F^2 T^5$ is the rate of *active* neutrino production (G_F is the Fermi constant), and $\theta(T)$ is the temperature dependent mixing angle

$$\theta_1 \to \theta_M \simeq \frac{\theta_1}{1 + 2.4(T/200\ \mathrm{MeV})^6(\mathrm{keV}/\mathrm{M}_1)^2} \,. \tag{11}$$

The rate Γ_N peaks roughly at [16]

$$T_{peak} \sim 130 \left(\frac{M_1}{1\ \mathrm{keV}}\right)^{1/3} \mathrm{MeV} \,, \tag{12}$$

corresponding to the temperature of the QCD cross-over for keV scale sterile neutrinos and is strongly suppressed at $T > 100$ MeV, $\Gamma_N \propto T^{-7}$. At temperatures above the electroweak scale the rate is determined mainly by the decays and inverse decays of the Higgs boson to sterile neutrino and a lepton and is of the order

$$\Gamma_N \sim \frac{\theta_1^2 M_1^2 T}{v^2} \,. \tag{13}$$

Since the maximal production occurs near the temperature of the QCD cross-over, where the quark-gluon plasma is strongly coupled and the dilute quark or hadron gas pictures are not valid, an exact computation of the number of produced sterile neutrinos happens is a difficult task. At the same time, the fact that DM sterile neutrinos were never in thermal equilibrium allows to express their abundance through certain equilibrium finite temperature Green's functions, to be computed in the Standard Model *without inclusion of right-handed fermions* [24, 25, 26].

Let us discuss the results of the computations. The creation of DM sterile neutrinos depends crucially on lepton asymmetries

$$\Delta_L^\alpha = \frac{(n_L^\alpha - \bar{n}_L^\alpha)}{(n_L^\alpha + \bar{n}_L^\alpha)} \,, \quad \Delta_L = \sum \Delta_L^\alpha = 244 \frac{n_{\nu e}}{s(T)} \,, \tag{14}$$

which may exist during the QCD epoch (here $n_{\nu e}$ is the asymmetry in electron neutrino flavour). It has been shown in [27] that a substantial lepton

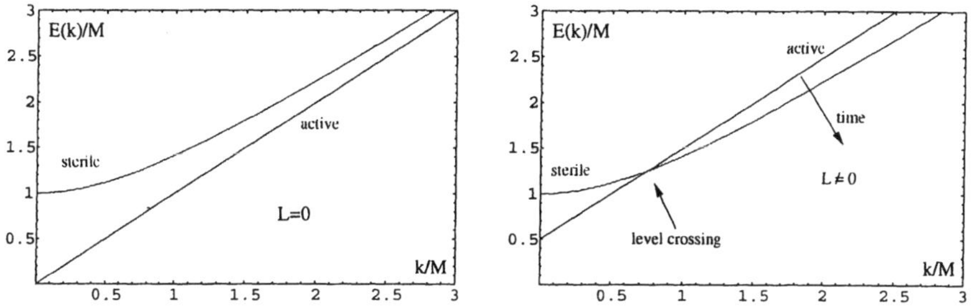

Figure 3: Non-resonant transitions $\nu \longrightarrow N_1$, $\Delta_L = 0$ (left). Resonant transitions $\nu \longrightarrow N_1$, $\Delta_L \neq 0$ (right)

asymmetry can be created in reactions with heavier singlet fermions of the νMSM *below the temperature of the sphaleron freeze-out.* It can be as large as $\Delta_L \simeq 0.2$. Note that the constraints on lepton asymmetry coming from Big Bang Nucleosynthesis are weaker than that [28].

The parameter Δ_L plays an important role in the theory of sterile dark matter production. In Fig. 3 we show the dispersion relations (dependence of energy of active and sterile neutrino excitations in plasma as a function of momenta) for zero (left) and non-zero (right) leptonic asymmetries. For the case of non-zero asymmetries the dispersion relations for active and sterile neutrinos cross each other at some point, leading to resonance production of sterile neutrinos [17, 26]. As a result, the number of created sterile neutrinos and their spectrum depend essentially on three parameters: their mass, the mixing angle θ_1 and the lepton asymmetry. Many plots which show different quantities can be found in [26].

If non-zero leptonic asymmetry is present, it is transferred to the population of dark matter neutrinos in a resonant way, see Fig. 4. In this case both baryon asymmetry of the Universe and dark matter are produced by essentially one and the same mechanism, providing an explanation of similar abundances of dark and baryonic matter in the Universe. The right panel in Fig. 4 gives the values of parameters necessary to reproduce the observed abundance of dark matter.

The combined constrains are shown in Fig. 5. The allowed region of parameters for DM sterile neutrinos produced via mixing with the active ones corresponds to the unshaded region. Two thick black lines bounding this region are production curves for non-resonant production (upper line, $L_6 = 0$, where L_6 is the lepton asymmetry defined by $L_6 \equiv 10^6(n_{\nu_e} - n_{\bar{\nu}_e})/s$,

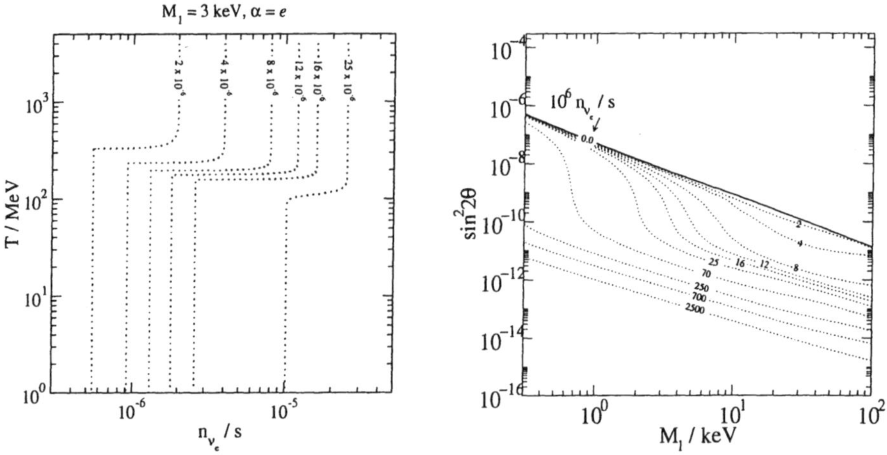

Figure 4: Left: evolution of lepton asymmetry as a function of temperature. The depletion of Δ_L is due to the resonant transfer of leptonic excess to dark matter sterile neutrinos. Right: A combination of the sterile neutrino mass, mixing angle and of lepton asymmetry, parametrised by $\frac{n_{\nu e}}{s(T)}$, which leads to correct dark matter abundance.

s is the entropy density) and for resonant production (RP) with the maximal lepton asymmetry, attainable in the νMSM (lower line, $L_6^{max} = 700$). The thin coloured curves between these lines represent (from top to bottom) production curves for $L_6 = 8, 12, 16, 25, 70$. The red shaded region in the upper right corner represents X-ray constraints (rescaled by a factor of 2 to account for possible systematic uncertainties in the determination of DM content). The black dashed-dotted line shows approximately the RP models with minimal $\langle q \rangle$ for each mass, i.e., the family of models with the largest cold component. The black filled circles along this line are compatible with the Lyman-α bounds, while those with $M_1 \leq 4$ keV are also compatible with X-ray bounds. Region below 1 keV is ruled out from the phase-space density arguments [20, 21].

The fact that the constraints exist from all sides makes the model testable with the use of X-ray observations. Unfortunately, the new data from *Chandra* and *XMM-Newton* can hardly improve the constraints by more than a factor 10 because these instruments have the energy resolution exceeding greatly the expected width of the DM line. To go much further one would need an improvement of spectral resolution up to the natural line width

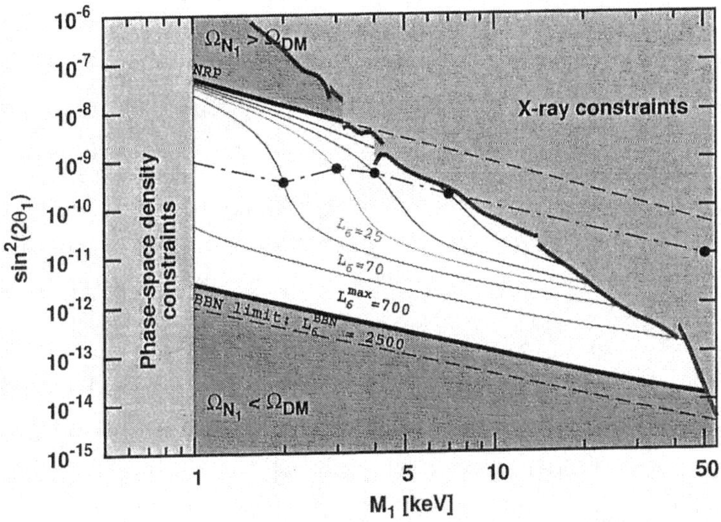

Figure 5: Constraints on the mixing angle of DM sterile neutrinos.

($\Delta E/E \sim 10^{-3}$), have a reasonably wide field of view $\sim 1°$ (size of a dSph) and perform a wide energy scan, from $\mathcal{O}(100)$ eV to $\mathcal{O}(100)$ keV.

5 Baryon asymmetry

Our Universe is baryon asymmetric – it does not contain antimatter in amounts comparable with matter. The baryon-to-entropy ratio gives a measure of baryon asymmetry,

$$\eta_B \equiv \frac{n_B}{s} = (8.8 \pm 0.2) \times 10^{-11} , \tag{15}$$

where n_B (s) is the baryon number (entropy) density respectively. To produce baryon asymmetry of the Universe three conditions should be satisfied [29]: *(a)* Conservation of the baryon number should be violated, *(b)* CP symmetry should be broken, and *(c)* the corresponding processes should be out of thermal equilibrium. Although the SM fulfils all three requirements [12], it cannot produce BAU of the required magnitude because the SM model interactions are strong enough to keep the system in a state very close to thermal equilibrium. In principle, strong deviations from thermal equilibrium could appear as a result of the bubble nucleation during the first-order electroweak phase transition converting the symmetric phase of the EW theory to the Higgs phase. However, the computations have shown [30] that the phase transition is absent for experimentally allowed Higgs boson masses.

Let us see what happens in the νMSM. The baryon (B) and lepton (L) numbers are not conserved in the νMSM. The lepton number is violated by the Majorana neutrino masses, while $B + L$ is broken by the electroweak anomaly. As a result, the sphaleron processes with baryon number non-conservation are in thermal equilibrium for $T_{\rm sph} < T < 10^{12}$ GeV, where $T_{\rm sph} \sim 100$ GeV depending on the Higgs mass [31]. As for CP-breaking, the νMSM (with practically decoupled DM sterile neutrino) contains 3 CP-violating phases in the lepton sector and a Kobayashi-Maskawa phase in the quark sector. This makes two of the Sakharov conditions for baryogenesis satisfied. Similarly to the SM, this theory does not have an electroweak phase transition with allowed values for the Higgs mass, making impossible the electroweak baryogenesis, associated with the non-equilibrium bubble expansion. However, the νMSM contains extra degrees of freedom - sterile neutrinos - which may be out of thermal equilibrium exactly because their Yukawa couplings to ordinary fermions are very small. The latter fact is a key point for the baryogenesis in the νMSM, ensuring the validity of the third Sakharov condition.

Let us consider the thermal history of the singlet fermions. They equilibrate due to reactions of the type $Nt \leftrightarrow \nu t$, $H \leftrightarrow N\nu$ or $N \leftrightarrow H\nu$ with the rate of the order $F^2 f_t^2 T/(4\pi)$ (f_t is the top Yukawa coupling, t is t-quark and H is the Higgs boson) and enter in thermal equilibrium at[1]

$$T_{\rm eq} \simeq \frac{m_{\rm atm} M_0}{4\pi v^2} M_I \simeq 100 M_I \ , \tag{16}$$

where M_0 is a parameter entering the temperature-time relation, $t = \frac{M_0}{2T^2}$, $M_0 \simeq M_{Pl}/1.66\sqrt{g_{eff}}$, with g_{eff} being the number of effectively massless degrees of freedom.

If M_I is larger than the mass of intermediate vector boson M_W then the singlet fermions decay at temperatures

$$T_{\rm decay} \simeq \left(\frac{m_{\rm atm} M_0}{24\pi v^2}\right)^{\frac{1}{3}} M_I \simeq 3 M_I \ . \tag{17}$$

[1]Our discussion here will be based on order-of-magnitude estimates. For a refined treatment see a comprehensive review [13] for the case of heavy Majorana leptons and and [27] for the case $M_I < M_W$.

If, on the contrary, $M_I < M_W$, the Majorana fermions decay at temperatures below the electroweak scale.

The densities of Majorana fermions deviate from equilibrium concentrations for $T > T_{\rm eq}$ or for $T < T_{\rm decay}$, pointing out to two regions of temperatures where the baryogenesis can happen.

If $T_{eq} > T_{\rm sph}$, then baryon asymmetry, created at $T > T_{\rm eq}$ will be destroyed in sphaleron processes and in equilibrium reactions with singlet fermions. So, in this case baryogenesis can only happen at temperature $T_{\rm decay}$. Since no baryon asymmetry can be produced directly in decays of Majorana particles (but lepton asymmetry can appear), the decays must happen during the epoch when sphaleron processes are still active, i.e. it is required that $T_{\rm decay} > T_{\rm sph}$. This may only be satisfied if $M_I > M_W$, i.e. when the Majorana leptons are sufficiently heavy. In this range of the parameter-space of the model one has "thermal leptogenesis" [11]. It will not be discussed here.

Now, if $T_{eq} < T_{\rm sph}$, baryogenesis will take place at $T > T_{\rm sph}$, when Majorana leptons are still out of thermal equilibrium and sphaleron processes operate. This regime can only be realized with the Majorana leptons which are light enough, see eq. (16), $M_I \lesssim T_{\rm sph}/100 \sim \mathcal{O}(1)$ GeV. It leads to baryogenesis via singlet fermion oscillations [32, 33]. This is exactly the case of the νMSM: N_I only thermalize at temperatures below M_W, providing a source of thermal non-equilibrium for all $T > M_W$: concentrations of N_I do not coincide with equilibrium ones, $n_N \neq n_{eq}$.

The computation of baryon asymmetry in this case *does* depends on the initial conditions for Majorana fermions in the high-temperature phase, in the contrary to the case of thermal leptogenesis. However, as was found in [22] at the end of inflation one can safely put their concentration to zero, as the production of these particles at reheating is strongly suppressed by the small values of Yukawas. This removes the uncertainty.

The smallness of Yukawa couplings suppresses the CP-violating effects. They can only be amplified if N_μ and N_τ are nearly degenerate. Also, the smallness of the Majorana masses of the singlet fermions in comparison with the sphaleron freeze-out temperature implies that the total lepton number $L_{\rm tot}$, defined as the lepton number in active species plus helicity in singlet fermions, is effectively conserved. At quantitative level, baryogenesis occurs as follows. The singlet fermions are created in the early Universe in Yukawa interactions with the particles of the SM and oscillate in a coherent way with CP-breaking. As a result, the total lepton number is zero but gets unevenly distributed between active and sterile neutrinos. The lepton number of active left-handed neutrinos is transferred to baryons due to equilibrium sphaleron processes.

Let us estimate the temperature T_B of baryogenesis in this mechanism [32, 33]. The frequency of oscillations between N_2 and N_3 is given by

$$\omega \sim \frac{|M_2^2 - M_3^2|}{E_I} \sim \frac{M_2 \Delta M(T)}{T}, \tag{18}$$

where we took into account that the typical energy of a singlet fermion $E_I \sim T$ and that $\Delta M(T) \ll M_2 \approx M_3$. Here $\Delta M(T)$ is the mass difference between N_2 and N_3 (temperature dependent due to Higgs' vev T-dependence and radiative corrections at high temperatures). If the rate of oscillations exceeds the rate of the Universe expansion $H(T)$, the system is out of resonance, and no amplification of CP-violation happens. In the opposite case CP-violation has no time to develop. This allows to determine the temperature T_B at which baryogenesis happens: $\omega \sim H(T_B)$, leading to

$$T_B \sim \left(M_I \Delta M(T) M_0\right)^{1/3}. \tag{19}$$

The described mechanism turns out to be quite effective. The *maximal* baryon asymmetry $\Delta \equiv \frac{n_B - n_{\bar{B}}}{n_B + n_{\bar{B}}} \sim 1$ is produced if $T_B \sim T_{\text{sph}} \sim T_{\text{eq}}$ leading to $\Delta M(T_B) \sim 0.01\ \text{eV} \left(\frac{1\ \text{GeV}}{M_I}\right)$. It is important for these considerations that interactions with plasma do not destroy quantum-mechanical nature of oscillations of $N_{2,3}$. i.e. that $T_{\text{eq}} \lesssim T_B$.

The kinetics of sterile neutrino oscillations and of the transfers of leptonic number between active and sterile neutrino sectors has been worked out in [33]. The effects to be taken into account include oscillations, creation and destruction of sterile and active neutrinos, coherence in sterile neutrino sector and its lost due to interaction with the medium, dynamical asymmetries in active neutrinos and charged leptons. The corresponding equations are written in terms of the density matrix for sterile neutrinos and concentrations of active neutrinos and are rather lengthy. They can be found in the original work [33].

The value of baryon to entropy ratio $\frac{n_B}{s}$ for $T_B \gg T_{sph}$, $T_{\text{eq}} < T_{sph}$ is given by [33]

$$\frac{n_B}{s} \simeq 1.7 \cdot 10^{-10}\, \delta_{\text{CP}} \left(\frac{10^{-5}}{\Delta M(T)/M_2}\right)^{\frac{2}{3}} \left(\frac{M_3}{10\ \text{GeV}}\right)^{\frac{5}{3}},$$

where the CP-breaking factor $\delta_{\rm CP}$ is expressed through the different mixing angles and CP-violating phases, parametrizing the neutrino mixing matrix and couplings of N_I to active leptons. The CP violation parameter δ_{CP} can be $\mathcal{O}(1)$, given the present experimental data on neutrino oscillations. Interestingly, δ_{CP} is non-zero even in the limit when θ_{13} angle in active neutrino mixing matrix is zero. In other words, baryon asymmetry is generated even if there is no CP-violation in neutrino oscillations. The observable value of baryon asymmetry can be obtained in a wide range of the parameters of the νMSM.

To summarize, if $M_I < M_W$, the model (1) can explain the baryon asymmetry of the Universe, provided the pair of singlet fermions $N_{\mu,\tau}$ is sufficiently degenerate (they a forming almost a Dirac 4-component spinor in this case). The temperature of baryogenesis can be as small as that of the sphaleron freezing, $T_{\rm sph} \sim 100$ GeV. At the same time, this model can explain neutrino masses and mixings and provide the candidate for dark matter particle. The Higgs mass in the νMSM is stable against radiative corrections coming from new particles. The relatively small masses of the Majorana fermions are protected from large radiative corrections by a symmetry – lepton number conservation, which is exact in the limit $M_I \to 0$. The ratio between the masses on $N_{\mu,\tau}$ and N_e – the dark matter candidate in the νMSM can be of the order of 10^5 – a typical number for the maximal mass ratio in the quark sector.

6 Testing the baryogenesis directly

The baryogenesis in the νMSM falls into the category of *experimentally testable* scenarios. By experimentally testable, we mean those in which particle physics experiments could potentially determine *all* parameters of the theory, necessary for theoretical computation of the *amplitude* and of the *sign* of the baryon asymmetry of the Universe[2]. The reason is that the mass of the Majorana fermions is small, so that they can be produced at existing accelerators.

In Fig. 6, we present different constraints on singlet fermion mixing angle (defined as $\theta = M_D/M_2$ and showing the suppression of their interactions in comparison with the weak interactions of ordinary leptons) versus their

[2]Yet another testable scenario is electroweak baryogenesis, for a review see Ref. [34] and the recent work [35].

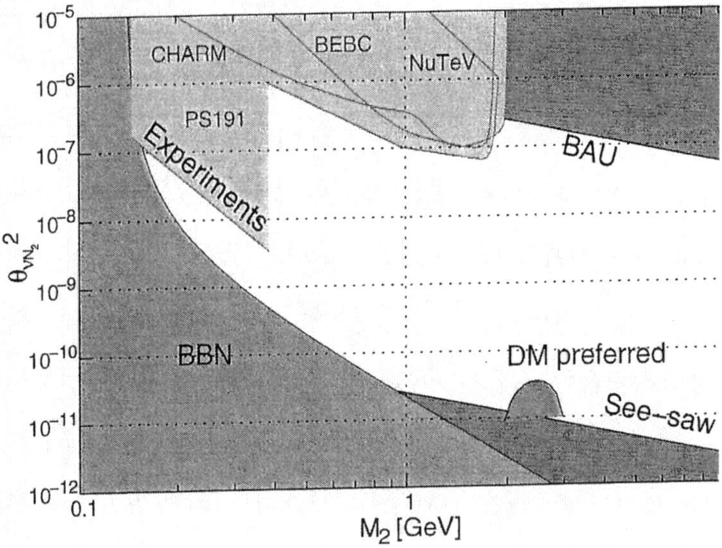

Figure 6: Constraints on the mixing angle of BAU generating singlet fermions.

mass, coming from different considerations. Above the line marked "BAU", baryogenesis is not possible: here, the coupling of $N_{2,3}$ to active neutrinos is so large that they come to thermal equilibrium above the EW temperature (this curve is derived by computation [27] rather than order-of-magnitude estimate presented in 5). Below the line marked "See-saw", the data on neutrino masses and mixings cannot be explained using eq. (4). The region noted as "BBN" is disfavoured by the considerations of Big Bang Nucleosynthesis - the decays of $N_{2,3}$ must not spoil the standard picture. A small region with the capture "DM preferred" in the domain of masses $2-3$ GeV is quite peculiar: here, the generation of BAU above the EW scale and the production of sterile neutrino DM well below T_{EW} are due to essentially the same mechanism, giving a hint as to why the DM abundance is similar to that of baryonic matter [27]. Finally, the region marked "Experiment" shows the part of the parameter space excluded by direct searches for singlet fermions (more detailed bounds for different decay modes of singlet fermions are summarized in Ref. [36], where BBN constraints are analysed as well). The analysis of the published works of different collaborations reveals that for the mass of the neutral lepton $M > 450$ MeV, the experiments BEBC [37],

CHARM [38], and NuTeV [39] have entered into the interesting for νMSM region below the line "BAU" (see Fig. 5), whereas the NOMAD [40] and L3 LEP experiments [41] give weaker constraints. The constraints in the small mass region, $M < 450$ MeV, are coming from the CERN PS191 experiment [42, 43].

Naturally, several distinct strategies can be used for the experimental search of these particles. The first one is related to their production (θ^2 effect). The singlet fermions participate in all the reactions the ordinary neutrinos do with a probability suppressed roughly by a factor θ_2^2. Since they are massive, the kinematics of, say, two body decays $K^\pm \to \mu^\pm N$, $K^\pm \to e^\pm N$ or three-body decays $K_{L,S} \to \pi^\pm + e^\mp + N_{2,3}$ changes when $N_{2,3}$ is replaced by an ordinary neutrino. Therefore, the study of *kinematics* of rare K, D, and B meson decays can constrain the strength of the coupling of heavy leptons. This strategy has been used in a number of experiments for the search of neutral leptons in the past [44, 45], where the spectrum of electrons or muons originating in decays π and K mesons has been studied. The precise study of kinematics of rare meson decays is possible in Φ (like KLOE), charm, and B factories, or in experiments with kaons where their initial 4-momentum is well known (like NA48 or E787 experiments).

The second strategy is to use the proton beam dump (θ^4 effect). As a first step, the proton beam heating the fixed target creates K, D or B mesons, which decay and produce $N_{2,3}$. The second step is a search for decays of N in a near detector, looking for the processes "nothing" $\to$ leptons and hadrons [42, 43, 39, 40]. To this end, quite a number of already existing or planned neutrino facilities (related, e.g., to CERN SPS, MiniBooNE, MINOS or J-PARC), complemented by a near *dedicated* detector[3], can be used. Finally, these two strategies can be unified, so that the production and the decay occurs inside the same detector [41].

For the mass interval $M_N < M_K$, both strategies can be used. Moreover, further constraints on the couplings of singlet fermions can be derived from the reanalysis of the *already existing but never considered from this point of view* experimental data of KLOE collaboration and of the E787 experiment. In addition, the NA62 experiment at CERN would allow the finding or exclusion of singlet fermions with the mass below that of the kaon.

If $m_K < M_{2,3} < m_D$, the search for the missing energy signal, potentially

[3]To minimize the background from normal neutrino interactions, the density of the detector must be much smaller than what is used in "standard" neutrino programs.

possible at beauty, charm, and τ factories, is unlikely to gain the necessary statistics and is very difficult if not impossible at hadronic machines like LHC. Thus, the search for decays of neutral fermions is the most effective opportunity. The dedicated experiments on the basis of the proton beam NuMI or NuTeV at FNAL, SPS at CERN, or J-PARC can touch a very interesting parameter range for $M_N < 1.8$ GeV. Experiments like NuSOnG [46] and HiResMν [47] should be able to enter in a cosmologically interesting region for masses and mixing angles of singlet fermions.

Going above D-meson but still below B-meson thresholds is very hard if not impossible with the present or planned proton machines or B-factories. To enter into a cosmologically interesting parameter space would require the increase in the present intensity of, say, CERN SPS beam by two orders of magnitude or to produce and study the kinematics of more than 10^{10} B-mesons.

7 Inflation

The fact that our Universe is almost flat, homogeneous and isotropic is often considered as a strong indication that the Standard Model (SM) of elementary particles is not complete. Indeed, these puzzles, together with the problem of generation of (almost) scale invariant spectrum of perturbations, necessary for structure formation, are most elegantly solved by inflation [48, 49, 50, 51, 52, 53]. The majority of present models of inflation require an introduction of an additional scalar—the "inflaton". This hypothetical particle may appear in a natural or not so natural way in different extensions of the SM, involving Grand Unified Theories (GUTs), supersymmetry, string theory, extra dimensions, etc. A number of inflaton properties are constrained by the observations of fluctuations of the Cosmic Microwave Background (CMB) and the matter distribution in the Universe. Though the mass and the interaction of the inflaton with matter fields are not fixed, the well known considerations prefer a heavy scalar field with a mass $\sim 10^{13}$ GeV and extremely small self-interacting quartic coupling constant $\lambda \sim 10^{-13}$ [54]. This value of the mass is close to the GUT scale, which is often considered as an argument in favour of existence of new physics between the electroweak and Planck scales.

According to the current theory of inflation (for a recent review see [55]) the early evolution of the Universe can be roughly divided into three parts.

During the first, inflationary stage, the Universe expands exponentially and becomes nearly flat. At this stage matter perturbations, leading to structure formation, are generated. During the second, reheating stage, the energy stored in the inflaton field is transferred to the fields of the Standard Model. The third stage is the radiation dominated Universe in nearly thermal equilibrium for most of the SM particles. The starting moment of this stage t_r corresponds to a maximal temperature of the Universe $T_{\rm max}$, and this is the onset of the standard hot Big Bang.

In [56] it was proposed that the Higgs boson of the SM can play the role of the inflaton and make the Universe flat, homogeneous and isotropic, produce the primordial fluctuations, necessary for structure formation, and heat up the Universe making the Big Bang. In other words, no new special particle is needed for inflation.

To describe the main idea of SM Higgs-inflation, let us consider Lagrangian of the SM *non-minimally* coupled to gravity,

$$L_{\rm tot} = L_{\rm SM} - \frac{M^2}{2}R - \xi H^\dagger H R\,, \tag{20}$$

where $L_{\rm SM}$ is the SM part, M is some mass parameter, R is the scalar curvature, H is the Higgs field, and ξ is an extra constant, characterising the strength of coupling of the Higgs field to gravity. The third term in (20) is in fact required by the renormalization properties of the scalar field in a curved space-time background [57]. If $\xi = 0$, the coupling of the Higgs field to gravity is said to be "minimal". Then M can be identified with Planck scale M_P related to the Newton's constant as $M_P = (8\pi G_N)^{-1/2} = 2.4\times10^{18}$ GeV. The parameter ξ cannot be fixed within the theory (20), it will be determined from the requirement of successful inflation.

For large Higgs backgrounds $\xi h^2 \gtrsim M_P^2$ (here $h^2 = 2H^\dagger H$) the masses of all the SM particles *and* the induced Planck mass $[M_P^{\rm eff}]^2 = M_P^2 + \xi h^2$ are proportional to one and the same parameter, leading to independence of physical effects on the magnitude of h. In other words, the Higgs potential in the large-field region is effectively flat and can result in successful inflation. This is not the case for the theory with minimal coupling, when $\xi = 0$.

The flatness of the physical scalar potential is better seen in the Einstein frame, in which the gravity is described by the standard Einstein-Hilbert action. Note that eq. (20) is written in the so-called Jordan frame, where the gravitational constant is field-dependent. The conformal transformation from the Jordan frame to the Einstein frame is given by

$$\hat{g}_{\mu\nu} = \Omega^2 g_{\mu\nu}\,, \quad \Omega^2 = 1 + \frac{\xi h^2}{M_P^2}\,. \tag{21}$$

This transformation leads to a non-minimal kinetic term for the Higgs field. It is convenient to make the change to the new scalar field χ with

$$\frac{d\chi}{dh} = \sqrt{\frac{\Omega^2 + 6\xi^2 h^2/M_P^2}{\Omega^4}}\,. \tag{22}$$

Finally, the action in the Einstein frame is

$$S_E = \int d^4x\sqrt{-\hat{g}}\left\{ -\frac{M_P^2}{2}\hat{R} + \frac{\mathrm{DM}\chi\partial^\mu\chi}{2} - U(\chi)\right\}, \tag{23}$$

where $\hat{R}$ is calculated using the metric $\hat{g}_{\mu\nu}$ and the potential is

$$U(\chi) = \frac{1}{\Omega(\chi)^4}\frac{\lambda}{4}\left(h(\chi)^2 - v^2\right)^2\,. \tag{24}$$

For small field values $h \simeq \chi$ and $\Omega^2 \simeq 1$, so the potential for the field χ is the same as that for the initial Higgs field. However, for large values of $h \gg M_P/\sqrt{\xi}$ (or $\chi \gg \sqrt{6}M_P$) the situation changes. In this limit $h \simeq M_P/\sqrt{\xi}\exp(\chi/(\sqrt{6}M_P))$. This means that the potential for the Higgs field is exponentially flat and has the form

$$U(\chi) = \frac{\lambda M_P^4}{4\xi^2}\left(1 - e^{-\frac{2\chi}{\sqrt{6}M_P}}\right)^2\,. \tag{25}$$

The full effective potential in the Einstein frame is presented in Fig. 7. It is the flatness of the potential at $\chi \gg M_P$ which makes the successful (chaotic) inflation possible.

If the inflaton potential is known, one can make predictions and compare them with observations. The basic inflationary parameters, which can be extracted from the analysis of anisotropies of cosmic microwave background are:

(i) The amplitude of the temperature fluctuation $\delta T/T$ at the WMAP normalization scale ~ 500 Mpc.

(ii) The value of spectral index n_s of scalar density perturbations

$$\left\langle \frac{\delta T(x)}{T}\frac{\delta T(y)}{T}\right\rangle \propto \int \frac{d^3k}{k^3} e^{ik(x-y)} k^{n_s-1}\,. \tag{26}$$

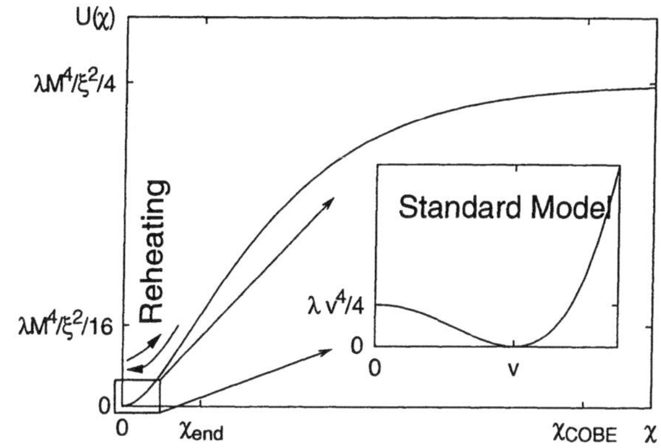

Figure 7: Effective potential in the Einstein frame.

(iii) The amplitude of tensor perturbations $r = \frac{\delta\rho_t}{\delta\rho_s}$.
Since in the Higgs inflation we have got one new parameter, we can fix ξ from (i) and make predictions of n_s and r.

The analysis can be performed in standard way using the slow-roll approximation (for a review see [58]). The condition (i) leads to the relation between the Higgs mass and the parameter ξ,

$$\xi \simeq 47000\sqrt{\lambda}\,. \qquad (27)$$

Since the Higgs self-coupling constant is of the order of one, ξ must be large enough. As anticipated, the Higgs-inflation predicts the specific values for spectral indexes describing scalar (n_s) and tensor (r) perturbations. They are in accordance with the WMAP-5 observations, see Fig. 8 (left panel).

The consideration carried out above tells nothing about the Higgs mass: change λ and ξ^2 in such a way that the ratio λ/ξ^2 stays constant - cosmological predictions do not change (see eq. (27)). This is not true any longer if quantum effects are taken into account. In particular, the Higgs self-coupling constant λ is not a constant as it depends on energy through renormalisation group equations. Since the typical inflationary energy scale is $M_P/\sqrt{\xi}$, for Higgs inflation to work, the SM must be a valid quantum field theory up to the inflation scale. A somewhat stronger requirement is the validity of the SM all the way up to the Planck scale, which leads to constraints on the Higgs mass:

$$m_{\min} < M_H < m_{\max}\,, \qquad (28)$$

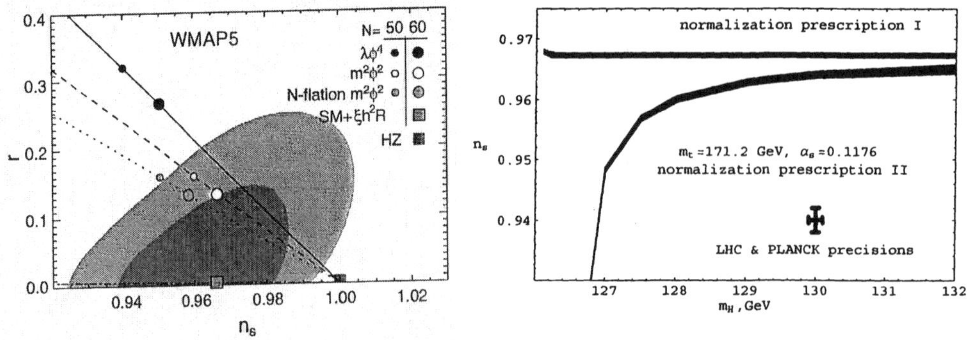

Figure 8: **Left.** The allowed WMAP region for inflationary parameters (r, n_s). The green box is our prediction for Higgs inflation. Black and white dots are predictions of usual chaotic inflation with $\lambda\phi^4$ and $m^2\phi^2$ potentials, HZ is the Harrison-Zeldovich spectrum. **Right.** Dependence of the spectral index of scalar perturbations on the Higgs mass in two different renormalisation prescriptions, related to the computations in the Jordan and Einstein frames. The cross indicates the accuracy to be achieved in the measurements of the Higgs mass at the LHC and of the spectral index n_s with the Planck satellite.

where

$$m_{\min} = [126.3 + \frac{m_t - 171.2}{2.1} \times 4.1 - \frac{\alpha_s - 0.1176}{0.002} \times 1.5]\ \text{GeV} \tag{29}$$

$$m_{\max} = [173.5 + \frac{m_t - 171.2}{2.1} \times 1.1 - \frac{\alpha_s - 0.1176}{0.002} \times 0.3]\ \text{GeV} \tag{30}$$

Here α_s is the strong coupling at the Z-mass, and theoretical uncertainty in $m_{\min}$ is equal to ± 2.2 GeV. These numbers are taken from the recent two-loop analysis of. [59] (see also [60, 61] and earlier computations in [62, 63, 64, 65]). The value of $m_{\max}$ corresponds to the (somewhat arbitrary) criterion $\lambda(M_P) < 6$, but changes only very little for arbitrarily large $\lambda(M_P)$. If $M_H > m_{\max}$ the quantum field theory is inconsistent for energies $E < M_P$ due to the presence of Landau pole in λ. If $M_H < m_{\min}$ there exists a lower additional ground state at $h < M_P$ and the electroweak vacuum is unstable.

This window for the Higgs mass can be refined accounting for radiative corrections in the inflationary region. They were computed in a number of papers [66, 67, 68, 59, 69], with different results. In Fig. 8 (right panel) we present the dependence of the spectral index n_s on the mass of the Higgs boson, found in [59]. It reveals the non-trivial relation between the Higgs mass and the properties of cosmological perturbations. The Higgs inflation is possible in a somewhat wider region of the Higgs masses: $m_{\min}$ stays practically the same as in (29), and $m_{\max}$ goes up to 194 GeV, accounting for the fact that the inflationary scale is smaller that the Planck mass. The computation of inflationary indexes has an extra uncertainty, related to renormalisation procedure [67, 59], which is shown in Fig. 8. The inflationary range of Higgs masses lies within the region allowed by the electroweak precision tests and by the direct LEP and FNAL searches for the Higgs boson. The combination of the future Planck measurements of n_s and r with the coming LHC data on the Higgs boson would allow to test the predictions of the Higgs inflation.

Remarkably, the Higgs inflation automatically solves the problem of the graceful exit from inflation. Roughly, for the Higgs fields $h > \frac{M_P}{\sqrt{\xi}}$ the Universe is inflating, for $\frac{M_P}{\xi} < h < \frac{M_P}{\sqrt{\xi}}$ it is in the matter dominated phase (the role of matter is played by the oscillating Higgs field), and at $h < \frac{M_P}{\xi}$ it enters into the radiation dominated phase. At $h \simeq \frac{M_P}{\xi}$ the energy stored in the Higgs field is transferred rapidly to other fields of the SM, leading to the Big Bang. The detailed discussion of these processes can be found in [22] (see also [70].)

8 Conclusions

New physics, responsible for neutrino masses and mixings, for dark matter, and for baryon asymmetry of the Universe may hide itself *below* the EW scale. This possibility can be offered by the the νMSM - a minimal model, explaining simultaneously *all well-established observational* drawbacks of the SM. Yet other problems of the SM, related to the cosmological constant puzzle, to the existence of Dark Energy (late Universe acceleration), and to the problem of stability of the Higgs mass against radiative corrections may be related to quantum scale invariance, and not to the existence of any intermediate energy scales between the Fermi and Planck scales. Due to the lack of time and space these problems were not discussed in these lectures and I refer to the original papers [71, 72] at this point.

This new physics (a pair of new neutral leptons, creating the baryon asymmetry of the Universe) can be searched for in dedicated experiments with the use of existing intensive proton beams at CERN, FNAL and neutrino facilities in Japan (J-PARC). An indirect evidence in favour of this proposal will be given by LHC, if it discovers the Higgs boson within the mass interval discussed above and nothing else. Moreover, the νMSM gives a hint on how and where to search for new physics in this case. It tells, in particular, that in order to uncover new phenomena in particle physics one should go towards high intensity proton beams or very high intensity charm or B-factories, rather than towards high energy electron-positron accelerators.

To search for DM sterile neutrino in the Universe one needs an X-ray spectrometer in Space with good energy resolution $\delta E/E \sim 10^{-3} - 10^{-4}$ getting signals from our Galaxy and its dwarf satellites. The laboratory search for this particle would require an extremely challenging detailed analysis of kinematics of β-decays of different isotopes [73].

To conclude, we give here a short summary of predictions of the νMSM.

The first prediction is the absolute values of masses of active neutrinos. One of the active neutrinos must be very light, $m_1 \lesssim \mathcal{O}(10^{-6})$ eV. This fixes the masses of two other active neutrinos: $m_2 \simeq 9 \cdot 10^{-3}$ eV, $m_3 \simeq 5 \cdot 10^{-2}$ eV for normal hierarchy or $m_{2,3} \simeq 5 \cdot 10^{-2}$ eV for the inverted hierarchy. As a result, an effective Majorana mass for neutrinoless double beta decay can be determined [74]. For normal (inverted) hierarchy the constraints read: 1.3 meV $< m_{\beta\beta} <$ 3.4 meV (13 meV $< m_{\beta\beta} <$ 50 meV). A very conservative bound on the mass of DM sterile neutrino comes from analysis of rotational curves of dwarf galaxies and reads $M_1 > 0.4$ keV (it is weaker than the one coming from Lyman-α observations, see [75, 76, 77]). Direct experimental searches and BBN require $M_{2,3} > 140$ MeV, whereas baryogenesis due to sterile neutrino oscillations is possible if $\Delta M = |M_2 - M_3| < 800\ m_{atm}\,(M/\text{GeV})^2$.

With quite a weak assumption about the initial conditions for the Big Bang (no sterile neutrinos at the beginning (this assumption is realized in the νMSM where the Higgs field plays the role of the inflaton) the predictions and constraints can be strengthened further. Namely the DM sterile neutrino mass should be in the interval 1 keV $<$ M$_1$ $<$ 50 keV (the lowest bound is related to phase density constraints), the DM sterile neutrino mixing angle is predicted to be in the region $2 \times 10^{-15} < \theta_1^2 < 2 \times 10^{-10}$. To produce the DM and BAU in correct amounts, the mass of heavier neutral leptons should be in the region $M_2 \sim 2$ GeV, their level of degeneracy is constrained as $\Delta M \lesssim 10^{-4} m_{atm}$, and their mixing angle should be $\theta_2^2 \simeq 10^{-11}$. The CP asymmetry in $N_{2,3}$ decays should be on the level of 1% .

Higgs inflation is only possible in a specific interval of the Higgs boson masses, discussed above, with theoretical uncertainty of ± 2 GeV. Moreover, the inflationary spectral indices have definite values in the Higgs inflation, what can be tested by the Planck satellite.

Acknowledgements. This work was supported in part by the Swiss National Science Foundation.

References

[1] M. Shaposhnikov, (2007), 0708.3550.

[2] A. Boyarsky, O. Ruchayskiy, and M. Shaposhnikov, Ann. Rev. Nucl. Part. Sci. **59**, 191 (2009), 0901.0011.

[3] R. Foot, G. C. Joshi, H. Lew, and R. R. Volkas, Mod. Phys. Lett. **A5**, 2721 (1990).

[4] P. Minkowski, Phys. Lett. **B67**, 421 (1977).

[5] T. Yanagida, Prog. Theor. Phys. **64**, 1103 (1980).

[6] P. Ramond, (1979), hep-ph/9809459.

[7] R. N. Mohapatra and G. Senjanovic, Phys. Rev. Lett. **44**, 912 (1980).

[8] F. Vissani, Phys. Rev. **D57**, 7027 (1998), hep-ph/9709409.

[9] G. 't Hooft, NATO Adv. Study Inst. Ser. B Phys. **59**, 135 (1980).

[10] A. Strumia and F. Vissani, (2006), hep-ph/0606054.

[11] M. Fukugita and T. Yanagida, Phys. Lett. **B174**, 45 (1986).

[12] V. A. Kuzmin, V. A. Rubakov, and M. E. Shaposhnikov, Phys. Lett. **B155**, 36 (1985).

[13] S. Davidson, E. Nardi, and Y. Nir, Phys. Rept. **466**, 105 (2008), 0802.2962.

[14] A. de Gouvea, Phys. Rev. **D72**, 033005 (2005), hep-ph/0501039.

[15] A. de Gouvea, J. Jenkins, and N. Vasudevan, Phys. Rev. **D75**, 013003 (2007), hep-ph/0608147.

[16] S. Dodelson and L. M. Widrow, Phys. Rev. Lett. **72**, 17 (1994), hep-ph/9303287.

[17] X.-D. Shi and G. M. Fuller, Phys. Rev. Lett. **82**, 2832 (1999), astro-ph/9810076.

[18] A. D. Dolgov and S. H. Hansen, Astropart. Phys. **16**, 339 (2002), hep-ph/0009083.

[19] K. Abazajian, G. M. Fuller, and M. Patel, Phys. Rev. **D64**, 023501 (2001), astro-ph/0101524.

[20] A. Boyarsky, O. Ruchayskiy, and D. Iakubovskyi, (2008), 0808.3902.

[21] D. Gorbunov, A. Khmelnitsky, and V. Rubakov, JCAP **0810**, 041 (2008), 0808.3910.

[22] F. Bezrukov, D. Gorbunov, and M. Shaposhnikov, JCAP **0906**, 029 (2009), 0812.3622.

[23] R. Barbieri and A. Dolgov, Phys. Lett. **B237**, 440 (1990).

[24] T. Asaka, M. Laine, and M. Shaposhnikov, JHEP **06**, 053 (2006), hep-ph/0605209.

[25] T. Asaka, M. Laine, and M. Shaposhnikov, JHEP **0701**, 091 (2007), hep-ph/0612182.

[26] M. Laine and M. Shaposhnikov, JCAP **0806**, 031 (2008), 0804.4543.

[27] M. Shaposhnikov, JHEP **08**, 008 (2008), 0804.4542.

[28] V. Simha and G. Steigman, JCAP **0808**, 011 (2008), 0806.0179.

[29] A. D. Sakharov, Pisma Zh. Eksp. Teor. Fiz. **5**, 32 (1967).

[30] K. Kajantie, M. Laine, K. Rummukainen, and M. E. Shaposhnikov, Phys. Rev. Lett. **77**, 2887 (1996), hep-ph/9605288.

[31] Y. Burnier, M. Laine, and M. Shaposhnikov, JCAP **0602**, 007 (2006), hep-ph/0511246.

[32] E. K. Akhmedov, V. A. Rubakov, and A. Y. Smirnov, Phys. Rev. Lett. **81**, 1359 (1998), hep-ph/9803255.

[33] T. Asaka and M. Shaposhnikov, Phys. Lett. **B620**, 17 (2005), hep-ph/0505013.

[34] M. Trodden, Rev. Mod. Phys. **71**, 1463 (1999), hep-ph/9803479.

[35] M. Carena, G. Nardini, M. Quiros, and C. E. M. Wagner, Nucl. Phys. **B812**, 243 (2009), 0809.3760.

[36] D. Gorbunov and M. Shaposhnikov, JHEP **10**, 015 (2007), 0705.1729.

[37] WA66, A. M. Cooper-Sarkar *et al.*, Phys. Lett. **B160**, 207 (1985).

[38] CHARM, F. Bergsma *et al.*, Phys. Lett. **B166**, 473 (1986).

[39] NuTeV, A. Vaitaitis *et al.*, Phys. Rev. Lett. **83**, 4943 (1999), hep-ex/9908011.

[40] NOMAD, P. Astier *et al.*, Phys. Lett. **B506**, 27 (2001), hep-ex/0101041.

[41] L3, P. Achard *et al.*, Phys. Lett. **B517**, 75 (2001), hep-ex/0107015.

[42] G. Bernardi *et al.*, Phys. Lett. **B166**, 479 (1986).

[43] G. Bernardi *et al.*, Phys. Lett. **B203**, 332 (1988).

[44] T. Yamazaki *et al.*, IN *LEIPZIG 1984, Proceedings, High Energy Physics, Vol. 1*, 262.

[45] M. Daum *et al.*, Phys. Rev. Lett. **85**, 1815 (2000), hep-ex/0008014.

[46] NuSOnG, T. Adams *et al.*, Int. J. Mod. Phys. **A24**, 671 (2009), 0803.0354.

[47] S. R. Mishra, R. Petti, and C. Rosenfeld, PoS **NUFACT08**, 069 (2008), 0812.4527.

[48] A. A. Starobinsky, JETP Lett. **30**, 682 (1979).

[49] A. A. Starobinsky, Phys. Lett. **B91**, 99 (1980).

[50] V. F. Mukhanov and G. V. Chibisov, JETP Lett. **33**, 532 (1981).

[51] A. H. Guth, Phys. Rev. **D23**, 347 (1981).

[52] A. D. Linde, Phys. Lett. **B108**, 389 (1982).

[53] A. Albrecht and P. J. Steinhardt, Phys. Rev. Lett. **48**, 1220 (1982).

[54] A. D. Linde, Phys. Lett. **B129**, 177 (1983).

[55] A. Linde, Lect. Notes Phys. **738**, 1 (2008), 0705.0164.

[56] F. L. Bezrukov and M. Shaposhnikov, Phys. Lett. **B659**, 703 (2008), 0710.3755.

[57] N. D. Birrell and P. C. W. Davies, Cambridge, Uk: Univ. Pr. (1982) 340p.

[58] D. H. Lyth and A. Riotto, Phys. Rept. **314**, 1 (1999), hep-ph/9807278.

[59] F. Bezrukov and M. Shaposhnikov, JHEP **07**, 089 (2009), 0904.1537.

[60] J. R. Espinosa, G. F. Giudice, and A. Riotto, JCAP **0805**, 002 (2008), 0710.2484.

[61] J. Ellis, J. R. Espinosa, G. F. Giudice, A. Hoecker, and A. Riotto, Phys. Lett. **B679**, 369 (2009), 0906.0954.

[62] G. Altarelli and G. Isidori, Phys. Lett. **B337**, 141 (1994).

[63] J. A. Casas, J. R. Espinosa, and M. Quiros, Phys. Lett. **B342**, 171 (1995), hep-ph/9409458.

[64] J. A. Casas, J. R. Espinosa, and M. Quiros, Phys. Lett. **B382**, 374 (1996), hep-ph/9603227.

[65] T. Hambye and K. Riesselmann, Phys. Rev. **D55**, 7255 (1997), hep-ph/9610272.

[66] A. O. Barvinsky, A. Y. Kamenshchik, and A. A. Starobinsky, JCAP **0811**, 021 (2008), 0809.2104.

[67] F. L. Bezrukov, A. Magnin, and M. Shaposhnikov, Phys. Lett. **B675**, 88 (2009), 0812.4950.

[68] A. De Simone, M. P. Hertzberg, and F. Wilczek, Phys. Lett. **B678**, 1 (2009), 0812.4946.

[69] A. O. Barvinsky, A. Y. Kamenshchik, C. Kiefer, A. A. Starobinsky, and C. Steinwachs, JCAP **0912**, 003 (2009), 0904.1698.

[70] J. Garcia-Bellido, D. G. Figueroa, and J. Rubio, Phys. Rev. **D79**, 063531 (2009), 0812.4624.

[71] M. Shaposhnikov and D. Zenhausern, Phys. Lett. **B671**, 187 (2009), 0809.3395.

[72] M. Shaposhnikov and D. Zenhausern, Phys. Lett. **B671**, 162 (2009), 0809.3406.

[73] F. Bezrukov and M. Shaposhnikov, Phys. Rev. **D75**, 053005 (2007), hep-ph/0611352.

[74] F. Bezrukov, Phys. Rev. **D72**, 071303 (2005), hep-ph/0505247.

[75] U. Seljak, A. Makarov, P. McDonald, and H. Trac, Phys. Rev. Lett. **97**, 191303 (2006), astro-ph/0602430.

[76] M. Viel, J. Lesgourgues, M. G. Haehnelt, S. Matarrese, and A. Riotto, Phys. Rev. Lett. **97**, 071301 (2006), astro-ph/0605706.

[77] A. Boyarsky, J. Lesgourgues, O. Ruchayskiy, and M. Viel, Phys. Rev. Lett. **102**, 201304 (2009), 0812.3256.

CHAIRMAN: M. SHAPOSHNIKOV

Scientific Secretaries: A. Agocs, A. Kurek

DISCUSSION I

- *Preghenella:*

Looking at the decay of the sterile neutrino into a photon and an active neutrino: N → $\gamma\nu$. It sounds to me like the process should occur via the electromagnetic interaction, but there is no electromagnetic charge, so how can this be possible?

- *Shaposhnikov:*

This process occurs through a loop correction, see Figure 1.

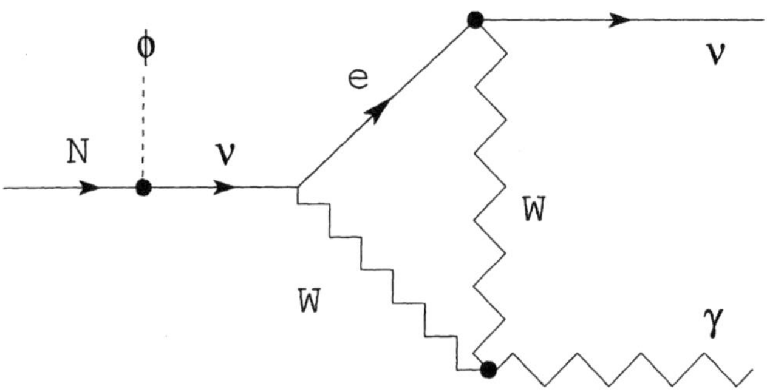

Figure 1: Nonzero Higgs vacuum expectation value induces the mixing between ordinary neutrino and sterile neutrino. Ordinary neutrino interacts with charged particles such as electron and W boson. The photon is attached then to W or to charged lepton lines. The probability of this decay is suppressed in comparison with decay into three neutrinos by the fine structure constant α.

- *Badziak:*

In your model you assume that Majorana masses are of the order of the Higgs mass. This implies extremely small Yukawa coupling to fit the correct mass of active neutrinos. Do you have any idea about the origin of such small Yukawa couplings?

- *Shaposhnikov:*

No, I even don't have any idea why electron is much lighter than the top quark. In the νMSM the smallness of Yukawa couplings plays the crucial role in explanation of dark matter and of the baryon asymmetry of the Universe.

- *Inguglia:*

You introduced 3 new light leptons (right-handed neutrinos). We have 18 new parameters, 15 of which are new Yukawa couplings, and other 3 are connected with 3 Majorana masses. The mechanism to give the mass to ordinary neutrinos is the see-saw, and the PMNS matrix, under see-saw, is not a unitary matrix any more. Are these 18 new parameters related to the loss of unitarity of PMNS?

- *Shaposhnikov:*

Indeed, the active neutrino mass matrix, which is given by the see-saw formula, can be diagonalized with the help of unitary matrix. There are corrections to the see-saw formula, which are suppressed by Yukawa couplings and by Majorana neutrino masses. As a result, the active neutrino mixing matrix is not unitary in general case. However, for the choice of νMSM parameters which leads to explanation of dark matter and baryon asymmetry of the universe these corrections are negligible. At this point there is no difference between electroweak see-saw and the traditional see-saw in which Majorana masses are of order of grand unified scale.

- *Inguglia:*

You said that Big Bang Nucleosynthesis is a evidence of the Dark Matter (DM), can you explain this relation?

- *Shaposhnikov:*

Big Bang Nucleosynthesis is an evidence in favour of non-baryonic DM. The logic is as follows. One can weight the amount of DM in the Universe by considering the rotational curves of different galaxies and from other cosmological observations. One finds that the abundance of DM is around $\Omega_{DM} \sim 22\%$ of the total energy density of the Universe. At the same time, the abundances of light elements (such as deuterium, helium and lithium), which are produced at nucleosynthesis, depend on the amount of the baryons in the Universe. The correct description is achieved if $\Omega_{baryons} \sim 5\%$. Given the fact that $\Omega_{baryons} < \Omega_{DM}$ one may conclude that DM is not baryonic.

- *Inguglia:*

One evidence for DM is the rotation curve of the galaxy. Law of gravitation was made by Newton by studying motions of comets, and simple objects as Earth. Also Keplerian distribution of velocity is based only on observations of objects inside the solar system. At that time none had an idea of what a galaxy was. Those laws were made only for the solar system. Today we know that there are lots of publications which do not include DM, for example the MOND theory. This one fits perfectly all the rotation curves of the (analysed) galaxies. So if today we assume that Quantum Mechanics is the theory which describes the smallest objects, and relativity is one which describes very fast objects, don't you think that gravity is the law which describes system of extension comparable with the solar system and that in order to describe the biggest structures as galaxies and clusters we have to improve the theory? Also owing to the fact that today, somehow, we have an idea of what a galaxy is. The problem is that I'm not a scientist, I'm a student and I do not understand how it is possible that there is a theory published about the same topic, which produces results which are completely different.

- *Shaposhnikov:*

For those who do not know: MOND is an abbreviation of "MOdified Newtonian Dynamics". This theory proposes that at very small accelerations $a < a_0$ the Newton's law $ma = F$ (where m is mass and F is force, and a_0 is of order of the present Hubble constant) must be replaced by another one, $ma^2/a_0 = F$. Indeed, this type of modification allows to describe quite a number of rotational curves of different galaxies, which is amazing. However, this theory does not fit a number of observations, such as the structure of dwarf galaxies or collision of clusters of galaxies, where DM hypothesis works well. Contrary to MOND (at least in it's present stage), the Einstein General Relativity (GR), which is a relativistic generalisation of the Newton gravity, allows to address in a consistent way the history of the Universe starting from inflation till present time, giving extra arguments in favour of GR and DM.

- *Marienfeld:*

How can one make an estimation of the entropy of the Universe?

- *Shaposhnikov:*

There are several contributions to the entropy of the Universe. An unknown contribution is associated with the black holes. Known contributions are related to photons, neutrinos and baryons. Entropy density in photons is given by $\frac{4\pi^2}{45}T^3$, where $T = 2.73$ K is the temperature of cosmic microwave background radiation. Another contribution is coming from neutrinos. In the early Universe, neutrinos decouple from plasma at the temperature of few MeV. This temperature is larger than the mass of electron. This tells that when electrons annihilate, the entropy sitting in electrons will go to photons rather than to neutrinos. As a result, neutrino temperature is smaller than that of photons. Accounting for this physics one can find that the entropy in all neutrinos and anti-neutrinos is roughly the same as that in photons. The entropy in baryons is much smaller.

- *Chkareuli:*

What about the mass scale of your Majorana neutrinos?

- *Shaposhnikov:*

The particle which is associated with DM has a mass from 3 keV to 50 KeV, and other two Majorana fermions are likely to have masses in GeV region, but even 140 MeV is not excluded.

- *Chkareuli:*

Do you have any restrictions from the flavour-changing processes like as $\mu e + \gamma e$ etc.

- *Shaposhnikov:*

No, absolutely nothing.

- *Chkareuli:*

I think there are no problems in the Standard Model, but these processes will definitely appear in the Super Symmetric Standard Model, where the scalar partners of your neutrinos naturally get SUSY scale masses.

- *Shaposhnikov:*

True. This is a very good argument in favour of the model as it stands, i.e. without supersymmetry.

- *'t Hooft:*

There are numerous "alternative" theories purporting to explain dark matter. I won't try to defend them all, but one charming idea, I thought, was that the electron-neutrino and muon-neutrino could have masses in the eV domain. In that case, neutrinos cooling after they were formed in the early universe could condense into low temperature gases representing a considerable amount of mass. Could you comment on this theory?

- *Shaposhnikov:*

If the masses of neutrinos (say of electron or muon type) are few eV, this can be consistent with all neutrino experiments. Then the neutrinos are required to be degenerate. However, if you don't change interactions of neutrinos with other particles, the hypothesis that they constitute all DM contradicts to the pattern of structures observed in the Universe. The structures in the Universe start to form roughly at matter radiation equality. If DM particles are relativistic at this time (and this is the case for active neutrinos – they give rise to the so-called hot DM), the primordial inhomogeneities will be smoothed at small scales due to neutrino free streaming. Though the large structures such as clusters of galaxies can be formed, this is not the case for galaxies and dwarf galaxies. One needs cold or warm DM to get the structure formation at smaller scales.

- *Badziak:*

Could you explain once more the link between the origin of dark matter and baryon asymmetry of the universe?

- *Shaposhnikov:*

In the νMSM the DM sterile neutrinos in the early Universe are produced through interactions with other particles. It happens to be crucial, whether the Universe at the moment of creation of dark matter is neutral with respect to lepton number or contains lepton asymmetry. If the universe contains net lepton number, the production rate of DM neutrinos is enhanced due to Mikheev-Smirnov-Wolfenstein resonance between active and sterile neutrinos (the same phenomenon in the Sun leads to efficient neutrino flavour transformations). Most of the lepton asymmetry is transformed to DM population.Now, due to CP-violation in neutrino sector and deviations from thermal equilibrium in the course of Universe expansion there is creation of lepton asymmetry. The lepton asymmetry, which is produced above the electroweak scale, at temperatures ~100 GeV, is converted to baryon asymmetry due to electroweak anomaly and sphalerons. This asymmetry is quite small, $n_B/s \sim 10^{-10}$ (here n_B and s are the baryon number and entropy densities respectively). However, the production of leptonic asymmetry continues after this point, and the created lepton asymmetry is transferred to DM as described above at temperatures of the order of 100 MeV. The correct relation between the amounts of baryon asymmetry and DM takes place in a particular corner of the parameter space of the νMSM.

- *Tannenbaum:*

Your last comment triggered a question on me. Everybody talks about it, I've heard hundreds of talks on baryon asymmetry of the universe. Well, there is obviously also a lepton asymmetry of the universe. How did that happen?

- *Shaposhnikov:*

That's what I just said. You create lepton asymmetry from the very beginning till pretty small temperature.

- *Tannebaum:*

No, I didn't mean in your model. I mean in all previous models before your model.

- *Shaposhnikov:*

If you assume that there are no low-energy sources of baryon and lepton number non-conservation then everything happens above the electroweak scale. In this case lepton asymmetry is related to baryon asymmetry. They are proportional to each other and are roughly of the same order of magnitude (i.e. small). The lepton asymmetry that small has no influence on the Universe evolution and cannot be detected. However, in the νMSM the process of creation of lepton asymmetry can occur at smaller temperatures. The lepton asymmetry thus can be much larger and can lead to observable consequences, such as DM production.

- *Inguglia:*

Can you please tell the order of magnitude of the age of the universe at the time of the lepton asymmetry?

- *Shaposhnikov:*

The age of the Universe at baryon asymmetry generation is roughly 10^{-10} seconds, and for lepton asymmetry is about 10^{-4} seconds. In terms of temperature: ~ 100 GeV and ~ 100 MeV.

- *Inguglia:*

Is it after the inflation?

- *Shaposhnikov:*

Yes, this all happens after the inflation, long time after the inflation. Temperature after inflation is likely to be of the order of 10^{13} GeV or so.

CHAIRMAN: M. SHAPOSHNIKOV

Scientific Secretaries: M. Ajaz, M.G. Chiu

DISCUSSION II

- *Kovetz:*

In the graph of the spectral index ns as a function of the Higgs mass, what is the dependency on the number of e-folds N? The margin of error depends on the number of e-folding s, right?

- *Shaposhnikov*:

The uncertainty in ns is associated mostly with the experimental accuracy in the measurement of the mass of the top-quark and strong gauge coupling constant. Also, the renormalization prescription at the Planck scale plays a role. The uncertainty in the number of e-foldings is less essential, since N depends only weakly (logarithmically) on the value of the reheating temperature, which can be estimated within a factor of 3 or 4 in the Higgs inflation. The reason is that when you describe inflation in the SM, you know quite a lot about the physics which is going on there. Numerically, the uncertainty in the spectral indexes due to uncertainty in the number of e-foldings is given by the thickness of the lines in the plots I have shown.

- *Kovetz:*

But how is the calculation done? Do you put in by hand the number of e-foldings?

- *Shaposhnikov:*

No, this is a result of computation. You should look at the evolution of the Higgs field, and compute the rate of energy transfer from the Higgs field oscillations to particles of the SM. From here you can find the time at which the radiation-dominated epoch starts and estimate the reheating temperature.

- *Burda:*

About the adding this extra term to the action, I am a little bit confused, this term changes the classical equation of motions for this scalar h, so why it isn't a problem? What is the correspondence between theories with or without this extra term?

- *Shaposhnikov:*

True, this term (non-minimal coupling of the Higgs field to gravity) does change the equations of motion. In particular, without this term there is no inflation in the Standard Model. So the physics with and without this interaction is very much different. The necessity of this type of coupling comes from the following argument. Suppose that you want to have a renormalizable field theory in fixed gravitational background. This type of action is not quite consistent because gravity is not quantized, but this is the best we can do at the moment. Now, if you start with the action without non-minimal coupling, it will be generated through quantum loops of matter fields. Moreover, quantum loops are divergent in the ultraviolet, telling that counter-terms with the same structure have to be introduced. Then, according to standard rules of quantum field theory, the term having this structure has to be added with a new coupling constant (ξ in this case). This constant cannot be computed within the theory under consideration. If ξ is small, then the Higgs field cannot play the role of the inflaton, but if it is large, then inflation takes place.

- *Inguglia:*

I would like to know which kind of effects were produced by inflation on early matter?

- *Shaposhnikov:*

If you talk about the effects which we can observe and test, then there are not that many (flatness, isotropy and homogeneity of the Universe, and almost flat spectrum of initial density perturbations). Most of the measurements are related to the properties of CMB, such as the value of different spectral indices (such as ns and r). If you talk about the physical picture of what is going on, there are plenty of effects, but many of them are washed away after the Big Bang. For instance, for chaotic inflation in the SM the Higgs field was misplaced from the origin. Therefore, the electroweak theory was in the Higgs phase, with huge values of masses for W and Z bosons.

- *Peters:*

How does the MWuncertainty go into the mmin and mmax calculation of the Higgs mass?

- *Shaposhnikov:*

The uncertainties in the W and Z mass have very small influence on mminand mmax, so they were not put in.

- *Peters:*

Are you sure, because the W mass uncertainty contribution is much larger than the top mass for the SM Higgs?

- *Shaposhnikov:*

Yes, I am sure. The W mass uncertainty cannot lead to any substantial change in mmin and mmax. Let me tell you how, for instance, mmin is computed. The constraint m > mmin comes from the requirement that the scalar self coupling must be positive for scales changing from the W mass up to the Planck energy. There is a known system of renormalization group equations which show the running of different couplings in the SM. There are four coupling constants which are essential here: scalar self-coupling λ, αs, weak gauge coupling g, and the U(1) gauge coupling g ′. This system of equations is supplemented by the initial conditions at small energy scale, which you get from experiment. To fix the initial conditions, one should use the procedure which relates the physical parameters (such as top quark mass, Higgs mass, Weinberg angle, etc.) to the coupling constants at this scale. When you solve the system of differential equations (i.e. "run" the couplings to high energy scale) you will find that the largest uncertainties are related to error bars in the top quark mass and in the αs. There are theoretical uncertainties (about ±2 GeV) in this computation. They can be reduced by a factor of 4 if the so-called pole MS matching is performed with 2-loop accuracy (the state of the art is 1-loop computation) and β functions for the SM are computed at 3 loops (only 2-loop results are available at present). There is an uncertainty coming from error bars in the masses of W and Z , but it is smaller than others and cannot exceed few tens of MeV.

- *Shcherbakov:*

Is the model νMSM super symmetric?

- *Shaposhnikov:*
Of course not.

- *Shcherbakov:*
What were the criteria for non-minimal generalization of general relativity to explain inflation?

- *Shaposhnikov:*
The criteria were the following. We wanted to have a theory which contains the same number of degrees of freedom as the SM and GR. Also, the field theory part of the action was required to be renormalizable. These requirements fix the action completely. For instance, if you add the R2 term, it contains higher derivatives, and thus there is one extra degree of freedom.

- *Shcherbakov:*
You considers only linear contributions in the curvature.

- *Shaposhnikov:*
Yes, we consider only the terms which are linear in curvature. If you have R2 term, you will get an extra scalar particle. I can add, by the way, that the gravity Lagrangian with R and R2 terms leads also to successful inflation, without adding any other scalar fields. This was found by A. Starobinsky. The values of inflationary indexes ns and r in this theory are close to those in Higgs inflation. However, one has an extra scalar particle which has a mass of the order of 1013 GeV or so.

- *Badziak:*
You computed the effective potential in the inflationary region in 2-loops. Are there any new qualitative effects as compared to tree level results?

- *Shaposhnikov:*
Yes, the qualitative difference is the appearance of the window for the Higgs mass. If you just consider the tree potential, and don't include any radiative corrections, you can get inflation for any value of the Higgs mass.

- *Badziak:*
If you could compute 3 loops, do you expect there could be some significant effects?

- *Shaposhnikov:*
I don't think so. I would expect that if we compute to three loops we wouldn't get any more qualitative differences, but we would get smaller uncertainties. Say, if we knew exactly the top massand alpha strong, then the uncertainties in the minimal value of the Higgs mass were ±2 GeV. If you add 2-loop effects to the so-called pole matching procedure and consider 3-loop running, I'd expect these uncertainties to be reduced to ±(0.4 – 0.5) GeV.

- *Cerchiai:*

Could you elaborate here on the spontaneous symmetry breaking of the scale invariance you mentioned at the end of your lecture, since you did not have enough time?

- *Shaposhnikov:*

Well, I have miserably miscalculated the timing of my second lecture, and, since I have only six minutes left now, I can say just a few words. If you have scale invariance in the theory, the only hope to describe the real world is to have spontaneous breaking of the scale invariance. Otherwise there are no scales (such as the proton mass or Newton's gravity constant) that we observe in Nature. The theory with exact (but spontaneously broken) quantum scale invariance is quite interesting. In particular, cosmological constant is zero, and quantum corrections to the Higgs mass are kept small due to the global dilatational symmetry. This is similar to the reason (gauge symmetry) why the mass of the photon is not modified by quantum corrections and is equal to zero in all orders of perturbation theory. One of the physical consequences of this scheme is the existence of a massless scalar particle dilaton, which is a Goldstone boson of the broken scale invariance. It has only derivative couplings to matter and is different from Brans-Dicke scalar, escaping thus numerous observational constraints. This particle can lead naturally to existence of Dark Energy, provided the gravity has a unimodular character.

- *Badziak:*

In a typical model of chaotic inflation there is large contribution of tensor perturbation. Do you have any understanding why in the use of non minimal coupling to gravity, tensor contribution is suppressed?

- *Shaposhnikov:*

The first part of your statement is not entirely correct. The chaotic inflation scenario doesn't fix the values of indices. The chaotic inflation is an assumption about initial condition for the inflaton field (Higgs in our case). If you are free to choose the inflaton potential, then you can get any values for the spectral indices, and in particular for tensor to scalar ratio.

- *Badziak:*

When I say chaotic inflation I mean quadratic or quintic potential.

- *Shaposhnikov:*

The steeper is the potential, the higher is the output of gravitational waves. The flatter is the potential, the smaller is the tensor to scalar ratio. In the Higgs-inflation the potential is flatter than a polynom in scalar field.

- *Badziak:*

In other models which are not Higgs models, but other models which predicts large tensor perturbations, this effect of non-minimal coupling to gravity it also can reduce the tensor perturbations?.

- *Shaposhnikov:*

Yes. For example, if you take $\lambda\varphi 4$ theory, and add non minimal coupling to gravity, even with very small coefficient ξ like 0.001, you will get the amplitude of tensor modes, consistent with CMB observations.

- *Nesterov:*

Unfortunately my question was totally similar to previous one. So, just to clarify: what new things do 2-loop calculations give compared to 1-loop calculations?

- *Shaposhnikov:*

The difference between 1 loop and 2 loop computation is a change of the admitted Higgs mass interval for the inflation. In the one-loop computation mmin is about 10 GeV larger than in 2-loop computation.

- *Nesterov:*

And what loop did you take into account? Only matter loops? What about graviton loops?

- *Shaposhnikov:*

Only matter loops. The gravitational loops are suppressed by the Planck mass, and give smaller contributions than the matter loops. This is a power counting argument, and it would be nice to make a real computation.

Results from RHIC with Implications for LHC

M. J. Tannenbaum *
Physics Department, 510c,
Brookhaven National Laboratory,
Upton, NY 11973-5000, USA
mjt@bnl.gov

October 26, 2010

1 Introduction

High energy nucleus-nucleus collisions provide the means of creating nuclear matter in conditions of extreme temperature and density [1, 2, 3]. The kinetic energy of the incident projectiles would be dissipated in the large volume of nuclear matter involved in the reaction. At large energy or baryon density, a phase transition is expected from a state of nucleons containing confined quarks and gluons to a state of "deconfined" (from their individual nucleons) quarks and gluons, in chemical and thermal equilibrium, covering a volume that is many units of the confining length scale. This state of nuclear matter was originally given the name Quark Gluon Plasma(QGP) [4], a plasma being an ionized gas. However the results at RHIC [2] indicated that instead of behaving like a gas of free quarks and gluons, the matter created in heavy ion collisions at nucleon-nucleon c.m. energy $\sqrt{s_{NN}}$ = 200 GeV appears to be more like a *liquid*. This matter interacts much more strongly than originally expected, as elaborated in peer reviewed articles by the 4 RHIC

*Research supported by U.S. Department of Energy, DE-AC02-98CH10886.

experiments [5, 6, 7, 8], which inspired the theorists [9] to give it the new name "sQGP" (strongly interacting QGP).

In the terminology of high energy physics, the QGP or sQGP is called a "soft" process, related to the QCD confinement scale

$$\Lambda_{\rm QCD}^{-1} \simeq (0.2\ {\rm GeV})^{-1} \simeq 1\,{\rm fm} \quad . \tag{1}$$

With increasing temperature, T, in analogy to increasing Q^2, the strong coupling constant $\alpha_s(T)$ becomes smaller, reducing the binding energy, and the string tension, $\sigma(T)$, becomes smaller, increasing the confining radius, effectively screening the potential[10]:

$$V(r) = -\frac{4}{3}\frac{\alpha_s}{r} + \sigma\, r \rightarrow -\frac{4}{3}\frac{\alpha_s}{r} e^{-\mu r} + \sigma\,\frac{(1-e^{-\mu r})}{\mu} \tag{2}$$

where $\mu = \mu(T) = 1/r_D$ is the Debye screening mass [10]. For $r < 1/\mu$ a quark feels the full color charge, but for $r > 1/\mu$, the quark is free of the potential and the string tension, effectively deconfined.

There has been considerable work over the past three decades in making quantitative predictions for the QGP [2]. The predicted transition temperature from a state of hadrons to the QGP varies, from $T_c \sim 150$ MeV at zero baryon density, to zero temperature at a critical baryon density roughly 1 GeV/fm^3, ~ 6.5 times the normal density of cold nuclear matter ($\rho_0 = 0.14\,$nucleons/fm^3, $\mu_B \simeq 930$ MeV), where μ_B is the Baryon chemical potential. A typical expected phase diagram of nuclear matter [11] is shown in Fig. 1. Not distinguished on Fig. 1 in the hadronic phase are the liquid self-bound ground state of nuclear matter and the gas of free nucleons [12].

A nice feature of the search for the QGP is that it requires the integrated use of many disciplines in Physics: High Energy Particle Physics, Nuclear Physics, Relativistic Mechanics, Quantum Statistical Mechanics, and, recently, AdS/CFT string theory [13, 14]. From the point of view of an experimentalist there are two major questions in this field. The first is how to relate the thermodynamical properties (temperature, energy density, entropy, viscosity ...) of the QGP or hot nuclear matter to properties that can be measured in the laboratory. The second question is how the QGP can be detected.

One of the major challenges in this field is to find signatures that are unique to the QGP so that this new state of matter can be distinguished from the "ordinary physics" of relativistic nuclear collisions. Another more

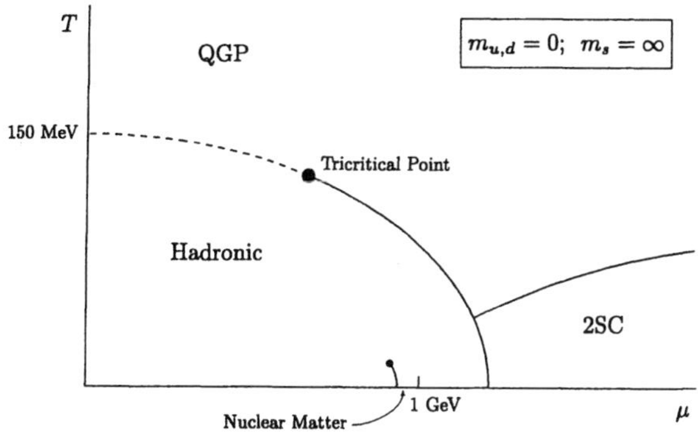

Figure 1: (left) A proposed phase diagram for nuclear matter [11]: Temperature, T, vs Baryon Chemical Potential, μ.

general challenge is to find effects which are specific to A+A collisions, such as collective or coherent phenomena, in distinction to cases for which A+A collisions can be considered as merely an incoherent superposition of nucleon-nucleon collisions [15, 16, 17].

2 Issues in Relativistic Heavy Ion Physics

2.1 J/Ψ suppression—the original "gold-plated" QGP signature

Since 1986, the 'gold-plated' signature of deconfinement was thought to be J/Ψ suppression. Matsui and Satz [18] proposed that J/Ψ production in A+A collisions will be suppressed by Debye screening of the quark color charge in the QGP. The J/Ψ is produced when two gluons interact to produce a $c, \bar{c}$ pair which then resonates to form the J/Ψ. In the plasma the $c, \bar{c}$ interaction is screened so that the $c, \bar{c}$ go their separate ways and eventually pick up other quarks at the periphery to become *open charm*. "Anomalous suppression" of J/Ψ was found in Pb+Pb collisions at the CERN SpS $\sqrt{s_{NN}} = 17.2$ GeV [19] (e.g. see Fig. 18 below) . This is the CERN fixed target heavy ion program's main claim to fame: but the situation is complicated because J/Ψ are suppressed in p+A collisions [20].

The search for J/Ψ suppression and thermal photon/dilepton radiation from the QGP drove the design of the RHIC experiments. My summary of the different views of dilepton resonances in the High Energy[21] and Relativistic Heavy Ion[18] Physics communities since the mid 1980's is shown in Fig. 2.

Success in HEP   Success in RHI

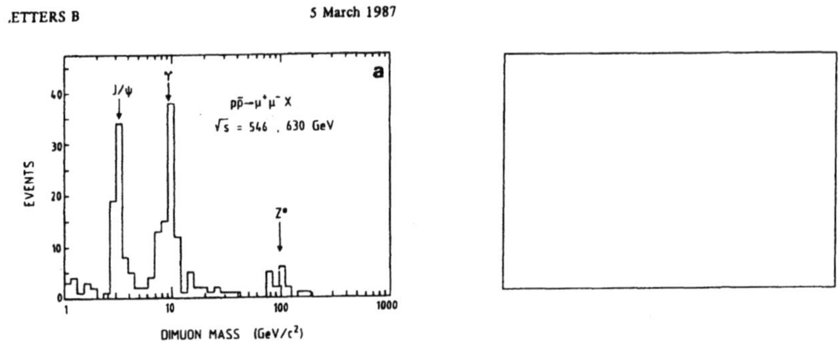

Figure 2: "The road to success": In High Energy Physics (left) a UA1 measurement[21] of pairs of muons each with $p_T \geq 3$ GeV/c shows two Nobel prize winning dimuon peaks and one which won the Wolf prize. Success for measuring these peaks in RHI physics is shown schematically on the right.

2.2 Detector issues in A+A compared to p-p collisions

Another main concern of experimental design in RHI collisions is the huge multiplicity in A+A central collisions compared to p-p collisions. A schematic drawing of a collision of two relativistic Au nuclei is shown in Fig. 3a. In the center of mass system of the nucleus-nucleus collision, the two Lorentz-contracted nuclei of radius R approach each other with impact parameter b. In the region of overlap, the "participating" nucleons interact with each other, while in the non-overlap region, the "spectator" nucleons simply continue on their original trajectories and can be measured in Zero Degree Calorimeters (ZDC), so that the number of participants can be determined. The degree of overlap is called the centrality of the collision, with $b \sim 0$, being the most central and $b \sim 2R$, the most peripheral. The maximum time of overlap is $\tau_\circ = 2R/\gamma c$ where γ is the Lorentz factor and c is the velocity of light. The energy of the inelastic collision is predominantly dissipated by multiple particle production, where $n_{\rm ch}$, the number of charged particles produced, is directly proportional [8] to the number of participating nucleons ($N_{\rm part}$) as sketched on Fig. 3a. Thus, $n_{\rm ch}$ or the total transverse energy E_T in central

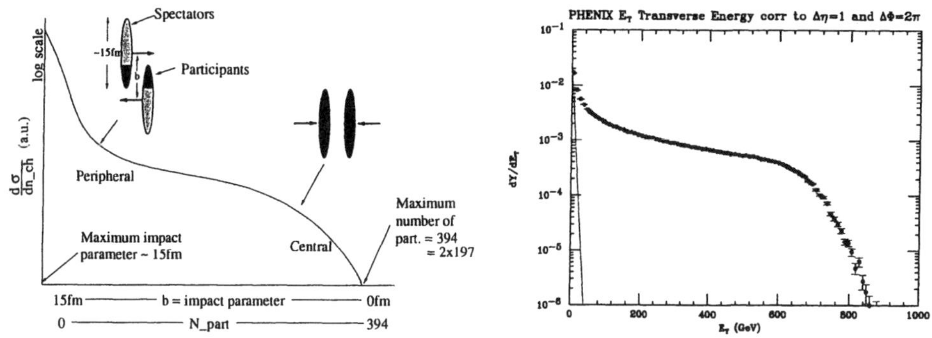

Figure 3: a) (left) Schematic of collision of two nuclei with radius R and impact parameter b. The curve with the ordinate labeled $d\sigma/dn_{\rm ch}$ represents the relative probability of charged particle multiplicity $n_{\rm ch}$ which is directly proportional to the number of participating nucleons, $N_{\rm part}$. b)(right) Transverse energy (E_T) distribution in Au+Au and p-p collisions at $\sqrt{s_{NN}} = 200$ GeV from PHENIX [22].

Au+Au collisions is roughly A times larger than in a p-p collision, as shown in the measured transverse energy spectrum in the PHENIX detector for Au+Au compared to p-p (Fig. 3b) and in actual events from the STAR and PHENIX detectors at RHIC in Fig. 4.

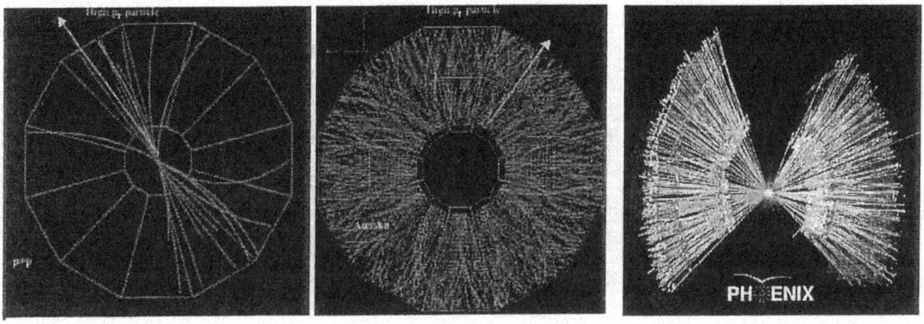

Figure 4: a) (left) A p-p collision in the STAR detector viewed along the collision axis; b) (center) Au+Au central collision at $\sqrt{s_{NN}} = 200$ GeV in the STAR detector; c) (right) Au+Au central collision at $\sqrt{s_{NN}} = 200$ GeV in the PHENIX detector.

As it is a daunting task to reconstruct all the particles produced in such events, the initial detectors at RHIC [23] concentrated on the measurement of single-particle or multi-particle inclusive variables to analyze RHI collisions, with inspiration from the CERN ISR which emphasized those techniques before the era of jet reconstruction. There are two major detectors in operation at RHIC, STAR and PHENIX, and there were also two smaller detectors, BRAHMS and PHOBOS, which have completed their program. As may be surmised from Fig. 4, STAR, which emphasizes hadron physics, is most like a conventional general purpose collider detector, a TPC to detect all charged particles over the full azimuth ($\Delta\phi = 2\pi$) and ± 1 units of pseudo-rapidity (η), while PHENIX is a very high granularity high resolution special purpose detector covering a smaller solid angle at mid-rapidity, together with a muon-detector at forward rapidity [24]. PHENIX is designed to measure and trigger on rare processes involving leptons, photons and identified hadrons at the highest luminosities with the special features: i) a minimum of material (0.4% $X_\circ$) in the aperture to avoid photon conversions; ii) possibility of zero magnetic field on axis to prevent de-correlation of e^+e^- pairs from photon conversions; iii) Electro-Magnetic Calorimeter (EMCal) and Ring Imaging Cherenkov Counter (RICH) for $e^\pm$ identification and level-1 $e^\pm$ trigger; iv) a finely segmented EMCal ($\delta\eta$, $\delta\phi = 0.01\times\ 0.01$) to avoid overlapping showers due to the high multiplicity and for separation of single-γ and π^0 up to $p_T \sim 25$ GeV/c; v) EMCal and precison Time of Flight measurement for particle identification.

In addition to the large multiplicity, there are two other issues in RHI physics which are different from p-p physics: i) space-time issues, both in momentum space and coordinate space—for instance what is the spatial extent of fragmentation? is there a formation time/distance?; ii) huge azimuthal anisotropies of particle production in non-central collisions (colloquially collective flow) which are interesting in their own right but can be troublesome.

2.3 Collective Flow

A distinguishing feature of A+A collisions compared to either p-p or p+A collisions is the collective flow observed. This effect is seen over the full range of energies studied in heavy ion collisions, from incident kinetic energy of 100A MeV to c.m. energy of $\sqrt{s_{NN}} = 200$ GeV [25]. Collective flow, or simply flow, is a collective effect which can not be obtained from a superposition of independent N-N collisions.

Immediately after an A+A collision, the overlap region defined by the

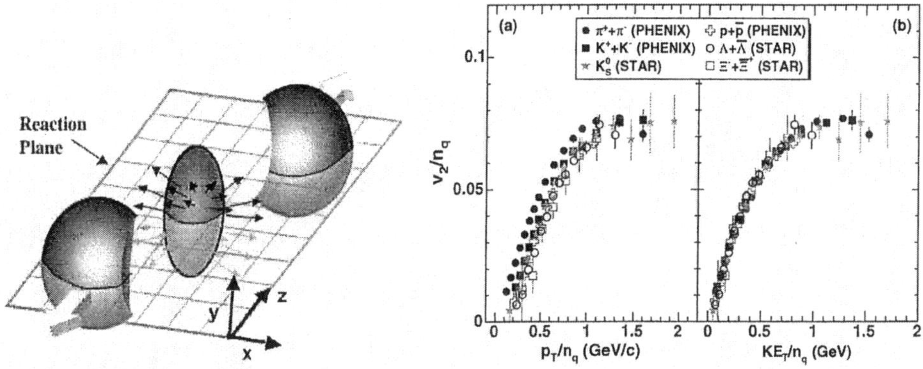

Figure 5: (left) Almond shaped overlap zone generated just after an A+A collision where the incident nuclei are moving along the $\pm z$ axis. The reaction plane by definition contains the impact parameter vector (along the x axis) [26] (see discussion session). (right) Measurements of elliptical-flow (v_2) for identified hadrons plotted as v_2 divided by the number of constituent quarks n_q in the hadron as a function of (a) p_T/n_q, (b) KE_T/n_q [27].

nuclear geometry is almond shaped (see Fig 5) with the shortest axis along the impact parameter vector. Due to the reaction plane breaking the ϕ symmetry of the problem, the semi-inclusive single particle spectrum is modified by an expansion in harmonics [28] of the azimuthal angle of the particle with respect to the reaction plane, $\phi - \Phi_R$ [29], where the angle of the reaction plane Φ_R is defined to be along the impact parameter vector, the x axis in Fig. 5:

$$\frac{Ed^3N}{dp^3} = \frac{d^3N}{p_T dp_T dy d\phi} = \frac{d^3N}{2\pi\, p_T dp_T dy}\left[1 + \sum_n 2v_n \cos n(\phi - \Phi_R)\right] . \quad (3)$$

The expansion parameter v_2, called elliptical flow, is predominant at mid-rapidity. In general, the fact that flow is observed in final state hadrons shows that thermalization is rapid so that hydrodynamics comes into play before the spatial anisotropy of the overlap almond dissipates. At this early stage hadrons have not formed and it has been proposed that the constituent quarks flow [30], so that the flow should be proportional to the number of constituent quarks n_q, in which case v_2/n_q as a function of p_T/n_q would represent the constituent quark flow as a function of constituent quark transverse momentum and would be universal. However, in relativistic hydrodynamics, at mid-rapidity, the transverse kinetic energy, $m_T - m_0 = (\gamma_T - 1)m_0 \equiv KE_T$, rather than p_T is the relevant variable, and in fact v_2/n_q as a function of KE_T/n_q seems to exhibit nearly perfect scaling [27] (Fig. 5b).

The fact that the flow persists for $p_T > 1$ GeV/c implies that the viscosity is small [31], perhaps as small as a quantum viscosity bound from string theory [32], $\eta/s = 1/(4\pi)$ where η is the shear viscosity and s the entropy density per unit volume. This has led to the description of the "sQGP" produced at RHIC as "the perfect fluid" [9].

2.4 Test of Hydrodynamics at LHC

A test of hydrodynamics at the LHC concerns the possible increase of the anisotropic flow v_2 beyond the 'hydrodynamic limit'. Wit Busza's extrapolation [33] of v_2 to the LHC energy is shown in Fig. 6a, a factor of 1.6 increase from RHIC. A previous paper by NA49 [34] which compared v_2

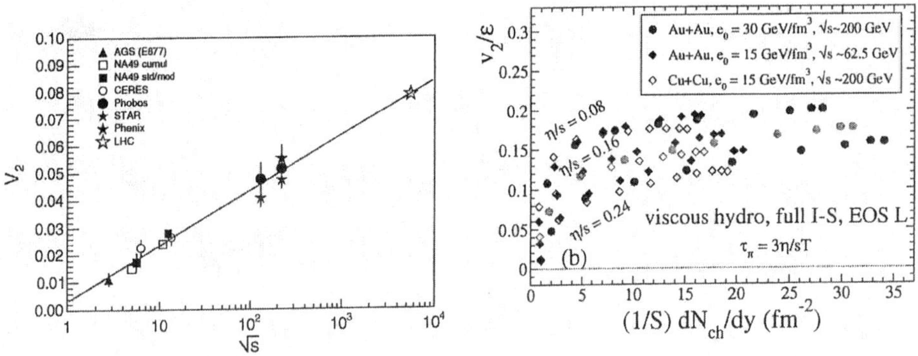

Figure 6: a) (left) Busza's extrapolation of v_2 to LHC [33]. b) (right) 'Hydro Limit' for v_2/ε vs 'Bjorken multiplicity density', $(1/S)dN_{\rm ch}/dy$ calculated in viscous hydrodynamics for several values of the initial energy density e_0 and $\eta/s \geq 1/(4\pi)$ [35].

measurements from AGS and CERN fixed target experiments to RHIC as a function of the 'Bjorken multiplicity density', $(1/S)dN_{\rm ch}/dy$, where $S =$ is the overlap area of the collision zone, showed an increase in v_2/ε from fixed target energies leading to a "hydro limit" at RHIC, where ε is the eccentricity of the collision zone. This limit was confirmed in a recent calculation using viscous relativistic hydrodynamics [35] which showed a clear hydro limit of $v_2/\varepsilon = 0.20$ (Fig. 6b). This limit is sensitive to the ratio of the viscosity/entropy density, the now famous η/s, but negligibly sensitive to the maximum energy density of the collision or to $(1/S)dN_{\rm ch}/dy$. Thus, I assume that this calculation would give a hydro limit at the LHC not too different from RHIC, $v_2/\varepsilon \approx 0.20$. Busza's extrapolation of a factor of 1.6 increase in v_2 from RHIC to LHC gives $v_2/\varepsilon = 0.32$ at LHC. In my opinion this is a measurement which can be done to high precision on the first day of Pb+Pb collisions at the LHC, since it is high rate and needs no p-p comparison data. Personally, I wonder what the hydro aficionados would say if both Heinz [35] and Busza's predictions were correct?

3 Measurements in p-p collisions at RHIC

In addition to being the first heavy ion collider, RHIC is also the first polarized proton collider. Proton-proton collisions are performed with both beams either longitudinally or transversely polarized [23, 36]. The bunch-by-bunch polarization is arranged so that the spin averaged cross section is obtained to high accuracy if polarization information is ignored. The emphasis on precision EM calorimetry allows PHENIX to excel in the measurement of reactions producing photons, such as direct-single-photon production, or particles which decay to photons, $\pi^0 \to \gamma + \gamma$, $\eta \to \gamma + \gamma$, etc.

In order to understand whether an effect observed in A+A collisions exhibits a sensitivity to collective effects or to the presence of a medium such as the QGP it is important to establish a precise baseline measurement in p-p collisions at the same value of nucleon-nucleon c.m. energy $\sqrt{s_{NN}}$. PHENIX measurements of the invariant cross section, $Ed^3\sigma/dp^3$, for π^0 and direct-single-γ production in p-p collisions at $\sqrt{s} = 200$ GeV are shown in Fig. 7a [37] and Fig. 7b [38], respectively. The inset on Fig. 7a shows that the π^0 cross section is exponential $\sim e^{-6p_T}$ for $p_T < 2$ GeV/c, as originally paramaterized by Cocconi [39, 40], which is the region of soft-multiparticle physics. For $p_T > 2$ GeV/c the spectrum is a power law which is indicative of the hard-scattering of the quark and gluon constituents of the proton. The excellent agreement of the measurements with theory is rewarding, although not surprising, since, after all, the discovery of π^0 production at large transverse momentum at the CERN-ISR proved that the partons of deeply inelastic scattering (DIS) interacted strongly with each other [40, 41].

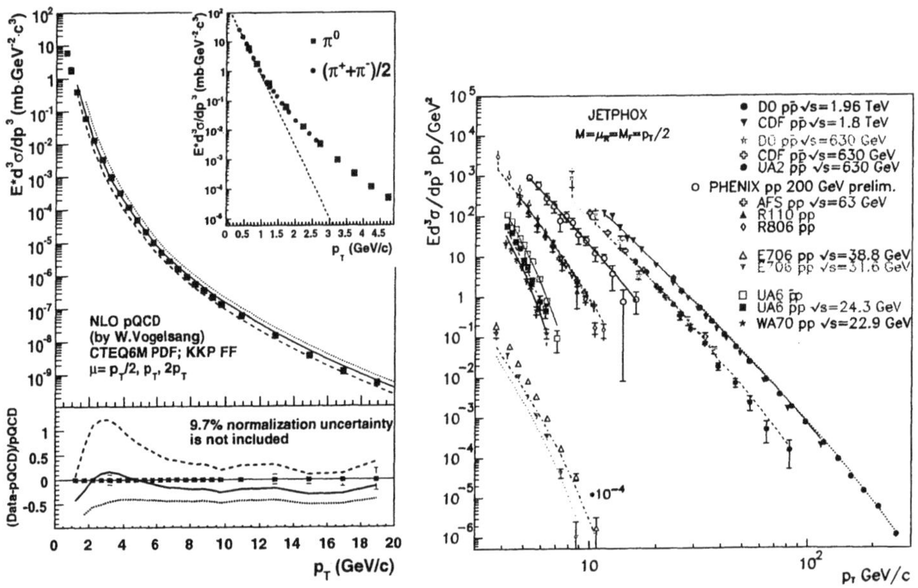

Figure 7: a) (left) PHENIX measurement of invariant cross section of π^0 vs. p_T at mid-rapidity in p-p collisons at $\sqrt{s} = 200$ GeV. [37]. b) (right) PHENIX measurement of inclusive direct-single γ in p-p collisons at $\sqrt{s} = 200$ GeV ($\circ$), together with all previous data compared to the theory. [38]

3.1 The influence of the CERN-ISR

The ISR discovery [41] (Fig. 8a) showed that the e^{-6p_T} dependence at low p_T breaks to a power law with characteristic $\sqrt{s}$ dependence for $p_T > 2$ GeV/c, which is more evident from the log-log plot of subsequent data [42] (Fig. 8b) as a function of $x_T = 2p_T/\sqrt{s}$. This plot exhibits that the cross section for hard-processes obeys the scaling law:

$$E\frac{d^3\sigma}{d^3p} = \frac{1}{p_T^{n_{\rm eff}}} F(\frac{p_T}{\sqrt{s}}) = \frac{1}{\sqrt{s}^{n_{\rm eff}}} G(x_T) \tag{4}$$

where $n_{\rm eff}(x_T, \sqrt{s}) \sim 4 - 6$ gives the form of the force-law between constituents as later predicted by Quantum Chromodynamics (QCD) with non-scaling structure and fragmentation functions and running coupling constant [43]. The more familiar equation for the constituent reaction $a + b \to$

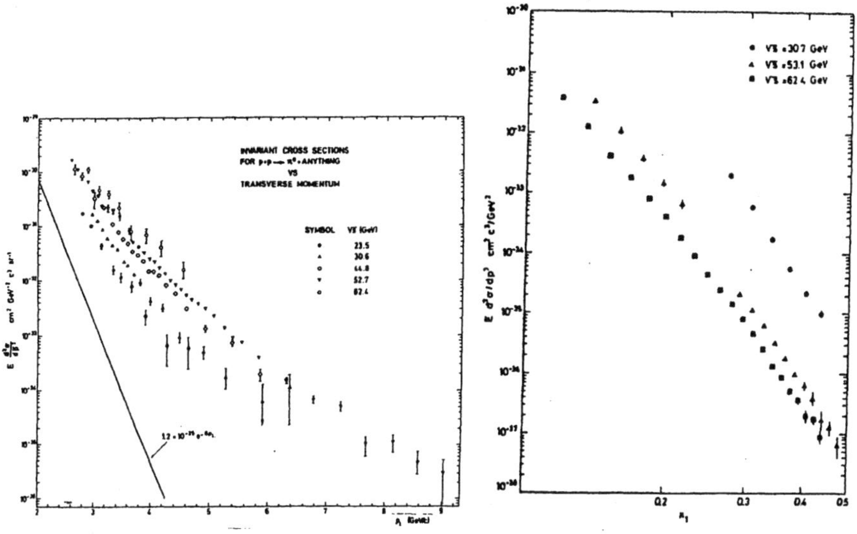

Figure 8: a) (left) CCR [41] measurement of the invariant cross section of π^0 vs. p_T at mid-rapidity in p-p collisons for 5 values of $\sqrt{s}$. b) (right) Later ISR measurement of invariant cross section of π^0 vs. $x_T = 2p_T/\sqrt{s}$ at mid-rapidity in p-p collisons for 3 values of $\sqrt{s}$ [42]

$c + d$ (e.g. $g + q \to g + q$) at parton-parton center-of-mass (c.m.) energy $\sqrt{\hat{s}}$ in "leading logarithm" pQCD [44] is:

$$\frac{d^3\sigma}{dx_1 dx_2 d\cos\theta^*} = \frac{s d^3\sigma}{d\hat{s} d\hat{y} d\cos\theta^*} = \frac{1}{s}\sum_{ab} f_a(x_1) f_b(x_2) \frac{\pi\alpha_s^2(Q^2)}{2x_1x_2}\Sigma^{ab}(\cos\theta^*) \quad (5)$$

where $f_a(x_1)$, $f_b(x_2)$, are parton distribution functions, the differential probabilities for partons a and b to carry momentum fractions x_1 and x_2 of their respective protons (e.g. $u(x_2)$), and where θ^* is the scattering angle in the parton-parton c.m. system. The parton-parton c.m. energy squared is $\hat{s} = x_1 x_2 s$, where $\sqrt{s}$ is the c.m. energy of the p-p collision. The parton-parton c.m. system moves with rapidity $\hat{y} = 1/2\ln(x_1/x_2)$ in the p-p c.m. system and the transverse momentum of a scattered parton is $p_T = p_T^* = \frac{\sqrt{\hat{s}}}{2}\sin\theta^*$. Only the characteristic subprocess angular distributions, $\Sigma^{ab}(\cos\theta^*)$ and the coupling constant, $\alpha_s(Q^2) = 12\pi/(25\ln(Q^2/\Lambda^2))$, are fundamental predictions of QCD [45, 46].

Subsequent ISR measurements utilizing inclusive single or pairs of hadrons established that high p_T particles in p-p collisions are produced from states with two roughly back-to-back jets which are the result of scattering of constituents of the nucleons as described by Quantum Chromodynamics (QCD), which was developed during the course of those measurements. These techniques have been used extensively and further developed at RHIC since they are the only practical method to study hard-scattering and jet phenomena in Au+Au central collisions at RHIC energies.

The di-jet structure of events triggered by a high p_T π^0, measured via two-particle correlations at the ISR, is shown in Fig 9 [47, 48]. The peaks

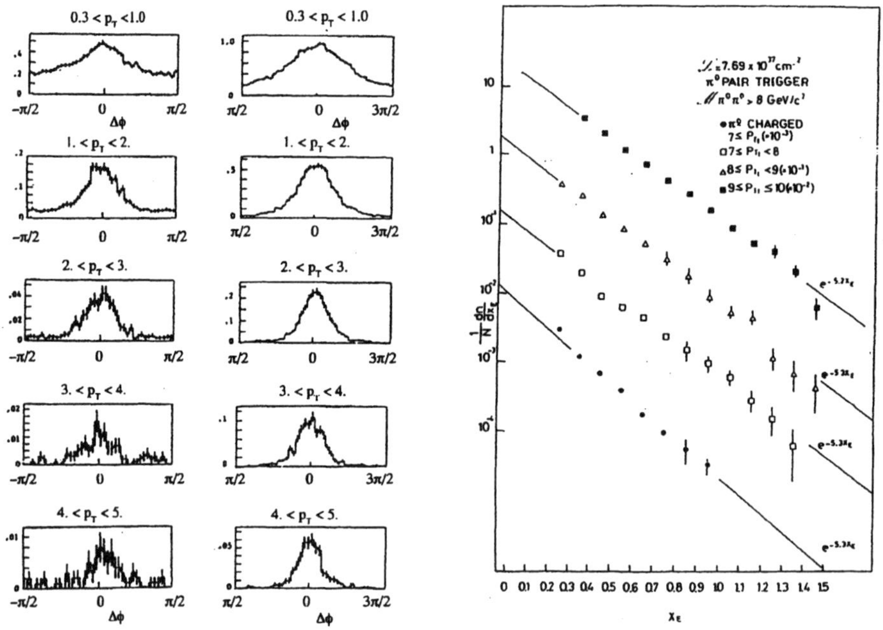

Figure 9: CCOR [47, 48] measurements at $\sqrt{s} = 62.4$ GeV. a,b)Distributions of azimuthal angle ($\Delta\phi$) of associated charged particles of transverse momentum p_{T_a}, with respect to a trigger π^0 with $p_{T_t} \geq 7$ GeV/c, for 5 intervals of $p_{T(a)}$: a) (left-most panel) for $\Delta\phi = \pm\pi/2$ rad about the trigger particle, and b) (middle panel) for $\Delta\phi = \pm\pi/2$ about π radians (i.e. directly opposite in azimuth) to the trigger. The trigger particle is restricted to $|\eta| < 0.4$, while the associated charged particles are in the range $|\eta| \leq 0.7$. c) (right panel) x_E distributions (see text) corresponding to the data of the center panel.

on both the same side (Fig. 9a) as the trigger π^0 and opposite in azimuth (Fig. 9b) are due to the correlated charged particles from jets. The integrated (in $\Delta\phi$) yield of the away side-particles as a function of the variable $x_E \equiv -p_{T_a}\cos(\Delta\phi)/p_{T_t} \approx z_a/z_t$, where $z_t = p_{T_t}/\hat{p}_{T_t}$ is the fragmentation variable of the trigger jet (with $\hat{p}_{T_t}$) and $z_a = p_{T_a}/\hat{p}_{T_a}$ is the fragmentation variable of the away jet (with $\hat{p}_{T_a}$), was thought in the ISR era to measure the fragmentation function of the away jet (Fig. 9c) but was found at RHIC to be sensitive, instead, to the ratio of the transverse momenta of the away-jet to the trigger jet, $\hat{x}_h \equiv \hat{p}_{T_a}/\hat{p}_{T_t}$ [49].

The QCD subprocess angular distribution $\Sigma^{ab}(\cos\theta^*)$ was also first measured with two-particle correlations of π^0 pairs of large invariant mass at the CERN-ISR [50, 51] (Fig. 10), in agreement with QCD [45, 46] at a fundamental level.

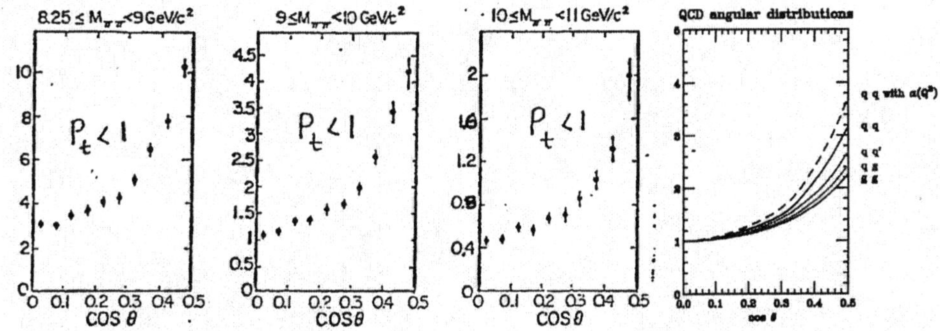

Figure 10: a) (left 3 panels) CCOR measurement [50, 51] of polar angular distributions of π^0 pairs with net $p_T < 1$ GeV/c at mid-rapidity in p-p collisions with $\sqrt{s} = 62.4$ GeV for 3 different values of $\pi\pi$ invariant mass $M_{\pi\pi}$. b) (rightmost panel) QCD predictions for $\Sigma^{ab}(\cos\theta^*)$ for the elastic scattering of gg, qg, qq', qq, and $q\bar{q}$ with $\alpha_s(Q^2)$ evolution.

3.2 Other ISR discoveries important at RHIC

Two other ISR discoveries, direct single-γ production and direct-single $e^\pm$ production, and one near miss, J/Ψ production, are important components of physics at RHIC.

Direct single-γ production via the inverse QCD-compton process [52] $g + q \to \gamma + q$ is an important probe in A+A collisions because the γ is a direct participant in the reaction (at the constituent level), which emerges from the medium without interacting and can be measured precisely. The cross sections for direct single-γ production at $\sqrt{s} = 62.4$ GeV [53] are shown in Fig. 11a. Two-particle azimuthal correlations of charged hadrons with neutral mesons (π^0), compared to direct-γ (Fig. 11b), show that direct-γ are isolated, with no accompanying same-side particles, while π^0 have accompanying particles since they are fragments of jets from high p_T partons.

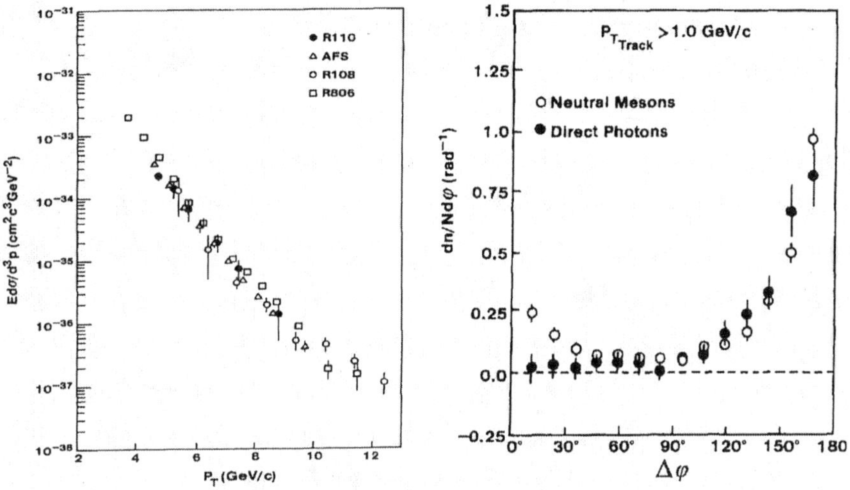

Figure 11: a)(left) Compilation of invariant cross sections of direct-γ production at ISR [53]; (right) azimuthal correlations of neutral mesons and direct-γ with $h^{\pm}$ [53].

Direct single-$e^{\pm}$ at a level of $e^{\pm}/\pi^{\pm} \approx 10^{-4}$ for all values of $\sqrt{s}$ at the CERN-ISR were discovered before either the J/Ψ or open-charm [54] (Fig. 12). After the discovery of the J/Ψ in 1974, it was demonstrated that the J/Ψ was not the source of the single-$e^{\pm}$ (Fig. 13) and two years later, when open charm was discovered, it was shown that the direct $e^{\pm}$ were due to the semi-leptonic decay of charm mesons [55]. Fig. 13a [56] shows the first J/Ψ at the ISR [56], Fig. 13b shows the best J/Ψ measurement at the ISR [57] while Fig. 13c [54] shows that the direct electrons (Fig. 12) are not the result of J/Ψ decay since $\langle p_T \rangle = 1.1 \pm 0.05$ GeV/c [57].

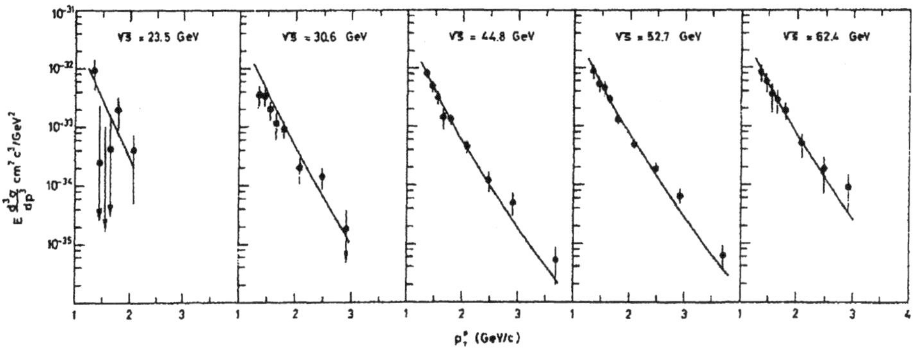

Figure 12: Invariant cross sections at mid-rapidity: $(e^+ + e^-)/2$ (points); $10^{-4} \times (\pi^+ + \pi^-)/2$ (lines) [54].

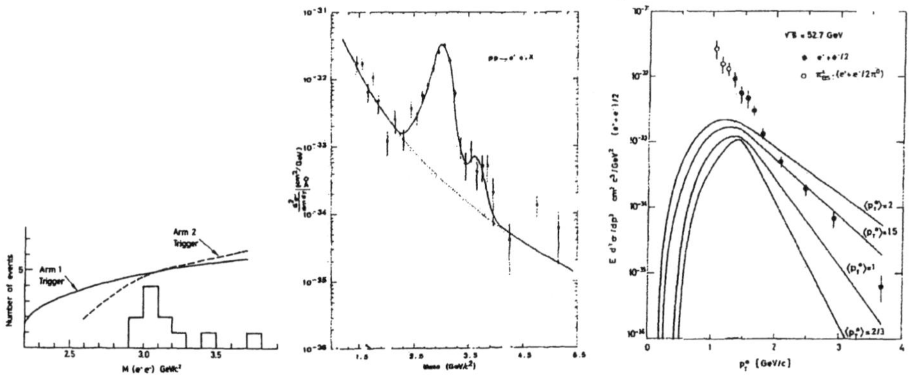

Figure 13: a)(left) First J/Ψ at ISR [56]; b) (center) Best $d\sigma_{ee}/dm_{ee}dy|_{y=0}$ [57]; c)(right) direct-$e^{\pm}$ data at $\sqrt{s}$ = 52.7 GeV (Fig. 12) with calculated $e^{\pm}$ spectrum for J/Ψ for several values of $\langle p_T \rangle$ [54].

4 From ISR p-p to RHIC A+A physics

Since hard-scattering at high $p_T > 2$ GeV/c is point-like, with distance scale $1/p_T < 0.1$ fm, the cross section in p+A (A+A) collisions, compared to p-p, should be larger by the relative number of possible point-like encounters, a factor of A (A^2) for p+A (A+A) minimum bias collisions. When the impact parameter or centrality of the collision is defined, the proportionality factor becomes $\langle T_{AA} \rangle$, the average overlap integral of the nuclear thickness functions.

4.1 Jet quenching from inclusive π^0 production

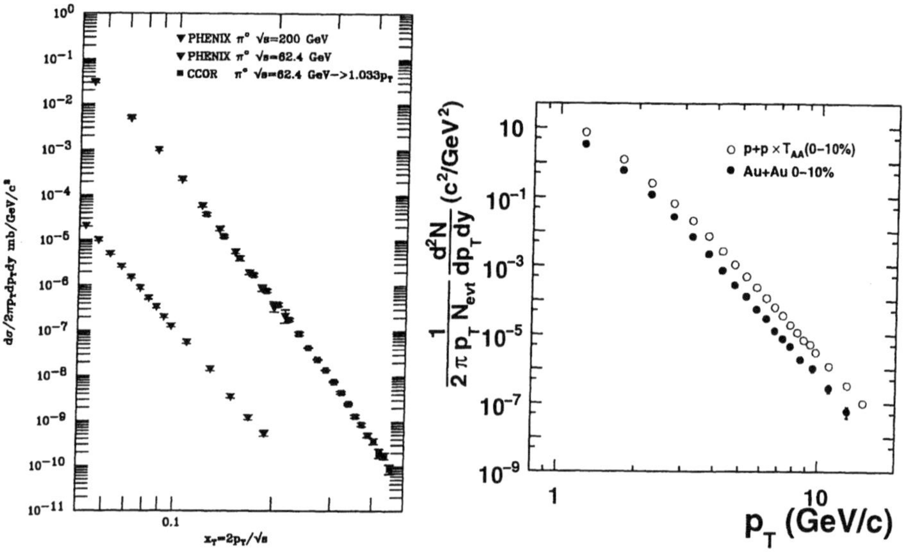

Figure 14: a) (left) $Ed^3\sigma/dp^3$ vs. x_T for PHENIX mid-rapidity π^0 at $\sqrt{s} = 200$ GeV in p-p collisions [37] plus PHENIX [58] and CCOR-ISR [42] measurements at $\sqrt{s} = 62.4$ GeV, where the absolute p_T scale of the ISR measurement has been corrected upwards by 3% to agree with the PHENIX data. b) (right) π^0 p-p data vs. p_T at $\sqrt{s} = 200$ GeV from a) multiplied by $\langle T_{AA} \rangle$ for Au+Au central (0-10%) collisions compared to semi-inclusive π^0 invariant yield in Au+Au central (0-10%) collisions at $\sqrt{s_{NN}} = 200$ GeV.

The discovery, at RHIC, that π^0 are suppressed by roughly a factor of 5 compared to point-like scaling of hard-scattering in central Au+Au collisions is arguably *the* major discovery in Relativistic Heavy Ion Physics. In Fig. 14a), the PHENIX measurement of $Ed^3\sigma/dp^3$ for π^0 production in p-p collisions at $\sqrt{s} = 62.4$ GeV [58] is in excellent agreement with the ISR data and the PHENIX π^0 data follow the same trend as the lower energy data, with a pure power law, $Ed^3\sigma/dp^3 \propto p_T^{-8/1\pm0.1}$ for $p_T > 3$ GeV/c at $\sqrt{s} = 200$ GeV. In Fig. 14b), the 200 GeV p-p data, multiplied by the point-like scaling factor $\langle T_{AA} \rangle$ for (0-10%) central Au+Au collisions are compared to the semi-inclusive invariant π^0 yield in central (0-10%) Au+Au collisions at $\sqrt{s_{NN}}$ = 200 GeV and, amazingly, the Au+Au data follow the same power-law as the p-p data but are suppressed from the point-like scaled p-p data by a factor of ~ 5, independent of p_T. The suppression is represented quantitatively by the "nuclear modification factor", $R_{AA}(p_T)$, the ratio of the measured semi-inclusive yield in A+A collisions to the point-like scaled p-p cross section at a given p_T:

$$R_{AA}(p_T) = \frac{d^2N^{\pi}_{AA}/dp_T dy N_{AA}}{\langle T_{AA}\rangle d^2\sigma^{\pi}_{pp}/dp_T dy} \quad . \tag{6}$$

In Fig. 15a, $R_{AA}(p_T)$ is shown for π^0, η mesons and direct-γ for $\sqrt{s_{NN}}$ = 200 GeV Au+Au central (0-10%) collisions. The π^0 and η mesons, which are fragments of jets from outgoing partons are suppressed by the same amount while the direct-γ which do not interact in the medium are not suppressed. This indicates a strong medium effect on outgoing partons. Fig. 15b shows

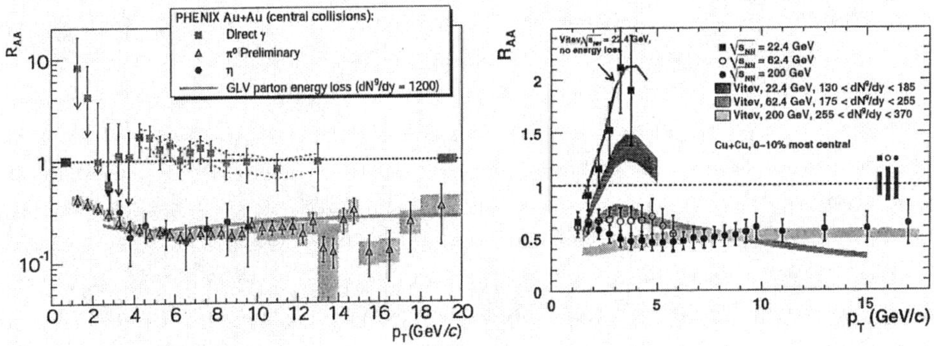

Figure 15: a) (left) Nuclear modification factor, R_{AA} for direct-γ, π^0 and η in Au+Au (0-10%) central collisions at $\sqrt{s_{NN}}$ = 200 GeV [59], together with GLV theory curve [60]. b) PHENIX R_{AA} for π^0 in Cu+Cu central collisions at $\sqrt{s_{NN}}$ = 200, 62.4 and 22.4 GeV [61], together with Vitev theory curves [62].

that R_{AA} for central (0-10%) Cu+Cu collisions is comparable at $\sqrt{s_{NN}}$ = 62.4 and 200 GeV, but that there is no suppression, actually a Cronin enhancement [63], at $\sqrt{s_{NN}}$ = 22.4 GeV. This indicates that the medium which suppresses jets is produced somewhere between $\sqrt{s_{NN}}$ = 22.4 GeV, the SpS Fixed Target highest c.m. energy, and 62.4 GeV.

The measurements at RHIC appear to be in excellent agreement with the theoretical curves [60, 62]. The suppression can be explained by the energy loss of the outgoing partons in the dense color-charged medium due to coherent Landau-Pomeranchuk-Migdal radiation of gluons, predicted in QCD [64], which is sensitive to properties of the medium. Measurements of two-particle correlations (discussed below, Sec. 7) confirm the loss of energy of the away-jet relative to the trigger jet in Au+Au central collisions compared to p-p collisions. However, lots of details remain to be understood.

5 Direct photons at RHIC: thermal photons?

5.1 Internal Conversions—the first measurement anywhere of direct photons at low p_T

Internal conversion of a photon from π^0 and η decay is well-known and is called Dalitz decay [65]. Perhaps less well known in the RHI community is the fact that for any reaction (e.g. $q + g \to \gamma + q$) in which a real photon can be emitted, a virtual photon (e.g. e^+e^- pair of mass $m_{ee} \geq 2m_e$) can also be emitted. This is called internal-conversion and is generally given by the Kroll-Wada formula [66, 67]:

$$\begin{aligned} \frac{1}{N_\gamma}\frac{dN_{ee}}{dm_{ee}} &= \frac{2\alpha}{3\pi}\frac{1}{m_{ee}}\left(1-\frac{m_{ee}^2}{M^2}\right)^3 \quad \times \\ & |F(m_{ee}^2)|^2\sqrt{1-\frac{4m_e^2}{m_{ee}^2}}\left(1+\frac{2m_e^2}{m_{ee}^2}\right) \quad , \end{aligned} \tag{7}$$

where M is the mass of the decaying meson or the effective mass of the emitting system. The dominant terms are on the first line of Eq. 7: the characteristic $1/m_{ee}$ dependence; and the cutoff of the spectrum for $m_{ee} \geq M$ (Fig. 16a) [67]. Since the main background for direct-single-γ production is a photon from $\pi^0 \to \gamma + \gamma$, selecting $m_{ee} \gtrsim 100$ MeV/c^2 effectively reduces the background by an order of magnitude by eliminating the background from π^0 Dalitz decay, $\pi^0 \to \gamma + e^+ + e^-$, at the expense of a factor ~ 1000 in rate. This allows the direct photon measurements to be extended (for the first time in both p-p and Au+Au collisions) below the value of $p_T \sim 4$ GeV/c, possible with real photons, down to $p_T = 1$ GeV/c (Fig. 16b) [67], which is a real achievement. The solid lines on the p-p data are QCD calculations which work down to $p_T = 2$ GeV/c. The dashed line is a fit of the p-p data to the modified power law $B(1 + p_T^2/b)^{-n}$, used in the related Drell-Yan [68] reaction, which flattens as $p_T \to 0$.

The relatively flat, non-exponential, spectra for the direct-γ and Drell-Yan reactions as $p_T \to 0$ is due to the fact that there is no soft-physics production process for them, only production via the partonic subprocesses, $g+q \to \gamma + q$ and $\bar{q}+q \to e^+ + e^-$, respectively. This is quite distinct from the case for hadron production, e.g. π^0, where the spectra are exponential as $p_T \to 0$ in p-p collisions (Fig. 7a) due to soft-production processes, as well as in Au+Au collisions. Thus, for direct-γ in Au+Au collisions, the exponential spectrum of excess photons above the $\langle T_{AA}\rangle$ extrapolated p-p fit is unique and therefore suggestive of a thermal source.

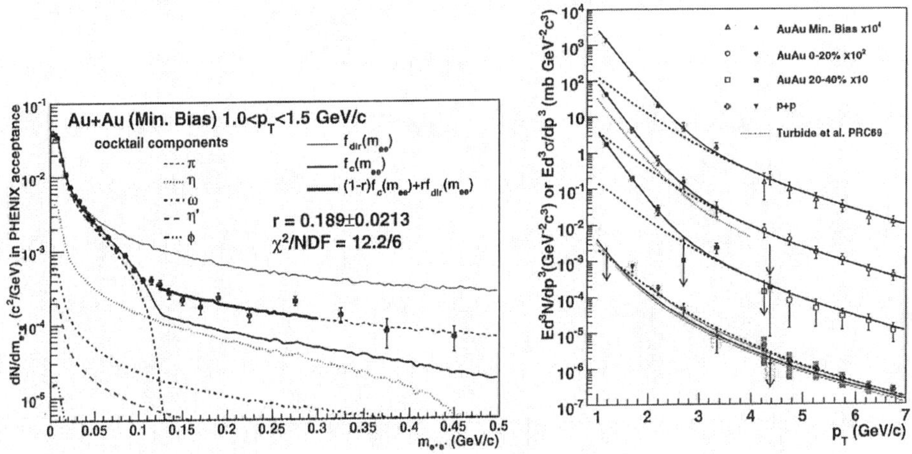

Figure 16: a) (left) Invariant mass (m_{ee}) distribution of e^+e^- pairs from Au+Au minimum bias events for $1.0 < p_T < 1.5$ GeV/c [67]. Dashed lines are Eq. 7 for the mesons indicated. Blue solid line is $f_c(m)$, the total di-electron yield from the sum of contributions or 'cocktail' of meson Dalitz decays; Red solid line is $f_{dir}(m)$ the internal conversion m_{ee} spectrum from a direct-photon ($M >> m_{ee}$). Black solid line is a fit of the data to the sum of cocktail plus direct contributions in the range $80 < m_{ee} < 300$ MeV/c^2. b) (right) Invariant cross section (p-p) or invariant yield (Au+Au) of direct photons as a function of p_T [67]. Filled points are from virtual photons, open points from real photons.

5.2 Low p_T vs high p_T direct-γ—Learn a lot from a busy plot

The unique behavior of direct-γ at low p_T in Au+Au relative to p+p compared to any other particle is more dramatically illustrated by examining the R_{AA} of all particles measured by PHENIX in central Au+Au collisions at $\sqrt{s_{NN}} = 200$ GeV (Fig. 17) [69]. For the entire region $p_T \leq 20$ GeV/c so far measured at RHIC, apart from the $p + \bar{p}$ which are enhanced in the region $2 \leq p_T \lesssim 4$ GeV/c ('the baryon anomaly'), the production of *no other particle* is enhanced over point-like scaling. The behavior of R_{AA} of the low

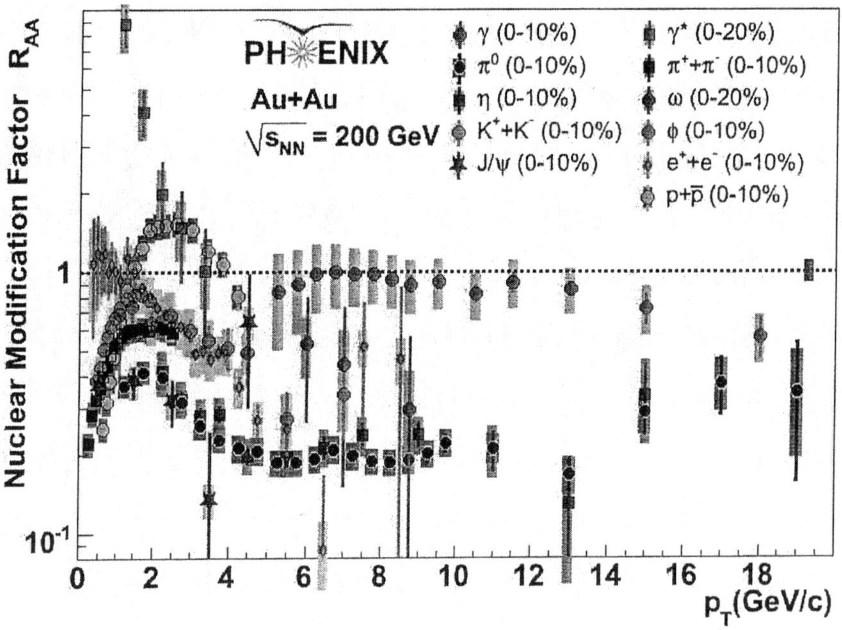

Figure 17: Nuclear Modification Factor, $R_{AA}(p_T)$ for all identified particles so far measured by PHENIX in central Au+Au collisions at $\sqrt{s_{NN}} = 200$ GeV. [69]

$p_T \leq 2$ GeV/c direct-γ is totally and dramatically different from all the other particles, exhibiting an order of magnitude exponential enhancement as $p_T \rightarrow 0$. This exponential enhancement is certainly suggestive of a new production mechanism in central Au+Au collisions different from the conventional soft and hard particle production processes in p-p collisions and its unique behavior is attributed to thermal photon production by many authors (e.g. see citations in reference [67]).

5.2.1 Direct photons and mesons up to $p_T = 20$ GeV/c

Other instructive observations can be gleaned from Fig. 17. The π^0 and η continue to track each other to the highest p_T. At lower p_T, the ϕ meson tracks the $K^{\pm}$ very well, but with a different value of $R_{AA}(p_T)$ than the π^0, while at higher p_T,the ϕ and ω vector mesons appear to track each other. Interestingly, the J/Ψ seems to track the π^0 for $0 \leq p_T \leq 4$ GeV/c; and it will be important to see whether this trend continues at higher p_T.

6 J/Ψ suppression, still golden?

The dramatic difference in π^0 suppression from SpS to RHIC c.m. energy (Fig. 15b) is not reflected in J/Ψ suppression, which is nearly identical at mid-rapidity at RHIC compared to the NA50 measurements at SpS (Fig. 18b) [70, 71]. This casts new doubt on the value of J/Ψ suppression as a probe of deconfinement in addition to the previous complication that J/Ψ are already suppressed (compared to point-like scaling) in p+A and B+A collisions (Fig. 18a). One possible explanation is that c and $\bar{c}$ quarks in the QGP recombine to regenerate J/Ψ, miraculously making the observed R_{AA} equal at SpS and RHIC c.m. energies (Fig. 19a) [71, 72]. The good news is that such models predict the vanishing of J/Ψ suppression or even an enhancement ($R_{AA} > 1$) at LHC energies [73, 74, 75], which would be spectacular, if observed.

Even without the LHC startup, there has been progress this past year when, after $\sim$ 20 years (!), p+A comparison data for the J/Ψ from the CERN fixed target program at 158A GeV/c finally became available [76]. The cold nuclear matter effect of J/Ψ suppression in p+A collisions is parameterized by an effective absorption cross section $\sigma_{abs}^{J/\Psi}$ which had been previously measured to be 4.3 ± 1.0 mb at 400 GeV/c proton beam energy and *"assumed to be independent of beam energy"*. The actual measurement for 158 GeV p+A collisions gives $\sigma_{abs}^{J/\Psi} = 7.6 \pm 0.9$ mb which considerably reduces the "anomalous suppression" effect shown in Fig. 18a to such an extent that there is now a clear difference between the CERN SpS and RHIC

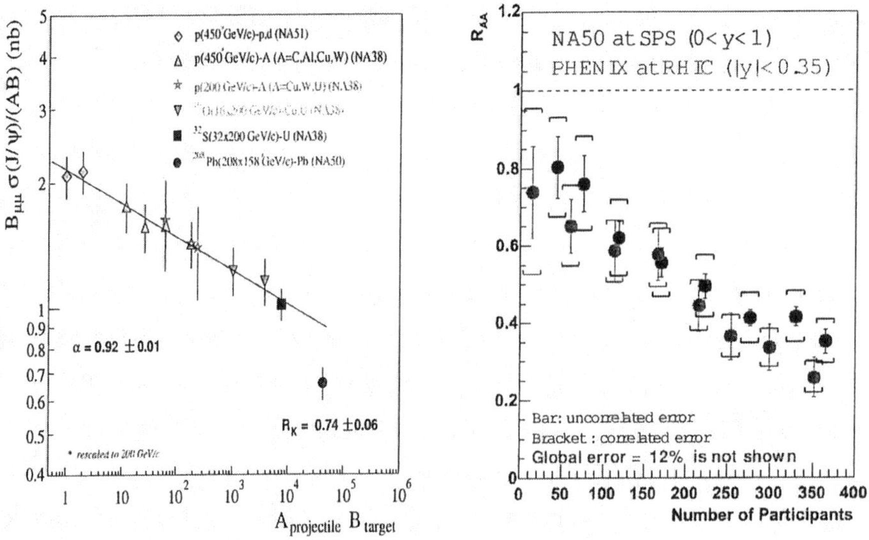

Figure 18: a) (left) Total cross section for J/Ψ production divided by AB in A+B collisions at 158–200A GeV [19]. b) (right) J/Ψ suppression relative to p-p collisions (R_{AA}) as a function of centrality ($N_{\rm part}$) at RHIC [70, 71] and at the CERN/SPS [19].

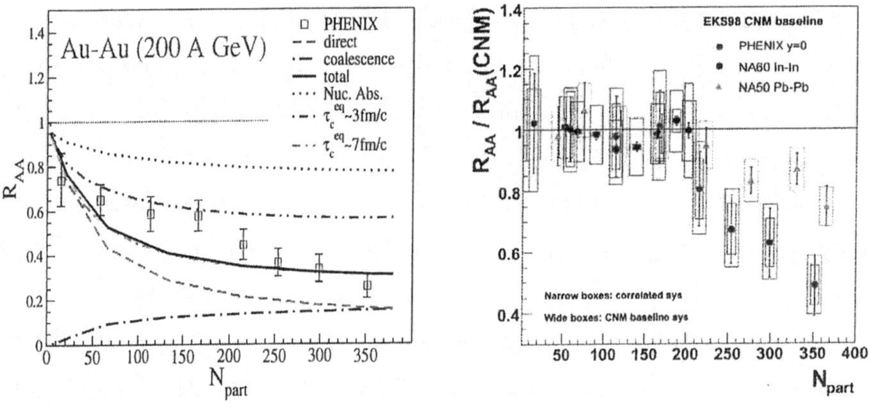

Figure 19: a) (left) PHENIX measurement of R_{AA} as a function of centrality from Fig. 18b together with prediction from a coalescence model [72]; b) (right) R_{AA} of J/Ψ at SpS and RHIC c.m. energies normalized to the measured $R_{AA}(CNM)$ from cold nuclear matter [77]

J/Ψ suppression for the most central A+A collisions relative to the measured Cold Nuclear Matter effect (Fig. 19b) [77]. Maybe there is still some hope for J/Ψ suppression as a QGP signature, but there is an important lesson for LHC. Comparison data for p-p and p+A MUST be taken and must be at the same $\sqrt{s_{NN}}$ as the A+A data.

7 Two-particle correlations

If the π^0 suppression shown in Fig. 15 is in fact explained by the energy loss of the outgoing partons in the dense color-charged medium, this can be confirmed by measurements of two-particle correlations. These measurements are sensitive to the ratio of the energy of the away-jet to the trigger jet, which can be compared in Au+Au collisions and p-p collisions. In analogy to Fig. 9 (above), the two-particle correlations in Au+Au collisions (Fig. 20a) show clear di-jet structure in both peripheral and central collisions. The away-side correlation in central Au+Au collisions is much wider than in peripheral Au+Au and p-p collisions and is further complicated by the large multiparticle background which is a modulated in azimuth by the v_2 collective flow of a comparable width to the jet correlation. After the v_2 correction, a double peak structure $\sim \pm 1$ radian from π is evident, with a dip at π radians. This may indicate a reaction of the medium to a passing parton in analogy to a "sonic-boom" [78] and is under active study both theoretically and experimentally. The energy loss of the away-parton is indicated by the fact that the x_E distribution in Au+Au central collisions (Fig. 20b) is steeper than that from p-p collisions. As noted above, we found in PHENIX [49, 79] that the x_E distribution did not measure the fragmentation function of the away-jet but is sensitive instead to $\hat{x}_h$, the ratio of the transverse momentum of the away-parton to that of the trigger parton, specifically [49]:

$$\left.\frac{dP}{dx_E}\right|_{p_{T_t}} = N(n-1)\frac{1}{\hat{x}_h}\frac{1}{(1+x_E/\hat{x}_h)^n} \tag{8}$$

where N is a normalization factor, and n (=8.1 at 200 GeV) is power of the inclusive invariant p_{T_t} distribution.

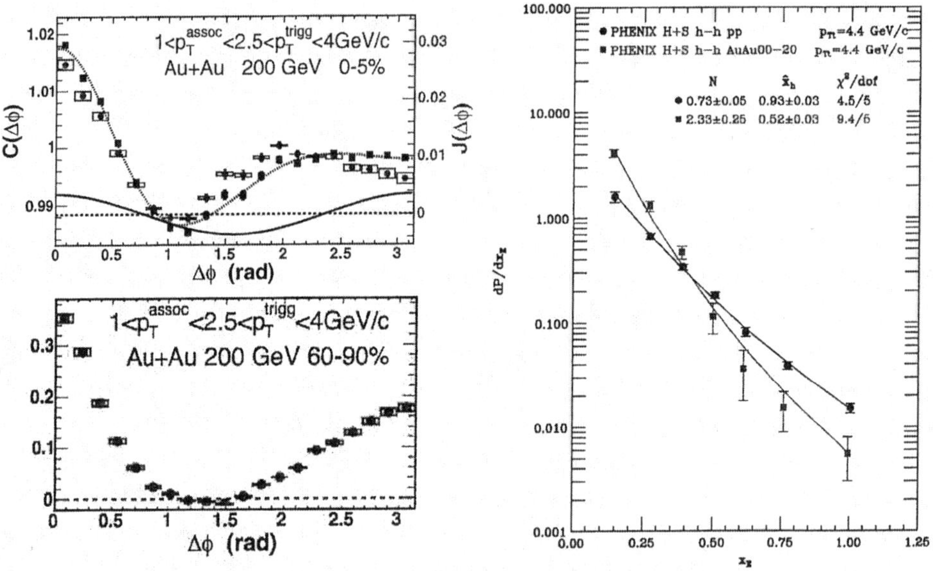

Figure 20: a) (left) Azimuthal correlation $C(\Delta\phi)$ of $h^{\pm}$ with $1 \leq p_{T_a} \leq 2.5$ GeV/c with respect to a trigger $h^{\pm}$ with $2.5 \leq p_{T_t} \leq 4$ GeV/c in Au+Au: (top) central collisions, where the line with data points indicates $C(\Delta\phi)$ before correction for the azimuthally modulated (v_2) background, and the other line is the v_2 correction which is subtracted to give the jet correlation function $J(\Delta\phi)$ (data points); (bottom)-same for peripheral collisions. b) (right) $x_E \approx p_{T_a}/p_{T_t}$ distribution for the Au+Au-central data compared to p-p.

8 A charming surprise

We designed PHENIX specifically to be able to detect charm particles via direct-single $e^{\pm}$ since this went along naturally with $J/\Psi \to e^{+} + e^{-}$ detection and since the single particle reaction avoided the huge combinatoric background in Au+Au collisions. We thought that the main purpose of open charm production, which corresponds to a hard-scale ($m_{c\bar{c}} \gtrsim 3$ GeV/c^2), would be a check of our centrality definition and $\langle T_{AA} \rangle$ calculation since the total production of c quarks should follow point-like scaling. In fact, our first measurement supported this beautifully [80]. However, our subsequent measurements proved to be much more interesting and even more beautiful. Figure 21a shows our direct-single-$e^{\pm}$ measurement in p-p collisions at

$\sqrt{s} = 200$ GeV [81] in agreement with a QCD calculation of c and b quarks as the source of the direct-single-$e^{\pm}$ (also called non-photonic $e^{\pm}$ at RHIC). The total yield of direct-$e^{\pm}$ for $p_T > 0.3$ GeV/c was taken as the yield of

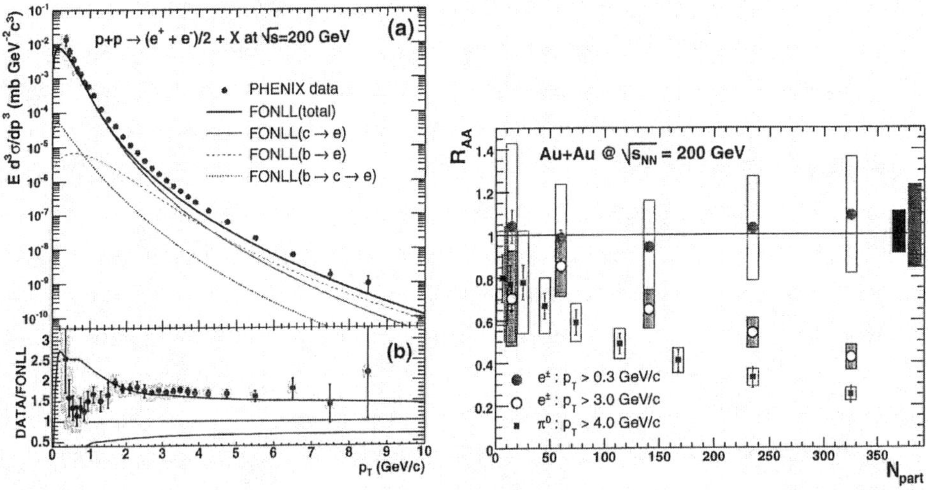

Figure 21: a) (left) Invariant cross section of direct $e^{\pm}$ in p-p collisions [81] compared to theoretical predictions from c and b quark semileptonic decay. b) (right) R_{AA} as a function of centrality (N_{part}) for the total yield of $e^{\pm}$ from charm ($p_T > 0.3$) GeV/c, compared to the suppression of the $e^{\pm}$ yield at large $p_T > 3.0$ GeV/c which is comparable to that of π^0 with ($p_T > 4$ GeV/c) [81]

c-quarks in p-p and Au+Au collisions. The result, $R_{AA} = 1$ as a function of centrality (Fig. 21b), showed that the total $c-(\bar{c})$ production followed point-like scaling, as expected. The big surprise came at large p_T where we found that the yield of direct-single-$e^{\pm}$ for $p_T > 3$ GeV/c was suppressed nearly the same as the π^0 from light quark and gluon production. This strongly disfavors the QCD energy-loss explanation of jet-quenching because, naively, heavy quarks should radiate much less than light quarks and gluons in the medium; but opens up a whole range of new possibilities including string theory [82].

The suppression of direct-single-$e^{\pm}$ is even more dramatic as a function of $p_T \gtrsim 5$ GeV/c (Fig 22a) which indicates suppression of heavy quarks as large as that for π^0 in the region where the $m \gtrsim 4$ GeV b-quarks dominate.

Figure 22b shows that heavy quarks exhibit collective flow (v_2), another indication of a very strong interaction with the medium.

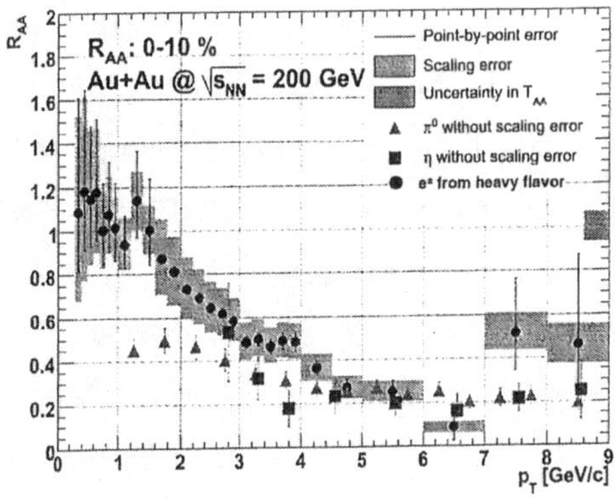

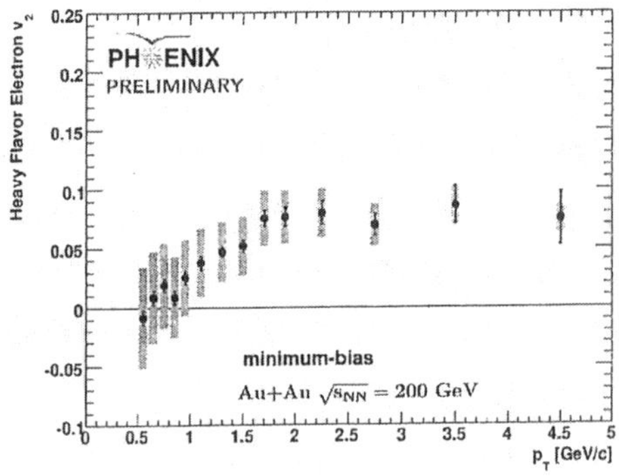

Figure 22: a) (left) R_{AA} (central Au+Au) b) (right) v_2 (minimum bias Au+Au) as a function of p_T for direct-$e^{\pm}$ at $\sqrt{s_{NN}} = 200$ GeV [81].

9 Zichichi to the rescue?

In September 2007, I read an article by Nino, "Yukawa's gold mine" in the CERN Courier taken from his talk at the 2007 International Nuclear Physics meeting in Tokyo, Japan, in which he proposed: "We know that confinement produces masses of the order of a giga-electron-volt. Therefore, according to our present understanding, the QCD colourless condition cannot explain the heavy quark mass. However, since the origin of the quark masses is still not known, it cannot be excluded that in a QCD coloured world, the six quarks are all nearly massless and that the colourless condition is 'flavour' dependent."

Nino's idea really excited me even though, or perhaps because, it appeared to overturn two of the major tenets of the Standard Model since it seemed to imply that: QCD isn't flavor blind; the masses of quarks aren't given by the Higgs mechanism. Massless b and c quarks in a color-charged medium would be the simplest way to explain the apparent equality of gluon, light quark and heavy quark suppression indicated by the equality of R_{AA} for π^0 and direct single-$e^{\pm}$ in regions where both c and b quarks dominate. Furthermore RHIC and LHC-Ions are the only place in the Universe to test this idea.

It may seem surprising that I would be so quick to take Nino's idea so seriously. This confidence dates from my graduate student days when I checked the proceedings of the 12th ICHEP in Dubna, Russia in 1964 to see how my thesis results were reported and I found several interesting questions and comments by an "A. Zichichi" printed in the proceedings. One comment about how to find the W boson in p+p collisions deserves a verbatim quote because it was exactly how the W was discovered at CERN 19 years later: "We would observe the μ's from W-decays. By measuring the angular and momentum distribution at large angles of K and π's, we can predict the corresponding μ-spectrum. We then see if the μ's found at large angles agree with or exceed the expected numbers."

Nino's idea seems much more reasonable to me than the string theory explanations of heavy-quark suppression (especially since they can't explain light-quark suppression). Nevertheless, just to be safe, I asked some distinguished theorists what they thought, with these results:

- Stan Brodsky: "Oh, you mean the Higgs field can't penetrate the QGP."
- Rob Pisarski: " You mean that the propagation of heavy and light quarks through the medium is the same."
- Chris Quigg (Moriond 2008): "The Higgs coupling to vector bosons γ, W, Z is specified in the standard model and is a fundamental issue. One big question to be answered by the LHC is whether the Higgs gives mass to fermions or only to gauge bosons. The Yukawa couplings to fermions are put in by hand and are not required." "What sets fermion masses, mixings?"
- Bill Marciano: "No change in the t-quark, W, Higgs mass relationship if there is no Yukawa coupling: but there could be other changes."

Nino proposed to test his idea by shooting a proton beam through a QGP formed in a Pb+Pb collision at the LHC and seeing the proton 'dissolved' by the QGP. My idea is to use the new PHENIX VTX detector, to be installed in 2010, to map out, on an event-by-event basis, the di-hadron correlations from identified $b-\bar{b}$ di-jets, identified $c-\bar{c}$ di-jets, which do not originate from the vertex, and light quark and gluon di-jets, which originate from the vertex and can be measured with π^0-hadron correlations. A steepening of the slope of the x_E distribution of heavy-quark correlations as in Fig. 20b will confirm in detail (or falsify) whether the different flavors of quarks behave as if they have the same energy loss (hence mass) in a color-charged medium. If Nino's proposed effect is true, that the masses of fermions are not given by the Higgs particle, and we can confirm the effect at RHIC or LHC-Ions, this would be a case where we Relativistic Heavy Ion Physicists may have something unique to contribute at the most fundamental level to the Standard Model, which would constitute a "transformational discovery." Of course the LHC could falsify this idea by finding the Higgs decay to $b - \bar{b}$ at the expected rate in p-p collisions. Clearly, there are exciting years ahead of us!

References

[1] *Report of the Workshop on BeV/Nucleon Coliisions of Heavy Ions—How and Why*, Bear Mountain, NY, 29 November–1 December 1974. (BNL-50445, Upton NY, 1975).

[2] See Ref. [3] for a more extensive list of references.

[3] M. J. Tannenbaum, Rep. Prog. Phys. **69** (2006) 2005–2059.

[4] E. V. Shuryak, Phys. Repts. **61** (1980) 71–158.

[5] I. Arsene *et al.*, BRAHMS Collab. Nucl. Phys. A**757** (2005) 1–27.

[6] B. B. Back *et al.*, PHOBOS Collab. Nucl. Phys. A**757** (2005) 28–101.

[7] J. Adams *et al.*, STAR Collab. Nucl. Phys. A**757** (2005) 102–183.

[8] K. Adcox *et al.*, PHENIX Collab. Nucl. Phys. A**757** (2005) 184–283.

[9] D. Rischke and G. Levin, eds., Nucl. Phys. A**750** (2005) 1–171.

[10] H. Satz Rep. Prog. Phys. **63** (2000) 1511-1574.

[11] K. Rajagopal, Nucl. Phys. A**661** (1999) 150c–161c.

[12] M. D'Agostino *et al.*, Nucl. Phys. A**749** (2005) 55c–64c.

[13] G. Policastro, D. T. Son and A. O. Starinets Phys. Rev. Lett. **87** (2001) 081601.

[14] H. Nastase "The RHIC fireball as a dual black hole" hep-th/0501068.

[15] It should be noted that the requirement of specificity to A+A collisions immediately rules out the QGP in p-p collisions, which is disputable. See Refs. [16, 17].

[16] R. M. Weiner, Int. J. Mod. Phys. E**15** (2006) 37–70.

[17] T. Alexopoulos, *et al.*, Phys. Lett. B**528** (2002) 43–48.

[18] T. Matsui and H. Satz, Phys. Lett. B**178** (1986) 416.

[19] B. Alessandro, *et al.*, NA50 Collab., Eur. Phys. J. C**39** (2005) 335–345. See also, F. Prino, Proc. XXX International Symposium on Multiparticle Dynamics, hep-ex/0101052.

[20] D. M. Alde, *et al.*, E772 Collab., Phys. Rev. Lett. **66** (1991) 2285–2288. See, also, M. J. Leitch, Eur. Phys. J. C**43** (2005) 157–160 and references therein.

[21] C. Albajar, *et al.*, UA1 Collab., Phys. Lett. B**186** (1987) 237.

[22] S. S. Adler, *et al.* PHENIX Collab., Phys. Rev. C**71** (2005) 034908.

[23] *The Relativistic Heavy Ion Collider Project: RHIC and its Detectors*, Nucl. Instrum. Methods A**499** (2003) 235–880.

[24] The detector is so non-conventional that it made the cover of Physics Today, October 2003.

[25] R. A. Lacey, Nucl. Phys. A**774** (2006) 199–214.

[26] M. Kaneta *et al.*, PHENIX Collab., J. Phys. G**30** (2004) S1217–S1220.

[27] A. Adare, *et al.*, PHENIX Collab., Phys. Rev. Lett. **98** (2007) 162301.

[28] J.-Y. Ollitrault, Phys. Rev. D**46** (1992) 229–245; Nucl. Phys. A**638** (1998) 195c–206c.

[29] H. Heiselberg and A.-M. Levy, Phys. Rev. C**59** (1999) 2716–2727.

[30] S. A. Voloshin, Nucl. Phys. A**715** (2003) 379c–388c.

[31] D. Teaney, Phys. Rev. C**68** (2003) 034913.

[32] P. K. Kovtun, D. T. Son, and A. O. Starinets Phys. Rev. Lett. **94** (2005) 111601.

[33] N. Armesto, N. Borghini, S. Jeon, U. A. Wiedemann, *et al.*, J. Phys. G**35** (2008) 054001.

[34] C. Alt, *et al.* NA49 Collab., Phys. Rev. C**68** (2003) 034903.

[35] H. Song and U. Heinz, Phys. Rev. C**78** (2008) 024902.

[36] Also, see, for example, M. J. Tannenbaum, Proc. Adriatico Research Conference on Trends in Collider Spin Physics, ICTP, Trieste, Italy, Dec. 5–8, 1995, Eds. Y. Onel, N. Paver, A. Penzo (World Scientific, Singapore, 1997) pp 31-53.

[37] S. S. Adler, *et al.* PHENIX Collab., Phys. Rev. D**76** (2007) 051006(R).

[38] P. Aurenche, *et al.*, Phys. Rev. D**73** (2006) 094007.

[39] G. Cocconi, L. J. Koester, and D. H. Perkins, Technical Report No. UCRL-10022 (1961), Lawrence Radiation Laboratory, (unpublished), p. 167.

[40] S. M. Berman, J. D. Bjorken and J. B. Kogut, Phys. Rev. D**4** (1971) 3388.

[41] F. W. Büsser, et al. (CCR), Phys. Lett. B**46** (1973) 471.

[42] A. L. Angelis, et al. (CCOR), Phys. Lett. B**79** (1978) 505.

[43] See M. J. Tannenbaum, Nucl. Phys. A**749** (2005) 219c for references.

[44] J. F. Owens, Rev. Mod. Phys. **59** (1987) 465.

[45] R. Cutler and D. Sivers, Phys. Rev. D**17** (1978) 196; Phys. Rev. D**16** (1977) 679.

[46] B. L. Combridge, J. Kripfganz and J. Ranft, Phys. Lett. B**70** (1077) 234.

[47] A. L. S. Angelis, *et al.*, Physica Scripta **19** (1979) 116.

[48] Data from Ref. [47] shown by M. Jacob, Proc. EPS Int. Conf. High Energy Physics, Geneva, 27 June–4 July, 2009, Ed. A. Zichichi (CERN, Geneva, 1980), p 512.

[49] S. S. Adler, *et al.* PHENIX Collab., Phys. Rev. D**74** (2006) 072002.

[50] Proc. 21st Int'l Conf. HEP, Paris, 1982, eds P. Petiau, M. Porneuf, J. Phys. C**3** (1982): see J. P. Repellin, p. C3-571; also see M. J. Tannenbaum, p. C3-134, G. Wolf, p. C3-525.

[51] A. L. S. Angelis, *et al.*, Nucl. Phys. B**209** (1982) 284.

[52] H. Fritzsch and P. Minkowski, Phys. Lett. B**69** (1977) 316.

[53] A. L. S. Angelis, et al (CMOR), Nucl. Phys. B**327** (1989) 541-568.

[54] F. W. Büsser, et al. (CCRS), Nucl. Phys. B**113** (1976) 189-245.

[55] I. Hinchliffe and C. H. Llewellyn Smith, Nucl. Phys. B**114** (1976) 45-60.

[56] F. W. Büsser, et al. (CCRS), Phys. Lett. B**56** (1976) 482-486.

[57] A. G. Clark, et al. (CSZ), Nucl. Phys. B**142** (1978) 29-52.

[58] A. Adare, *et al.* PHENIX Collab., Phys. Rev. D**79** (2009) 012003.

[59] Y. Akiba, *et al.* PHENIX Collab., Nucl. Phys. A**774** (2006) 403–408.

[60] M. Gyulassy, P. Levai and I. Vitev, Phys. Rev. Lett. **85** (2000) 5535-5538; I. Vitev and M. Gyulassy, Phys. Rev. Lett. **89** (2002) 252301.

[61] A. Adare, *et al.* PHENIX Collab., Phys. Rev. Lett. **101** (2008) 162301.

[62] I. Vitev, Phys. Lett. B**639** (2006) 38–45, and private communication to Ref. [61].

[63] D. Antreasyan, *et al.*, Phys. Rev. D**19** (1979) 764-778.

[64] See R. Baier, D. Schiff and B. G. Zakharov, Ann. Rev. Nucl. Part. Sci. **50** (2000) 37–69, and references therein.

[65] e.g. see N. P. Samios, Phys. Rev. **121** (1961) 275–281.

[66] N. M. Kroll and W. Wada, Phys. Rev. **98** (1955) 1355–1359.

[67] A. Adare, *et al.* PHENIX Collab., Phys. Rev. Lett. **104** (2010) 132301.

[68] A. S. Ito, *et al.*, Phys. Rev. D**23** (1981) 604–633.

[69] Thanks to Sasha Milov for the plot of $R_{AA}(p_T)$ for all PHENIX published and preliminary measurements. With the exception of the internal-conversion direct-γ where the fit to the p-p data is used to compute R_{AA}, all the other values of R_{AA} are computed from Eq. 6 using the measured Au+Au and p-p data points.

[70] A. Adare, *et al.*, PHENIX Collab., Phys. Rev. Lett. **98** (2007) 232301.

[71] T. Gunji, *et al.*, PHENIX Collab., J. Phys. G**34** (2007) S749–S752, `http://www.sinap.ac.cn/qm2006/parallel.htm#p21` .

[72] X. Zhao and R. Rapp, Phys. Lett. B**664** (2008) 253–257

[73] P. Braun-Munzinger and J. Stachel, Phys. Lett. B**490** (2000) 196–202.

[74] R. L. Thews, M. Schroedter and J. Rafelski, Phys. Rev. C**63** (2001) 054905.

[75] A. Andronic, P. Braun-Munzinger, K. Redlich and J. Stachel, Nucl. Phys. A**789** (2007) 334–356.

[76] R. Arnaldi, *et al.*, NA60 Collab., Nucl. Phys. A**830** (2009) 345c–352c.

[77] R. Arnaldi, presented at Heavy Quarkonia Production in Heavy Ion Collisions, Trento, Italy, 25-29 May 2009, `http://www.ect.it/Meetings/ConfsWksAndCollMeetings/ConfWksDocument/2009/Talks/WORKSHOP_25May09/arnaldi.ppt`

[78] J. Casalderrey-Solana, E. V. Shuryak and D. Teaney, J. Phys: Conf. Series **27** (2005) 22-31.

[79] M. J. Tannenbaum, PoS(CFRNC2006)001 (2006)

[80] S. S. Adler *et al.*, PHENIX Collab., Phys. Rev. Lett. **94** (2005) 082301.

[81] A. Adare, *et al.*, PHENIX Collab., Phys. Rev. Lett. **98** (2007) 172301.

[82] e.g. see Ref. [81] for a list of references.

CHAIRMAN: M.J. TANNENBAUM

Scientific Secretaries: M.A. Ahmad, V.N. Uvais

DISCUSSION I

- *Klopot:*

Could you please give some details on how in heavy ion collisions one can extract the ratio of shear viscosity to entropy density?

- *Tannenbaum:*

I don't wish to discuss this today. Ask me again tomorrow.

- *Perrey*:

In the PHENIX what difference would it make to be able to measure at full azimuth, 2π instead of π?

- *Tannenbaum*:

For detection of a single particle near mid-rapidity, simply a factor of 2 for the azimuthal acceptance. For two-particle detection like the J/ψ, or for correlations, we might gain a larger factor, possibly 4 by extending to full azimuth, depending how correlated the particles of interest are in azimuth—the less correlated back-to-back, the more we would gain. In rapidity, if we could have afforded to build our detector to cover $|\eta|<1.05$, like STAR, we could have gained another factor of 3 in acceptance for single particles. We also have a forward detector for muons which covers 2π in azimuth and $1.2<|\eta|<2.2$, primarily for J/ψ, Drell-Yan and W boson measurements. We also have two small electromagnetic calorimeters covering 2π in azimuth and $3.1<|\eta|<3.7$ for measurements in the forward cone. Personally, I like to measure at mid-rapidity because this emphasizes hard-scattering which occurs at large p_T, while in the forward direction p_T is limited by conservation of energy; and hard-scattering occurs with particles of large energy=($\sqrt{}$s-hat)/2, where ($\sqrt{}$s-hat=$2p_T/\sin\theta^*$) is the c.m. energy and θ^* the scattering angle for the parton-parton scattering. At mid rapidity, a parton has p_T=E=($\sqrt{}$s-hat)/2, and the kinematics of the parton are simple even with a single particle detected: $x_1 \approx x_2 \approx x_T = 2p_T/\sqrt{s}$, where $\sqrt{s}$ is the c.m. energy of the p-p collision. In the forward direction, the limited p_T and large energy forced by the kinematics puts the particles from hard-scattering in the region of low p_T dominated by soft processes such as diffraction dissociation. Also the so-called 'higher twist' QCD process come into play and complicate things further.

In general, I personally don't like to measure at forward rapidity—it's too complicated, with one exception. We can measure the gluon structure function in nuclei at low values of parton x by this method. The reaction we use is $g+q \rightarrow \gamma+q$. This will be my first slide tomorrow. The beauty of this reaction is that the γ comes straight out of the medium in a heavy ion collision without interacting, in distinction to the ougtoing q (or g) from hard-scattering. We have discovered that the q and g interact with the medium produced in A+A collisions at RHIC and lose energy, which suppresses the spectrum of particles, such as π^0, from the parton fragmentation by a factor of 5 relative to binary-scaling from p-p collisions. By contrast, the direct-γ are not suppressed, which shows that the suppression is a final state effect, caused by the medium.

The binary-scaling assumes that the probability of finding a g (and q) in a nucleus A is simply A times that in a nucleon. We know that for small values of x, the probability is less than A—this is called shadowing. It is possible that direct γ will be suppressed in A+A collisions at LHC due to initial state shadowing thus vastly complicating the study of the medium (QGP). For instance, for a 10 GeV direct γ at mid-rapidity at RHIC, the parton $x \approx x_T = 10/100 = 0.1$, while the same 10 GeV photon at the LHC will have $x = 10/5500 = 0.0018$. We can achieve a low x at RHIC by detecting both the direct γ and its quark jet at forward rapidity, $y_1 = y_2 = y$, in which case $x_1 = x_T\, e^{-y}$ and $x_2 = x_T\, e^{y}$. So, for $y = 4.0$ we can reduce x_1 by a factor of 55 (kinematics permitting), making up the ratio of energies of RHIC to the LHC and enabling us at RHIC to measure the gluon structure function in Au, at a value of x which is useful at the LHC.

- *Preghenella*:

I have a rather technical question about elliptical flow measurement at PHENIX. How do you determine the reaction plane? Do you have a dedicated detector?

- *Tannenbaum*:

Yes, we have three detectors to determine the reaction plane. The first detector is the beam-beam counter (BBC). We have two identical counters with 64 channels of quartz cerenkov counters located on each side of the interaction point at $3.0<|\eta|<3.9$. We measure the reaction plane in the detector by looking at the asymmetry of the azimuthal distribution of counts. To measure the resolution we compare the upstream and downstream counters. We also have a forward EM calorimeter that Mickey Chiu built which is just behind the BBC and measures the asymmetry of the energy. We recently added RxNP which is a pair of highly segmented scintillator Pb sandwich counters covering $1.2<|\eta|<2.8$. Using the 3 sets of counters gives us a factor of four better resolution on the angle of the reaction plane than our early measurements using the BBC only.

- *Preghenella*:

You told us about N_{ch} scaling with N_{part} using different models which work at different energies. What is the role of QGP quenching effects on this? Could it be the reason why one has to change model?

- *Tannenbaum*:

In some sense it is possible that the QGP could change the total multiplicity; so in principle, you are absolutely correct. However, I don't know what effect it will make, how the multiplicity would or should be changed. People who make calculations just assume that the multiplicity is the entropy. The measured multiplicity is consistent with simple scaling from p-p to Au-Au collisions. If you're right, there might be some effects. We mostly look for the non-statistical fluctuations in multiplicity to see such effects, but we observe very little of such fluctuations, they're tiny—mostly due to the Bose-Einstien effect. So, the answer is yes. On should expect effects on the final multiplicity due to

QGP formation, but so far we have not seen anything.

- *Yasnopolskiy:*

Using PbGl crystals at EmCal of PHENIX, do you have to keep them at low temperatures for better performance as it was originally designed for PbWO crystals at PHOS of ALICE experiment? If yes, have you also had long lasting problems with the cooling system?

- *Tannenbaum*:

Our PbGl calorimeter comes from the WA98 Experiment. We keep all our calorimeters at room temperature, including the PbWO crystals of the MPC. We have forced air cooling to dissipate the heat from the electronics for the central calorimters which in general works fine. For the MPC we can't apply cooling and they are kept at a relatively constant temperature by the large mass of the iron piston of the Muon spectrometer in which they are embedded. We don't have any major problems with our forced air cooling system.

- *Dixon*:

What causes the transition from n = 8 to n = 4 or 5 in the index of the single hadron inclusive p_T spectra you showed?

- *Tannenbaum*:

Going to larger p_T than the original CCR measurement and using only the two largest values of $\sqrt{s}$=52.7 and 62.4 GeV caused the effective index n_{eff} from x_T scaling to change from 8 to 5. Even at large x_T>0.30, including the $\sqrt{s}$=30.7 GeV in the calculation gives values of n_{eff}=7 (and 8 at lower x_T) rather than 5. Staying away from the lowest p_T and lowest $\sqrt{s}$ did the trick.

CHAIRMAN: M.J. TANNENBAUM

Scientific Secretaries: M.Kurkov, S.Yasnopolskiy

DISCUSSION II

- *Tannenbaum*:

Before taking questions, I would like to make a few comments on some unresolved issues between STAR and PHENIX that may not be clear from the lectures. Paul Sorensen presented STAR's view on the J/ψ suppression in Cu+Cu. On a semi-log plot, STAR presents only two points at p_T=5 and 7 GeV/c which show R_{AA}>1 with huge errors with the claim that the data are consistent with no J/ψ suppression at high p_T. However since Quark Matter 2009, PHENIX has added two more points to this plot at p_T=7 and 9 GeV/c which are consistent with all the PHENIX measurements at lower p_T and indicate a constant value of R_{AA}~0.6 from 0< p_T < 9 GeV/c. I think that STAR's emphasis on the rising R_{AA} here was misleading. Predictions show both a rising and falling $R_{AA}(p_T)$ while the PHENIX and STAR data together are consistent with a constant $R_{AA}(p_T)$.

STAR's view

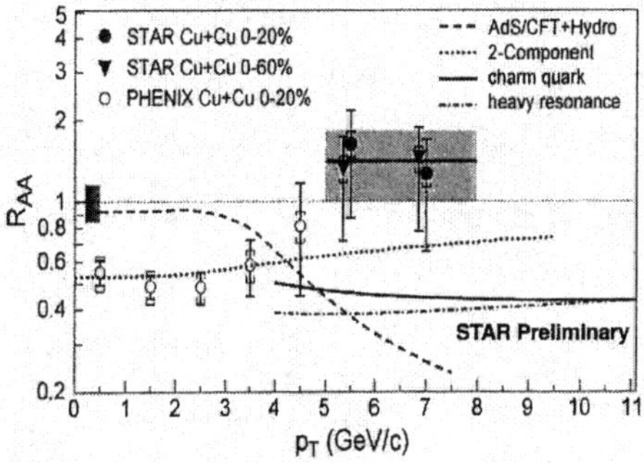

PHENIX' view post QM2009

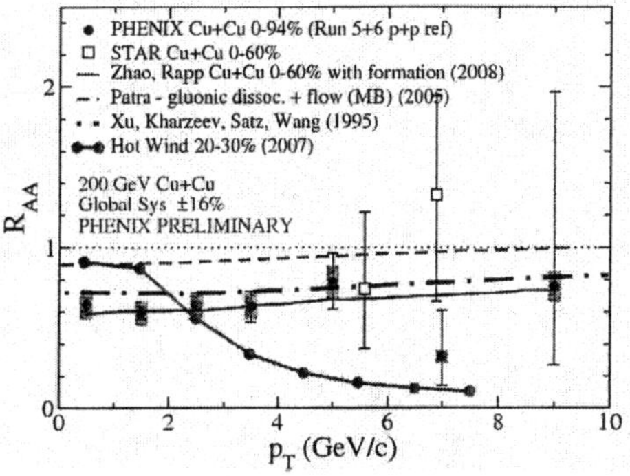

- *Sorensen*:

There are different points here.

- *Tannenbaum*:

I'm talking about these new PHENIX data points at 7 and 9 GeV/c.

- *Sorensen*:

So we are not ignoring these two data STAR points.

- *Tannenbaum*:

I'm not ignoring them, but whichever ones you use the STAR and PHENIX points are consistent, but STAR has huge errors.

- *Sorensen*:

Yes, but those are different than the ones on the final plot.

- *Tannenbaum*:

I don't know, I did not make this plot. Anyway, just look at the (PHENIX) red points.

- *Sorensen*:

I have it zero to sixty.

- *Tannenbaum*:

Anyway, it's PHENIX's view of this.

Second comment – a really interesting point made by Paul [Sorensen] is Stan Brodsky's idea about color transparency. I show the PHENIX measurement of two-particle correlations for identified mesons associated to identified baryon and meson triggers in the region of the baryon anomaly 2.5< p_{Tt}< 4.0 GeV/c which I did not discuss in my lecture but which Paul mentioned. For this range of p_{Tt}, the p/π ratio is 1 in central Au+Au collisions instead of the value of 0.3 measured in p-p collisions. For this plot the mesons are the sum of π and K and the baryons are protons + antiprotons. With the exception of the most central point for the near-side, there is no difference in the correlations to a baryon or a meson trigger. Since we know that the inclusive π mesons come from jets and satisfy the correct x_T scaling, while the inclusive protons have anomalous x_T scaling, the fact that the associated mesons are the same for trigger baryons and mesons means that the baryons also come from jets with the exception of the most central point where there are definitely fewer mesons associated to the baryon trigger than to the meson trigger. This could imply that some protons come out unacompanied for the most central collisions, possibly consistent with Brodsky's prediction of the higher twist process u+u→p+d-bar, where the proton comes out unacompanied and doesn't interact with the medium due to `color transparency' but the d-bar fragments into an away-jet. I like the Paul's point: this prediction can be further tested by measuring the v_2, since the 'color transparent' protons which don't interact with the medium should show zero v_2. This should reduce the total proton v_2 relative to the pion v_2 as N_{part}→350.

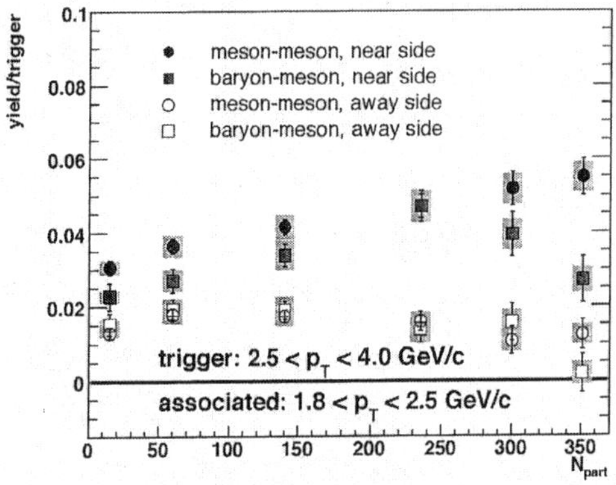

- *Tannenbaum*:

I have another point. Consider the plot from STAR which shows the Baryon to Meson ratio (B/M) of associated same-side particles to a non-identified hadron trigger, $h^{\pm}$, with p_{Tt}>4.0 GeV/c in both the ridge region (large $\Delta\eta$) and the jet region (small $\Delta\eta$) in comparison to the inclusive B/M ratio in p-p and Au+Au collisions (the dramatic difference in this ratio in Au+Au compared to p-p is the Baryon Anomaly which I previously mentioned). From this plot, STAR states that: the baryon to meson ratio in the ridge region is close to inclusive Au+Au and in the jet region is close to pp. However what they [STAR] really mean is that the baryon to meson ratio in the conditional yield of same side correlations to an $h^{\pm}$-trigger with p_{Tt}>4.0 GeV/c is close to the inclusive yield for p-p in the jet region and close to Au+Au in the ridge. They are not making a claim for the inclusive B/M ratio in the jet region, only the associated B/M ratio to a non-identified hadron trigger. When said in this precise way, the STAR result is clearly in agreement with PHENIX observation that both the meson and baryon triggers in the region of the baryon anomaly have associated particles which are consistent with the meson and baryon triggers both coming from jets (except maybe for the most central point). STAR ignored the large proton component in the trigger and did not add it to the associated particle yield to get the correct comparison to the inclusive yield. The STAR result for the associated same side B/M ratio in the jet region for an $h^{\pm}$-trigger with p_{Tt}~4.0 GeV/c is in reasonable agreement with a recent PHENIX publication [Phys. Rev. Lett. 101 (2008) 082301.]

- *Tannenbaum*:

Yesterday somebody asked me the question: "How to measure η/s?" which I didn't wish to answer then. Today, Paul Sorensen showed how this was derived by PHENIX in Phys. Rev. Lett. 98 (2007) 172301 from the $R_{AA}(p_T)$ and $v_2(p_T)$ of non-photonic electrons from heavy quark decay. Thanks Paul. However, please note that the PHENIX derivation is totally model dependent, too model dependent for my taste; that's why I prefer to leave it to the theorists. What one needs is a full hydro calculation which describes the initial state E_T or multiplicity distribution, the p_T spectra, and $v_2(p_T)$ for all identified particles. Also, at present there is still lots of model dependence for the eccentricity as well as the color charge density of the medium (QGCW as Zichichi calls it) as a function of time. Also we don't know the detailed physics of energy loss in the QGP. So there is a lot to learn and theorists are welcome to bring their models.

- *Tapia Takaki*:

What are the future plans at PHENIX concerning very low p_T (less than 3 GeV) direct photons?

- *Tannenbaum*:

The publication was submitted, but one referee had many criticisms with which we didn't agree, which caused us lots of additional work to respond. In about a month we are going to resubmit it to the journal and I hope it will have been published by the end of the year.

Every particle species produced in a p-p-collision has an exponential p_T distribution as $p_T \to 0$ except for direct-γ. The exponential distribution comes from soft processes such as diffraction dissociation; but we determined that in p-p collisions direct-γ are are not exponential as $p_T \to 0$, they are only produced by the hard process $g+q \to \gamma+q$. In other words, if you measure π^0 and direct-γ in p-p collisions, they both have a power law at high p_T, but as $p_T \to 0$ the π^0 becomes exponential but the direct-γ does not—the slope is nearly flat. Some people believe that p-p collisions are thermal because of the exponential; but I don't believe that an exponential distribution implies a thermal system—it is necessary, but not sufficient. Thus, an exponential distribution in Au-Au-collisions doesn't necessarily mean thermal, either. However, if the exponential distribution of low p_T direct-γ in Au+Au is really thermal emission, this implies radiation from the medium and thus these γ should have the v_2 that is characteristic of the medium, while direct-γ from hard-processes have zero v_2. So we plan to measure the v_2 of direct photons between 1 and 3 GeV. If the low-p_T photons are thermal they should have the same v_2 as the medium.

- *Marienfeld:*

Could you comment on the reasons for using the specific element in heavy ion collisions. For example why did you use gold, why will LHC use lead?

- *Tannenbaum*:

We started using gold in AGS experiments because the ordinary (24 karat) gold that you buy in a jewelry store is isotopcally pure; and since we used gold at AGS we know how to inject it, we know how to strip it of all its electrons, we know how to do everything. That's why we use it at RHIC. It will be interesting when we go to uranium, because uranium has a shape of a rugby ball, and hence when you make a central collision you have an asymmetric overlap which gives huge v_2 (normally v_2 equals 0 for central collisions for spherical nuclei); we are looking forward to it.

- *Marienfeld:*

What element would you favor if there were no problems of purity or availability?

- *Tannenbaum*:

Gold is fine. In fact I wanted element 256, but the nuclear physicists couldn't do this for me.

- *Chiu:*

For the "thermal photons" at low p_T, are there any measurements which one can do to constrain many models which now describe the data so that we can use the thermal photons to extract quantitative information about the QGP? The photons are very important since they would carry the information about the early stages without being biased by final state effects.

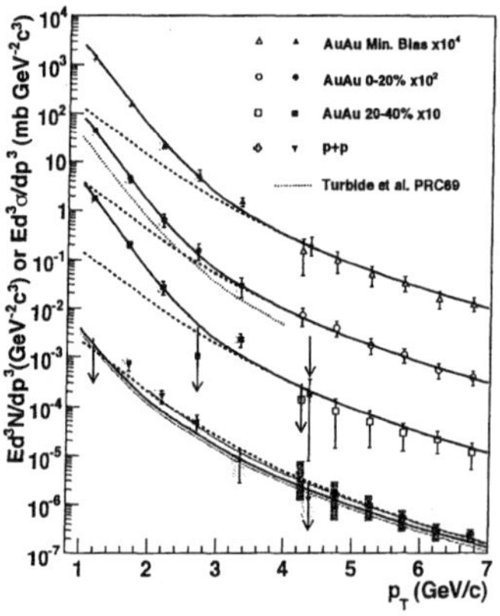

- *Tannenbaum*:

I don't think we can do it yet for the photon signal, because the exponential slope doesn't vary much for the three cases - peripheral, central and inclusive – which is rather suspicious. Maybe we will get a better idea if we have a better measurement versus centrality so that we could see the difference. The other test that I discussed above is the measurement as a function of angle to the reaction plane, v_2, which is zero for the direct-γ from hard-scattering which do not interact with the medium, while thermal γ are generated in the medium and so should exhibit the v_2 of the medium, which should be similar to the v_2 observed for hadrons. These are the only two ways I know, so far. I think we are stuck with them, because I don't see any measurements that we can make to pick out photons coming from the original time, which is ~10^{-23} sceconds in comparison to the 16-17 nsec for photons to get to out detectors which we measure with ~100 psec resolution.

- *Chiu:*

Can't you do something like use HBT?

- *Tannenbaum*:

That's an excellent point. People are trying to do HBT-correlations for photons, but they are extremely difficult.

- *Ichou:*

What do you think is the best measurement at LHC to help clarify the open issues at RHIC in the π^0 or jet quenching sector?

- *Tannenbaum*:

The good thing about ALICE is that you have identified particles. But your calorimeter is unfortunately nearly the same as ours, so I do not think you will be able to do better than we do at 20 GeV. We see that the direct-γ seems to be suppressed at 20 GeV/c approaching the π^0

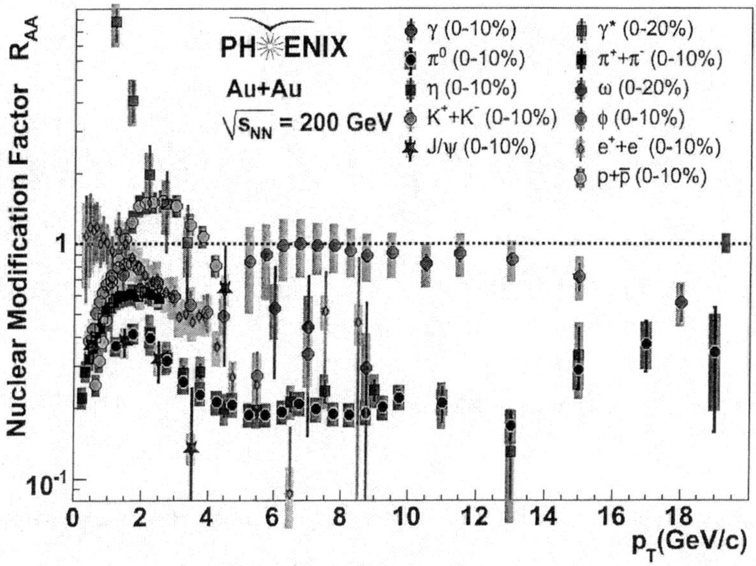

suppression, but it could easily be a detector effect. We have problems with π^0's because the photons merge and I am not sure about the efficiency of our cuts. However, suppression of direct-γ also could be due to the initial state structure function; we are working on that. The equality of R_{AA} for π^0 and γ at 20 GeV at RHIC would imply an initial state effect, which would mean that the medium effect is insignificant for p_T~20 GeV/c. I think the measurements from 10 to 20 GeV at LHC would be interesting, and going above 20 GeV even more interesting. However, because of the low x values for this p_T range at the LHC, direct photons will likely be suppressed due to shadowing in the structure function, so comparison with π^0's to show suppression by the medium will not be as obvous as at RHIC; and besides, the direct-γ measurement is very difficult. Thus you will need extensive p-p and p+Pb collision data at the same $\sqrt{s}$ for the π^0 comparison. A measurement of v_2 is probably easier because you don't need comparison data. Another thing you could do is a p/π ratio vs p_T and centrality. Typically, from jet fragmentation it is 0.3 and you will probably get lots of data in p-p collisions to verify this. The baryon anomaly [peak at 2 GeV on figure] is due to the fact that the p/π ratio is 1 in Au+Au at RHIC in this range instead of 0.3. This is still not explained so the behavior of the baryon anomaly at the LHC is important and should clarify our understanding. Also p/pbar tells you the baryon chemical potential straightaway.

- *Aschieri:*

Given the importance of the impact geometry in obtaining the properties of the quark-gluon plasma, is there the possibility of varying this impact geometry?

- *Tannenbaum*:

In a fixed target experiment it is very easy, but in a collider the situation is different because charged fragments get swept away from the collision axis. The only thing you can detect in a zero-degree calorimeter are free neutrons. We tried to detect all the charged fragments in an external calorimeter, but not very successfully. I don't know if it is possible to do it at LHC. I agree with your point: we don't have a direct measurement of the overlap region, what the ellipse is. In principle we know how to do it at RHIC, but we haven't succeeded in practice. However, I am confident that our measurement of collision centrality (hence impact parameter) by the percentiles of the multiplicity distribution is accurate to within our stated errors, so quite good.

- *Aschieri:*

I was thinking about varying the ion's shape via e.g. polarizing them.

- *Tannenbaum*:

If we take a nucleus like U^{238}, which has a football shape and spin, we could in principle have collisions with the U^{238} nuclei both transversely or longutidinally polarized. In RHIC, we have spin rotators so we can have either transverse or longitudinal polarized beams and we have the so-called Siberian Snakes to preserve the polarization. Each half turn around the ring, we flip the spin, so whatever imperfection develops in one half turn is canceled out in the other. Thus we preserve polarization with protons exceedingly well. I don't know what the spin of U^{238} is but I am sure that it is much less than 238/2. Unless something has roughly the magnetic moment of a proton, it's very hard to preserve its polarization, but if you find the right nucleus, in principle you can do it.

- *Klopot*

You told us about away side and near side correlations. How are these quantities measured in an actual experiment?

- *Tannenbaum*:

Here is a typical plot of the azimuthal angle difference $\Delta\phi$ of two particles, a trigger particle

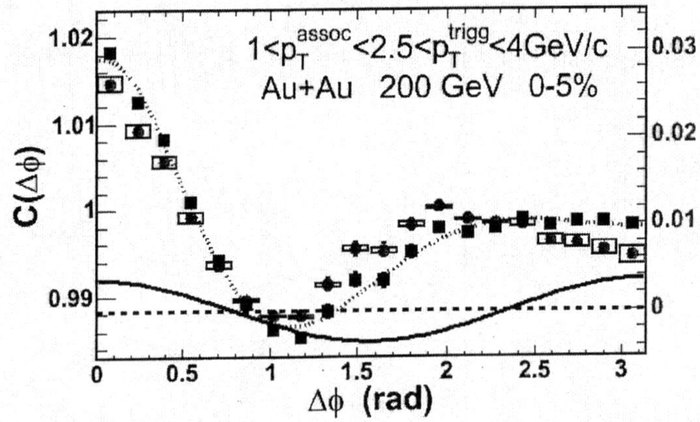

with p_{Tt} and an associated particle with p_{Ta}. This angle difference ranges from 0 to 2π radians and we fold the distribution around π radians in the plot. The filled squares are the actual conditional yield $C(\Delta\phi)$ of associated $h^{\pm}$ with $1< p_{Ta}< 2.5$ GeV/c per trigger $h^{\pm}$ with $2.5< p_{Tt}< 4.0$ GeV/c, corrected for the azimuthal acceptance, but only for particles in the range $|\eta|<0.35$. The red filled circles are the correlation due to the di-jet nature of the events, what we call the jet function, $J(\Delta\phi)$, which is $C(\Delta\phi)$ corrected for v_2 (the cosine curve), which is subtracted since v_2 is the correlation of each particle to the event plane, not to each other. This correction can be measured using correlations of particles from different events which have the same event plane angle. We define the same-side correlation as the region $|\Delta\phi|< \pi/2=1.57$ radians from the trigger, the left region of the plot, and the away side as $|\Delta\phi- \pi|< \pi/2$. For the same-side yield we simply integrate the yield in the peak. However, note that in Au+Au collisions the away-side peak is wide and spills over to the same side. For the away-side yield we quote the region of integration in $\Delta\phi$, sometimes over the whole "away" peak, sometimes only over the away-side, $|\Delta\phi- \pi|< \pi/2$. In both cases we only measure the yield in the limited η range, $|\eta|<0.35$, of our detector. The absolute yield is not correct (STAR is much better on this issue); but the relative yield for Au+Au and p-p is fine.

ITP-UU-09/38
SPIN-09/35

Quantum Gravity without Space-time Singularities or Horizons*

Gerard 't Hooft

Institute for Theoretical Physics
Utrecht University
and

Spinoza Institute
Postbox 80.195
3508 TD Utrecht, the Netherlands
e-mail: g.thooft@uu.nl
internet: http://www.phys.uu.nl/~thooft/

Abstract

In an attempt to re-establish space-time as an essential frame for formulating quantum gravity – rather than an "emergent" one –, we find that exact invariance under scale transformations is an essential new ingredient for such a theory. Use is made of the principle of "black hole complementarity", the notion that observers entering a black hole describe its dynamics in a way that appears to be fundamentally different from the description by an outside observer. These differences can be boiled down to conformal transformations. If we add these to our set of symmetry transformations, black holes, space-time singularities, and horizons disappear, while causality and locality may survive as important principles for quantum gravity.

Version September 18, 2009

*Erice School of Subnuclear Physics 2009

1. Introduction: scales in general relativity

Due to quantum effects, a black hole emits particles of all sorts. Thus, if left entirely by itself, a black hole gradually looses mass, eventually ending its life with a gigantic explosion. This is the Hawking effect[1]. For an intuitive understanding of our world, the Hawking effect seems to be quite welcome. It appears to imply that black holes are just like ordinary forms of matter: they absorb and emit things, they have a finite temperature, and they have a finite lifetime. One would have to admit that there are still important aspects of their internal dynamics that are not yet quite understood, but this could perhaps be considered to be of later concern. Important conclusions could already be drawn: the Hawking effect implies that black holes come in a denumerable set of distinct quantum states. This also adds to a useful and attractive picture of what the dynamical properties of space, time and matter may be like at the Planck scale: black holes seem to be a natural extension of the spectrum of elementary physical objects, which starts from photons, neutrinos, electrons and all other elementary particles.

In such a picture, however, what happens to the horizon and the space-time singularities? An answer sometimes suggested by string theorists, as well as others, is that all of space-time is just "emergent"[2][3][4]; the theory should first be formulated without space-time altogether. Or, perhaps, time alone is an emergent concept[5]. It was argued that, at least, locality would have to be abandoned[6][7]. In this paper, however, we dismiss all such options. In particular, we insist that any satisfactory theory should have built in a strong form of causality, as well as locality, in order to explain why cause precedes effect, and why events separated at some distance from one another appear to evolve independently. For this, space-time appears to be indispensable.

Something has to give, and in this paper we claim to have found a good candidate for that: the definition of *scales* in space-time. It should be done in a way that differs from conventional wisdom.

In what follows it is useful to write the metric tensor $g_{\mu\nu}(x)$ as:

$$\begin{aligned} g_{\mu\nu}(x) &= \omega^2(x)\,\hat g_{\mu\nu}(x)\,, \qquad \text{where} \\ \det(\hat g_{\mu\nu}) &= -1\,, \qquad \omega = (-\det(g_{\mu\nu}))^{1/8}\,. \end{aligned} \tag{1.1}$$

Since the light cone is defined by solutions of the equation

$$\mathrm{d}s^2 = g_{\mu\nu}\mathrm{d}x^\mu \mathrm{d}x^\nu\,, \tag{1.2}$$

it is $\hat g_{\mu\nu}$ that determines the positions of light cones, whereas $\omega(x)$ defines the scales for rulers and clocks in the theory.

As for general coordinate transformations, the split (1.1) simply implies that $\hat g_{\mu\nu}(x)$ and $\omega(x)$ transform anomalously. When $x^\mu \to u^\mu(x)$, then

$$\begin{aligned} \hat g_{\mu\nu}(x) &\to \tilde{\hat g}_{\mu\nu}(u) = \frac{\partial x^\alpha}{\partial u^\mu}\frac{\partial x^\beta}{\partial u^\nu}\,\hat g_{\alpha\beta}(x)\,\det\left(\frac{\partial x}{\partial u}\right)^{-1/2}\,; \\ \omega(x) &\to \tilde\omega(u) = \omega(x)\,\det\left(\frac{\partial x}{\partial u}\right)^{1/4}\,. \end{aligned} \tag{1.3}$$

2. Black hole complementarity

An observer entering a black hole would experience quantized fields $\phi(x)$ that are defined in a space-time expanding across the horizon. Indeed, this observer would not even know where exactly the horizon is, and her description should certainly not depend on that. Assuming the horizon to be a plane that we put at the origin of our coordinate frame, we can regard space-time as to be split into two pieces:

$$\text{region } I\ , \quad z > 0,\ \phi(x) = \phi_I(\varrho, \tau)\ , \quad \text{with} \quad z = \varrho \cosh \tau\ , \tag{2.1}$$

$$\text{region } II\ , \quad z < 0,\ \phi(x) = \phi_{II}(\varrho, \tau)\ , \quad \text{with} \quad z = -\varrho \cosh \tau\ . \tag{2.2}$$

So here, $\phi(x)$ is the field used by the ingoing observer, while the outside observer has only ϕ_I to his disposal. One derives that the vacuum state $|\text{vac}\rangle$ experienced by the ingoing observer, splits up as

$$|\text{vac}\rangle = \sum_n |n\rangle_I |n\rangle_{II} e^{-\pi E_n}\ , \tag{2.3}$$

where $|n\rangle_I$ is the complete set of states of the outside observer in region I, while $|n\rangle_{II}$ is a complete set of states that all escape the attention of this observer. E_n are the energies in these states, and τ is the normalized time coordinate for the outside observer. The probability that a state n is detected is therefore

$$W(n) = C\, e^{-2\pi E_n}\ , \tag{2.4}$$

so that we have a temperature T which, in the units used, obeys $kT = 1/2\pi$.

Black hole complementarity[8][9][10][11] refers to the fact that observers who stay outside the black hole can see the Hawking particles, including the effect they have on the space-time metric, while observers entering the black hole can neither see these particles, nor the effect they have on the metric. On the other hand, these observers do have to add all particles that entered, and will enter, the black hole to obtain a meaningful description of what is going on there. These differences are caused by the fundamental difference in the way the two sets of observers experience the notion of time.

We now bring a slight twist in this notion of complementarity. Rather than distinguishing observers who go into a black hole from the ones staying out, we note that the ingoing observer sees everything that went into the black hole, whereas the second observer sees all objects going out – the Hawking radiation. Thus reformulated, complementarity really refers to the mapping from the in-states to the out-states.

It is instructive then, to consider the special case that the black hole was formed by the collapse of a single shell of massless non interacting particles moving in radially, because this case has a simple exact analytical solution: outside the shell we have the Schwarzschild metric, inside the shell space-time is flat, and the two regions are glued together in such a way that the metric is continuous. In particular, the transverse components of the metric must match, which means that the r coordinate coincides inside and outside the shell. Furthermore, we take the case that the outgoing Hawking particles also form a single shell of matter. Since, actually, the Hawking particles have a thermal spectrum, it is highly unlikely that these particles emerge as a single shell, but it is not impossible. The shell is suppressed by large Boltzmann factors, but this just means that it occupies a very tiny portion of phase space; that is not relevant to our procedure. Later, one may replace the in- and outgoing matter configurations by more realistic ones. Thus, for simplicity, the in-configuration is described by Fig. 1 *a*, and the out-configuration by 1 *b*.

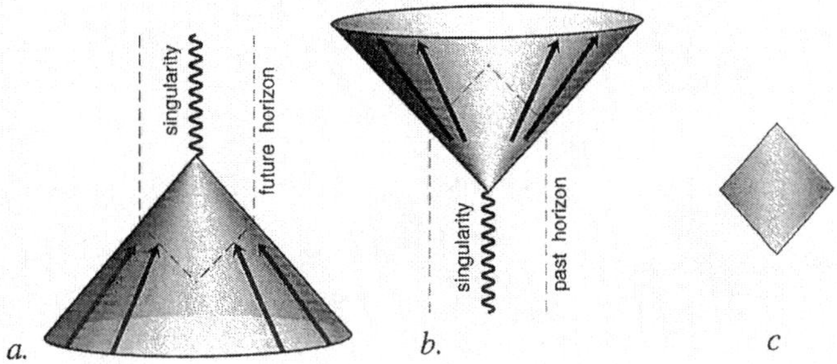

Figure 1: *a*) The in-configuration of a black hole; *b*) the out-configuration. Dotted lines are the horizons in both pictures. In the case both the in-going matter and the out-going matter forms a single shell, there is a region at the center that both pictures seem to have in common (*c*).

According to the complementarity principle, the ingoing observer thinks that only the in-going matter in Fig. 1 *a* is responsible for the matter contribution to the space-time curvature, while the Hawking radiation is essentially invisible. In Fig. 1 *b*, space-time is shown as it is recorded by the late outside observer: *only* the Hawking particles are responsible for the space-time curvature, while all in-going matter that formed the black hole in turn is invisible to him[1]. Naturally, one asks how these differences can be implemented in one all-embracing theory. Are there transformations from one picture to the other? How complex may these transformations be? How can these be reconciled with causality?

3. Causality and locality

We now insist that the complementarity transformation that relates Fig. 1 *a* with Fig. 1 *b* should be a local transformation of dynamical physical degrees of freedom, and furthermore, that this transformation preserve the causal order of events, so that an evolution law obeying obvious causality can be phrased in either one of the two systems. This necessarily means that the light cones should not be affected by this transformation. Therefore,

[1]This picture differs somewhat from the older versions of "complementarity", where the outside observer does see the matter going in; we presently take the viewpoint that if an observer registers the effect of Hawking radiation on the metric, all ingoing matter becomes invisible, which is the exact time-reverse of what the ingoing observer sees.

in Eqs. (1.1) and (1.2), $\hat{g}_{\mu\nu}(x)$ should remain the same, but $\omega(x)$ may be different. To be precise, $\omega(x)$ is observer dependent. The value of $\omega(x)$ depends on whether the point x is seen through a curtain of Hawking particles or directly. Therefore, according to the complementarity principle, the physical events taking place at some space-time point x should not depend explicitly on ω. We end up with a picture where $\omega(x)$ is entirely unobservable locally, much like a local gauge parameter $\Lambda(x)$ in a gauge theory.

The fields $\hat{g}_{\mu\nu}(x)$ must obey equations of their own. Formulating these equations will be difficult, and we postpone this to future work. By first studying the physical consequences of this idea, we must cultivate some feeling about how to proceed.

At this point, we can already identify some of these consequences. Suppose we have a space-time where we are free to choose ω. Suppose that *cosmic censorship*[12][13] holds. This means that, in terms of the conventional metric $g_{\mu\nu}$, any naked singularity is hidden behind a horizon. Then, whenever we encounter such a singularity, we can adjust ω in such a way that a clock shows infinite time when approaching this singularity. Therefore, it now occurs at $t = \infty$. In practical examples, the singularity will also have been smeared out so that it effectively disappears, and then also the horizon disappears, as we will demonstrate.

4. Conformal transformations

In the case described in Section 2, where both in- and out-going matter forms a single shell, there is an internal flat region that both descriptions have in common. Since it is flat, they have the same, flat value $\hat{g}_{\mu\nu}(x) = \eta_{\mu\nu}$ for the metric. So here, the only choice we have for the transformation is the value for the scales $\omega(x)$. As is well-known, the only transformations that leave $\hat{g}_{\mu\nu} = \eta_{\mu\nu}$ intact, are the conformal transformations. If in both Figures 1 a and 1 b, we choose the tip where the singularity starts as the origin of space-time, then this mapping is

$$x^\mu_{\text{in}} = (-1)^{\delta^{\mu o}} \frac{x^\mu_{\text{out}}}{|(x_{\text{out}})^2|} , \tag{4.1}$$

where the square is of course the Lorentz invariant square of the coordinate x_{out}. We chose the sign here such that the timelike causal order is preserved. All vectors in the relevant region are timelike with respect to the origin chosen.

The transformation (4.1) is illustrated in Figure 2, where we see that indeed lightlike geodesics stay lightlike. We see cosmic censorship at work: the singularity in both frames are moved to $t \to \pm\infty$ in the other frames. As for the region outside the collapsing and the expanding shells, we might decide to choose no change in ω. At the lightlike shells themselves, the mapping is singular, but this may be attributed to the fact that we are dealing with macroscopic black holes in this example, so that, in both coordinate frames, matter tends to take divergent values for their stress-energy-momentum tensors. In microscopic black holes, the transformation may be less singular at these regions.

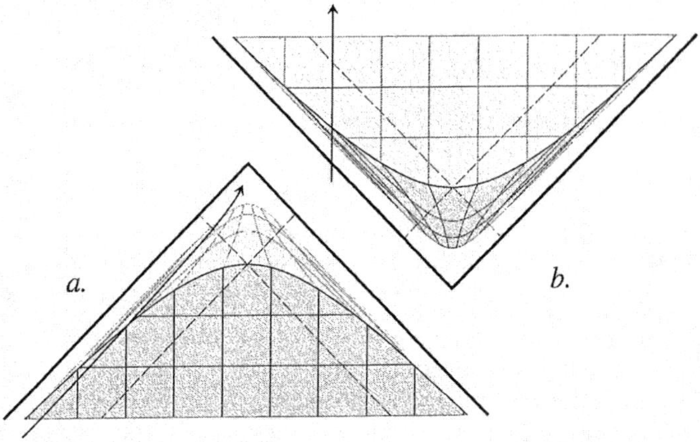

Figure 2: The conformal transformation for the inside of the black hole of Fig. 1. Bold lines show matter going in (a), and out (b). The dotted lines show how lightlike geodesics transform. The arrowed line is the same timelike trajectory in the two frames.

5. Scale invariance and the stress-energy-momentum tensor

The above considerations imply that the scale factor $\omega(x)$ is ambiguous, yet it is needed to describe clocks and rulers in the macroscopic world, and it is also needed if we wish to compute the Riemann and the Ricci curvature, because they depend on the entire metric $g_{\mu\nu}$, not just $\hat{g}_{\mu\nu}$. Clearly, the world that is familiar to us is not scale invariant. Without ω we cannot define distances and we cannot define matter, but we can define the geometry of the light cones, and, if $\hat{g}_{\mu\nu}$ is non-trivial, it should describe some of the physical phenomena that are taking place. It should be possible to write down equations for it without directly referring to ω.

To what extent will $\omega(x)$ be determined by $\hat{g}_{\mu\nu}(x)$, if we impose some physical conditions? For instance, we can impose the condition that the space-time singularities are all moved to $t \to \infty$, so that the interiors of black holes at the moment of collapse, are re-interpreted as the exteriors of decaying black holes, as in Figs. 1 and 2. After that is done, conformal transformations are no longer possible, because they would move infinities from the boundary back to finite points in space-time, which we do not allow. But we do wish to impose some extra "gauge" condition on $\omega(x)$, so as to fix its value, allowing us to define what clocks, rulers, and matter are.

A natural thing to impose is that all particles appear to be as light as possible. If we would have only non-interacting, massless particles, then $T^\mu_\mu = 0$. Therefore, a natural gauge to choose is the vanishing of the induced Ricci scalar:

$$R(\omega^2 \hat{g}_{\mu\nu}) = 0 \,. \tag{5.1}$$

When $\hat{g}_{\mu\nu}(x)$ is given, we can compute (in 4 space-time dimensions):

$$\begin{aligned} R_{\mu\nu} &= \hat{R}_{\mu\nu} + \omega^{-2}(4\,\partial_\mu\omega\,\partial_\nu\omega - \hat{g}_{\mu\nu}\hat{g}^{\alpha\beta}\,\partial_\alpha\omega\,\partial_\beta\omega) \\ &\quad + \omega^{-1}(-2D_\mu\,\partial_\nu\omega - \hat{g}_{\mu\nu}\hat{g}^{\alpha\beta}D_\alpha\,\partial_\beta\omega)\ ; \end{aligned} \tag{5.2}$$

$$\omega^2 R = \hat{R} - 6\,\omega^{-1}\hat{g}^{\mu\nu}D_\mu\partial_\nu\omega\ . \tag{5.3}$$

Here, we write $R_{\mu\nu}(\omega^2\hat{g}) = R_{\mu\nu}$, and $R_{\mu\nu}(\hat{g}_{\mu\nu}) = \hat{R}_{\mu\nu}$, and similarly the Ricci scalars. The covariant derivative D_μ treats the field ω as a scalar with respect to the metric $\hat{g}_{\mu\nu}$.

The condition that Eq. (5.3) vanishes gives us the equation

$$\hat{g}^{\mu\nu}D_\mu\partial_\nu\omega(x) = \tfrac{1}{6}\hat{R}\,\omega(x)\ . \tag{5.4}$$

This happens to be a linear equation for $\omega(x)$, which is the reason for having included the square in its definition, in Eq. (1.1), since ω is not restricted to be infinitesimal.

Now let $\omega(x)$ obey Eq. (5.4) in a region where $\hat{g}_{\mu\nu} = \eta_{\mu\nu}$, so that $\hat{R} = 0$, and consider the Fourier transform $\omega(k)$ of $\omega(x)$, which, at $k \neq 0$, we now take to be infinitesimal. The wave vector k obeys $k^2 = 0$. Let this wave vector be in the $+$ direction. Then, according to Eq. (5.2), the only non-vanishing component of the Ricci tensor, and hence also the only non-vanishing component of the stress-energy-momentum tensor, is the $++$ component. Thus, we have shells of massless particles traveling in the x^--direction. Adding all possible Fourier components, we see that, in the infinitesimal case, our transition to non-trivial $\omega(x)$ values leads us from the vacuum configuration to the case that massless particles are flying around. As long as $\omega(k)$ at $k \neq 0$ stays infinitesimal, these massless particles are non-interacting.

In the case of an evaporating black hole, these massless particles are the Hawking particles. Our complementarity transformation, the transformation that modifies the values of $\omega(x)$ while keeping Eq. (5.4) valid, switches on and off the effects these Hawking particles have on the metric[2].

Often, it will seem to be more convenient to impose that we have flat space-time, *i.e.* $\omega^2\hat{g}_{\mu\nu} = \eta_{\mu\nu}$ at infinity, rather than imposing $R = 0$ everywhere. In the case of an implosion followed by an evaporating black hole (for simplicity emitting a single shell of matter as in Section 2), this would allow us to glue the different conformal regions together, to obtain Figure 3. This then gives us a region of large Ricci scalar curvature near the horizon. Ricci scalar curvature is associated with a large burst of pressure in the matter stress-energy-momentum tensor. This means that an observer who uses this frame sees a large explosion near the horizon that sends the ingoing material back out, that is, the black hole explodes classically. The *local* observer would be wondering what causes this "unnatural" explosion, which, at the very last moment, avoided the formation of a permanent black hole; in our theory, however, the local observer would see no reason to use this function ω, so she would not notice any causality violation. Only the distant,

[2]It is often claimed that Hawking particles have little or no effect on the stress-energy-momentum operator on the horizon[14][15]. This, however, is true only for the expectation values in the states used by the ingoing observer (the Hartle-Hawking states). The outgoing observer uses Boulware states, for which the $T_{\mu\nu}$ diverges[16][17].

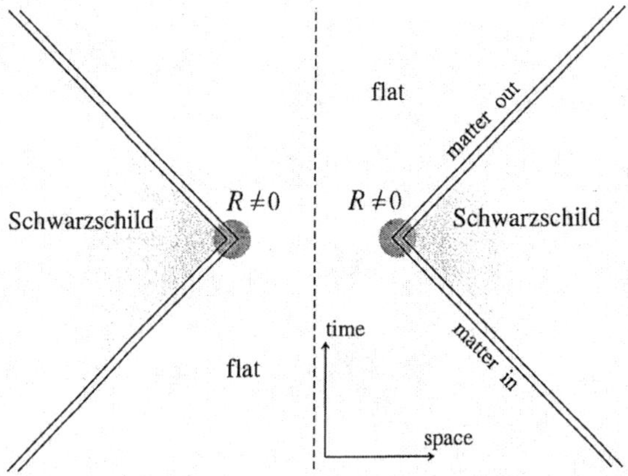

Figure 3: Combining the different conformal frames into one metric for the imploding and subsequently evaporating black hole, gives us a region of large Ricci scalar values R near the horizon (here, the two circles). Note that the metric outside the collapsing and evaporating shells is Schwarzschild, while inside the shells it is flat. Demanding ω to be continuous across the shells implies that the r coordinate should be matched there (see Ref. [18]).

outside observer would use this ω, concluding that, indeed, the black hole is not an eternal one since it evaporates.

The metric described in Figure 3 is related to the metrics described in Ref. [18]. There, care was taken that no conical singularity should arise when ingoing and outgoing shells meet. Here, we do allow this singularity, interpreting it as a region of large R values, which we now demand not to be locally observable.

6. Conclusions

Consider a large region of space-time, filled with light particles scattered here and there. Suppose we ask what this state looks like after a very large Lorentz boost. This question is significant for instance if we consider the neighborhood of a black hole horizon; a (large) time boost for an external observer corresponds to a (large) Lorentz boost for a local observer. A problem with that is that light particles in one frame transform to extremely energetic ones in the boosted frame. Their energies may become so large that the ensuing back reaction upon the metric may no longer be ignored. As we saw in Section 5, the effects on the space-time metric due to light particles moving nearly with the speed of light can be efficiently represented by a conformal factor in the metric[3]. This means that

[3]The configurations described in Section 5 are actually limited to superpositions of flat shells of massless matter; this would leave a problem for massless particles that are point-like. In a quantum theory this happens to be not such a bad restriction, if we assume that the particles form plane waves,

large Lorentz transformations may work as usual only on spaces with a $\hat{g}_{\mu\nu}$ metric, but generate delicate gravitational corrections in the scale function $\omega(x)$. This may be seen as a different way to phrase our motivation for claiming that $\omega(x)$ should be considered as locally unobservable, in particular when black hole horizons are considered. This way, the Lorentz group can be kept as a perfect symmetry near the Planck scale, but only for the $\hat{g}_{\mu\nu}$ part of the metric.

Our theory avoids the need to treat space-time as "emergent" [2][3]. Rather, we concentrate upon the demands of causality and locality. The light cones therefore come first, and the scale function ω is of secondary importance.

Describing matter in a $\hat{g}_{\mu\nu}$ metric will still be possible as long as we restrict ourselves to conformally invariant field theories, which may perhaps be not such a bad constraint when describing physics at the Planck scale. Of course that leaves us the question where Nature's mass terms come from, but an even more urgent problem is to find the equations for the gravitational field itself, considering the fact that Newton's constant G_N is not scale-invariant at all. The Einstein-Hilbert action is not scale-invariant. Here, we cannot use the Riemann curvature or its Ricci components, but the Weyl component is independent of ω, so that may somehow have to be used.

We suspect that, eventually, scales enter into our world in the following way. Information is now strictly limited to move along the light cones, since only light like geodesics are well-defined, not the time like or space like ones. It is generally believed that the amount of information moving around in Nature is limited to exactly one bit in each surface element of size $4 \ln 2$ Planck lengths squared. Turning this observation around, one might assume that, whatever the equations are, they define information to flow around. The density of this information flow may well define the Planck length locally, and with that all scales in Nature. Obviously, this leaves us with the problem of defining what exactly information is, and how it links with the equations of motion. The notion of information might not be observer-independent, as the scale factor ω isn't. Quantum mechanics will probably require that all these bits of information form distinct elements of a basis for Hilbert space, as described in Refs. [19][20].

References

[1] S.W. Hawking, *Particle creation by black holes*, Commun. Math. Phys. **43** (1975) 199, Erratum ibid. **46** (1976) 206.

[2] S. Liberati, M. Visser, and S. Weinfurtner, *Naturalness in emergent spacetime*, Phys. Rev. Lett. 96:151301, 2006, arXiv:gr-qc/0512139.

[3] T. Konopkal, F. Markopoulou and S. Severini, *Quantum Graphity: a model of emergent locality*, arXiv:0801.0861v2 [hep-th].

[4] F. Girelli, S. Liberati, and L. Sindoni, *On the emergence of Lorentzian signature and scalar gravity* ,arXiv:0806.4239v2 [gr-qc].

---

but we will not further expand on this issue here.

[5] J. Butterfield and C. J. Isham, *On the emergence of time in quantum gravity*, arXiv:gr-qc/9901024.

[6] S. B. Giddings, *Black hole information, unitarity, and nonlocality*, Phys. Rev. **D 74**, 106005 (2006) [arXiv:hep-th/0605196].

[7] S. B. Giddings, D. Marolf and J. B. Hartle, *Observables in effective gravity*, Phys. Rev. **D 74**, 064018 (2006) [arXiv:hep-th/0512200].

[8] G. 't Hooft, *On The quantum structure of a black hole*, Nucl. Phys. **B 256** (1985) 727.

[9] G. 't Hooft, *The black hole interpretation of string theory*, Nucl. Phys. **B 335** (1990) 138.

[10] G. 't Hooft, *The Black hole horizon as a quantum surface*, Phys. Scripta **T36** (1991) 247.

[11] L. Susskind, L. Thorlacius and J. Uglum, *The Stretched horizon and black hole complementarity*, Phys. Rev. **D 48** 3743 (1993) [arXiv:hep-th/9306069].

[12] R. Penrose, *The Question of Cosmic Censorship*, Chapter 5 in Black Holes and Relativistic Stars, R. Wald (editor), (1994) (ISBN 0-226-87034-0).

[13] R.M. Wald, *Gravitational collapse and cosmic censorship*, to appear in 'The Black Hole Trail', ed. by B. Iyer. In *Iyer, B.R. (ed.) et al.: Black holes, gravitational radiation and the universe*, p.69. e-Print: gr-qc/9710068.

[14] K. W. Howard, *Vacuum $\langle T_\mu^{\ \nu} \rangle$ in Schwarzschild space-time.*, Phys. Rev. D **30** (1984) 2532.

[15] K. W. Howard and P. Candelas, *Quantum stress tensor in Schwarzschild space-time*, Phys. Rev. Lett. **53** (1984) 403.

[16] B. P. Jensen, J. G. McLauglin and A. C. Ottewill, *Anisotropy of the quantum thermal state in Schwarzschild spacetime*, Phys. Rev. D **45** (1992) 3002.

[17] M. Visser, *Gravitational vacuum polarization II: Energy conditions in the Boulware vacuum*, Phys. Rev. **D54** (1996) 5116 [arXiv:gr-qc/9604008].

[18] T. Dray and G. 't Hooft, *The effect of spherical shells of matter on the Schwarzschild black hole*, Comm. Math. Phys. **99** (1985) 613.

[19] G. 't Hooft, *Emergent quantum mechanics and emergent symmetries*, presented at PASCOS 13, Imperial College, London, July 6, 2007; ITP-UU-07/39, SPIN-07/27; arXiv:0707.4568[hep-th].

[20] G. 't Hooft, *Entangled quantum states in a local deterministic theory*, 2nd Vienna Symposium on the Foundations of Modern Physics (June 2009), ITP-UU-09/77, SPIN-09/30; arXiv:0908.3408.

CHAIRMAN: G. 't HOOFT

Scientific Secretaries: I. Florakis, M. Marienfeld

DISCUSSION

- *Haidt:*

Our universe is expanding. Now the fact that it is expanding means that you create more volume. Is this volume vacuum and is its metric then a $\delta_{\mu\nu}$ or $g_{\mu\nu}$ or $\hat{g}_{\mu\nu}$?

- *'t Hooft:*

The question is what are the ultimate equations of nature. My claim is those should be phrased entirely in terms of $\hat{g}$. At this stage, the concept of volume is not yet defined. Once you have $\hat{g}$ you can say "now I want to fill in what I believe the local scale is for all practical purposes". So I have an ω, I put ω in, and my claim is that ω is not fixed locally - only globally. To fix ω you must make certain assumptions and one assumption is that there's practically no matter or a very little bit of matter, which is nothing compared to the Planck scale energies that you'd encounter at the Planck scale. So the amount of matter that you and I see is nothing compared to the matter at the Planck scale. So, assume for a moment that there's no matter at all, then I can fill in ω by this condition: $T^{\mu}{}_{\mu} \propto R \propto \hat{g}^{\mu\nu}\partial_{\mu}\partial_{\nu}\omega = 0$. And ω is 1 at infinity. This is how scales, including the notion of volume, enter into the picture.

- *Haidt:*

So, that means that you have a different concept of vacuum.

- *'t Hooft:*

Yes. In principle I want to leave as much as possible of ordinary physics untouched, because I don't want to throw away all the good things of the Standard Model and all the good things of General Relativity. So, I want to keep that as much as possible. So, whenever someone talks about the vacuum, I don't think you can even talk about the vacuum unless you know what the scale is. Normally in quantum field theory we have a $g_{\mu\nu}$ – not the $\hat{g}_{\mu\nu}$ – we have the $g_{\mu\nu}$ entire, before we can even phrase the question of a quantum field theory. So, the whole notion of vacuum doesn't exist if you only have $\hat{g}$.

- *Haidt:*

Thank you.

- *Ferrara:*

Wald gave a formula for the entropy function as a way to compute entropy in the general theory of relativity. What happens when you make this change of variables to the entropy - what is its meaning? What happens to the Bekenstein-Hawking formula, since there's no longer a horizon or singularity? The entropy should be something which has a meaning.

- *'t Hooft:*

Your question can be rephrased in a much more vicious way, which is "what happens to Quantum Mechanics altogether, what happens to quantum states". And I have no complete answer to that, yet. I think that the theory has to be quantized in a way which I've been advocating for some time. You can't use the notion of action – I don't even have an action principle, because actions require also a scale to be defined in, so the situation is much worse because I might have to throw away the whole principle of action and a lot of things on which Quantum Mechanics is based. But I see a very practical way of describing black holes here, so we have to re-think very hard what a 'state' is. It's a description such that if you leave ω out you don't know how many massless particles there are around. It's ill-defined. So, also, you have to redefine what you mean by a state, in which the system is sitting.

- *Ferrara:*

So essentially you object to the entropy as related to the counting of microstates. For instance, in String Theory, in some limit Vafa and Strominger have proven that if you compute counting microstates (which in that case are D-branes) you reproduce Hawking. So you just say that your formulation in some sense disagrees with this picture.

- *'t Hooft:*

I absolutely don't know how to reconcile this picture with String Theory, although it's very conceivable that there is a natural way to embed this idea in String Theory. As you know in String Theory you also very often end up with a scale-invariant theory. So scale transformations already play an important role in String Theory and I think that this has to be reconsidered and more brought to the extreme. The equations of nature don't even know what the scale is and the scale factor has to be taken out of the theory. But that forces us to reconsider what states are, what the Schroedinger equation in Quantum Mechanics is and all that. I admit that the notion of entropy, the logarithm of the number of states, appears to be very robust, yet I am not sure that it can be defined before having the scale factor.

- *Vikman:*

Did I understand you correctly, using this analogy with Rindler spacetime, that a freely falling observer into a black hole would not see any Hawking particles? Usually very close to the horizon vacuum polarization effects are as important as Hawking radiation. So what is the role of vacuum polarization in your approach?

- *'t Hooft:*

That, indeed, is a question I cannot answer at this point. That is a very important question. To embed this whole picture into Quantum Mechanics is very difficult, that is, in conventional Quantum Mechanics. So my belief is that what you have to do is to quantize things my way, which is to promote all physical configurations you can have into elements of a basis of a quantum Hilbert space and then superimpose those elements to get the quantum states back. So that would be the approach I would have to take.

- *Vikman:*

In your approach, the matter fields should couple to $\hat{g}_{\mu\nu}$ or $g_{\mu\nu}$?

- *'t Hooft:*

Eventually, in the complete description, all the physical degrees of freedom should be in terms of $\hat{g}$. It may be at a later stage that you introduce a scalar field, but this is very dangerous. If you introduce a scalar field that somehow fixes the scale it is very tempting to say that this spontaneously breaks scale invariance and you're back at square one. Now, I don't want to go back to square one. So I don't want to introduce such scalar fields. What I could introduce, however, is a connection field. That is something that is still on the back of my mind: if I introduce a connection field that at least gives me the complete Levi-Civita connection fields, I need something that connects the scale. Now that happens to be a U(1) connection field, very reminiscent of electromagnetism. So maybe while adding electromagnetism in the theory, you get back to something that indirectly allows you to arrive at the scale again. And then you can use that to go back to a more complete formalism.

- *Vikman:*

Thank you.

- *Aschieri:*

Is there any relation of your theory to unimodular gravity, where indeed one does transformations that leave the determinant of the metric invariant?

- *'t Hooft:*

I've tried such things myself many times earlier but I never before entered into the picture I have now. And I think those attempts you mention are more like the things I tried earlier, because I now realize how important it is that ω is unobservable. But, apart from that, I think that, yes, there must be a relation between many of these attempts. So the answer is yes, there must be relations but exactly which they are, I don't know.

- *Nesterov:*

Do you have in mind some concrete (particular) realization of scale invariance, at least at the classical level ?

- *'t Hooft:*

The question should be how to realize this invariance at the classical level. But that is exactly what I've been trying to say, which is: replace $g_{\mu\nu}$ by $\hat{g}_{\mu\nu}$. $\hat{g}_{\mu\nu}$ does not fix the scale. And so all you have is light-cones and prescriptions about massless particles - you don't have massive particles anymore. Only massless particles which move along their light-cones, and these particles may or may not hit each other. And there are rules about what they do next: they may produce other massless particles coming out. For that we need equations, but those equations are not allowed to have any scale. So that, in practice, will be very difficult. So, like I mentioned in my answer to the previous question, it might be possible to add a connection field for the scale, not unlike the electromagnetic field, maybe it is the electromagnetic field. That, I don't know.

- *Nesterov:*

In your conclusion you mentioned the exclusion of topological non-trivialities. Is it an extra input to all that you have said before? Or does it somehow follow from the philosophy of what you were talking about (scale invariance)?

- *'t Hooft:*

At least the black hole no longer has any topology in this picture. The black hole is replaced by a configuration with $\hat{g}_{\mu\nu}$, such that it is regular everywhere, it obeys equations and the rules are: if ever you encounter a horizon use another ω – so that the horizon disappears, whenever you encounter a singularity move it towards infinity. If you have nothing else at your disposal use a conformal transformation: a conformal transformation will move the point in question to infinity and you do that for all these singularities - you don't need a global conformal transformation, because if you don't obey the equations all the way you just introduce a curvature scalar, which is not a disaster but not locally observable.

- *Nesterov:*

Do you exclude peculiar topologies, such as wormhole topologies?

- *'t Hooft:*

Now the wormhole becomes completely invisible. So, for all practical purposes, there is no wormhole. I think it's a good thing to see that spacetime is just normal, so it allows us to try to formulate a good theory, in particular having causality as a foremost important principle. And for that, topological non-triviality has always been standing in the way of causality as an important principle of nature.

- *Kovetz:*

My question is if this totally contradicts the String Theory picture where we do have an intrinsic lowest scale.

- *'t Hooft:*

I would not go as far as to say that it contradicts String Theory. I'd say it puts things in a different light.

- *Kovetz:*

But how can you settle the two?

- *'t Hooft:*

I don't know. So far what I have is independent of String Theory. Maybe there's a way to reconcile all these pictures – actually I hope so, because String Theory is a beautiful construction by itself. So I don't want to throw String Theory away. I'll just say that there's work to be done.

- *Kovetz:*

At the very basic level, is it at all possible to reconcile these?

- *'t Hooft:*

I hope it will be.

- *Kovetz:*

But then you have to throw away the basic point, that there is a lowest string scale.

- *'t Hooft:*

If history of science is anything to go by, very often those totally conflicting views turn out to describe the same reality. For instance, think of the particle-wave discussion in the old days: one would contradict the other, "either one is true or the other but not both". Now we know both are true. I actually suspect that String Theory has a lot of truth in it. But there always have been things missing. I think I have been talking about one of the things missing.

- *S. Ferrara:*

You are essentially splitting the 10 degrees of freedom $g_{\mu\nu}$ in a part of constant volume plus the volume ω. What is the equation of motion for the volume?

- *'t Hooft:*

I mentioned the equation of motion for the volume factor ω, I'll just say that ω is not locally observable. So I could just about throw away that equation and replace with anything I like – it's not observable. I need the equation for $\hat{g}$ itself, which does not yet exist. So, that's why this theory is very far from finished and that's why I didn't want to talk about it at first. So I don't understand those equations for $\hat{g}$ at all, but you can also say "temporarily, locally put in the ω for the time being". Then you know what you're doing, write down the equations and subsequently, to obtain $\hat{g}$, remove ω – say that it's not locally observable. Having these black holes in mind, I'd say ω can't be locally observable. Because different observers disagree about the amount of Hawking particles. In-going observer and outside observer disagree about Hawking particles so they disagree about the scale factor ω. Apparently, they can't observe it locally.

- *Vikman:*

We do know one of the phenomena when we do observe it where certainly the world is not scale-symmetric, namely for the expansion of the universe we do have the scale factor. How can we reconcile it with your approach?

- *'t Hooft:*

That is the kind of question I've asked myself: how come the world we live in does not at all look scale-invariant. The molecules, the atoms, the universe, the cosmological constant, they all are not scale invariant. So, the only answer can be that we have defined the scale at infinity and we use equations without realizing it - we have been using equations all the time to define what the scale is elsewhere. Without realizing it, we're using a scale that is defined in such a way that there's no gigantic amount of massless particles sitting between you and me for instance. There could have been a curtain of massless particles between us and I'd be mistaking the scale you are at, but we are just assuming there are no such particles. We're just looking at the entire universe as if there are no massless particles between us. We should realize however that, if in today's vacuum the laws are transformed by a boost of a factor of a million you'd see an enormous number of particles – curtains of particles separating us. Those particles are there, but we are assuming they are not there. So we are assuming some things about ω and then we can formulate our laws of physics. And then it's seen as if everything has a specific scale. So, I think that the fact that we see our world having a scale, comes from our wish to describe the things around us. So it looks as if there is a scale, but it also looks as if there are no Hawking particles. The demand that the universe is not completely filled with Hawking particles gives us a special frame in which the scale is well-defined. But for the whole universe at large, this becomes a little bit more difficult. Actually this could be the seed of a possible explanation of the cosmological constant. The cosmological constant gives you a very small scalar curvature. And that is now an observer-induced thing: We think we see a cosmological constant, but it is just induced by our desire to see the universe as empty as possible: "things you don't see aren't there".

- *Vikman:*

So, as far as I remember, in this unimodular gravity, Λ plays the role of integration constant so it's a first integral. So in that sense, it should be fixed by initial conditions somehow.

- *'t Hooft:*

Precisely. However, one has to be very careful not to jump to conclusions too quickly. For instance, there obviously still is a hierarchy problem: why is there something we call a cosmological constant (we may call it something else, I don't care), with a value that is off by 120 orders of magnitude? That I have not explained. I haven't even come close to an explanation of that. So the cosmological constant is still a problem despite the fact that in my theory it's not a well-defined concept.

Diffraction in Deep Inelastic Electron Proton Scattering at HERA

G. Wolf
Deutsches Elektronen-Synchrotron, DESY, Hamburg, Germany

Abstract

At the electron-proton collider HERA diffraction by photon proton scattering has been studied for centre of mass energies up to 250 GeV and mass squared -Q^2 of the photon up to $Q^2 = 1600$ GeV^2. Results obtained by the experiments H1 and ZEUS on the production of vector mesons and of inclusive final states by deep inelastic scattering are summarized. Diffraction accounts for a substantial fraction of the total virtual photon proton cross section even at Q^2 values as large as 300 GeV^2.

This article is dedicated to Antonino Zichichi, on the occasion of his 80th birthday.

1 Introduction

The subject of these lectures is diffraction in deep inelastic electron proton scattering, where the virtual photon has a negative mass squared, Q^2, with values up to 1600 GeV^2. Such measurements became possible with the electron-proton collider HERA.

When HERA and the experiments H1 and ZEUS were in the planning phase the study of diffraction at substantial values of Q^2 was not on the list of topics to be pursued. The cross section was expected to be too small, since large values of Q^2 correspond to small transverse sizes of the photon such that the photon probing the proton will become a thin needle.

1.1 HERA and the experiments H1 and ZEUS

The DESY accelerator complex is shown in Fig. 1. Beams of electrons and protons were accelerated in the synchrotron DESY, then in PETRA, transferred to HERA (Fig.2) and accelerated to their final energies of 27.5 GeV for electrons and 820 (920) GeV for protons. The ring has a cricumference of 6 km. The proton ring (top) uses superconducting magnets, half of which were provided by Italy. The electron ring (bottom) uses normal conducting magnets. The electron-proton collider HERA was in operation for physics from 1992 until 2007.

The electron and proton beams collided in the detectors of H1 (Hall North), Fig. 3, and ZEUS (Hall South), Fig. 4. For the detection of forward scattered protons both experiments were equipped with spectrometers integrated into the magnet system of HERA. Figure 5 shows the leading proton spectrometer (LPS) of ZEUS.

2 General features of virtual photon proton scattering

The broad features of inclusive and diffractive virtual photon proton scattering are sketched in Fig. 6: (a) in the inclusive process the produced particles fill more or less uniformly the rapidity space between the incoming photon and proton; (b) for diffractive scattering two separate groups are emitted, one with mass M_X in the direction of the virtual photon (γ^*), and a second one, consisting of a proton or a low-mass nucleonic system (N), separated by a large gap in rapidity. Figure 7 shows a nondiffractive (bottom) and a diffractive event (top) as observed in the ZEUS detector. Also shown are the distributions of the produced particles as a function of their pseudorapidity $\eta = -\ln(tan\theta/2$, where θ is the polar angle measured w.r.t. the direction of the electron beam. For the nondiffractive event (bottom) particles are produced - more or less uniformly - over a wide space in η, while for the diffractive event (top) particle emission is grouped around the directions of the incoming photon and proton.

3 Total cross section for real and virtual photon proton scattering

The total photon proton cross section, $\sigma_{\gamma * p}^{tot}$ at $Q^2 = 0$ has been measured directly by ZEUS [1, 2] and H1 [3]. For virtual photons ($Q^2 > 0$) it has been deduced from the measurements of the proton structure function F_2 by ZEUS [1, 2, 4, 5] and H1 [6]:

$$\sigma_{\gamma * p}^{tot} = \frac{4\pi^2\alpha}{Q^2(1-x)} F_2(x, Q^2) \tag{1}$$

Figure 8 shows $\sigma_{\gamma * p}^{tot}$ as a function of the c.m. energy W of the virtual-photon proton system, for Q^2 between 0 and 320 GeV^2. The cross section is rising with W proportional to W^δ, where δ increases with Q^2 from $\delta = 0.2$ at $Q^2 = 0$ to $\delta \approx 0.8$ at $Q^2 = 320$ GeV^2. The latter value of δ implies an almost linear increase of $\sigma_{\gamma^* p}^{\text{tot}}$ with W, a behaviour not observed for hadron-hadron interactions.

The HERA data for $\sigma_{\gamma^* p}^{\text{tot}}$ can be represented in a compact form:

$$\sigma_{\gamma^* p}^{\text{tot}} = \frac{4\pi^2\alpha}{Q^2(1-x)} F_2(x, Q^2) \tag{2}$$

with

$$\begin{aligned}\log_{10} F_2(x, Q^2) = c_1 + c_2 \cdot \log_{10} x + c_3 \cdot (\log_{10} x) \cdot \log_{10}(Q^2/Q_0^2) \\ +c_4 \cdot (\log_{10} x) \cdot (\log_{10}(Q^2/Q_0^2))^2\end{aligned} \tag{2}$$

where Q^2 in units of GeV^2. A fit to the HERA data for $Q^2 \geq 2.7$ GeV^2 with 389 data points yielded $c(1) = -0.735 \pm 0.004$, $c(2) = -0.098 \pm 0.002$, $c(3) = -0.158 \pm 0.002$, and $c(4) = 0.023 \pm 0.001$, with $\chi^2 = 368$. The fit gives a good representation of the data, see the curves in Fig. 8.

4 Diffraction

First evidence for a substantial contribution from diffraction in deep-inelastic scattering was reported by ZEUS [7], observing events in which no energy is deposited close to the direction of the proton beam, i.e. at large angle θ, see Fig. 9. Described in terms of pseudorapidity η, where

$$\eta = -\ln(\tan\theta/2), \tag{3}$$

nondiffractive events show uniform particle production in η, see Fig. 6(a), while diffractive events exhibit a sizeable gap between the most forward calorimeter cells ($\theta = 1.5^0$, $\eta = 4.3$) and the pseudorapidity of the hadron(s) observed closest to the proton direction, see Fig. 6(b): in diffractive scattering particle emission is concentrated around the directions of the incoming photon and proton. Figure 10 shows the distribution of the maximum rapidity (η_{max}) observed in the calorimeter for events from deep-inelastic scattering with $Q^2 > 10$ GeV^2. There is a large excess of events at $\eta_{max} < 2$ compared to the expectation from nondiffractive deep-inelastic scattering (shaded histogram).

Different strategies have been followed by H1 and ZEUS for isolating diffractive contributions. In the case of vector meson production, $\gamma^* p \to VN$, resonance signals in the decay mass spectrum combined with the absence of other substantial activity in the detector have been used. The contribution from inclusive diffraction has been extracted using the presence of a large rapidity-gap, the detection of the leading proton or the hadronic mass spectrum observed in the central detector (M_X method [8,9]). Event samples selected on the basis of a large rapidity-gap or a leading proton may include additional contributions from Reggeon exchange. Such contributions are exponentially suppressed when using the M_X method.

5 Production of vector mesons

5.1 ρ^0 production

A study of $\gamma^* p \to \rho^0 p$ with a large data sample was presented by ZEUS [10], based on an integrated luminosity of 120 pb^{-1}. The $\pi^+\pi^-$ mass distribution (Fig. 11) is dominated by ρ^0 production[1]. The cross section for $\gamma^* p \to \rho^0 p$, as measured by H1 [11] and ZEUS, is presented in Fig. 12 as a function of W for Q^2 values between 2 and 19.5 GeV^2. For fixed W, $\sigma_{\gamma^* p \to \rho^0 p}$ is falling rapidly with Q^2. For fixed Q^2, the cross section rises with W. A parametrisation of the form $\sigma_{\gamma^* p \to \rho^0 p} \propto W^\delta$ yielded the δ values shown in Fig. 13 as a function of $(Q^2 + M_V^2)$: the power δ rises from around 0.15 at $Q^2 \approx 0$ to about 0.8 at $(Q^2 + M_V^2) = 30$ GeV^2.

[1] The skewing of the mass distribution is explained by the Söding model [12]: the excess of events observed at masses below 0.65 GeV is due to background from $\omega \to \pi^+\pi^-\pi^0$, where the π^0 was not detected, and from $\phi \to K^+K^-$, where the K's were misidentified as π's.

The Q^2 dependence of $\sigma(\gamma^* p \to \rho^0 p)$ is shown in Fig. 14 for $W = 90$ GeV [10]. There is an almost exponential decrease from $\sigma \approx 700$ nb at $Q^2 = 2$ GeV^2 to 0.04 nb at $Q^2 = 100$ GeV^2. The measurements were compared with several models which are based on the dipole representation of the virtual photon: the photon fluctuates into a $q\bar{q}$ pair - the colour dipole - which in turn interacts with the gluon cloud of the proton to produce the ρ^0, see e.g. [13]. For $Q^2 > 1$ GeV^2, the calculations of [14] (KMW) and [15] (DF) give a good description of the data while those of FSS [16] are somewhat low for Q^2 above 10 GeV^2.

As a function of $|t|$, the differential cross-section $d\sigma/d|t|$ at $W = 90$ GeV falls exponentially with $|t|$, $d\sigma/d|t| \propto e^{-b|t|}$ [10], see Fig. 15. The dependence of b on $(Q^2 + M_V^2)$, where M_V is the mass of the vector meson, is shown in Fig. 16, taken from [17]. For $(Q^2 + M_V^2) \leq 1$ GeV^2, b is of the order of $10 - 12$ GeV^{-2}, similar to what has been measured for elastic scattering at high energy: for pp scattering at $W = 31 - 62$ GeV, $b \approx 11$ GeV^{-2} [18]. At larger values of $(Q^2 + M_V^2)$ the slope b decreases rapidly, approaching a constant value of $b \approx 5$ GeV^{-2} for $(Q^2 + M_V^2) > 10$ GeV^2. In an optical model, b is proportional to the sum of the radii squared of the virtual photon and the proton, $b = [(R_{\gamma*})^2 + (R_p)^2]/4$. As Q^2 increases, the transverse extension of the virtual photon, which is expected to be proportional to $1/Q$, goes to zero, such that $b \to (R_p)^2/4$.

The W dependence of $d\sigma/d|t|$ at fixed values of $|t|$ was used to determine the parameters of the Pomeron trajectory, $\alpha_{I\!P} = \alpha_{I\!P}(0) + \alpha'_{I\!P} * |t|$, with the following result [10]:

$< Q^2 >= 3$ GeV^2:
$\alpha_{I\!P}(0) = 1.113 \pm 0.010(stat)^{+0.009}_{-0.012}(syst)$, $\alpha'_{I\!P} = 0.185 \pm 0.042(stat)^{+0.022}_{-0.057}(syst)$ GeV^{-2}
$< Q^2 >= 10$ GeV^2:
$\alpha_{I\!P}(0) = 1.152 \pm 0.011(stat)^{+0.006}_{-0.008}(syst)$, $\alpha'_{I\!P} = 0.114 \pm 0.043(stat)^{+0.026}_{-0.024}(syst)$ GeV^{-2}

The data suggest a small increase of $\alpha_{I\!P}(0)$ with increasing Q^2.

The decay angular distribution of the ρ^0 mesons has been used to determine the orientation of the ρ^0 spin [19, 20]. Exclusive electroproduction of ρ^0 mesons and their decay into $\pi^+\pi^-$ is described by three angles: Φ_h - the angle between the ρ^0 production plane and the electron scattering plane in the $\gamma^* p$ c.m. system, and θ_h, ϕ_h - the polar and azimuthal angles of the π^+ w.r.t. the proton direction, all determined in the ρ^0 rest system. The ρ^0 decay angular distribution, $W(\cos\theta_h, \phi_h, \Phi_h)$, can be parametrised by the ρ^0 spin-density matrix elements, $\rho^{\alpha}_{i,k}$, with $i,k =$ -1,0,1, and $\alpha = 0, 1, ., 6$. If no separation is made between the contributions from transverse and longitudinal photons, the ρ^0 s-channel helicity decay angular distribution w.r.t. θ_h and ϕ_h can be expressed in terms of the matrix elements

r_{ik}^{04} as follows [20]:

$$\frac{1}{N}\frac{dN}{d\cos\theta_h} = \frac{3}{4}[1 - r_{00}^{04} + (3r_{00}^{04} - 1) \cdot \cos^2\theta_h] \tag{4}$$

(5)

where the density matrix element r_{00}^{04} represents the probability that the ρ^0 is produced longitudinally polarised.

Figure 17 summarises the measurements of r_{00}^{04} for ρ^0, ϕ and J/Ψ production as a function of Q^2/M_V^2. The same behaviour is observed for all three vector meson species: r_{00}^{04} rises sharply from zero at $Q^2 = 0$ to a value of $0.7 - 0.8$ for $Q^2 > 5$ GeV^2. This shows that near $Q^2 = 0$ the vectormesons are transversely polarised while at large Q^2 they are predominantly longitudinally polarised. Under the assumption of SCHC their production at large Q^2 is dominated by longitudinal photons.

The ratio of the cross sections for production by either transverse or longitudinal photons, $R = \sigma_L/\sigma_T$, for $\gamma^* p \to \rho^0 p$ is shown in Fig. 18: R rises rapidly with Q^2 from zero at $Q^2 = 0$, to unity at $Q^2 \approx 2$ GeV^2 and values around 4 at $Q^2 = 20$ GeV^2: for $Q^2 > 2$ GeV^2 the dominant contribution to $\gamma^* p \to \rho^0 p$ comes from longitudinal photons. The measurements are well described by the models of [16] and [21], which are based on two-gluon exchange.

5.2 ω production

The data on ω production, $\gamma^* p \to \omega p$ are scanty, due mainly to the difficulty in reconstructing the π^0 of the final state $\gamma p \to \omega p \to p\pi^+\pi^-\pi^0$. A single measurement on ω production by virtual photons has been published [22]. It is based on an integrated luminosity of 37.7pb^{-1} and covers the kinematic region $3 < Q^2 < 20$ GeV^2, $40 < W < 120$ GeV and $|t| < 0.6$ GeV^2. The results are presented in Fig. 19. The $\pi^+\pi^-\pi^0$ mass spectrum shows clear signals for the production of ω and ϕ mesons. The cross section for $\gamma^* p \to \omega p$ at $Q^2 = 7$ GeV^2 is a factor of about 10 smaller than for $\gamma^* p \to \rho^0 p$ and shows, for fixed $W = 70$ GeV, a rapid decrease with Q^2, similar to the Q^2-behaviour of $\sigma(\gamma^* p \to \rho^0 p)$.

5.3 ϕ production

First results from HERA on ϕ production by *virtual* photons, $\gamma^* p \to \phi p$, have been published by ZEUS [23], based on an integrated luminosity of 2.6 pb^{-1}. Further results were presented by H1 [24] (integrated luminosity L = 3.1 pb^{-1}) and ZEUS [25,26] (L = 45 and 119 pb^{-1}, respectively). Figure 20 shows the results from ZEUS: the K^+K^- mass spectrum is completely dominated by ϕ production. The cross section for $\gamma^* p \to \phi p$ at

$Q^2 = 2.4, 6.5$ and 13.0 GeV^2 (Fig. 20a) is rising with W. The dependence on $(Q^2+M_\phi^2)$ is shown in Fig. 20b for $W = 75$ GeV, separately for the contributions from longitudinal (L) and transverse photons (T). At $(Q^2+M_\phi^2) = 3$ GeV^2, σ_T and σ_L are approximately equal; for larger values of $(Q^2+M_\phi^2)$, σ_T falls off more rapidly than σ_L. At $(Q^2+M_\phi^2) \approx 45$ GeV^2, σ_L is larger than σ_T by about a factor of 10.

The $|t|$ - dependence of ϕ production is displayed in Fig. 21 for different values of Q^2. The data yield $d\sigma/d|t| \propto e^{-b|t|}$, with $b \approx 5.5$ GeV^{-2} for $(Q^2+M_\phi^2) > 5$ GeV^2, see Fig. 16.

The distribution of the ϕ polar decay angle in the helicity system, θ_h, is of the form $dN/d\cos\theta_h \propto a + \cos^2\theta$, where a tends to zero as Q^2 increases from 2.4 to 19.7 GeV^2, see Fig. 22 (bottom). This implies that, at large Q^2, the ϕ mesons are predominantly longitudinally polarised - as has been observed for ρ^0 production (see above). Under the assumption of SCHC, ϕ production at $Q^2 \geq 2.4$ GeV^2 proceeds predominantly by longitudinal-photon-proton scattering.

5.4 J/Ψ production

The first signal for J/Ψ production in deep-inelastic scattering was reported by H1 [27]. Further results were presented by ZEUS [28, 29] and H1 [30, 31]. The final results stem from about 1200 (ZEUS) and 600 (H1) $J/\Psi p$ events with $Q^2 > 2$ GeV^2, see Figs. 23, 24. The W dependence of the cross-section $\sigma(\gamma^* p \to J/\Psi)$ is shown in Fig. 25 as a function of W for Q^2 values between 3.2 and 22.4 GeV^2. For fixed Q^2, $\sigma(\gamma^* p \to J/\Psi)$ is rising with W. A parametrisation of the form $\sigma(\gamma^* p \to J/\Psi p) \propto W^\delta$ yields the δ values shown as a function of Q^2 in Fig. 25(bottom); within errors, $\delta \approx 0.7$.

The differential cross-section $d\sigma(\gamma^* p \to J/\Psi p)/d|t|$ is proportional to $e^{-b|t|}$, where $b = 4.72 \pm 0.15 \pm 0.12$ GeV^{-2}, independent of Q^2, as shown by Fig. 26. The measured b-value of 4.7 GeV^{-2} corresponds to the size of the radius of the proton alone. Hence, the radius of the virtual photon must be small.

The decay $J/\Psi \to l^+l^-$ was studied in terms of the decay angles θ_h and $\psi_h = \phi_h - \Phi_h$, where θ_h and ϕ_h are the polar and azimuthal angles of the positively charged decay lepton, and Φ_h is the angle between the J/Ψ production plane and the lepton scattering plane. The distributions of $\cos\theta_h$ and of ψ_h are shown in Figs. 27a-f for Q^2 between 2 and 100 GeV^2. The full set of r_{ik}^α is shown in Fig. 28 as a function of Q^2 (left) and $|t|$ (right). The element r_{1-1}^{04} and the combinations $r_{00}^5 + 2r_{11}^5$ and $r_{00}^1 + 2r_{11}^1$ are zero, within errors, in agreement with SCHC. Assuming SCHC, the ratio of the longitudinal to transverse cross sections, $R = \sigma_L/\sigma_T$, was determined from r_{00}^{04}: $R = \frac{1}{\epsilon}\frac{r_{00}^{04}}{1-r_{00}^{04}}$, where ϵ is the polarisation parameter. The ratio R is shown in Fig 27(g): it rises proportional to Q^2: $R = \alpha \cdot (Q^2/M_{J/\Psi}^2)$ with $\alpha = 0.52 \pm 0.16$. For $Q^2 = 20$ GeV^2, σ_L and σ_T are approximately equal.

5.5 The Q^2 and t-dependences of vector meson production

Figure 29 shows the cross sections for $\gamma^* p \to Vp$ at $W = 75$ GeV, as a function of $(Q^2 + M_V^2)$, where $V = \rho^0, \omega, \phi, J/\Psi, \Upsilon$ [24]. The cross sections lie on a universal curve and fall by five orders of magnitude from $(Q^2 + M_V^2) = 1$ to $(Q^2 + M_V^2) = 100$ GeV^2. The curve shown is of the form $\sigma = a_1(Q^2 + M_V^2 + a_2)^{a_3}$, with $a_1 = 10689 \pm 165$ nb, $a_2 = 0.42 \pm 0.09$ GeV^2 and $a_3 = -2.37 \pm 0.10$ [24]. The cross sections for $V = \rho^0, \phi, J/\Psi$ are compared in Fig. 30 with the predictions of the bSat and gGGC models [32] [2], as provided by [34]. A good description of the data is obtained.

The differential cross-sections $d\sigma/d|t|$ for $\gamma^* p \to Vp$ are shown in Fig. 31 as a function of $|t|$, where $V = \rho^0, J/\Psi$ ($W = 90$ GeV) and $V = \phi$ ($W = 75$ GeV); Q^2 is fixed with values between 2.4 and 19.7 GeV^2. The relative size of these cross sections at large Q^2 can be understood by assuming that the virtual - predominantly longitudinal - photon couples to a vector meson V, which scatters elastically on the proton, see Fig 32. The cross section for $\gamma_L^* p \to V_L p$ is given by:

$$\sigma(\gamma_L^* p \to V_L^0 p) = \frac{\alpha}{4} \cdot \frac{1}{\gamma_V^2/4\pi} \cdot \frac{M_V^4}{(Q^2 + M_V^2)^2} \cdot \frac{Q^2}{M_V^2} \sigma(V_L^0 p \to V_L^0 p). \tag{6}$$

From the data presented in Fig. 31 one obtains for $|t| = 0.5$ GeV^2:

$$\frac{d\sigma(\phi_L p \to \phi_L p, W = 75, Q^2 = 19.7)/d|t|}{d\sigma(\rho_L^0 p \to \rho_L^0 p, W = 90, Q^2 = 19.7)/d|t|} = 1.03 \pm 0.05 \tag{7}$$

$$\frac{d\sigma(J/\Psi_L p \to J/\Psi_L p, W = 90, Q^2 = 19.7)/d|t|}{d\sigma(\rho_L^0 p \to \rho_L^0 p, W = 90, Q^2 = 19.7)/d|t|} = 0.59 \pm 0.15. \tag{8}$$

Within a factor of two, the cross sections for elastic $\rho_L p$, $\phi_L p$ and $J/\Psi_L p$ scattering are expected to be the same at W around 90 GeV and $|t| = 0.5$ GeV^2.

5.6 Inclusive diffraction by $\gamma^* p$ interactions

The early observations of inclusive diffractive production in deep-inelastic scattering by ZEUS [7,35] and subsequently by H1 [36], were followed by detailed studies.

[2] These models are based on the so called color glass concept [33]: the hadrons participating in the scattering process behave as a new form of matter (color glass condensate) which is made of small-x gluons. These gluons carry small fractions $x << 1$ of the total momentum and are created by slowly moving color sources - i.e. partons at large x - as for a glass where, on short time scales, the constituents appear to be frozen.

The first extraction of the diffractive cross section by ZEUS with the M_X-method [9] was based on data from an integrated luminosity of 0.54 pb^{-1}. The M_X and $\ln M_X^2$ distributions are shown in Fig. 34 at $Q^2 = 14$ GeV^2 for several bins of W. As a function of M_X (top) the diffractive contribution (shaded) is falling rapidly with increasing M_X, while the $\ln M_X^2$ spectra (bottom) show a plateau for the diffractive component once M_X is sufficiently above the kinematical threshold. This allows the extraction of the diffractive contribution by a fit to the $\ln M_X^2$ spectrum of the form:

$$\frac{dN}{d\ln M_X^2} = D + c \cdot exp(b \cdot \ln M_X^2) \tag{9}$$

The fit is applied in the interval $\ln M_X^2 \leq \ln W^2 - \eta_0$, where $\ln W^2 - \eta_0$ is the maximum value of $\ln M_X^2$ up to which the exponential behaviour of the nondiffractive part holds. The diffractive contribution is not taken from the fitted value of D; rather it is obtained by subtracting the nondiffractive contribution determined by the fit $(c \cdot exp(b \cdot \ln M_X^2))$ from the observed number of events.

In the following, the high-statistics results on inclusive diffraction as obtained by H1 and ZEUS are presented in terms of the diffractive cross section, σ^{diff}/dM_X, and the diffractive structure function, $x_{\mathbb{P}} F_2^{D(3)}(x_{\mathbb{P}}, \beta, Q^2)$, where $\beta = Q^2/(M_X^2 + Q^2)$ and $x_{\mathbb{P}} = (M_X^2 + Q^2)/(W^2 + Q^2)$.

5.7 Inclusive diffraction with high statistics: results from H1

The diffractive structure function has been measured by H1 [41] with integrated luminosities of 2.0 pb^{-1} ($3 < Q^2 < 13.5$ GeV^2), 10.6 pb^{-1} ($13.5 < Q^2 < 105$ GeV^2) and 61.6 pb^{-1} ($Q^2 > 133$ GeV^2). Events produced by diffraction were required to have a rapidity gap characterised by $\eta_{max} < 3.3$. In the H1 measurement, the mass of the nucleonic system Y escaping through the forward beam hole is limited to $M_Y < 1.6$ GeV.

Figure 35 shows the diffractive structure function as a function of $x_{\mathbb{P}}$ for β values between 0.01 and 0.9, and Q^2 values from 3.5 to 1600 GeV^2. Here, H1 denotes he diffractive structure function $x_{\mathbb{P}} F_2^{d(3)}$ by $x_{\mathbb{P}} \sigma_r^{D(3)}$. Figure 35 shows that $x_{\mathbb{P}} F_2^{D(3)}$ increases as $x_{\mathbb{P}} \to 0$, provided $\beta \geq 0.2$. In Figs. 36 and 37 measurements for $x_{\mathbb{P}} F_2^{D(3)}$ are shown as a function of Q^2 for fixed values of $x_{\mathbb{P}}$ and β. The dependence on Q^2 can be summarised as follows: for $\beta \leq 0.5$ the diffractive structure function rises with increasing Q^2, while for $\beta > 0.8$ it decreases with Q^2. H1 has performed a fit of their $x_{\mathbb{P}} F_2^{D(3)}$ measurements to a QCD-motivated model, which includes the contributions from Pomeron plus Reggeon exchanges in the t-channel. Diffractive parton distribution functions have been determined by a next-to-leading order DGLAP analysis of the data with $Q^2 \geq 8.5$ GeV^2, $\beta \leq 0.8$ and $M_X > 2$ GeV. The fit gives a good account of the data as shown by the curves in Figures 35 - 37.

The resulting quark and gluon distributions are presented in Fig. 38 as a function of z. Here z is the longitudinal momentum fraction of the parton entering the hard sub-process, such that $z = \beta$ for the lowest order quark-parton process, whereas $0 < \beta < z$ for higher order processes. Figure 38 shows that a large fraction of the momentum of the Pomeron exchanged in the t-channel is carried by gluons (dashed curves). In the model, the rise of $x_{I\!P} F_2^{D(3)}$ for $x_{I\!P} > 0.1$ is due to Pomeron-Reggeon interference.

5.8 Inclusive diffraction with high statistics: results from ZEUS

Inclusive diffraction has been measured by ZEUS for a wide range in Q^2, W and M_X, making use of the forward-plug calorimeter (FPC). The FPC limited the mass of the nucleonic system escaping undetected in the forward (= proton) direction to $M_N < 2.3$ GeV, on average. Data were collected for $Q^2 = 2.7 - 80$ GeV^2 and $Q^2 = 25 - 320$ GeV^2, with integrated luminosities of 4.2 [4] and 52.4 pb^{-1} [5], respectively.

The diffractive cross section $d\sigma^{diff}/dM_X$ is shown in Figs. 39, 40 as a function of W for the bins of Q^2 and M_X indicated. In all Q^2, M_X bins with sufficient W coverage, the diffractive cross section shows a strong rise with W.

The ratio of the diffractive contribution to the total $\gamma^* p$ cross section is displayed in Figs. 41, 42 as a function of W, for different values of Q^2 and M_X. For fixed Q^2, the ratio is approximately independent of W. It is substantial when $M_X^2 > Q^2$. The ratio $r = \sigma^{\rm diff}(0.28 < M_X < 35 \text{ GeV}, M_N < 2.3 \text{ GeV})/\sigma^{\rm tot}$, which includes most of the diffractive contribution, amounts to $15.8^{+1.1}_{-1.0}\%$ at $Q^2 = 4$ GeV^2, decreasing to $5.0 \pm 0.9\%$ at $Q^2 = 190$ GeV^2, see Fig. 43. A fit of the ratio r to the form $r = a - b \cdot \ln(1 + Q^2)$ yielded $a = 0.207 \pm 0.008$ and $b = 0.032 \pm 0.002$. The fit is shown by the curve.

Since the optical theorem relates the forward amplitude for elastic scattering to the total cross section - which behaves, say, proportional to W^δ - one would naively expect that the diffractive cross section is proportional to $W^{2\delta}$. However, the measured ratio of the diffractive contribution to the total $\gamma^* p$ cross section is approximately independent of W, as shown in Figs. 41, 42. This apparent contradiction is understood as the result of a strong growth of the gluon density ("color glass condensate") such that gluon-gluon interactions become important and damp the rise of the diffractive cross section [38].

The diffractive structure function $x_{I\!P} F_2^{D(3)}$ for $\gamma^* p \to XN$, $M_N < 2.3$ GeV, is shown in Figs. 44, 45 as a function of $x_{I\!P}$ for fixed Q^2, M_X, or equivqalently for fixed Q^2, β: $x_{I\!P} F_2^{D(3)}$ rises approximately proportional to $\ln 1/x_{I\!P}$ as $x_{I\!P} \to 0$, reflecting the rise of the diffractive cross-section $d\sigma^{\rm diff}/dM_X$ with increasing W. The rise is observed for most Q^2 values between 2.7 and 320 GeV^2.

The Q^2 dependence of $x_{I\!P} F_2^{D(3)}$ for fixed β and $x_{I\!P}$ is shown in Fig. 46. The data are dominated by positive scaling violations for $x_{I\!P}\beta = x < 10^{-3}$, by negative scaling violations for $x \geq 5 \cdot 10^{-3}$, and by constancy in between. The data contradict the assumption of Regge factorisation, namely that $x_{I\!P} F_2^{D(3)}(\beta, x_{I\!P}, Q^2)$ factorises into a term that depends only on $x_{I\!P}$, and a second term that depends only on β and Q^2.

5.9 Leading protons and the contribution from diffraction

Deep inelastic diffractive scattering has also been studied in measurements where the scattered proton was detected in a forward spectrometer. The first such measurement has been reported by ZEUS [39]. Figure 47 shows $d\sigma/d|t|$ as a function of the four-momentum-transfer squared t, for the range $5 < Q^2 < 20$ GeV^2, $50 < W < 270$ GeV and $0.15 < \beta < 0.5$. A fit of the form $d\sigma/d|t| = A \cdot e^{-b|t|}$ yielded for the slope $b = 7.2 \pm 1.1(\text{stat})^{+0.7}_{-0.9}(\text{syst})$ GeV^{-2}. Further results on b [40, 41] are also presented in Figure 47. For the Q^2-values selected, the transverse size of the virtual photon is negligible compared to the radius of the proton.

The final measurement [42], based on an integrated luminosity of 32.6 pb^{-1}, provided results for the kinematic region $Q^2 > 2$ GeV^2, $W = 40 - 240$ GeV, $M_X > 2$ GeV. The diffractive peak covered the region $x_L > 0.98$, where x_L is the fraction of the momentum of the incoming proton carried by the scattered proton. The differential cross section $d\sigma^{ep\to eXp}/d|t|$ is shown in Fig. 48 as a function of $|t|$ for $x_{I\!P} = 0.0002 - 0.01$ and $x_{I\!P} = 0.01 - 0.1$, respectively. The cross section falls exponentially with $|t|$, $d\sigma^{ep\to eXp}/d|t| \propto e^{-b|t|}$, where $b = 7.0 \pm 0.3$ GeV^{-2} for $x_{I\!P} = 0.0002 - 0.01$, and $b = 6.9 \pm 0.3$ GeV^{-2} for $x_{I\!P} = 0.01 - 0.1$. Within errors the value of b is independent of $x_{I\!P}$.

The cross section for $ep \to eXp$ was also studied in terms of the azimuthal angle Φ between the positron scattering plane and the proton scattering plane. Within errors, no significant dependence on Φ was observed for $x_{I\!P} = 0.0002 - 0.01$ and $x_{I\!P} = 0.01 - 0.1$, respectively. This is in contrast to ρ^0 production in deep-inelastic scattering, $ep \to e\rho^0 p$ [10], where a substantial dependence on Φ had been observed, viz. $\frac{d\sigma}{d\Phi} \propto 1 + A_{LT}\cos\Phi + A_{TT} cos 2\Phi$, $A_{LT} = \sqrt{2\epsilon(1+\epsilon)}(r^5_{00} + 2r^5_{11}) = -0.256 \pm 0.030^{+0.032}_{-0.022}$.

The structure function $x_{I\!P} F_2^{D(3)}$, here denoted by $x_{I\!P}\sigma_r^{D(3)}$, is shown in Figs. 49, 50 and compared with the FPCI,II measurements discussed before. The $x_{I\!P} F_2^{D(3)}$ data from the FPCI,II measurements have been scaled down by a factor of 0.83 to account for the extra contribution from $p \to N$ dissociation with $M_N < 2.3$ GeV. For most data points, there is good agreement between the FPC and LPS measurements.

5.9.1 Comparison with the BEKW parametrisation

Further insight into the behavior of the diffractive structure function can be gained with the help of the BEKW parametrisation [43], which is guided by QCD. It assumes that the incoming virtual photon fluctuates into a $q\bar{q}$ or $q\bar{q}g$ dipole, which in turn interacts with the target proton via the exchange of a two-gluon system, the Pomeron. The contributions from the transitions: transverse photon to $q\bar{q}$ and $q\bar{q}g$, respectively, and longitudinal photon to $q\bar{q}$, are taken into account. The β spectrum and the scaling behaviour in Q^2 are derived in the non-perturbative limit from the wave functions of the incoming transverse (T) and longitudinal (L) photons on the light cone. The $x_{I\!P}$ dependence of the cross section is not predicted by BEKW and has to be determined by experiment. The BEKW parametrisation reads as follows:

$$x_{I\!P} F_2^{D(3)}(\beta, x_{I\!P}, Q^2) \;=\; c_T \cdot F^T_{q\bar{q}} + c_L \cdot F^L_{q\bar{q}} + c_g \cdot F^T_{q\bar{q}g}, \tag{10}$$

where

$$F^T_{q\bar{q}} = \left(\frac{x_0}{x_{I\!P}}\right)^{n_T(Q^2)} \cdot \beta(1-\beta), \tag{11}$$

$$F^L_{q\bar{q}} = \left(\frac{x_0}{x_{I\!P}}\right)^{n_L(Q^2)} \cdot \frac{Q_0^2}{Q^2+Q_0^2} \cdot \left[\ln\left(\frac{7}{4} + \frac{Q^2}{4\beta Q_0^2}\right)\right]^2 \cdot \beta^3(1-2\beta)^2, \tag{12}$$

$$F^T_{q\bar{q}g} = \left(\frac{x_0}{x_{I\!P}}\right)^{n_g(Q^2)} \cdot \ln\left(1 + \frac{Q^2}{Q_0^2}\right) \cdot (1-\beta)^{\gamma}. \tag{13}$$

In the ZEUS analysis, $x_0 = 0.01$. The powers $n_{L,T,g}$ were assumed to be the same for the contributions from T, L and g, and of the form $n(Q^2) = n_1 \cdot \ln(1 + \frac{Q^2}{Q_0^2})$ with $Q_0^2 = 0.4$ GeV^2.

The contribution from longitudinal photons coupling to $q\bar{q}$ is limited to β values close to unity. The $q\bar{q}$ contribution from transverse photons is expected to have a broad maximum around $\beta = 0.5$, while the $q\bar{q}g$ contribution becomes important at small β, provided the power γ is large.

A fit to the ZEUS data [4,5] gave an excellent description of the data with the following result for the five parameters left free: $c_T = 0.118 \pm 0.002$, $c_L = 0.087 \pm 0.005$, $c_g = 0.0090 \pm 0.0003$, $n_1 = 0.062 \pm 0.002$ and $\gamma = 8.22 \pm 0.46$. The curves in Figs. 44, 45 represent the individual contributions from the $(q\bar{q})_T$, $(q\bar{q})_L$ and $(q\bar{q}g)_T$ terms, and their sum, as obtained by the fit. The contribution from $(q\bar{q})_L$ is important for $\beta \geq 0.96$; the $(q\bar{q})_T$ contribution is dominant for $0.14 < \beta < 0.9$, while gluon emission, $(q\bar{q}g)_T$, is important at small $\beta < 0.05-0.1$. The curves in Fig. 46 show the sum of the contributions.

The β dependence of $x_{I\!P}F_2^{D(3)}$ is shown in Figs. 51, 52, 53 for fixed values of $x_{I\!P}$ and Q^2. In general, the $x_{I\!P}F_2^{D(3)}$ values measured at different Q^2 values cluster around a narrow band of the form $\beta \cdot (1-\beta)$. However, for $x_{I\!P} > 0.005$ and low $\beta < 0.3$ or high $\beta > 0.9$ the data lie above the $\beta \cdot (1-\beta)$ band, a feature which becomes more prominent as $x_{I\!P}$ increases. The curves show the contributions from $(q\bar{q})_T$, $(q\bar{q}g)_T$ and $(q\bar{q})_L$. The dominant contribution is of the form $\beta(1-\beta)$ as predicted for $(q\bar{q})_T$; the rise towards small $\beta \leq 0.2$ is the result of gluon emission, $(q\bar{q}g)_T$; contributions from longitudinal photons, $(q\bar{q})_L$, are important near $\beta = 1$.

6 Summary and conclusions

A substantial fraction of diffractive scattering by virtual-photon proton interactions leads to the production of vector mesons, $\gamma^* p \to Vp$, where $V = \rho^0, \omega, \phi, J/\Psi$. For fixed Q^2, the cross sections rise with c.m. energy proportional to W^δ, with $\delta \approx 0.2$ at $Q^2 + M_V^2 = 1$ GeV^2, increasing to ≈ 0.8 at $(Q^2+M_V^2) = 35$ GeV^2. The integrated cross-sections at fixed $W = 75$ GeV lie on a universal curve, decreasing exponentially with $(Q^2+M_V^2)$ provided $(Q^2+M_V^2) > 3$ GeV^2. The differential cross-sections $d\sigma/d|t|$ decrease with $|t|$ proportional to $e^{-b|t|}$, where $b \approx 10$ GeV^{-2} for $(Q^2+M_V^2) < 1$ GeV^2, dropping rapidly to a constant value of $b \approx 5$ GeV^{-2}; a value which corresponds to the radius of the proton. Therefore, for $(Q^2+M_V^2) \geq 5$ GeV^2 the transverse size of the virtual photon is negligible. The Q^2 dependence of $\sigma(\gamma^* p \to \rho^0 p)$ is reproduced well by a QCD-based model where the photon fluctuates into a $q\bar{q}$ pair which interacts with the gluon cloud of the proton to produce the ρ^0. The decay angular distributions show that under the assumption of SCHC, ρ^0, ϕ and J/Ψ for $(Q^2+M_V^2) \geq 2$ GeV^2 are produced predominantly by longitudinal photons.

Inclusive diffraction, $\gamma^* p \to XN$, where N is a proton or a low-mass nucleonic system, behaves in many aspects similar to the production of vector mesons. For fixed Q^2, the cross section differential in the mass M_X of the system X, shows a rapid rise with W, $d\sigma^{diff}/dM_X \propto W^\delta$, where $\delta \approx 0.6-0.7$ for $Q^2 > 20$ GeV^2. The W dependence yields for the average of the Pomeron trajectory $\bar{\alpha_{I\!P}} = 1.23 \pm 0.02(stat) \pm 0.04(syst)$. The dependence on t is of the form $d\sigma/d|t| \propto e^{-b|t|}$ with $b = 7.2 \pm 1.4$ GeV^{-2}. The ratio of the diffractive cross section, integrated over $M_X < 35$ GeV, to the total cross section is approximately constant with W, being $15.8^{+1.1}_{-1.0}\%$ at $Q^2 = 4$ GeV^2 and decreasing to $5.0 \pm 0.9\%$ at $Q^2 = 320$ GeV^2.

The diffractive structure function of the proton, $x_{I\!P}F_2^{D(3)}(\beta, Q^2, x_{I\!P})$, where $\beta = Q^2/(Q^2 + M_X^2)$ and $x_{I\!P} = (Q^2+M_X^2)/(Q^2+W^2)$, has been measured for Q^2 values between 2.7 and 1600 GeV^2. For fixed values of β and Q^2, $x_{I\!P}F_2^{D(3)}$ rises as $x_{I\!P} \to 0$. The data have been used to determine diffractive parton distribution functions. A QCD-inspired model

(BEKW), in which the incoming photon fluctuates into a $q\bar{q}$ or $q\bar{q}g$ dipole that interacts with the target proton by the exchange of a two-gluon system (the "Pomeron"), gives an excellent description of the data for inclusive diffraction, after adjusting the five free parameters. The contribution from longitudinal photons via $q\bar{q}$ exchange is important for $\beta \geq 0.96$, while transverse photons dominate the regions $0.14 < \beta < 0.9$ by $q\bar{q}$ exchange, and $\beta < 0.1$ by $q\bar{q}g$ exchange.

References

[1] ZEUS Coll., M. Derrick et al., Z. Phys. **C 63**, 391 (1994).

[2] ZEUS Coll., S. Chekanov et al., Nucl. Phys. **B 627**, 3 (2002).

[3] H1 Coll., S. Aid et al., Z. Phys. **C 69**, 27 (1995).

[4] ZEUS Coll., S. Chekanov et al., Nucl. Phys. **B 713**, 3 (2005).

[5] ZEUS Coll., S. Chekanov et al., Nucl. Phys. **B 800**, 1 (2008).

[6] H1 Coll., C. Adloff et al. Eur. Phys. J. **C 21**, 33 (2001).

[7] ZEUS Coll., M. Derrick et al., Phys. Lett. **B 315**, 481 (1993).

[8] H. Kowalski, M. Roco and G. Wolf, ZEUS note 96-038 (18-12-1995), unpublished .

[9] ZEUS Coll., M. Derrick et al., Z. Phys. **C 70**, 391 (1996).

[10] ZEUS Coll., S. Chekanov et al., PMC **A 1**, 6 (2007).

[11] H1 Coll., C. Adloff et al., Eur. Phys. J. **C 13**, 317 (2000).

[12] P. Söding, Phys. Lett. **19**, 702 (1996).

[13] Henri Kowalski and Derek Teany, Phys. Rev. **D 68**, 114005 (2003).

[14] H. Kowalski, L. Motyka and G. Watt, Phys. Rev. **D 74**, 074016 (2006).

[15] H. G. Dosch and E. Ferreira, hep-ph/0610311 (2006).

[16] J. R. Forshaw, R. Sandapen and G. Shaw, Phys. Rev. **D 69**, 094013 (2004).

[17] ZEUS Coll., S. Chekanov et al., J. High Energy Phys. **05**, 108 (2009).

[18] A. Breakstone et al., Nucl. Phys. **B 248**, 253 (1984).

[19] K. Schilling, P. Seyboth and G. Wolf, Nucl. Phys. **B 15**, 397 (1973).

[20] K. Schilling and G. Wolf, Nucl. Phys. **B 61**, 381 (1970).

[21] A. D. Martin, M. G. Ryskin and T. Teubner, Phys. Rev. **D 55**, 4329 (1997).

[22] ZEUS Coll., J. Breitweg et al., Phys. Lett. **B 487**, 3 (2000).

[23] ZEUS Coll., M. Derrick et al., Phys. Lett. **B 380**, 220 (1996).

[24] H1 Coll., C. Adloff et al., Phys. Lett. **B 483**, 360 (2000).

[25] ZEUS Coll., S. Chekanov et al., Phys. Lett. **B 553**, 141 (2003).

[26] ZEUS Coll., S. Chekanov et al., Nucl. Phys. **B 718**, 3 (2003).

[27] H1 Coll., S. Aid et al., Nucl. Phys. **B 468**, 3 (1996).

[28] ZEUS Coll., J. Breitweg et al., Eur. Phys. J. **C 6**, 371 (2000).

[29] ZEUS Coll., S. Chekanov et al., Nucl. Phys. **B 695**, 3 (2004).

[30] H1 Coll., C. Adloff et al., Eur. Phys. J. **C 10**, 373 (1999).

[31] H1 Coll., A.Aktas et al., Eur. Phys. J. **C 46**, 585 (2006).

[32] E. Iancu, K. Itakura and S. Munier, Phys. Lett. **B 590**, 199 (2004).

[33] L. McLerran and R. Venugopolan, Phys. Rev. **D49**, 2233 (1994), **D50**, 2225 (1994).

[34] G. Watt and H. Kowalski, Phys. Rev. **D 78**, 014016 (2008).

[35] ZEUS Coll., M. Derrick et al., Phys. Lett. **B 338**, 483 (1994).

[36] H1 Coll., T. Ahmed et al., Nucl. Phys. **B 429**, 477 (1994).

[37] H1 Coll., A. Aktas et al., Eur. Phys. J. **C 48**, 715 (2007).

[38] J. Bartels, Eur. Phys. J. **C 43**, 3 (2005).

[39] ZEUS Coll., J. Breitweg et al., Eur. Phys. J. **C 1**, 81 (1998).

[40] ZEUS Coll., S. Chekanov et al., Eur. Phys. J. **C 38**, 43 (2004).

[41] H1 Coll., A. Aktas et al., Eur. Phys. J. **C 48**, 715 (2006).

[42] ZEUS Coll., S. Chekanov et al. Nucl. Phys. **B 816**, 1 (2009).

[43] J. Bartels, J. Ellis, H. Kowalski, M. Wüsthoff, Eur. Phys. J. **C 7**, 443 (1999)

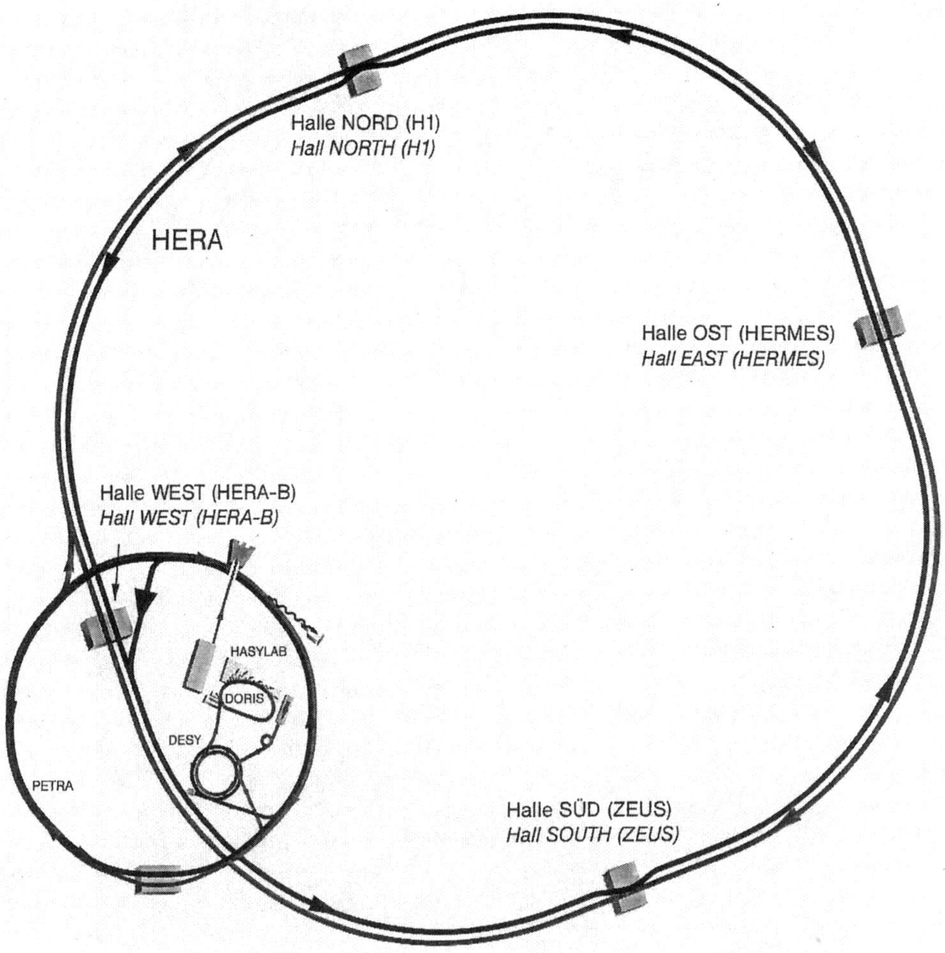

Figure 1: *Electron-Proton collider HERA at DESY.*

Figure 2: *HERA tunnel: proton ring with superconducting magnets (top) and electron ring with normal conducting magnets (bottom).*

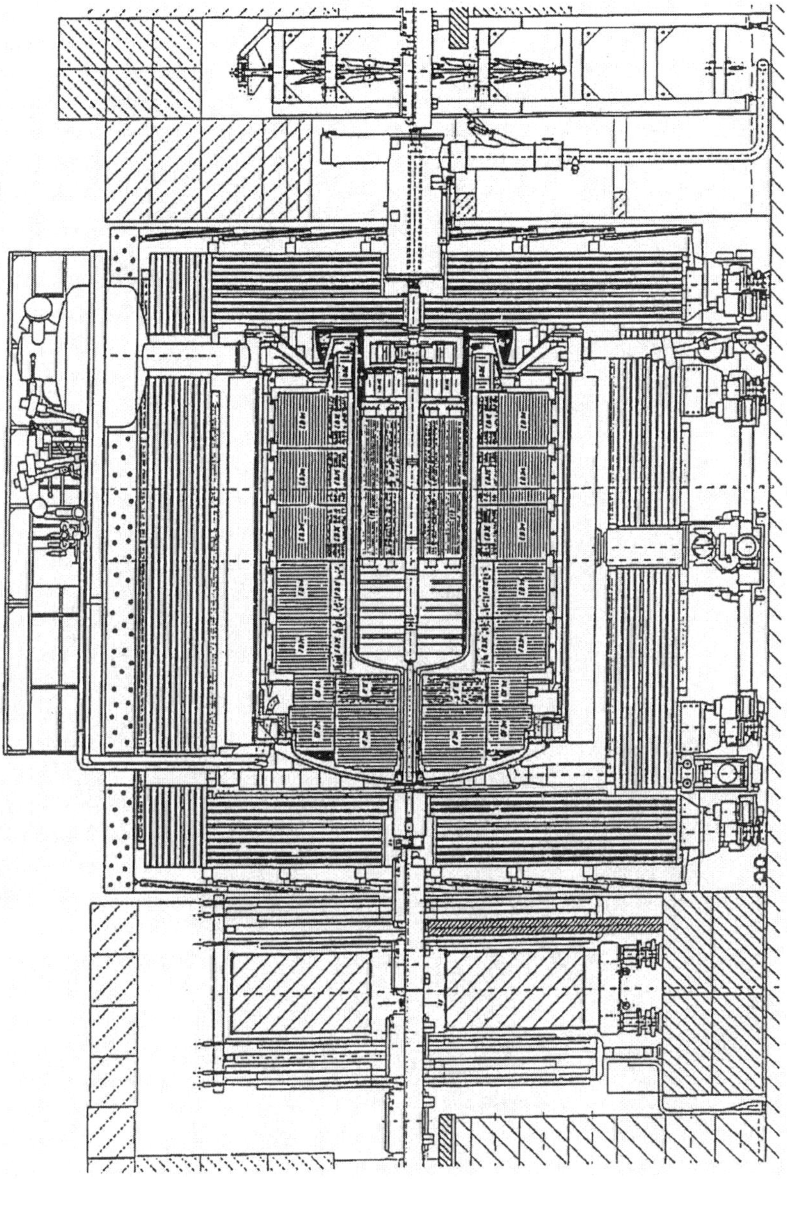

Figure 3: *Side view of the H1 detector. The proton (electron) beam enters from the right (left).*

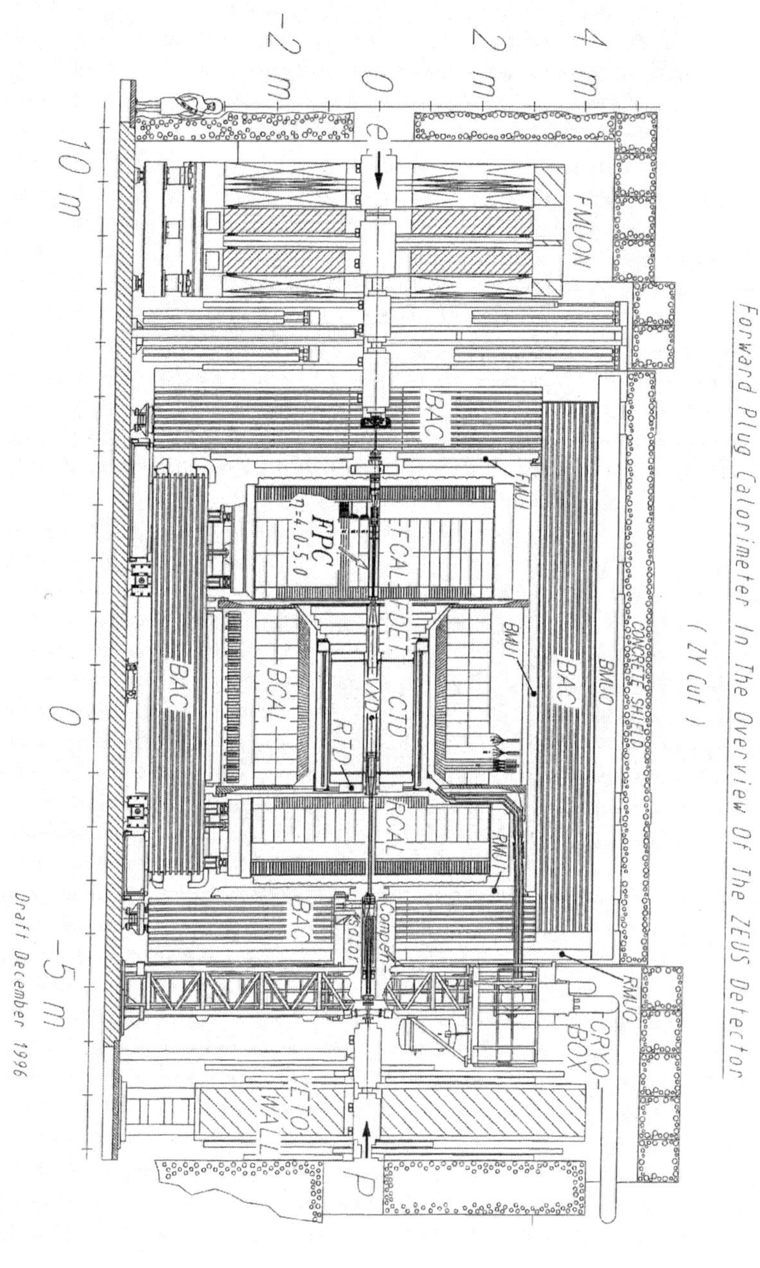

Figure 4: *Side view of the ZEUS detector with the FPC integrated in the ZEUS Calorimeter. The proton (electron) beam enters from the right (left).*

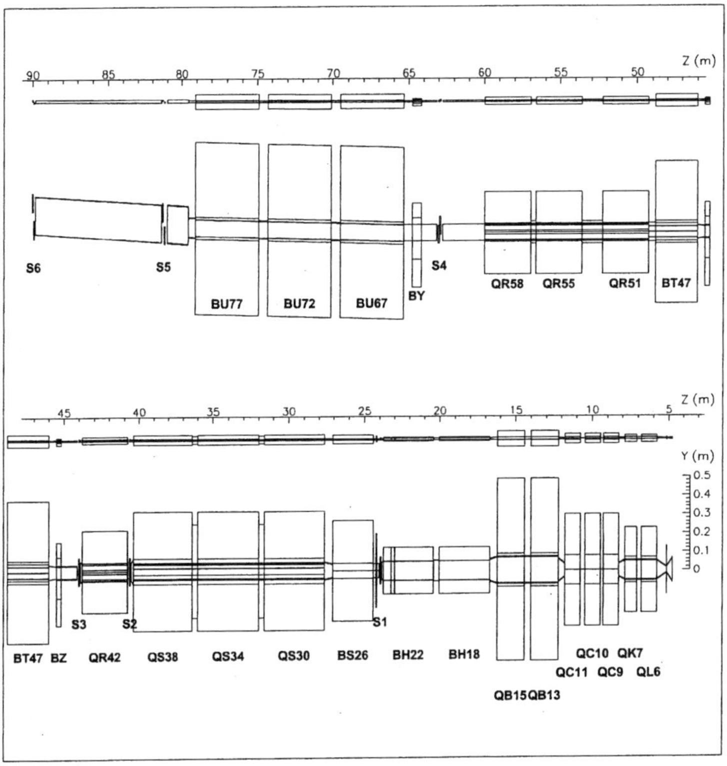

Figure 5: *The ZEUS system of detectors S1, ..,S6 for the detection of forward scattered protons; the interaction point is located at* $z = 0$.

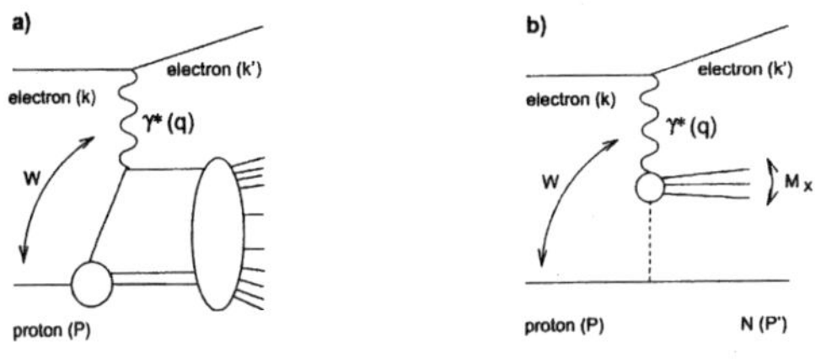

Figure 6: *Diagrams illustrating particle production (a) by deep inelastic inclusive scattering, (b) by deep inelastic diffractive scattering.*

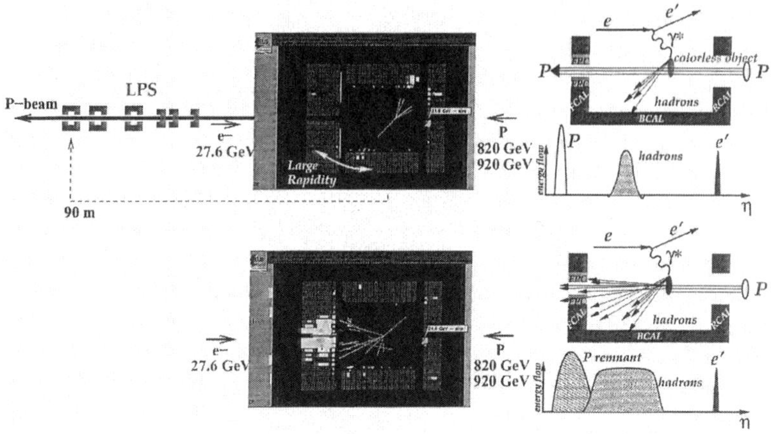

Figure 7: *Event topologies of diffractive (top) and non-diffractive deep inelastic scattering (bottom) as seen in the ZEUS detector. The top figure shows also the leading proton spectrometer (LPS) which detects forward scattered protons. Shown are the tracks of charges particles in the central tracking chamber and the energy deposits in the uranium-scintillator calorimeter.*

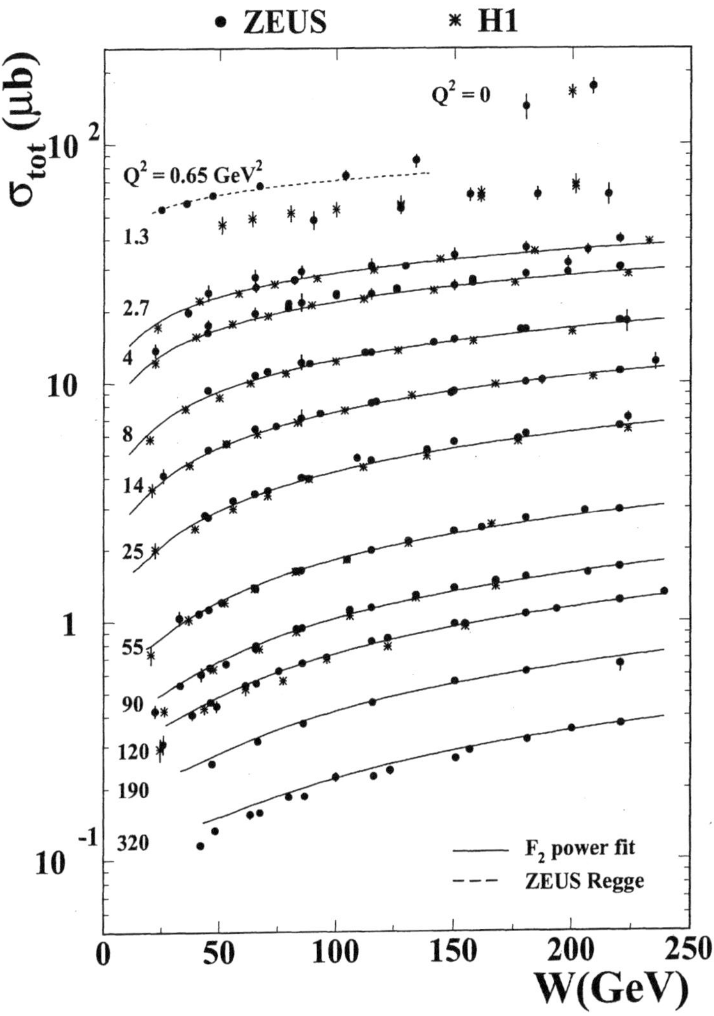

Figure 8: *The photon-proton total cross section for real ($Q^2 = 0$) and virtual photons ($Q^2 > 0$) as a function of the photon-proton center-of-mass energy, W, measured by H1 and ZEUS. Also shown is the ZEUS Regge fit for $Q^2 = 0.65$ GeV^2 (dashed curve), and the result of the F_2 power fit (solid curves) for $Q^2 \geq 2.7$ GeV^2.*

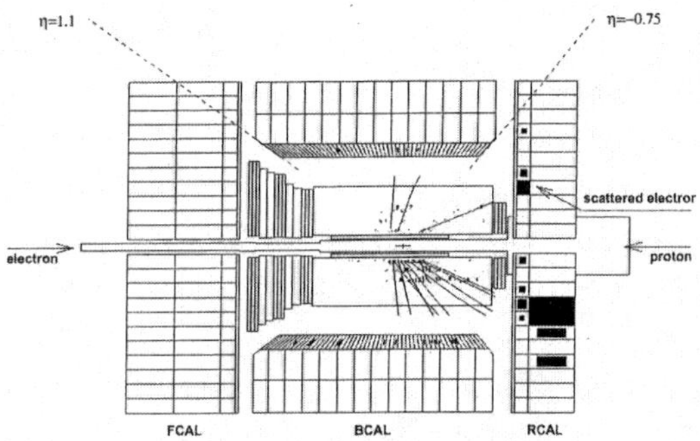

Figure 9: *One of the first events from deep inelastic ep scattering with a large rapidity gap, as recorded in the ZEUS detector.*

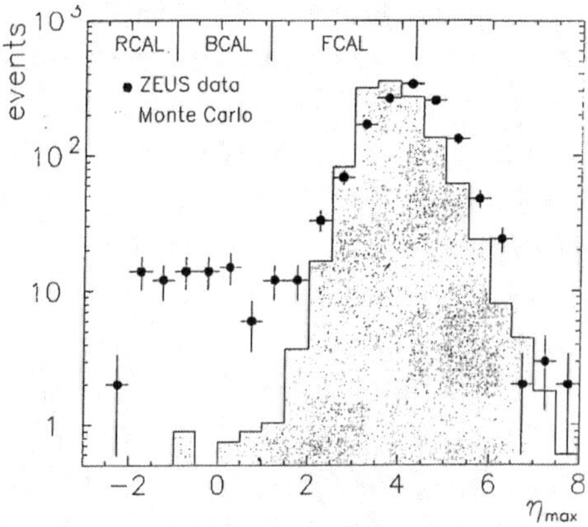

Figure 10: *Distribution of η_{max}, the maximum rapidity of a calorimeter cluster in an event, for data and for Monte Carlo events generated for non-diffractive scattering (shaded).*

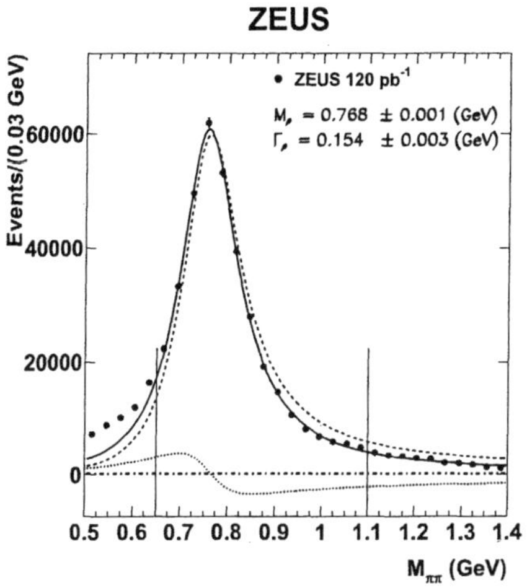

Figure 11: *Reaction $\gamma^* p \to \pi^+\pi^- p$: the $\pi^+\pi^-$ invariant-mass distribution. The solid line represents a fit to the Söding form, see text.*

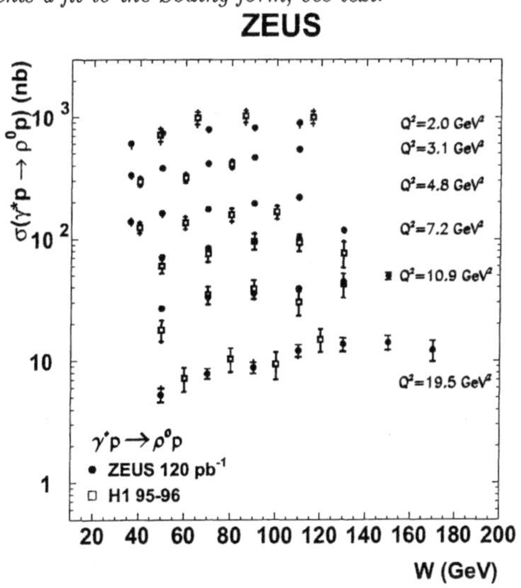

Figure 12: *The cross section for $\gamma^* p \to \rho^0 p$ as a function of W for different values of Q^2; from H1 and ZEUS.*

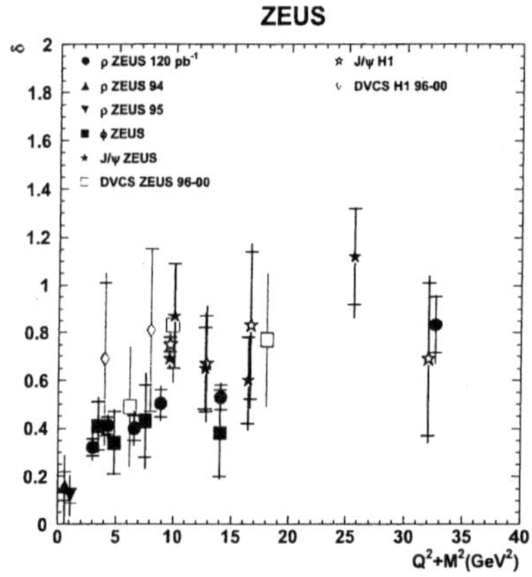

Figure 13: *The power δ determined from a fit of the cross section for $\sigma(\gamma^* p \to Vp) \propto W^\delta$ as a function of $(Q^2 + M_V^2)$, where M_V is the mass of the vector meson indicated; for DVCS: $M_V = 0$. The data stem from H1 and ZEUS.*

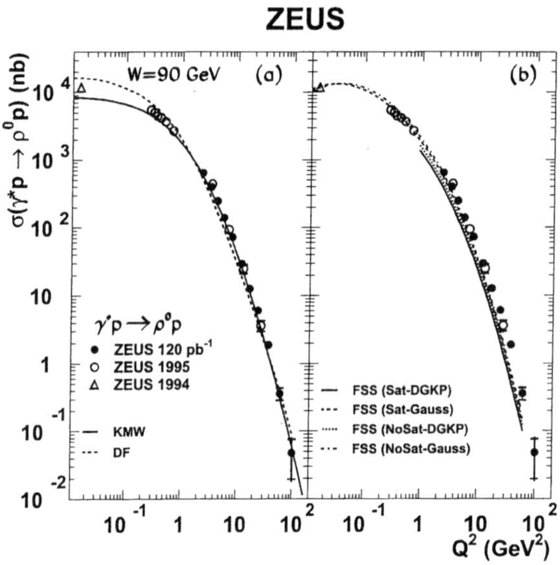

Figure 14: *The cross section for $\gamma^* p \to \rho^0 p$ as a function of Q^2 for $W = 90$ GeV, from ZEUS; the data presented in (a) and (b) are the same. Shown are also the predictions of several models, see text.*

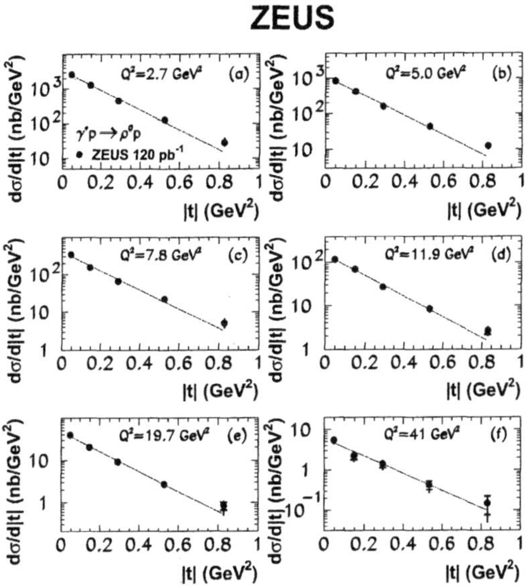

Figure 15: *The differential cross section $d\sigma/d|t|$ for $\gamma^* p \to \rho^0 p$ as a function of $|t|$ for $W = 90$ GeV and fixed values of Q^2. The lines represent exponential fits to the data; from ZEUS.*

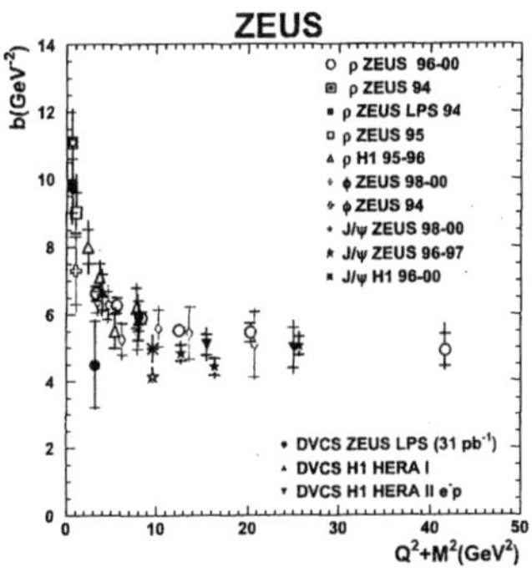

Figure 16: *The slope b characterizing the t dependence, $d\sigma/d|t| \propto e^{-b|t|}$, of photon and vector meson production by $\gamma, \gamma^* p$ interactions, shown for DVCS and for $V = \rho^0, \phi, J/\Psi$ production as a function of $(Q^2 + M_V^2)$, where $M_V = 0$ for DVCS and equal to the mass of the vector meson otherwise.*

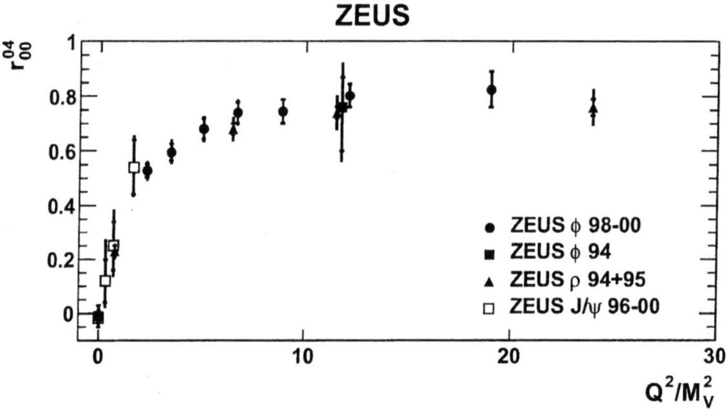

Figure 17: *The reaction $\gamma^* p \to Vp$: the density-matrix-element r^{04}_{00} as a function of Q^2/M_V^2 for $V = \rho^0, \phi, J/\Psi$ (bottom); from ZEUS.*

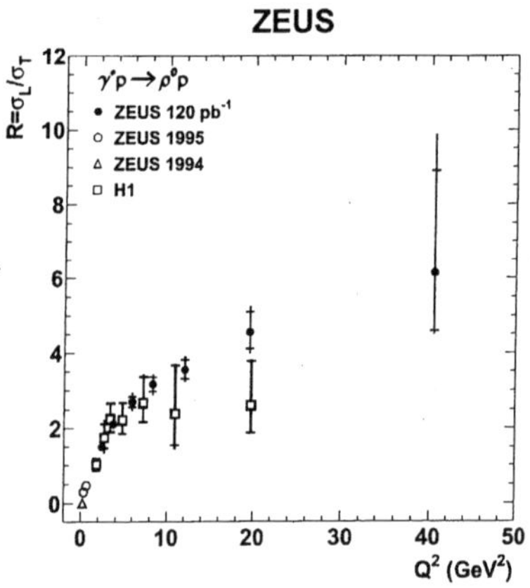

Figure 18: *For $\gamma^* p \to \rho^0 p$ the ratio $R = \sigma_L/\sigma_T$ as a function of Q^2; the data from H1 were determined for $W = 75$ GeV, those from ZEUS for $W = 90$ GeV.*

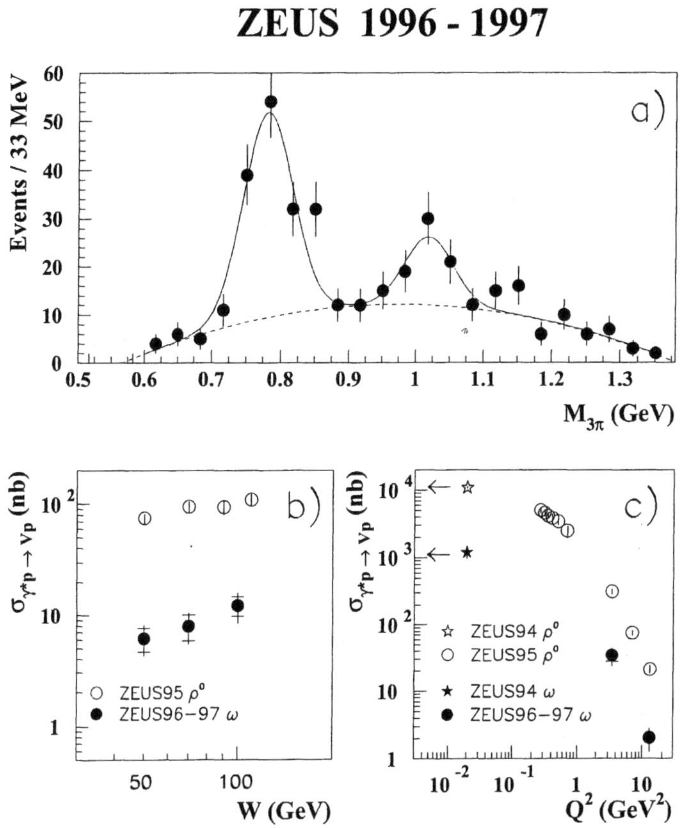

Figure 19: *The reaction $\gamma^* p \to \pi^+\pi^-\pi^0$: (a) distributions of the $\pi^+\pi^-\pi^0$ mass spectrum; (b) cross sections for $\gamma^* p \to \rho^0 p$ and $\gamma^* p \to \omega p$ as a function of the c.m. energy W at $Q^2 = 7$ GeV^2; (c) the same as a function of Q^2 at $W = 70$ GeV; from ZEUS.*

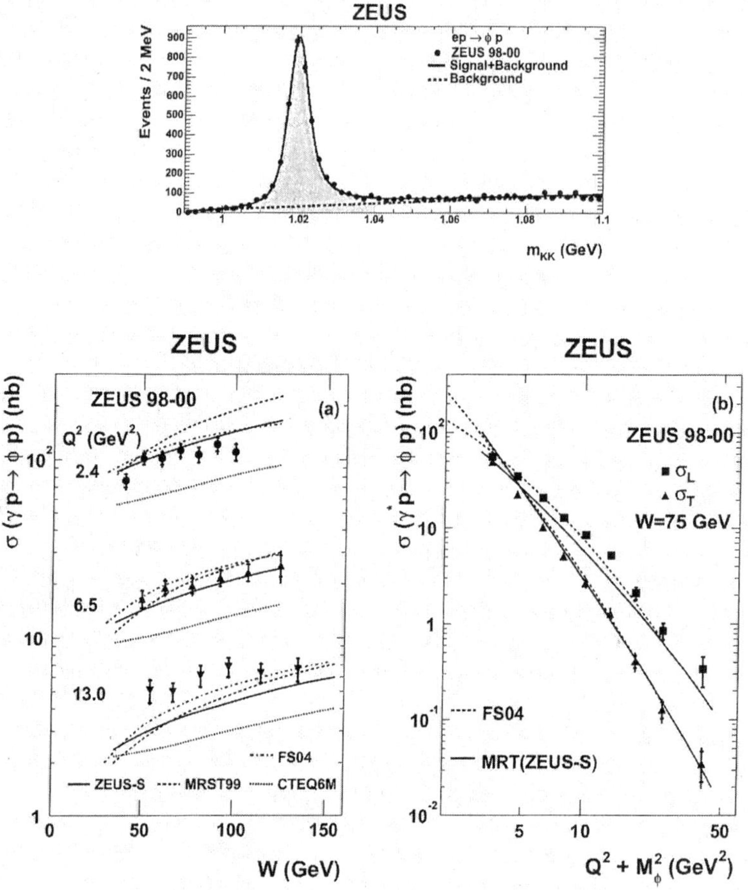

Figure 20: *The reaction $\gamma^* p \to \phi p$: (top) distribution of the K^+K^- mass spectrum ($L = 119\ pb^{-1}$); (bottom, left): cross section as a function of W for $Q^2 = 2.4, 6.5 and 13$ GeV^2; (bottom, right): cross section as a function of $(Q^2 + M_\phi^2)$ separately for the contributions from transverse (T) and longitudinal photons (L); from ZEUS.*

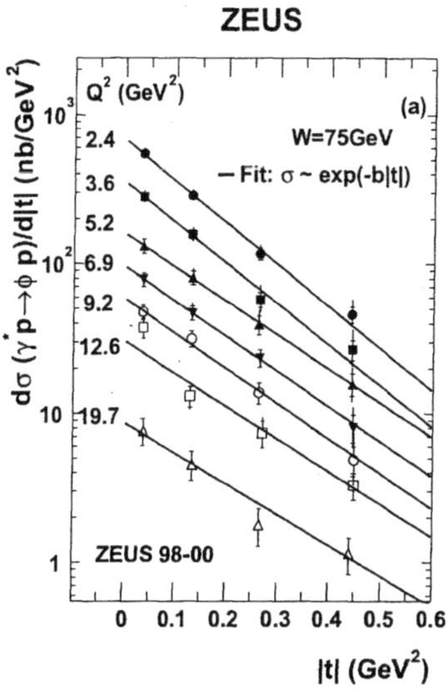

Figure 21: *Reaction $\gamma^* p \to \phi p$: differential cross-section $d\sigma/d|t|$ as a function of $|t|$; from ZEUS.*

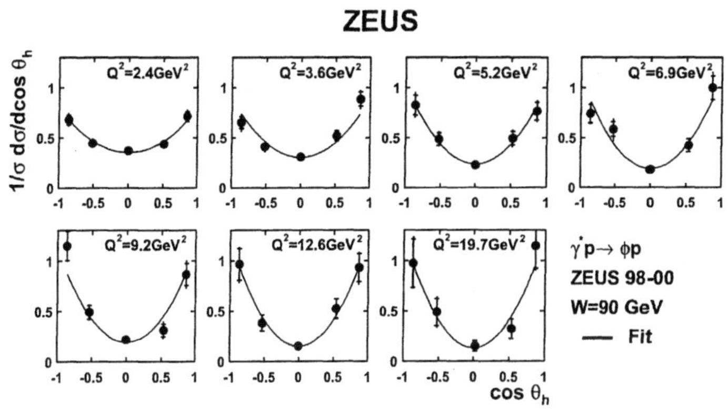

Figure 22: *Reaction $\gamma^* p \to \phi p$: distribution of $cos\theta_h$ for different values of Q^2; from ZEUS.*

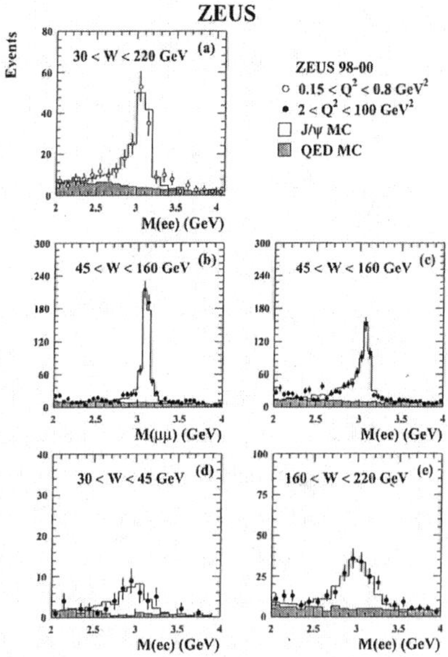

Figure 23: *The reaction $\gamma^* p \to J/\Psi p$: distributions of $M_{l^+l^-}$, $l = e, \mu$, for the W, Q^2 regions indicated; from ZEUS.*

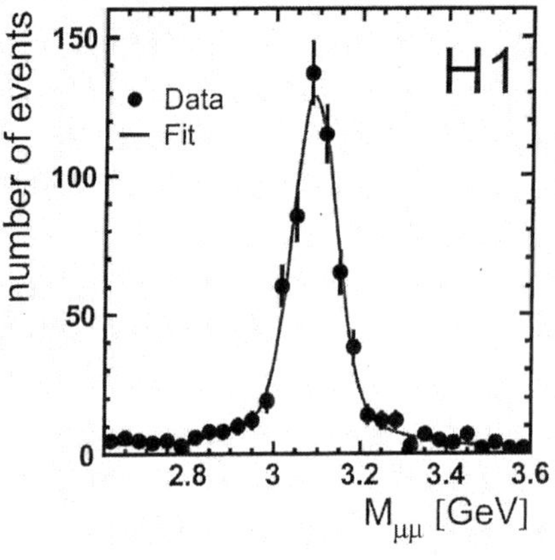

Figure 24: *The reaction $\gamma^* p \to J/\Psi p$: distribution of $M_{\mu^+\mu^-}$ for $Q^2 = 2-$ 80 GeV^2; from H1.*

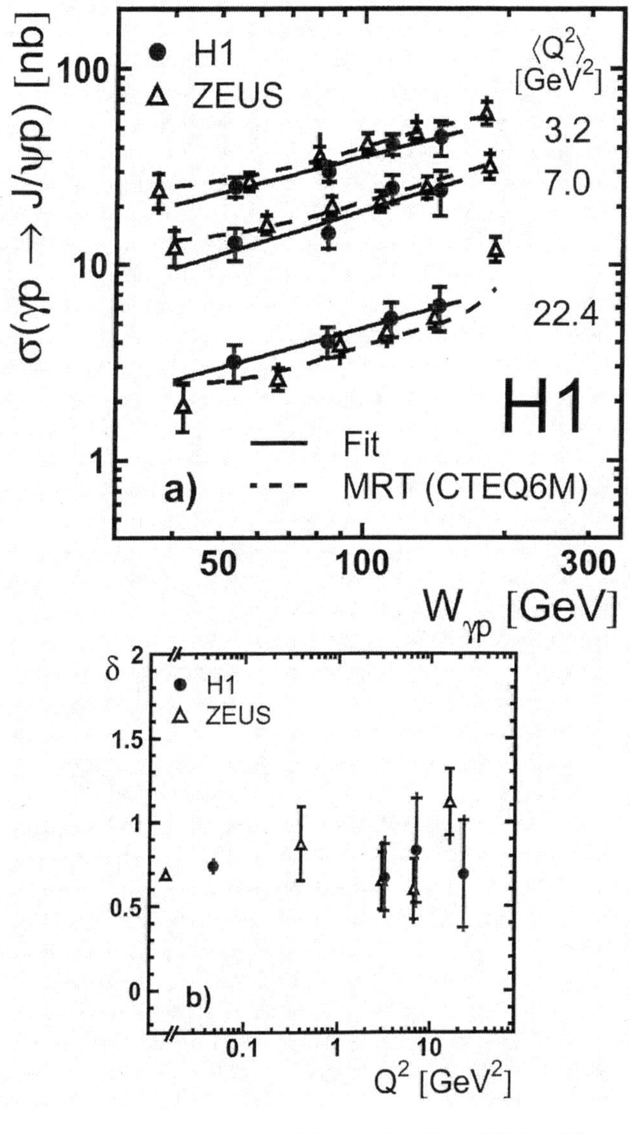

Figure 25: *The cross-section $\gamma^* p \to J/\Psi p$ as a function of W for different values of Q^2 as measured by H1 and ZEUS (top); parametrisation $\sigma(\gamma^* p \to J/\Psi p) \propto W^\delta$: the power δ as a function of Q^2 as determined by H1 (bottom); from H1.*

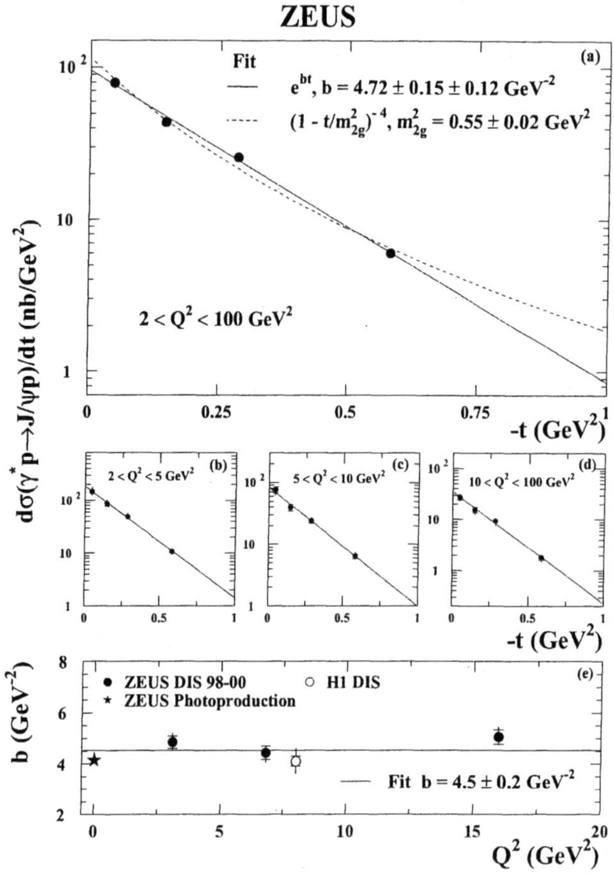

Figure 26: *The reaction $\gamma^* p \to J/\Psi p$: (a)-(d) differential cross section $d\sigma/dt$ as a function of $-t$ for $Q^2 = 2 - 100\ \mathrm{GeV}^2$ and $Q^2 = 2-5, 5-10, 10-100\ \mathrm{GeV}^2$; (e) assuming $\sigma(\gamma^* p \to J/\Psi p) \propto W^\delta$: δ as a function of Q^2; from ZEUS.*

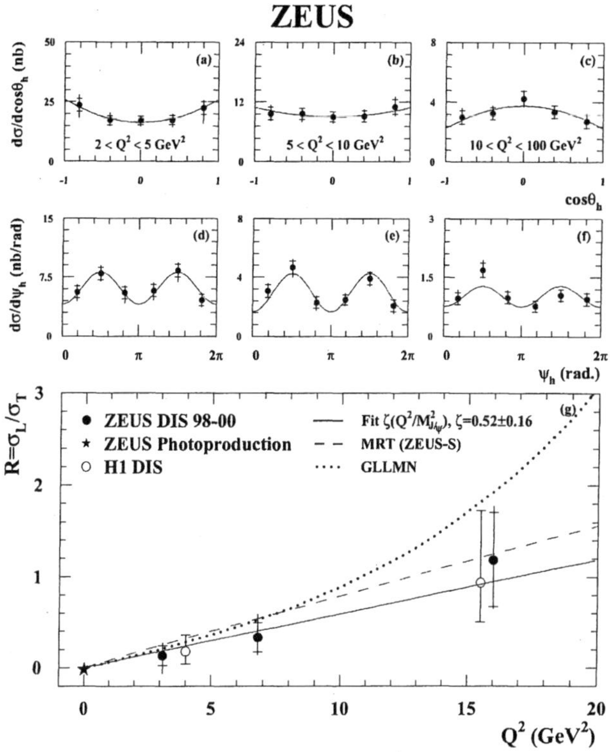

Figure 27: *The reaction $\gamma^* p \to J/\Psi p$: (a)-(f) decay angular distributions of the J/Ψ in the s-channel helicity system for the Q^2 bins indicated; (g) ratio $R = \sigma_L/\sigma_T$ as a function of Q^2; from ZEUS.*

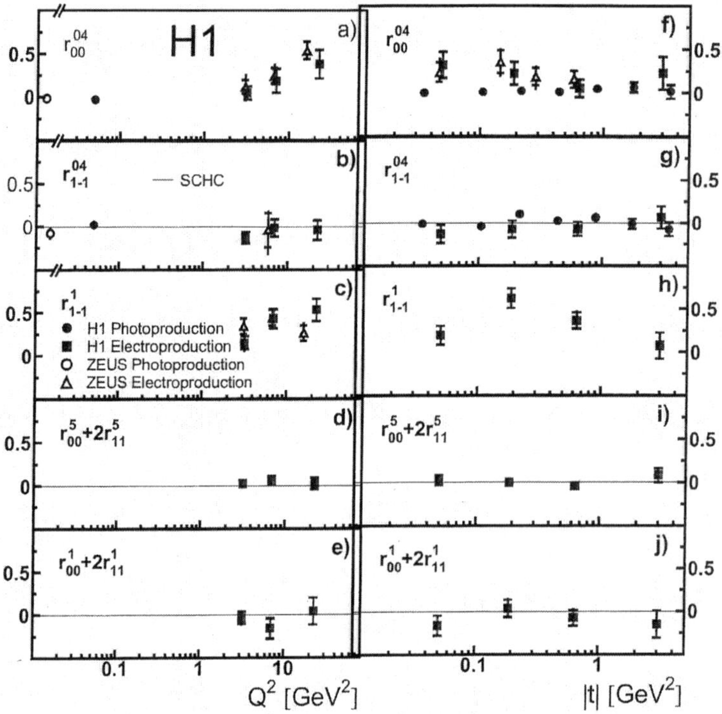

Figure 28: *Reaction $\gamma^* p \to J/\Psi p$: spin-density matrix elements as a function of Q^2 (left) and $|t|$ (right), with data from H1 and ZEUS; from H1.*

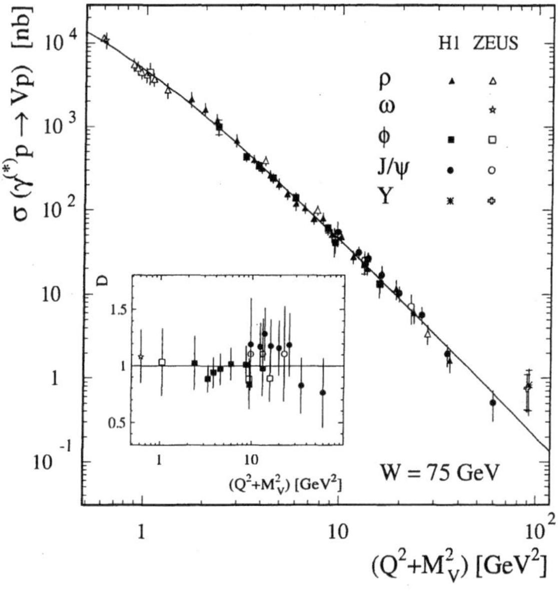

Figure 29: *The cross section $\sigma(\gamma^* p \to Vp)$ with $V = \rho^0, \omega, \phi, J/\Psi, Y$, as a function of $(Q^2 + M_V^2)$ for $W = 75$ GeV: results from H1 and ZEUS.*

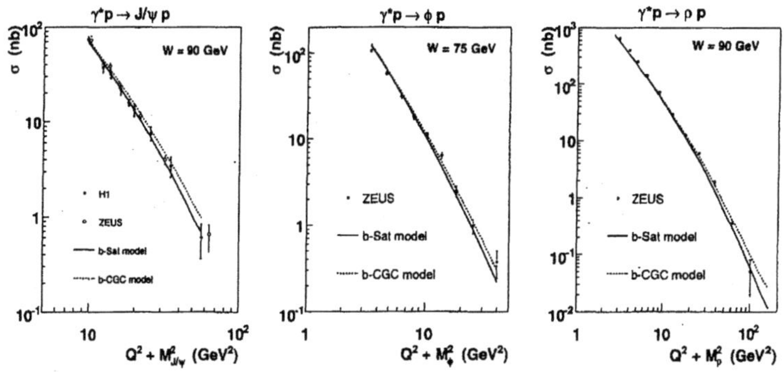

Figure 30: *The cross section $\sigma(\gamma^* p \to Vp)$ with $V = \rho^0, \phi, J/\Psi$, as a function of $(Q^2 + M_V^2)$ for $W = 75$ GeV: results from H1 and ZEUS. The curves have been computed with the b-Sat and b-CGC models; from Watt and Kowalski.*

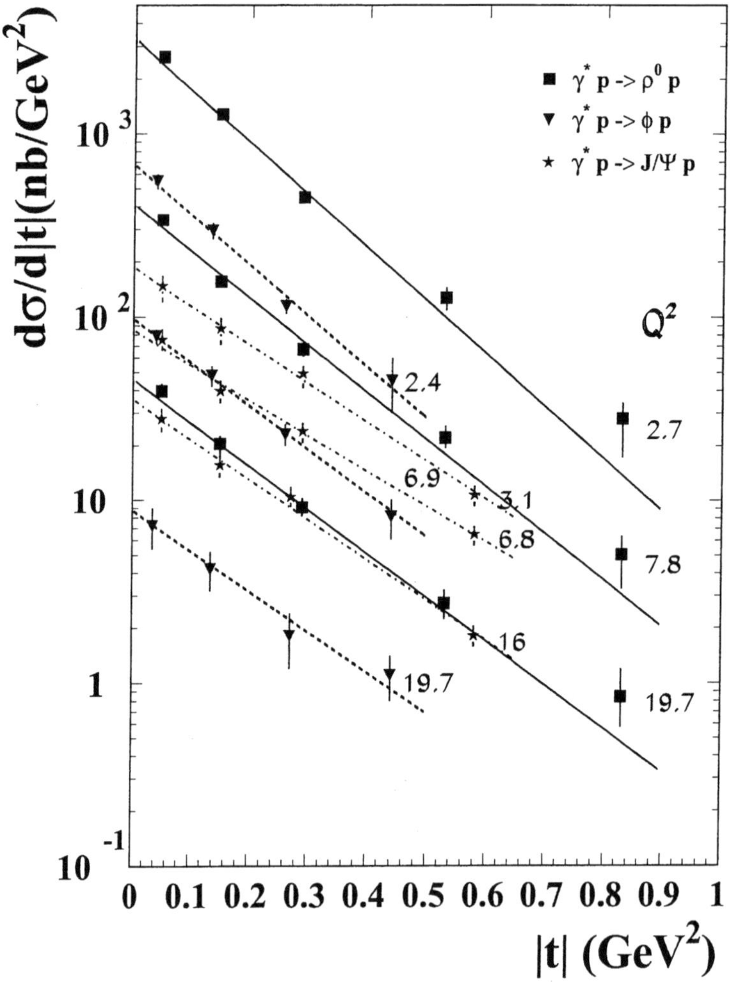

Figure 31: *The differential cross sections $d\sigma/d|t|$ as a function of $|t|$for vector meson production, $\gamma^* p \to Vp$, $V = \rho^0, \phi, J/\Psi$; at $W = 90$ GeVfor $\rho^0, J/\Psi$ and at $W = 75$ GeVfor ϕ; as determined by ZEUS. The lines represent the fits of the form $d\sigma/d|t| = c \cdot e^{-b|t|}$.*

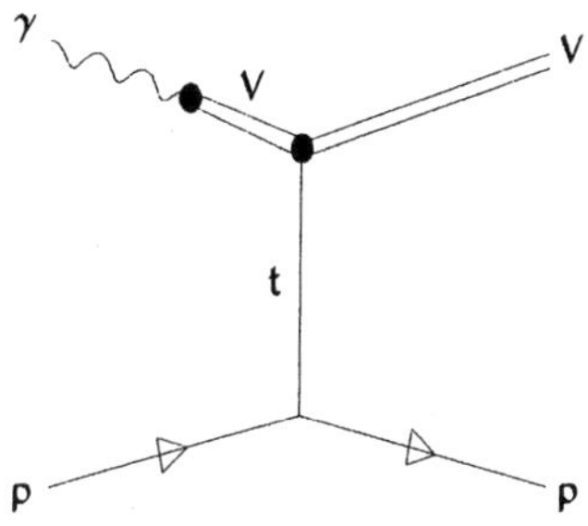

Figure 32: *Diagram for diffractive vector meson production,* $\gamma^* p \rightarrow Vp$.

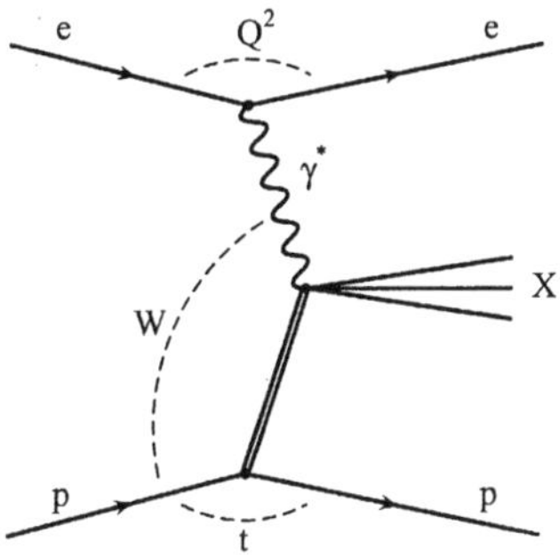

Figure 33: *Diagram for the reaction* $ep \rightarrow eXp$.

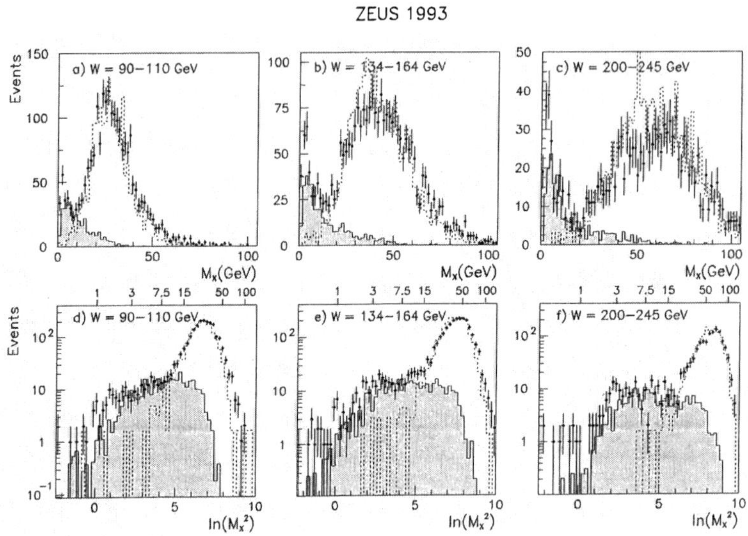

Figure 34: *Reaction $ep \to eX$: distribution of M_X (top) and $\ln M_X^2$ (bottom) for different regions of W at $Q^2 = 14$ GeV^2. The shaded distributions show the prediction of the NZ model for diffractive production; the dashed histograms show the expectations for nondiffractive production predicted by the CDMBGF model; from ZEUS.*

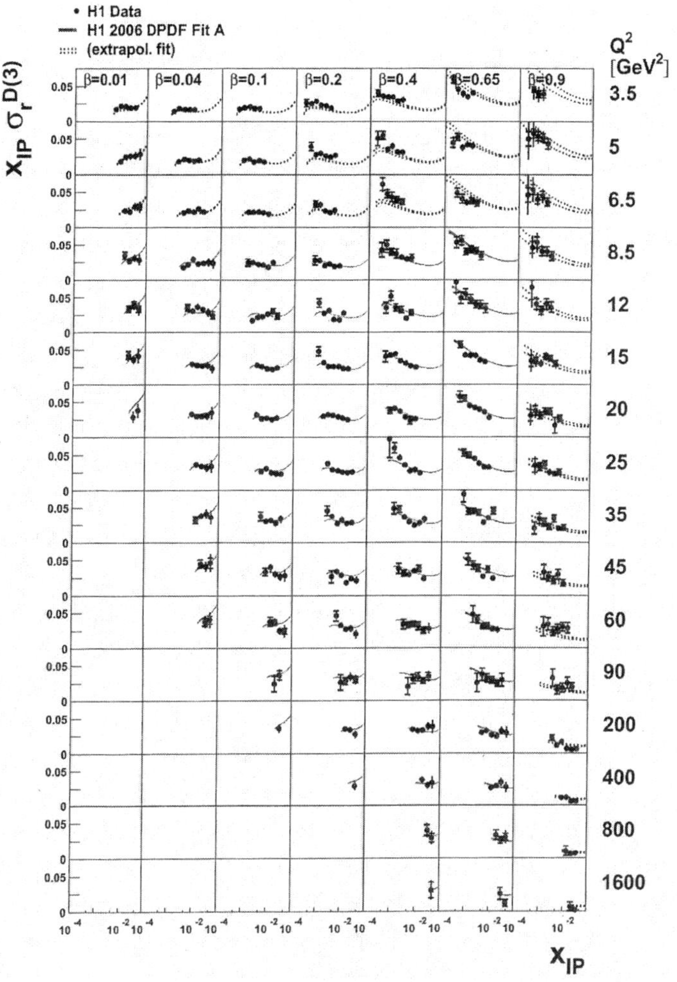

Figure 35: *The diffractive structure function* $x_{IP} F_2^{D(3)}$ *(here denoted by* $x_{IP}\sigma_r^{D(3)}$*), as a function of* x_{IP} *for the* β, Q^2 *values indicated. The curves show the H1 2006 DPDF Fit A to these data as a shaded error band in kinematic regions included in the fit, and as a pair of dashed lines in regions excluded from the fit; from H1.*

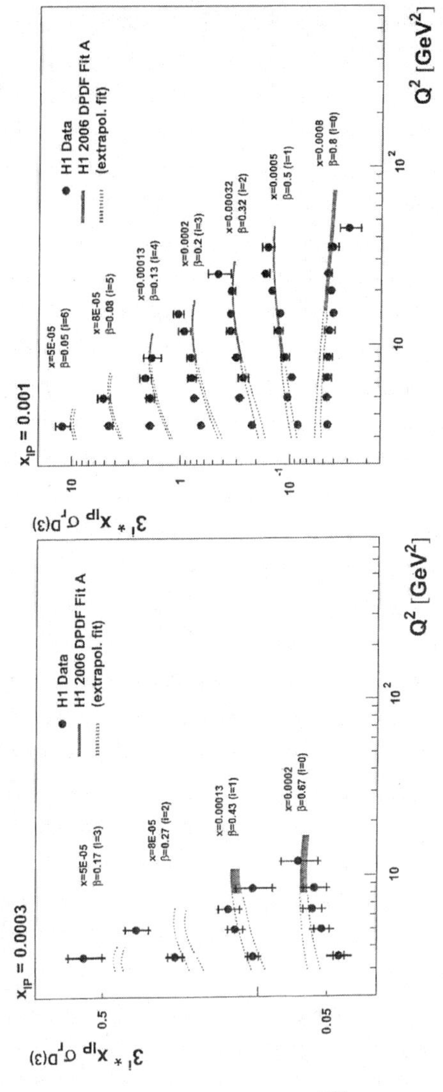

Figure 36: *The diffractive structure function $x_{\mathbb{P}} F_2^{D(3)}$ (here denoted by $x_{\mathbb{P}} \sigma_r^{D(3)}$) at $x_{\mathbb{P}} = 0.0003$ and 0.001 as a function of Q^2, for the x values indicated; from H1.*

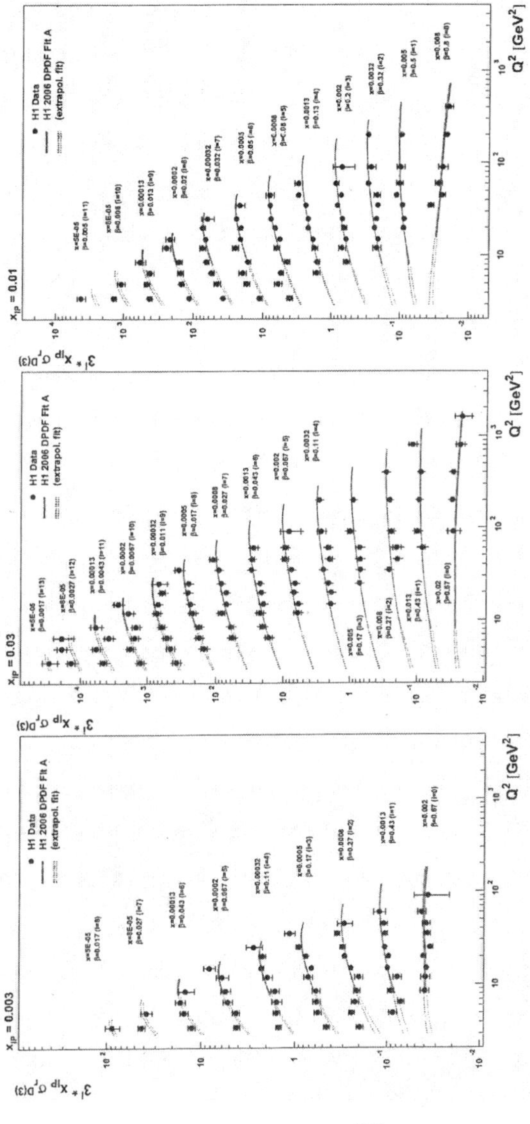

Figure 37: *The diffractive structure function* $x_{I\!P} F_2^{D(3)}$ *(here denoted by* $x_{I\!P}\sigma_r^{D(3)}$*) at* $x_{I\!P} = 0.003$, 0.01 *and* 0.03 *as a function of* Q^2, *for the x values indicated; from H1.*

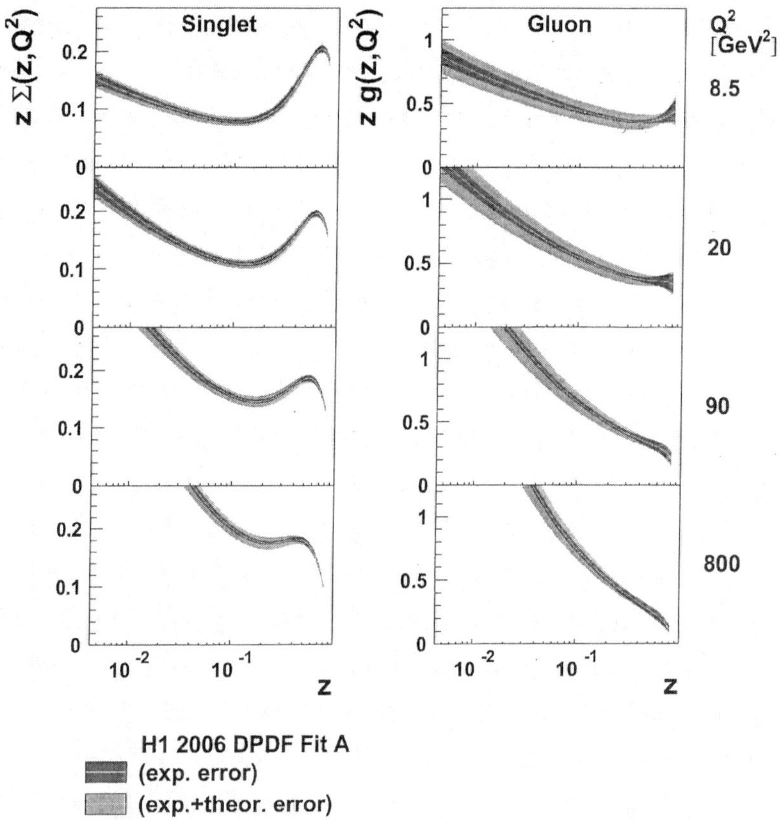

Figure 38: *The total quark singlet and gluon distributions as a function of z from an NLO QCD fit by H1 to the data for diffraction with* $Q^2 \geq 8.5$ GeV^2, $\beta \leq 0.8$ *and* $M_X > 2$ GeV; *z is the longitudinal momentum fraction of the parton entering the hard sub-process: for the lowest order quark-parton process* $z = \beta$, *whereas* $0 < \beta < z$ *for higher order processes; from H1.*

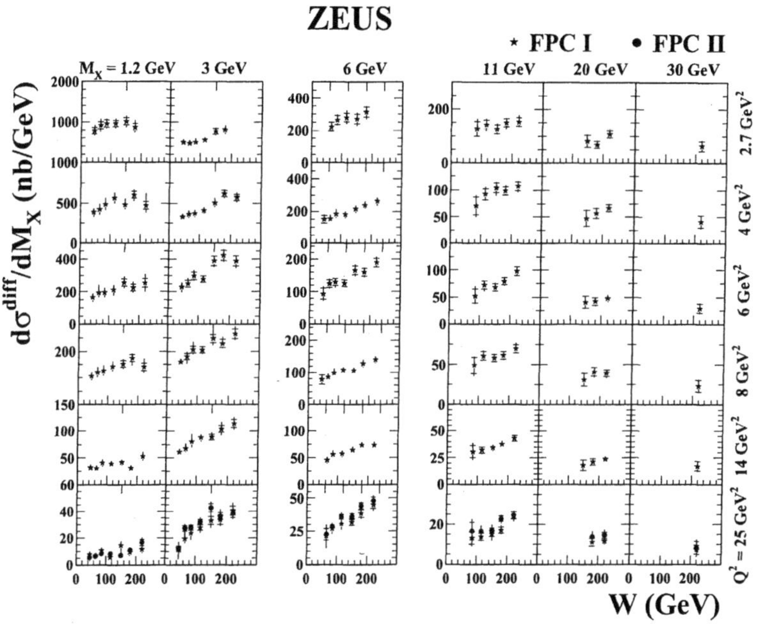

Figure 39: *The differential cross section, $d\sigma^{\rm diff}_{\gamma^* p \to XN}/dM_X$, $M_N < 2.3$ GeV, as a function of W for bins of Q^2, M_X, for FPC I data (stars) and FPC II data (dots), for Q^2 between 2.7 and 25* GeV^2*; from ZEUS.*

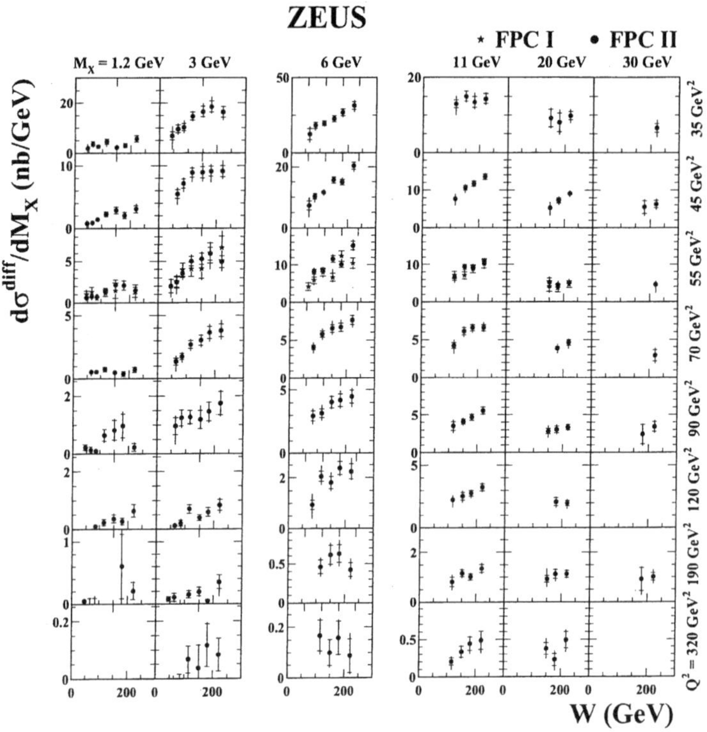

Figure 40: *The differential cross section, $d\sigma^{\rm diff}_{\gamma^* p\to XN}/dM_X$, $M_N < 2.3$ GeV, as a function of W for bins of Q^2, M_X, for FPC I data (stars) and FPC II data (dots), for Q^2 between 35 and 320* GeV^2*; from ZEUS.*

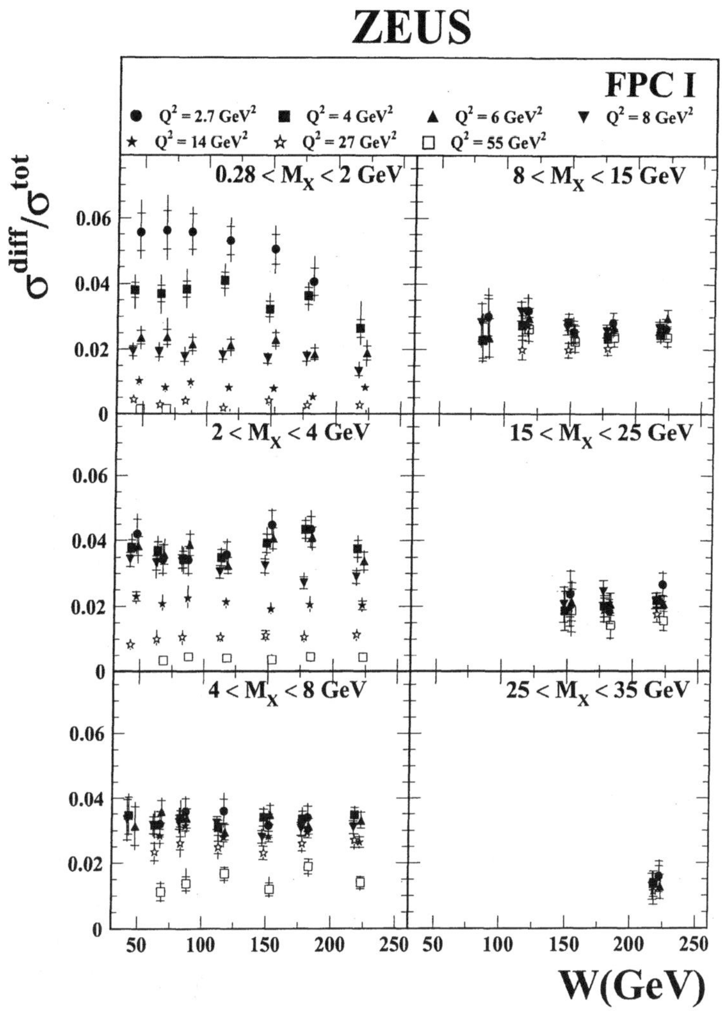

Figure 41: *The ratio of the diffractive cross section, integrated over the M_X intervals indicated, $\int_{M_a}^{M_b} dM_X d\sigma^{\rm diff}_{\gamma^* p\to XN, M_N<2.3{\rm GeV}}/dM_X$, to the total $\gamma^* p$ cross section, $r^{\rm diff}_{\rm tot} = \sigma^{\rm diff}/\sigma^{\rm tot}_{\gamma^* p}$, as a function of W for Q^2 between 2.7 and 55 ${\rm GeV}^2$. The inner error bars show the statistical uncertainties and the full bars the statistical and systematic uncertainties added in quadrature; from ZEUS.*

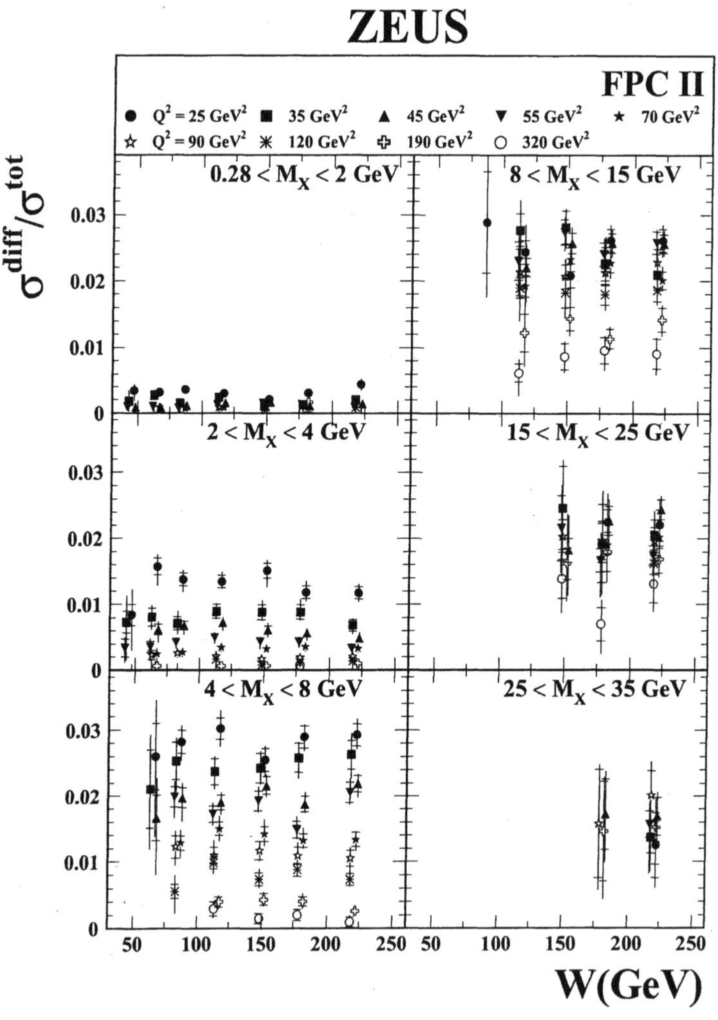

Figure 42: *The ratio of the diffractive cross section, integrated over the M_X intervals indicated, $\int_{M_a}^{M_b} dM_X d\sigma^{\rm diff}_{\gamma^* p\to XN, M_N<2.3{\rm GeV}}/dM_X$, to the total $\gamma^* p$ cross section, $r^{\rm diff}_{\rm tot} = \sigma^{\rm diff}/\sigma^{\rm tot}_{\gamma^* p}$, as a function of W for Q^2 between 25 and 320* GeV^2. *The inner error bars show the statistical uncertainties and the full bars the statistical and systematic uncertainties added in quadrature; from ZEUS.*

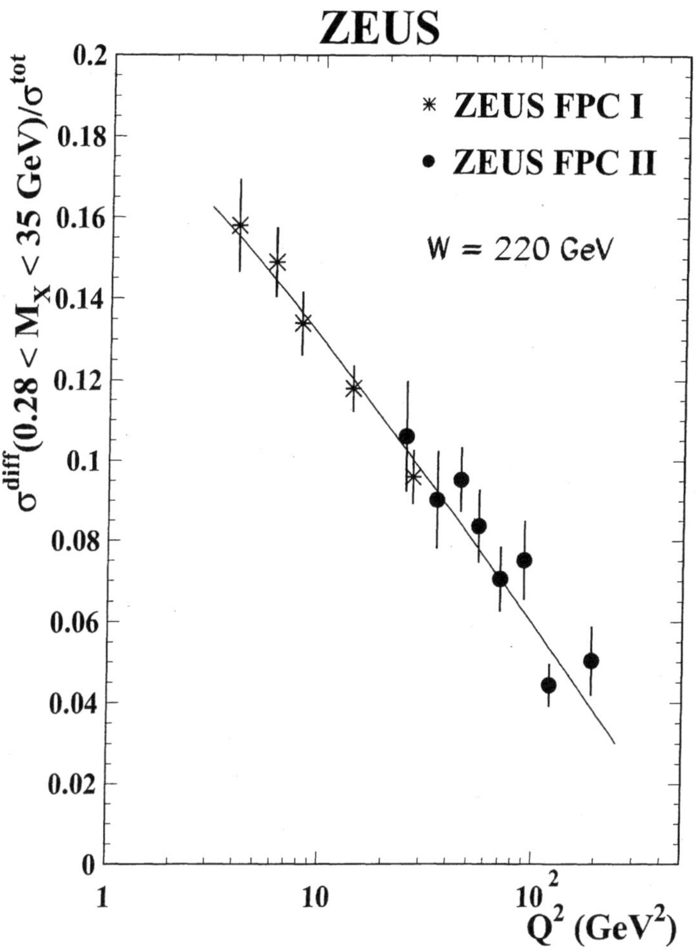

Figure 43: *The ratio of the diffractive cross section, integrated over* $0.28 < M_X < 35$ GeV, *to the total* $\gamma^* p$ *cross section, at* $W = 220$ GeV *as a function of* Q^2. *The line shows the result of fitting the data to the form* $r = a - b \cdot \ln(1 + Q^2)$, *see text; from ZEUS.*

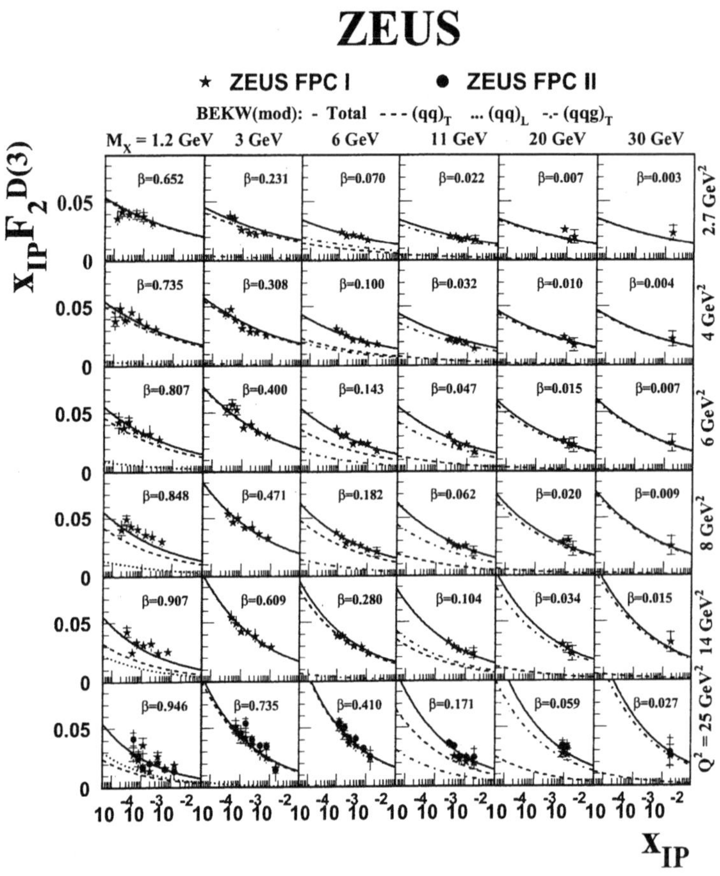

Figure 44: *The diffractive structure function of the proton multiplied by* $x_{I\!P}$, $x_{I\!P}F_2^{D(3)}$, *for* $\gamma^* p \to XN$, $M_N < 2.3$ GeV, *as a function of* $x_{I\!P}$ *for different regions of* β *and* $Q^2 \leq 25$ GeV^2. *The curves show the results of the BEKW(mod) fit for the contributions from* $(q\bar{q})$ *for transverse (dashed) and longitudinal photons (dotted) and for the* $(q\bar{q}g)$ *contribution for transverse photons (dashed-dotted) together with the sum of all contributions (solid); from ZEUS.*

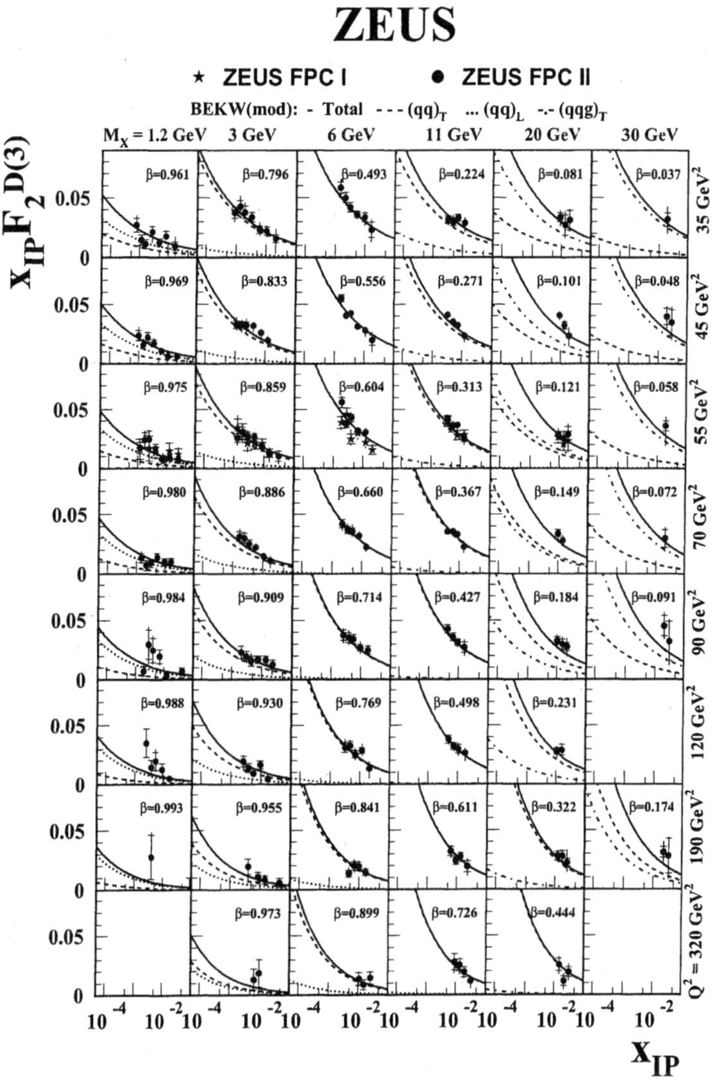

Figure 45: *The diffractive structure function of the proton multiplied by* $x_{I\!P}$, $x_{I\!P}F_2^{D(3)}$, *for* $\gamma^* p \to XN$, $M_N < 2.3$ GeV, *as a function of* $x_{I\!P}$ *for different regions of* β *and* $Q^2 \geq 35$ GeV^2. *The curves show the results of the BEKW(mod) fit for the contributions from* $(q\bar{q})$ *for transverse (dashed) and longitudinal photons (dotted) and for the* $(q\bar{q}g)$ *contribution for transverse photons (dashed-dotted) together with the sum of all contributions (solid); from ZEUS.*

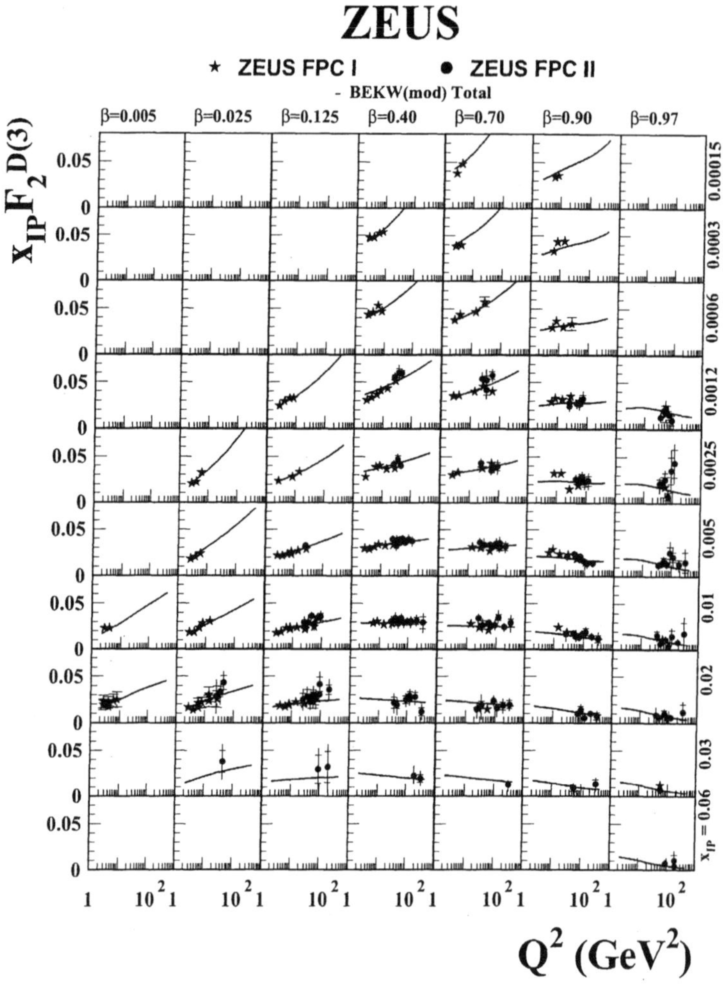

Figure 46: *The diffractive structure function of the proton multiplied by* $x_{\mathbb{P}}$, $x_{\mathbb{P}}F_2^{D(3)}$, *as a function of* Q^2 *for different regions of* β *and* $x_{\mathbb{P}}$. *The curves show the result of the BEKW(mod) fit to the data; from ZEUS.*

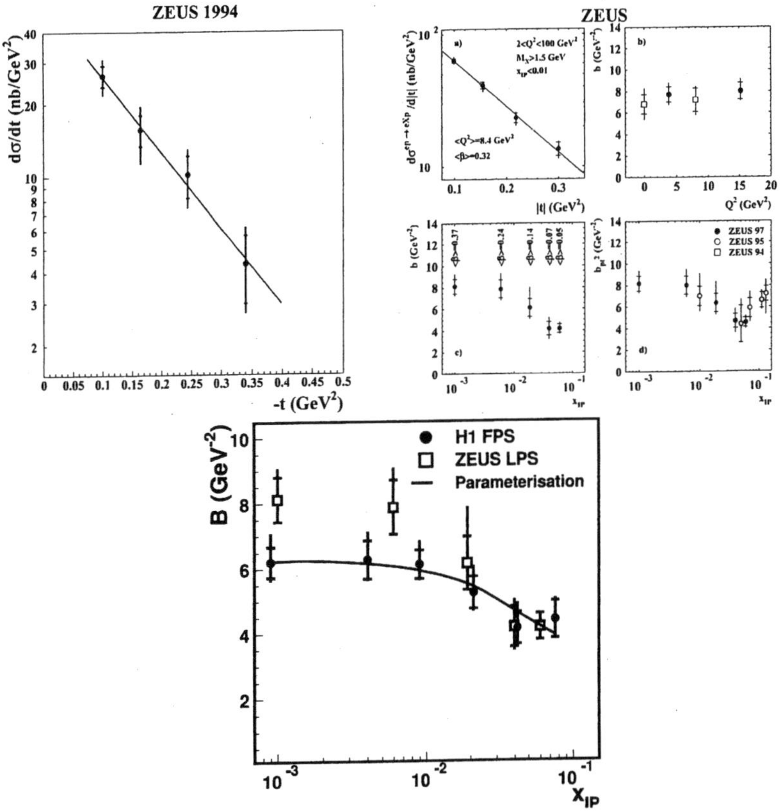

Figure 47: *Top left: The differential cross section* $d\sigma/d|t|$ *for diffractive events with a leading proton of* $x_L > 0.7$*, determined in the range* $5 < Q^2 < 20$ GeV^2, $50 < W < 270$ GeV *and* $0.15 < \beta < 0.5$. *Top right: a) The differential cross section* $d\sigma^{ep \to Xp}/d|t|$ *for* $x_{I\!P} < 0.01$, $Q^2 = 2 - 100$ GeV^2, $M_X > 1.5$ GeV; *b) the slope* b *as a function of* Q^2, *and c) as a function of* $x_{I\!P}$; *d) the slope* $b_{p_T^2}$ *as a function of* $x_{I\!P}$; *from ZEUS. Bottom: The slope* b *of the* $|t|$ *distribution - here denoted by* B *- as a function of* $x_{I\!P}$ *for* $Q^2 = 2 - 50$ GeV^2 *and* $y = 0.02 - 0.6$; *from H1.*

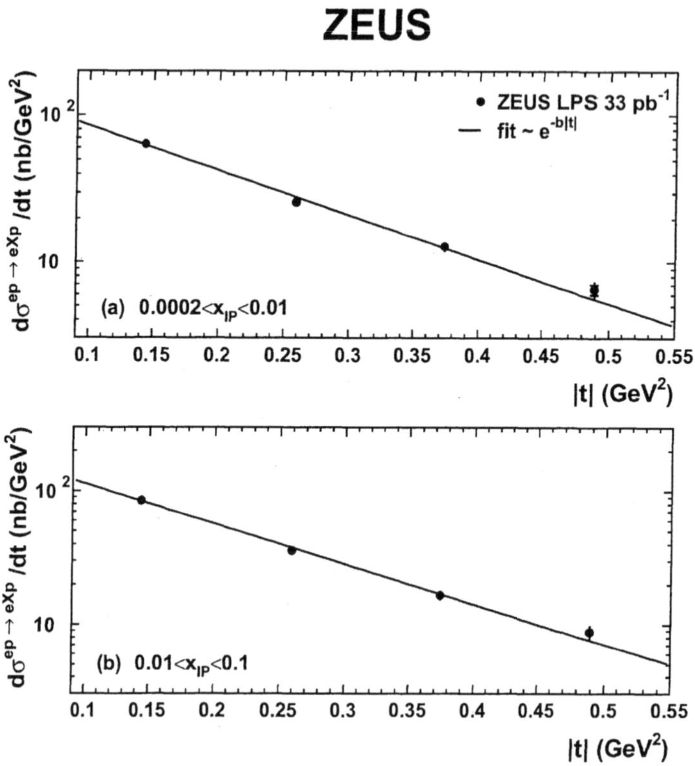

Figure 48: *The differential cross section* $d\sigma^{ep\to eXp}/d|t|$ *for (a)* $0.0002, x_{I\!P} < 0.01$ *and (b)* $0.01x_{I\!P} < 0.1$*; from ZEUS.*

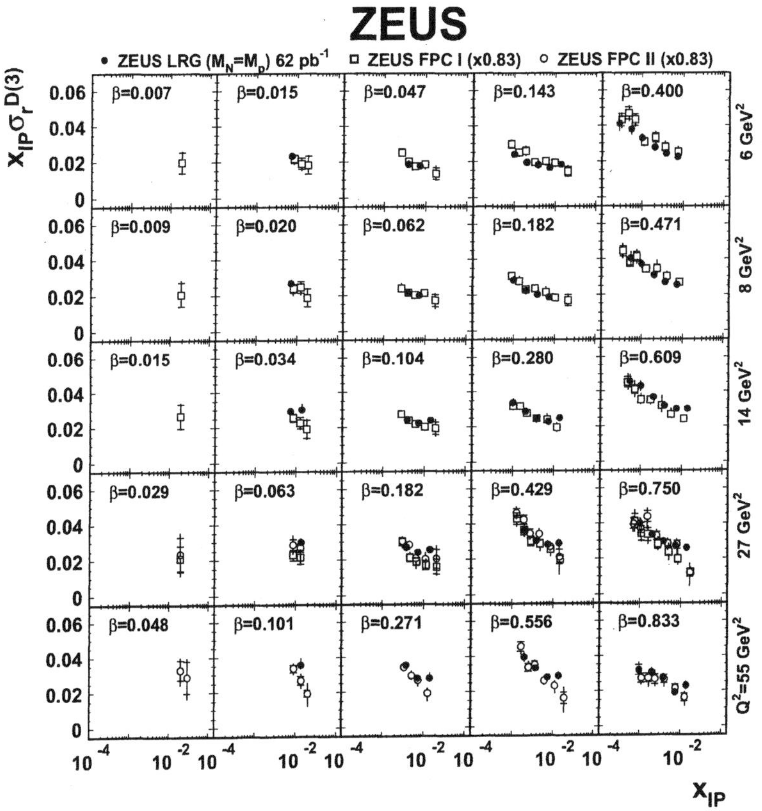

Figure 49: *The diffractive structure function of the proton multiplied by $x_{I\!P}$, $x_{I\!P}F_2^{D(3)}$, here denoted by $x_{I\!P}\sigma^{D(3)_r}$, as function of $x_{I\!P}$ for Q^2 values between 6 and 55 GeV^2; from ZEUS.*

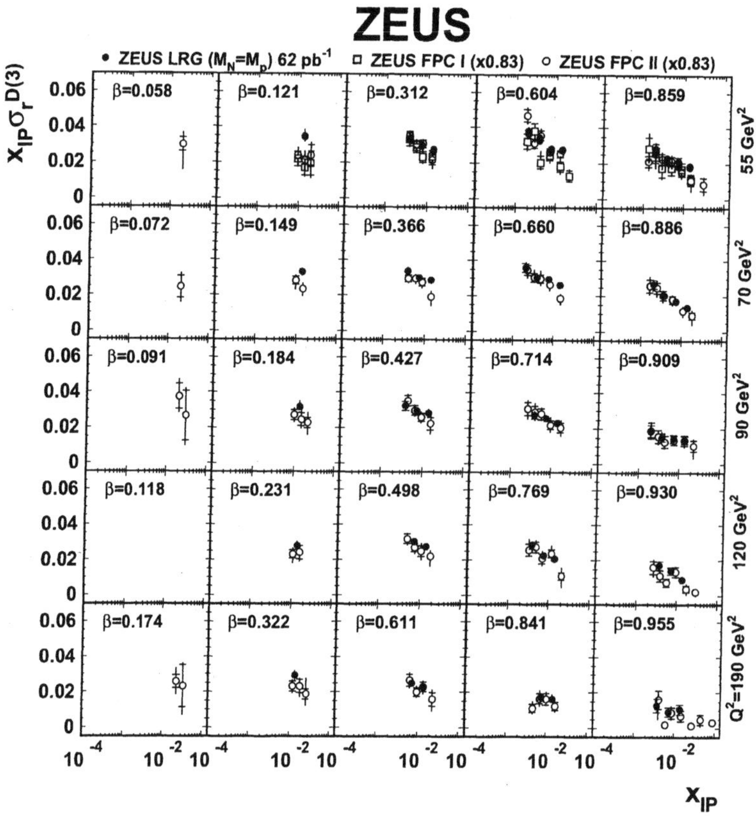

Figure 50: *The diffractive structure function of the proton multiplied by* $x_{I\!P}$*,* $x_{I\!P}F_2^{D(3)}$*, here denoted by* $x_{I\!P}\sigma^{D(3)_r}$*, as function of* $x_{I\!P}$ *for* Q^2 *values between 55 and 190 GeV*^{2}*; from ZEUS.*

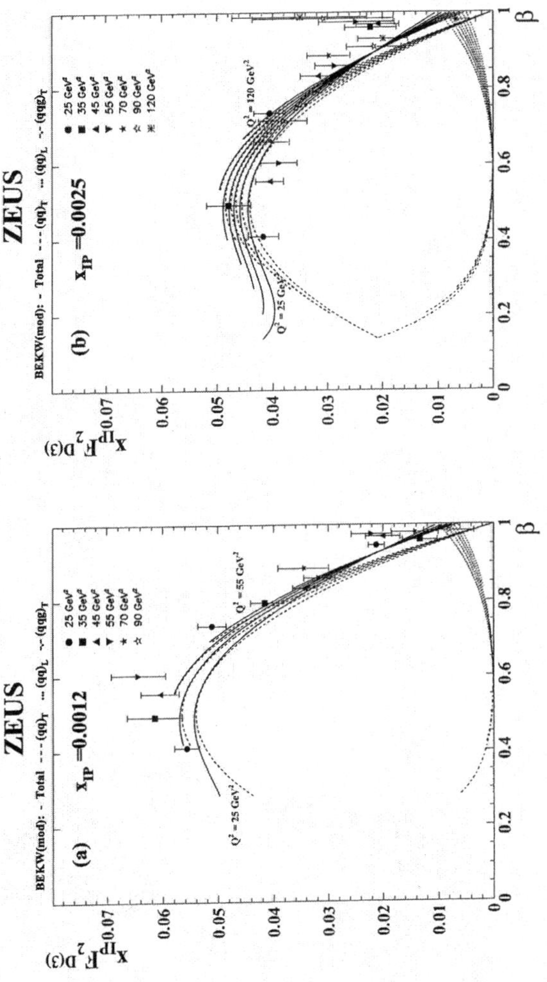

Figure 51: *The diffractive structure function of the proton multiplied by* $x_{\mathbb{P}}$*,* $x_{\mathbb{P}} F_2^{D(3)}$*, as a function of* β *for the* Q^2 *values indicated, at fixed* $x_{\mathbb{P}} = 0.0012$ *(a) and* $x_{\mathbb{P}} = 0.0025$ *(b). The curves show the results of the BEKW(mod) fit for the contributions from* $(q\bar{q})$ *for transverse (dashed) and longitudinal photons (dotted) and for the* $(q\bar{q}g)$ *contribution for transverse photons (dashed-dotted) together with the sum of all contributions (solid), for the* β*-region studied for diffractive scattering; from ZEUS.*

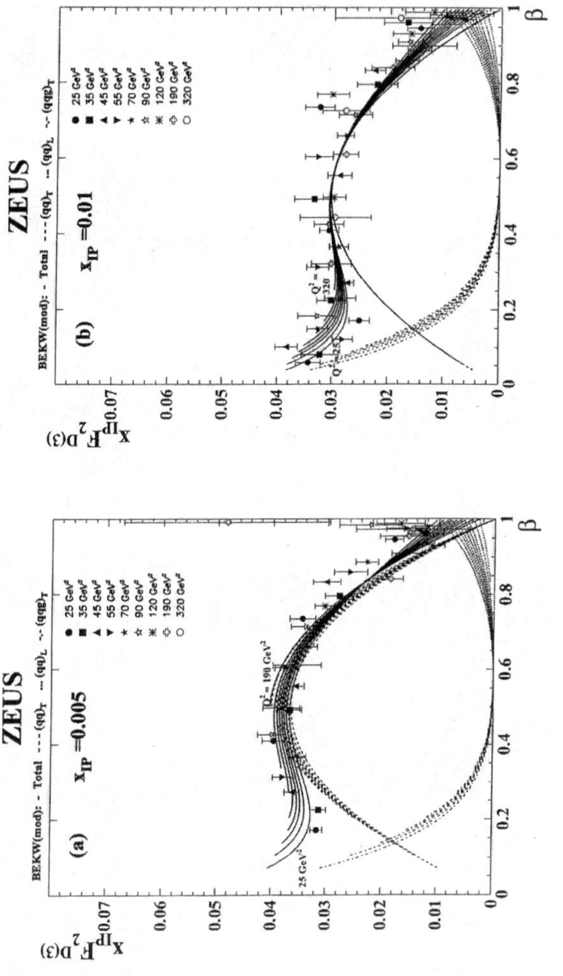

Figure 52: *The diffractive structure function of the proton multiplied by* $x_{I\!P}$, $x_{I\!P}F_2^{D(3)}$, *as a function of* β *for the* Q^2 *values indicated, at fixed* $x_{I\!P} = 0.005$ *(a) and* $x_{I\!P} = 0.01$ *(b). The curves show the results of the BEKW(mod) fit for the contributions from* $(q\bar{q})$ *for transverse (dashed) and longitudinal photons (dotted) and for the* $(q\bar{q}g)$ *contribution for transverse photons (dashed-dotted) together with the sum of all contributions (solid), for the* β*-region studied for diffractive scattering; from ZEUS*

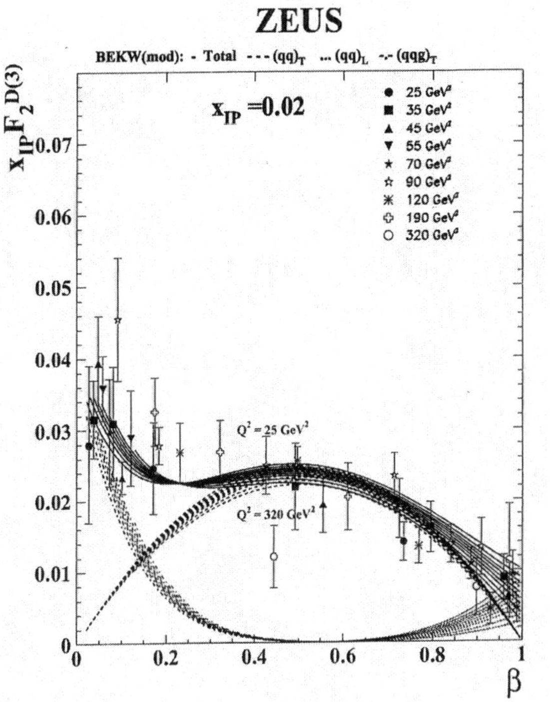

Figure 53: *The diffractive structure function of the proton multiplied by $x_{I\!P}$, $x_{I\!P}F_2^{D(3)}$, as a function of β for the Q^2 values indicated at fixed $x_{I\!P} = 0.02$. The curves show the results of the BEKW(mod) fit for the contributions from $(q\bar{q})$ for transverse (dashed) and longitudinal photons (dotted) and for the $(q\bar{q}g)$ contribution for transverse photons (dashed-dotted) together with the sum of all contributions (solid), for the β-region studied for diffractive scattering; from ZEUS.*

CHAIRMAN: G.WOLF

Scientific Secretaries: A. Calivà, C.Di Giglio

DISCUSSION I

- *Inguglia:*

Please, I would like to know the technical setup for the production and detection of (2S) bottomonium. I mean the energy of the proton beam, energy and source of photons and detectors used.

- *Wolf:*

The HERA energy of the electron beam was 27.6 GeV; the energy of the proton beam was 820 GeV for the first part and 920 GeV for the second part of running. Data have been taken simultaneously by the H1 and ZEUS experiments. As shown the Upsilon signal has been observed by ZEUS and by H1 with limited statistics and on top of a substantial background. Most of the events in the Upsilon mass region are attributed to the 1S state.

- *Pacetti:*

The question is about slide n. 15. Could you please say more about the non resonant background in $\pi+\pi-$ production?

- *Wolf:*

Slide 15 shows the pi+pi- mass distribution for the production of the final state: proton $\pi+\pi-$ by virtual photon scattering. The mass distribution is dominated by the signal from ρ^0 production. A careful inspection shows that the ρ mass peak is shifted slightly to a mass value lower than that of the nominal ρ mass. This shift can be understood in the Soeding model where the amplitude for ρ production, γ^* p -> ρ p -> $\pi+\pi-$ p, interferes with nonresonant production, i.e. with the final state proton $\pi+\pi-$. The contribution from the interference term is shown by the dotted curve.

- *Takaki:*

From your summary plot of photoproduction, could you explain what is the relevance of the scaling observed in pQCD.

- *Wolf:*

The summary figure shows the total cross section for photoproduction, together with the cross sections for the production of various vector mesons, as a function of the total c.m. energy W. The dependence on W is of the form W^δ, where \delta increases with the mass of the vector meson. It is conceivable that this increase with vector meson mass is a phase space effect, and that at much higher values of W, the dependence on W is the same for all vector mesons.

- *Baldini:*

Could you explain why the production of vector mesons should scale as $1/M^4$.

- *Wolf:*

Consider photoproduction of vector mesons, and assume that the photon turns into a vector meson V (V = rho, omega, phi, J/Ψ, ..), then the cross section is of the form $\sigma \propto 1/((M_V)^2)^2$ times the cross section for Vp -> Vp. Assume, that the cross section for Vp -> Vp is the same for all V species. Then $\sigma \propto 1/((M_V)^2)^2$.

CHAIRMAN: G. WOLF

Scientific Secretaries: J. Davey, A. Monin

DISCUSSION II

- *Baldini:*

It seems to me there is a lack of statistics in the data you showed in some cases. Is there any chance in the near future to have data from other accelerators?

- *Wolf:*

At the LHC, it may be possible to study some of the diffractive processes at higher energies compared to HERA. As an example, imagine that at LHC in pp collisions one of the proton emits a photon which then interacts with the other proton.

Two comments:

- The background from standard pp collisions will be very large: it might be possible to identify processes where one of the incoming protons emits a photon of energy $E\gamma$, which in turn, interacts with a proton of the other beam.

1) Under the assumption that the photon is real, photon-proton centre-of-mass energies of 0.5 to 1 TeV seem feasible, which would be a factor of 2 to 5 above what has been studied at HERA.

2) Because of the form factor of the proton, the rate of events with a large virtual mass squared, $-Q^2$, are strongly suppressed.

- *Cifarelli:*

Do you think it would be interesting to have an electron-proton collider at the LHC?

- *Wolf:*

Yes! Assuming 100 GeV electrons (= LEP ring) collide with 6 TeV protons the gain in real(virtual) photon proton c.m. energy over HERA would be a factor of 5.

- *Chiu:*

Now that HERA is turned off, are there any measurements that you wish you had been able to do at DESY with the same sort of energy but, for instance,with polarized protons or some new detector design? There is now an effort in the US to build an electron - proton (or ion) collider, where electrons and/or protons (ions?) would be polarised, but with c.m. energies much lower than at HERA. Do you think that effort will lead to new insights?

- *Wolf:*

I am not an expert in nuclear physics. I would suspect that much will be learned on the structure of nuclei.

- *Chiu:*

Suppose, HERA would have been operated for much longer, and the detectors would have been upgraded: which questions could have been addressed?

- *Wolf:*

An order of magnitude increase in integrated luminosity would have increased the sensitivity for new pieces in the charged and neutral currents, for instance for heavier W's, Z's.

- *Tannenbaum:*
I didn't understand how one measures the gluon structure function from diffraction.

- *Wolf:*
H1 did compare their measurements of the diffractive structure function, xpom.F2D(3), to a QCD model of diffraction. The diffractive cross sections are expressed in terms of diffractive parton densities which are evolved with Q^2, using DGLAP evolution equations. Those parameters not provided by other data, are determined by a fit to the data.

- *Tannenbaum:*
Oh, but that's like you do with just F2.

- *Wolf:*
Yes, but with more free parameters to play with.

«ETTORE MAJORANA» FOUNDATION AND CENTRE FOR SCIENTIFIC CULTURE

TO PAY A PERMANENT TRIBUTE TO GALILEO GALILEI, FOUNDER OF MODERN SCIENCE AND TO ENRICO FERMI, THE "ITALIAN NAVIGATOR", FATHER OF THE WEAK FORCES

INTERNATIONAL SCHOOL OF SUBNUCLEAR PHYSICS

47th Course:

THE MOST UNEXPECTED AT LHC AND THE STATUS OF HIGH ENERGY FRONTIER

THE LESSON NEEDED FOR THE FUTURE

Antonino Zichichi
INFN and University of Bologna, Italy
CERN, Geneva, Switzerland
Enrico Fermi Centre, Rome, Italy
World Federation of Scientists, Beijing, Geneva, Moscow, New York

THE LESSON NEEDED FOR THE FUTURE

Antonino Zichichi

Table of Contents

1 WE PHYSICISTS CANNOT REMAIN SILENT

We physicists **cannot remain silent** when the great public shows a vivid interest for topics such as:

- Global warming
- The energy crisis
- The information security
- The environment
- The Intelligent Design
- The Evolution

We have to convince the great public that the solution to all these **problems** requires **clarity** and **rigour** and that the best way to study these problems is through Science. Since Physics is the "Queen of all Sciences", the solution of these problems needs physicists. But the present trend is to study the topics mentioned above through the "new Science", the so called "Science of Complexity".

In fact the "Modern Culture" considers "Complexity" a source of new insights in physics, biology, geology, cosmology, social sciences and in all intellectual activities which look at the world through the lens of a standard analysis in terms of either "Platonic Simplicity" or "Complexity".

The reason why we have to care (see Chapter 3) about "Platonic Simplicity" is due to the emergence of this "new Science", which condemns "Reductionism" and promote "Holism". **But "Complexity**" is ill-defined, as shown by the existence of very many definitions (at least seventy) of Complexity.

We will see that we have nothing to learn from this "new Science". in fact it is thanks to "Reductionism" that we are able to reach the following conclusion: despite the seventy definitions of "Complexity" there are only two experimentally observable effects which exist when "Complexity" is at work. These two effects are the UEEC events and the AFB phenomena (see later). Both are present in Physics.

We start with UEEC.

2 UEEC (*UNEXPECTED EVENTS WITH ENORMOUS CONSEQUENCES*)

Let me show a (non exhaustive) synthesis of UEEC during the last Century (Figure 1). For the first 50 years, I have listed 22 totally unexpected discoveries, including the invention of electronic computers by Von Neumann, which no one could have imagined at the beginning of the XXth Century. Point no. 22 refers to the impressive list of Fermi discoveries: once again, all totally unexpected.

THE UEEC OF THE XXth CENTURY

FIRST 50 YEARS

1 Lorentz Invariance (Lorentz).
2 The quantum nature of the World (Planck).
3 Two worlds: "Time-like" and "Space-like" (Einstein).
4 The photon (Einstein).
5 Gauge Invariance (Weyl).
6 The Space-Time curvature (Einstein).
7 Bose-Einstein statistics (Bose-Einstein).
8 The structure of the atom (Bohr).
9 Wave nature of particles (de Broglie).
10 Wave function (Schrödinger) and its probabilistic interpretation (Born).
11 Nuclear Forces (Rutherford) and their "glue" (Yukawa).
12 The Weak Forces (Fermi).
13 The proof that Quantum Mechanics is self consistent (no contradictions) (Von Neumann).
14 The antiparticles (Dirac).
15 The Uncertainty Principle (Heisenberg).
16 The Exclusion Principle (Pauli).
17 Fermi statistics (Fermi).
18 Time Reversal Invariance (Wigner).
19 Other Invariance Laws (Wigner-Parity; Dirac, Weyl-charge conjugation; Pauli CPT).
20 The Stars are "nuclear-fusion" candles (Fermi-Bethe).
21 Electronic computing (Von Neumann).
22 The sequence of unexpected Fermi discoveries: *Fermi-coupling, Fermi-gas, Fermi-momentum, Fermi-temperature, Fermi-surface, Fermi-statistics, Fermi-transition, Fermi-length* (plus the other two quoted above).

Figure 1

The discoveries of the second 50 years are grouped into a unique chapter: the **"Standard Model and Beyond" SM&B** (Figure 2).

THE SECOND HALF OF THE XXth CENTURY
The Standard Model and Beyond (SM&B)

Figure 2

THE STANDARD MODEL AND BEYOND

The superb synthesis called the "Standard Model" is part of a more general structure, where many problems are open. We call this structure "The Standard Model and Beyond", "**SM&B**".

This Structure brings to the unification of all Fundamental Forces of Nature, suggests the existence of the Superworld which exists as a fascinating mathematical structure, but lacks the Galilean experimental proof. The Superworld produces the need for a non-point-like description of Physics processes (the so-called Relativistic Quantum String Theory: RQST), thus opening the way to quantize gravity.

All this is summarised in Figure 3.

THE FIVE BASIC STEPS IN UNDERSTANDING NATURE

Point ① in Figure 3 refers to the renormalization group equations (RGEs) which imply that the gauge couplings (α_i) and the masses (m_j) all run with k^2. It is this "running" with k^2 which allows to think of a Grand Unification Theory, GUT. The value of the grand unified gauge coupling turns out to be $\alpha_{GUT} \cong 1/24$. Here a problem arises.

The energy where the three gauge couplings ($\alpha_1, \alpha_2, \alpha_3$) converge into a unique point, α_{GUT}, is two orders of magnitude below the other energy level where the "string-unification" happens to be, i.e. $E_{SU} \cong 10^{18}$ GeV. Thus a problem arises if a gap exists [1] between these two energy levels of Grand Unification.

DETAILED INFORMATION ON THE STANDARD MODEL AND BEYOND (SM&B)

① RGEs ($\alpha_i\,(i \equiv 1, 2, 3)$; $m_j\,(j \equiv q,\ l,\ G,\ H)$) : $f\,(k^2)$.
- GUT ($\alpha_{GUT} \cong 1/24$) & GAP ($10^{16} - 10^{18}$) GeV.
- SUSY (to stabilize $m_F/m_P \cong 10^{-17}$).
- RQST (to quantize Gravity).

② Gauge Principle (hidden and expanded dimensions).
— How a Fundamental Force is generated: SU(3); SU(2); U(1) and Gravity.

③ The Physics of Imaginary Masses: SSB.
— The Imaginary Mass in SU(2) × U(1) produces masses ($m_{W^\pm}$; m_{Z^0}; m_q; m_l), including $m_\gamma = 0$.
— The Imaginary Mass in SU(5)⇒SU(3)×SU(2)×U(1) or in any higher Symmetry Group (not containing U(1)) ⇒ SU(3) × SU(2) × U(1) produces Monopoles.
— The Imaginary Mass in $SU(3)_c$ generates Confinement.

④ **Flavour Mixings & CP ≠ , T ≠ .**
— No need for it but it is there.

⑤ **Anomalies & Instantons.**
— Basic Features of all Non-Abelian Forces.

Note:

q	≡	quark and squark;	m_F	≡	Fermi mass scale;
l	≡	lepton and slepton;	m_P	≡	Planck mass scale;
G	≡	Gauge boson and Gaugino;	k	≡	quadrimomentum;
H	≡	Higgs and Shiggs;	C	≡	Charge Conjugation;
RGEs	≡	Renormalization Group Equations;	P	≡	Parity;
GUT	≡	Grand Unified Theory;	T	≡	Time Reversal;
SUSY	≡	Supersymmetry;	≠	≡	Breakdown of Symmetry Operators.
RQST	≡	Relativistic Quantum String Theory;			
SSB	≡	Spontaneous Symmetry Breaking.			

The five basic steps in our understanding of nature. ① The renormalization group equations (RGEs) imply that the gauge couplings (α_i) and the masses (m_j) all run with k^2. It is this running which allows GUT, suggests SUSY and produces the need for a non point-like description (RQST) of physics processes, thus opening the way to quantize gravity. ② All forces originate in the same way: the gauge principle. ③ Imaginary masses play a central role in describing nature. ④ The mass-eigenstates are mixed when the Fermi forces come in. ⑤ The Abelian force QED has lost its role of being the guide for all fundamental forces. The non-Abelian gauge forces dominate and have features which are not present in QED.

Figure 3

The "running" with k^2 which allows the Grand Unification suggests also the existence of a new invariance law for fermions and bosons, called Supersymmetry, SUSY.

This is the origin of the **Superworld** whose purpose is to stabilize the two energy levels, the Fermi level (m_F) and the Planck level (m_P), which are separated by 17 orders of magnitude.

It is at the Planck scale that gravity becomes relevant and the Relativistic Quantum String Theory (RQST) is supposed to solve the problem of quantizing it.

Point ② is the Gauge Principle. It implies the existence of hidden dimensions.

In fact a fundamental force is generated by the invariance of all physical phenomena when symmetry operations, controlled by the mathematics of SU(3), SU(2) and U(1), take place in a space whose dimensions are hidden, non-expanded.

Point ③ illustrates the fact that imaginary masses play a central role in describing nature.

Point ④ refers to another unexpected feature: the mass-eigenstates are mixed when the Fermi forces come in.

This is a totally unpredicted phenomenon. In all other interactions, the strength of a gauge force is fully given to a mass-eigenstate of a particle.

With SU(2), this strength is subdivided and fully given to a "mixture" of states. Furthermore the Symmetry Laws, C, P, CP and T, are violated (this is the meaning of the symbol ≠).

We recall that CP is the product of charge C and parity P Symmetry operators, while T stands for Time reversal invariance.

The last **point**, ⑤, refers to a discovery which needed a century. In fact during a century, the Abelian force QED was considered to be the guide for all fundamental forces.

After many decades of experimental searches in the field of high energy processes, it has been finally discovered that the Fermi forces and the strong subnuclear forces are non-Abelian: they have to obey the Symmetry properties dictated by the mathematics of SU(2) and SU(3).

We now know that the non-Abelian gauge forces dominate and have features which are not present in QED, again a totally unexpected series of discoveries.

The **SM&B** is the greatest synthesis of all times in the study of the fundamental phenomena governing the Universe in all its structures. The basic achievements of the **SM&B** have been obtained via an impressive series of UEEC events; moreover the **SM&B** could not care less about the existence of Platonic Simplicity. The reason why we have to care about "Platonic Simplicity" (see Chapter 3) is due to the emergence of a "new science": the so called "Science of Complexity", to which "modern Culture" attributes an enormous importance: **new insights** in physics, biology, geology, cosmology, social sciences should come from Complexity, as said before. We have also mentioned that despite the seventy definitions of Complexity, the experimentally observable effects for the **existence** of **Complexity are only two**:

1) The Anderson-Feynman-Beethoven-type phenomena (**AFB**) i.e. phenomena whose laws and regularities ignore the existence of the Fundamental Laws of Nature from which they originate, and
2) The Sarajevo-type effects, i.e. Unexpected Events of quasi irrelevant magnitude which produce Enormous Consequences (**UEEC**).

For details I refer the interested reader to Reference [2]. Here we limit ourselves to note that the Logic of Nature allows the existence of AFB phenomena. This corresponds to a large variety of structures with their regularities and laws which appear to be independent from the basic constituents and the Fundamental Laws of Nature which govern their interactions. But, without these constituents and these laws it would be impossible to have AFB phenomena and the real world which is in front of us and of which we are part of. The proof is in the series of complex systems shown in Figure 4 (not exhaustive).

As you can see, we go from traffic flux, to the internet network, to earthquakes and seismicity, to social and economic systems, to the behaviour of financial markets, to the study of cosmological structures, and so on.

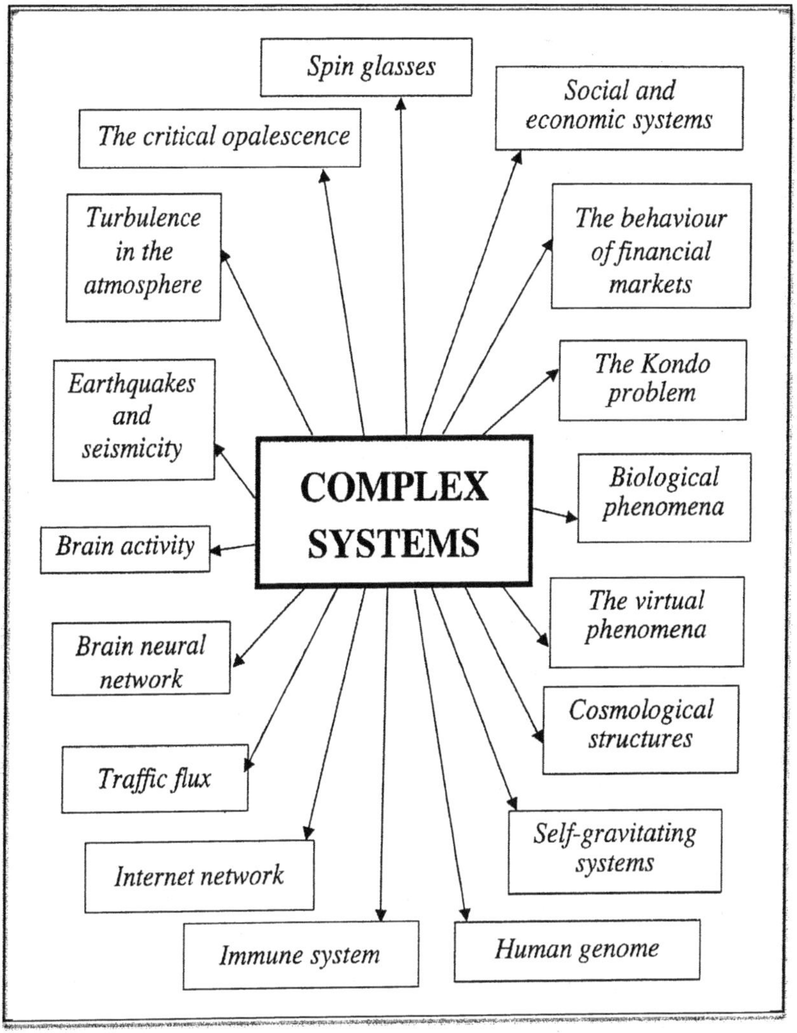

Figure 4

The only certainty about Complexity is the existence of the experimentally observable effects mentioned before: AFB and UEEC. Nature tell us that AFB and UEEC exist at all scales, and therefore Complexity exists at all scales, as illustrated in Figure 5.

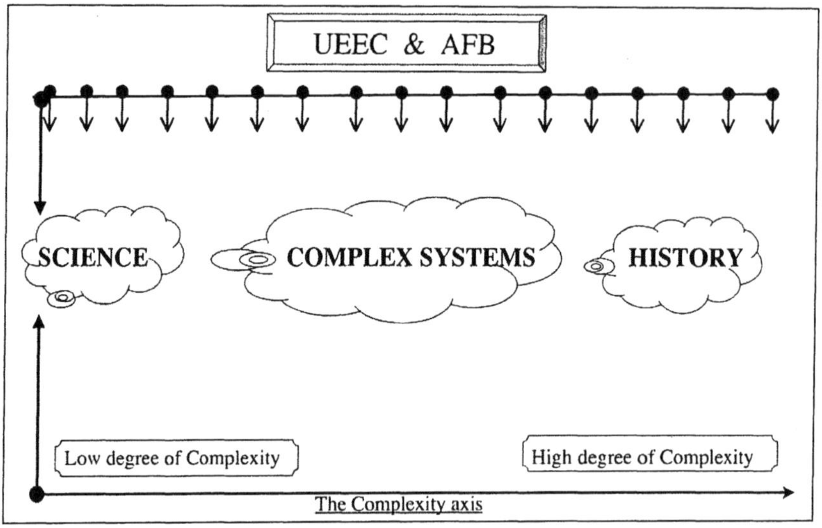

Figure 5

At the lowest level of Complexity there is Science, which we define **EBUS** ≡ Evolution of our **B**asic **U**nderstanding of the laws governing the world in its **S**tructure. At the highest degree of Complexity, there is History, that we define **EWRL** ≡ Evolution of the **W**orld in its **R**eal **L**ife. It is interesting to define ***Science*** and ***History*** in terms of this property, **probably the only one**, which they share; i.e. evolution.

The historians call "Sarajevo" type events our UEEC. In Sarajevo a man is assassinated. This is an event which happens all over the World, frequently, without producing World catastrophes. But the Sarajevo event is the origin of World War I and World War II. This is how historians elaborate the "Virtual History"; in other words, our World could not be as it is, if one or several of the Sarajevo type events (UEEC) would not have taken place.

This is not the case for Science. The world would have exactly the same laws and regularities, whether Galileo Galilei or somebody else had discovered $F = mg$ ($F \equiv$ force; $m \equiv$ mass; $g \equiv$ acceleration due to gravity), and so on for all the other scientific discoveries. This is illustrated in Table 1, where we compare EWRL and EBUS.

TABLE 1

'WHAT IF?'

	In History ≡ EWRL		In Science ≡ EBUS
I	What if Julius Caesar had been assassinated many years before?	*I*	What if Galileo Galilei had not discovered that $F = mg$?
II	What if Napoleon had not been born?	*II*	What if Newton had not discovered that $F = G\frac{m_1 \cdot m_2}{R_{12}^2}$?
III	What if America had been discovered few centuries later?	*III*	What if Maxwell had not discovered the unification of electricity, magnetism and optical phenomena, which allowed him to conclude that light is a vibration of the EM field?
IV	What if Louis XVI had been able to win against the 'Storming of the Bastille'?	*IV*	What if Planck had not discovered that $h \neq 0$?
V	What if the 1908 Tunguska Comet had fallen somewhere in Europe instead of Tunguska in Siberia?	*V*	What if Lorentz had not discovered that space and time cannot be both real?
VI	What if the killer of the Austrian Archduke Francisco Ferdinand had been arrested the day before the Sarajevo event?	*VI*	What if Einstein had not discovered the existence of time-like and space-like real worlds? Only in the time-like world, simultaneity does not change, with changing observer.
VII	What if Lenin had been killed during his travelling through Germany?	*VII*	What if Rutherford had not discovered the nucleus?
VIII	What if Hitler had not been appointed Chancellor by the President of the Republic of Weimar Paul von Hindenburg?	*VIII*	What if Hess had not discovered the cosmic rays?
IX	What if the first nuclear weapon had been built either by Japan before Pearl Arbour (1941) or by Hitler in 1942 or by Stalin in 1943?	*IX*	What if Dirac had not discovered his equation, which opens new horizons, including the existence of the antiworld?
X	What if Nazi Germany had defeated the Soviet Union?	*X*	What if Fermi had not discovered the weak forces?
XI	What if Karol Wojtyla had not been elected Pope, thus becoming John Paul II?	*XI*	What if Fermi and Dirac had not discovered the Fermi–Dirac statistics?
XII	What if the USSR had not collapsed?	*XII*	What if the 'strange particles' had not been discovered in the Blackett Lab?

It is in the consequences of "**What if ?**" that the two asymptotic limits of Simplicity and Complexity seem to diverge, despite the fact that the sequence of "**What if ?**" in Science belongs to the "totally unexpected events" (UEEC) exactly like the others listed in the column of History.

Let us now say few words on "Reductionism" and "Holism".

Reductionism and Holism

The world is characterized by two basic features, which are on the opposite side of one another: *Simplicity* and *Complexity*.

It is generally accepted that *Simplicity* is the outcome of *Reductionism*, while *Complexity* is the result of *Holism*.

Talking about asymptotic limits, as shown in Figure 5, the general trend is to consider ***History*** the asymptotic limit of *Holism* and of *Complexity*; ***Science*** as the asymptotic limit of *Reductionism* and of *Simplicity*. This is illustrated in Figure 6.

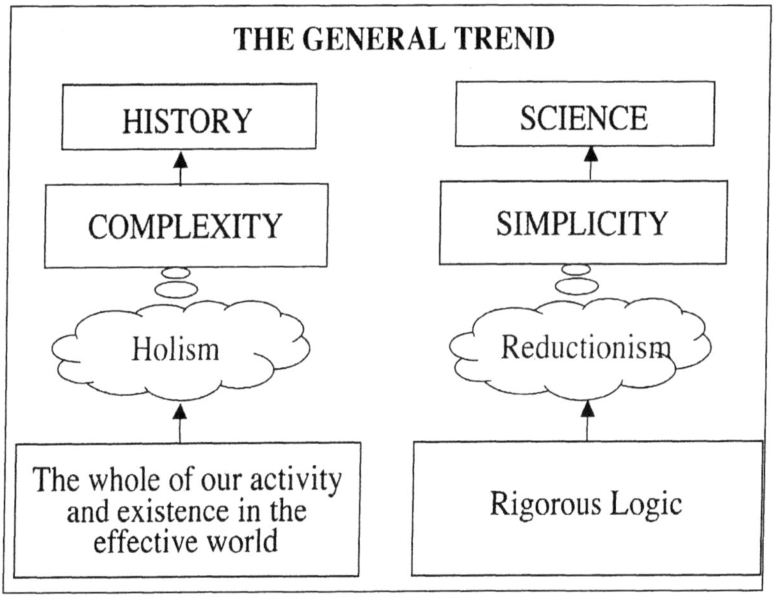

Figure 6

The new enemies of Science emphasize that the reductionistic studies (i.e. Physics) have to be abandoned in favour of Holism. It is forgotten that Holism has been going on for ten thousands years since the dawn of civilisation, as shown in Figure 7.

During this Holistic approach to understand the World, not a single Fundamental Law of Nature has been discovered.

It is thanks to reductionism (started by Galileo Galilei) that we have reached the unification of all Fundamental Forces of Nature.

And it is trough reductionism that we can face the problem of Science being NP-complete or not, as illustrated in Figure 7.

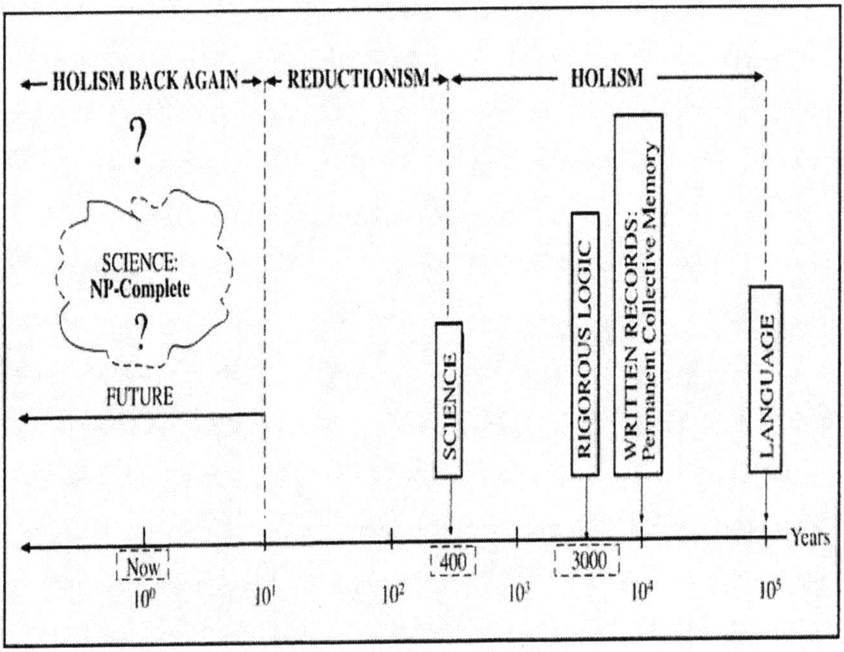

Figure 7

FROM PLANCK TO COMPLEXITY

The developments we have been talking about, started thirty years ago, when a great scientific novelty came: all experimental discoveries obtained with our powerful accelerators were to be considered only matters of extremely low energy.

The scale of energy on which to direct the attention to understand the Logic that rules the world, from the tiniest structures to the galactic ones, had to be shifted at a much higher level: to the mass-energy named after Planck, E_{Planck}, something like seventeen powers of ten above the Fermi scale, E_{Fermi}, that already seemed to be an extremely high level of energy.

The novelty now is the fact that Complexity exists at the Fundamental level, as illustrated in Figure 8.

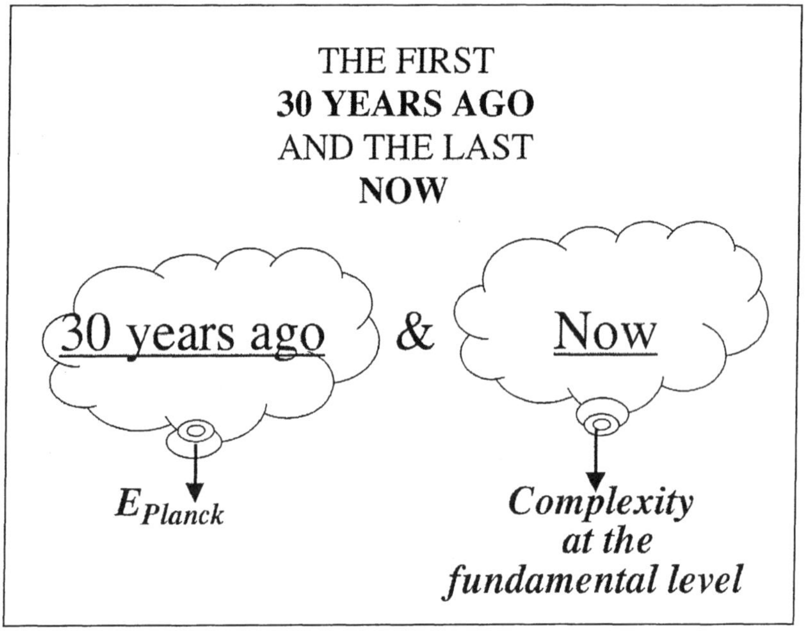

Figure 8

Let me now speak about the cases, seven in total, where I have been directly involved. They are given in Figure 9.

A FEW EXAMPLES WHERE I HAVE BEEN INVOLVED

① **The 3rd lepton**
despite *the abundance of neutrinos* (ν_e, ν_μ), *another* ***one*** ($\boldsymbol{\nu_{HL}}$).

② **Antimatter**
despite *S-matrix and C, P, CP breaking.*

③ **Nucleon Time-like** EM structure
despite *S-matrix.* **inst**

④ **No quarks** in violent (pp) collisions
despite *scaling.*

⑤ **Meson mixings**
$\theta_V \neq \theta_{PS}$: (51°) ≠ (10°) ≠ 0 ***despite*** $SU(3)_{uds}$. **inst**

---

⑥ **Effective energy:** *the QCD-light*
despite *QCD.*

---

⑦ **The running** of α_1 α_2 α_3 versus **energy** at ***a point*** E_{GU}.
(1979) (1991) ***despite*** *straight line convergence.*
EGM (Evolution Gaugino Masses – not only the three gauge couplings) lowers the threshold energy for the discovery of the first (the lightest) particle of the Superworld by nearly 3 orders of magnitude.

Figure 9

Point 1: The great novelty here was to search for a lepton heavier than the muon since the reason why muons are all over the place was due to it's incredible small mass difference with π–meson.

If a lepton heavier than the muon existed, for example in the GeV-mass region, no one would have been able to detect its existence. This is why in the early sixties at CERN the search for a third lepton with its own neutrino started to be implemented via its pair production $[(HL^+) + (HL^-)]$ in $(\bar{p}p)$ annihilation.

This is how a new type of technology able to detect electrons in a high hadronic background was invented and a set-up for the simultaneous production of acoplanar $(e\mu)$ pairs was implemented at CERN with the PAPLEP (Proton AntiProton into LEpton Pairs) experimental set-up.

Instead of finding $(e\mu)$ acoplanar pairs we discovered that the proton had a very big time-like electromagnetic form-factor. This is how it was realized that the effective reaction to produce a third lepton was (e^+e^-) annihilation since the electron had not a time-like electromagnetic form-factor and the cross-section for the pair production of a third lepton was "point-like".

At CERN with $(\bar{p}p)$ the cross section was found to be nearly three orders of magnitude lower that the (point-like) cross section.

The apparatus build at CERN for the detection of $(e\mu)$ acoplanar pairs from the reaction

$$(\bar{p}p) \rightarrow HL^+ + HL^-$$

was brought to Frascati where in 1967 the experimental proposal [3] to search for the production of a third lepton via the annihilation process

$$(e^+e^-) \rightarrow HL^+ + HL^-$$

giving as signal an acoplanar $(e\mu)$ pairs was presented.

The first limit on the production cross section for the new heavy lepton was published in [4], five years before the SLAC experiment where the HL was first observed, looking exactly at same mechanism of $(e\mu)$ acoplanar pairs. The only novelty was its name: τ^+ instead of $HL^{\pm}$.

All this is in the special volume where all details concerning the origin of the third lepton are reported [5].

Let me explain why, in Figure 9, I mention "despite the abundance of neutrinos (ν_e, ν_μ) another one (ν_{HL})". When the CERN-DG (Victor Weisskopf) decided to support the PAPLEP experiment whose purpose was to see if another lepton HL heavier than the muon was produced with its own neutrino ν_{HL}, the Head of the CERN Theory Division said: «*Nature can do all that is needed with a single neutrino why do you want Nature to be so stupid to have more than one neutrino?*». The proposal to search at CERN for a new heavy lepton $HL^{\pm}$ with its own neutrino was made before the ($\nu_{\mu} \neq \nu_{e}$) discovery by Danby, Gaillard, Goulianos, Lederman, Mistry, Schwartz and Steinberger at BNL in 1962 [6].

Talking about Nature and only one neutrino, it should not be forgotten that for the Electroweak force (EW), Nature has not chosen the simplest way out SU(2), but, unexpectedly, SU(2) × U(1). More details in Figure 10.

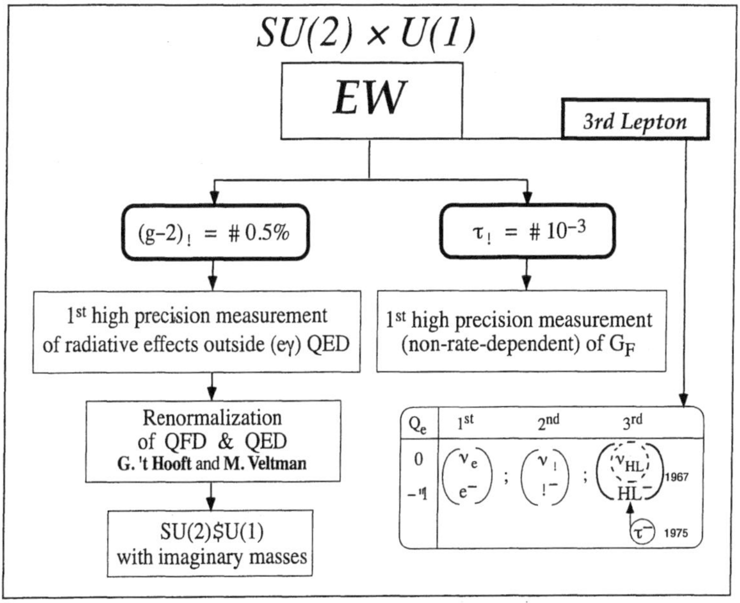

Figure 10

<u>Point 2</u>: **Antimatter despite** S-matrix and C, P, CP breaking.

The basic problem of distinguishing "mass" from "matter" is illustrated in Figure 11. The incredible series of unexpected events which originated with the problem of understanding the stability of matter is shown in Figure 12, together with the unexpected violation of the Symmetry Operators (C, P, T, CP) and the discovery of Matter-Antimatter Symmetry.

Mass ≠ Matter

$$|m_i\rangle \equiv \textbf{Mass} \equiv \textbf{Antimass} \equiv |\bar{m}_i\rangle$$

i ≡ **1** (Intrinsic); i ≡ **2** (Confinement); i ≡ **3** (Binding)

$$C\,|m_i\rangle = |m_i\rangle \qquad \boxed{***}$$

$$i \equiv 1, 2, 3$$

$$|m_i\, Q_j\rangle \equiv \textbf{Matter} \neq \textbf{Antimatter} \equiv |m_i\, \bar{Q}_j\rangle$$

Q_j ≡ Flavour Charges

$$j \equiv (u \;\; d \;\; c \;\; s \;\; t \;\; b) \equiv (1, 2, 3, 4, 5, 6)$$

$$(\nu_e \;\; e^- \;\; \nu_\mu \;\; \mu^- \;\; \nu_{HL} \;\; HL^-) \equiv (7, 8, 9, 10, 11, 12)$$

$$HL^- \downarrow \tau^-$$

$$C\,|m_i\, Q_j\rangle = |m_i\, \bar{Q}_j\rangle \qquad \boxed{***}$$

$$i \equiv 1, 2, 3\,; \;\; J \equiv 1, 2, 3, 4, 5, 6, 7, 8, 9, 10, 11, 12.$$

Figure 11

Many deviations from **simplicity are needed** to corroborate the **stability of matter: one charge** (the electric one) is not enough. Another unexpected discovery was the distinguish "colour" charge from "flavour" charge, with six colour charges and twelve flavour charges.

Six "Colour" charges (1 + 2 + 3) →1 for QED, 2 for QFD, 3 for QCD.

Twelve "Flavour" charges (6 + 6)→6 for quarks + 6 for leptons.

We can see in Figure 12 that there are **seven decades of developments**, started **from** the **antielectron** and **C-invariance, to** arrive at the experimental discovery of **Matter Antimatter Symmetry**. The detailed series of all these totally unexpected events are reported in the volume published on the occasion of the 30th anniversary of the discovery of antimatter [7].

THE INCREDIBLE STORY TO DISENTANGLE THE ORIGIN OF THE STABILITY OF MATTER
FROM THE ANTIELECTRON TO ANTIMATTER AND THE UNIFICATION OF ALL GAUGE FORCES

○ The validity of C invariance from 1927 to 1957.

After the discovery by Thomson in 1897 of the first example of an elementary particle, the Electron, it took the genius of Dirac to theoretically discover the Antielectron thirty years after Thomson.

1927 → Dirac equation [8]; the existence of the antielectron is, soon after, theoretically predicted. Only a few years were needed, after Dirac's theoretical discovery, to experimentally confirm (Anderson, Blackett and Occhialini [9]) the existence of the Dirac antielectron.

1930-1957 → <u>Discovery of the C operator</u> [(charge conjugation) H. Weyl and P.A.M. Dirac [10]]; discovery of the P Symmetry Operator [E.P. Wigner, G.C. Wick and A.S. Wightman [11, 12]]; discovery of the T operator (time reversal) [E.P. Wigner, J. Schwinger and J.S. Bell [13, 14, 15, 16]]; discovery of the CPT Symmetry Operator from RQFT (1955-57) [17].

1927-1957 → Validity of C invariance: e^+ [9]; $\bar{p}$ [18]; $\bar{n}$ [19]; $K_2^0 \to 3\pi$ [20] but see LOY [21].

○ The new era starts: C ≠ ; P ≠ ; CP ≠ (*).

1956 → Lee & Yang P ≠ ; C ≠ [22].

1957 → Before the experimental discovery of P ≠ & C ≠, Lee, Oehme, Yang (LOY) [21] point out that the existence of the second neutral K-meson, $K_2^0 \to 3\pi$, is proof neither of C invariance nor of CP invariance. Flavour antiflavour mixing does not imply CP invariance.

1957 → C.S. Wu et al. P ≠ ; C ≠ [23]; CP ok [24].

1964 → $K_2^0 \to 2\pi \equiv K_L$: CP ≠ [25].

1947-1967 → QED divergences & Landau poles.

1950-1970 → The crisis of RQFT & the triumph of S-matrix theory (i.e. the negation of RQFT).

1965 → Nuclear antimatter is (experimentally) discovered [26]. See also [7].

1968 → The discovery [27] at SLAC of Scaling (free quarks inside a nucleon at very high q^2) but in violent (pp) collisions no free quarks at the ISR are experimentally found [28]. Theorists consider Scaling as being evidence for RQFT not to be able to describe the Physics of Strong Interactions. The only exception is G. 't Hooft who discovers in 1971 that the β-function has negative sign for non-Abelian theories [29].

1971-1973 → β = − ; 't Hooft, Gross & Wilczek. The discovery of **non-Abelian** gauge theories. Asymptotic freedom in the interaction between quarks and gluons [29].

1974 → All gauge couplings $\alpha_1\ \alpha_2\ \alpha_3$ run with q^2 but they do not converge towards a unique point.

1979 → A.P. & A.Z. point out that the new degree of freedom due to SUSY allows the three couplings $\alpha_1\ \alpha_2\ \alpha_3$, **<u>to converge towards a unique point</u>** [30].

1980 → QCD has a "hidden" side: the multitude of final states for each pair of interacting particles: (e^+e^- ; $p\bar{p}$; πp; Kp; νp; pp; etc.)
The introduction of the Effective Energy allows to discover the Universality properties [31] in the multihadronic final states.

1992 → All gauge couplings converge towards a unique point at the gauge unification energy: $E_{GU} \cong 10^{16}$ GeV with $\alpha_{GU} \cong 1/24$ [32, 33].

1994 → The Gap [34] between E_{GU} & the String Unification Energy: $E_{SU} \cong E_{Planck}$.

1995 → **<u>CPT loses its foundations at the Planck scale (T.D. Lee)</u>** [35].

1995-1999 → **<u>No CPT theorem from M-theory (B. Greene)</u>** [36].

1995-2000 → A.Z. points out the need for new experiments to establish if matter-antimatter symmetry or asymmetry are at work.

(*) The symbol ≠ stands for "Symmetry Breakdown".

Figure 12

Let us skip the Point 3 and 4 and move to 5.

Point 5: In the Physics of Mesons the totally unexpected result was the difference existing between the two mesonic mixing angles, pseudoscalar θ_{PS} and vector θ_V. They should both be zero if $SU(3)_{uds}$ was a good Symmetry. The experimental results gave $\theta_V = 51°$ and $\theta_{PS} = 10°$ **despite** $SU(3)_{uds}$. This is illustrated in Figure 13. More details about mesonic mixings are given in Addendum 2. The Physics of Instantons in QCD is needed to explain θ_V and θ_{PS}, i.e. the mixing in the pseudoscalar and in the vector mesons. The existence of Instantons was not known. The Instantons came after the unexpected discovery that $\theta_{PS} \neq \theta_V \neq 0$.

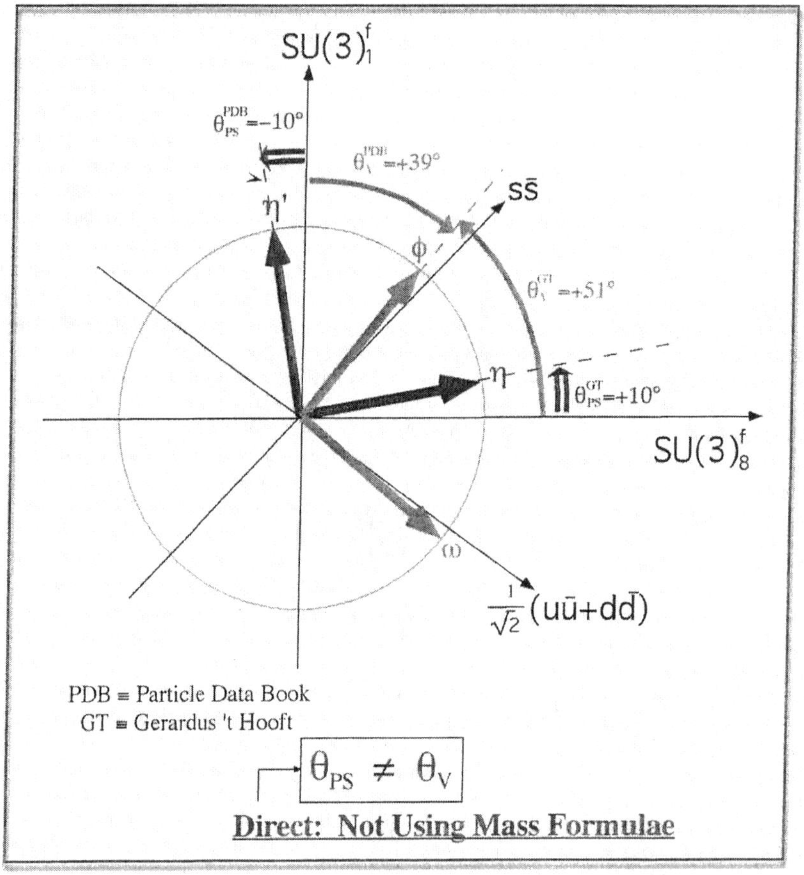

Figure 13

Point 6: **Effective energy:** the QCD-light ***despite*** *QCD*.

In **QCD** we have *quarks and gluons* interacting and producing *Jets* made of many pions pp → π + X as shown in Figure 14 for the case:

$$pp \rightarrow \pi^+ + X \text{ at } (\sqrt{s})_{pp} = 62 \text{ GeV}.$$

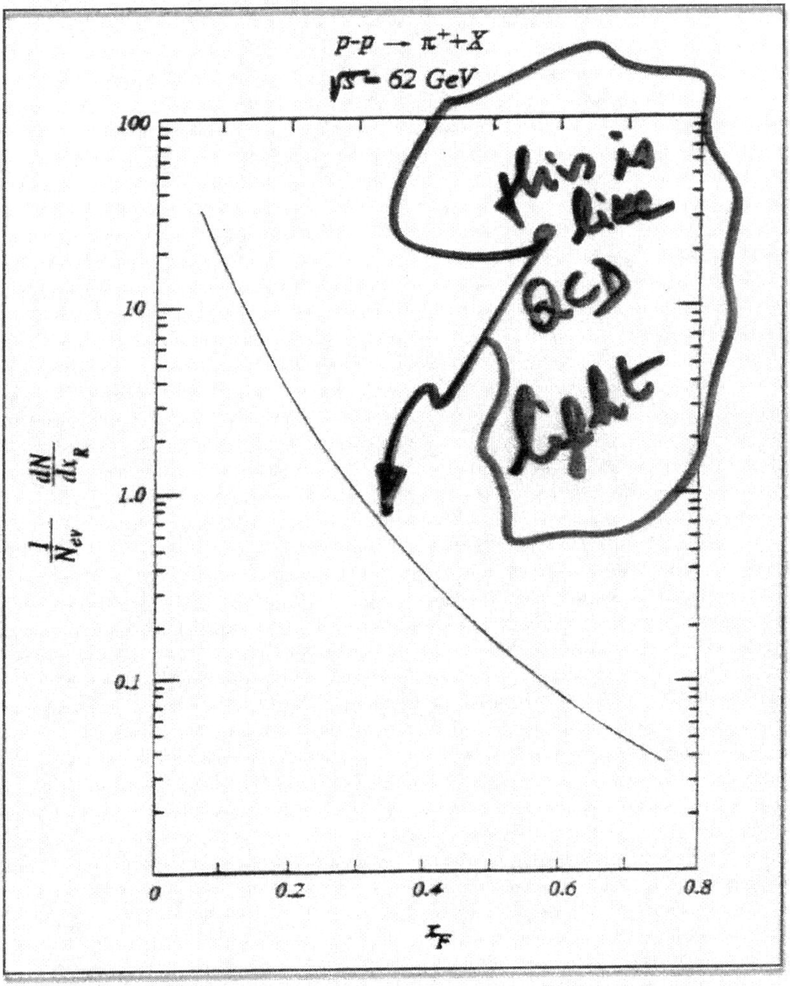

Figure 14

This is what Gribov defined: ***the QCD light.***

The Effective Energy shown in Figure 15 is at its origins, despite being totally unexpected.

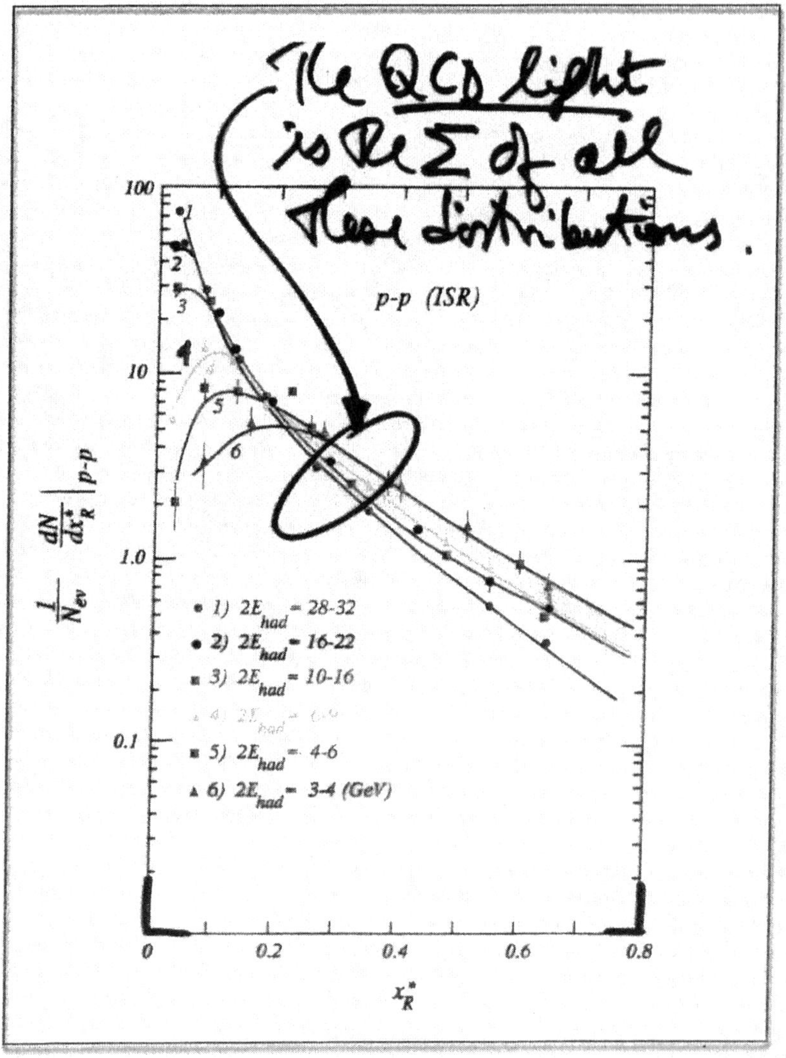

Figure 15

The non-Abelian nature of the Interaction describing quarks, gluons and the Effective Energy with the set of unexpected discoveries is illustrated in Figure 16.

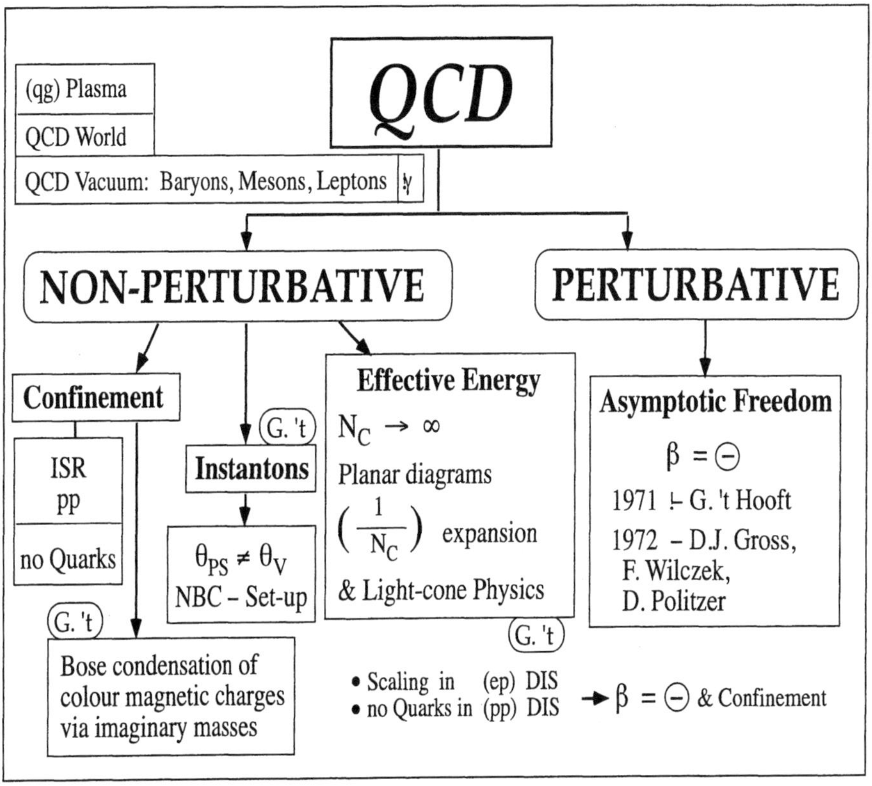

Figure 16

<u>Point 7</u>: **The running** of α_1 α_2 α_3 versus **energy (1979)** at **a point (1991)** E_{GU}; **despite** straight line convergence.

An example of "Platonic Simplicity" is shown in Figure 17, where the straight line, small dots, would be the Platonic simple solution towards the Unification of all Fundamental Forces. Nature follows the big dots.

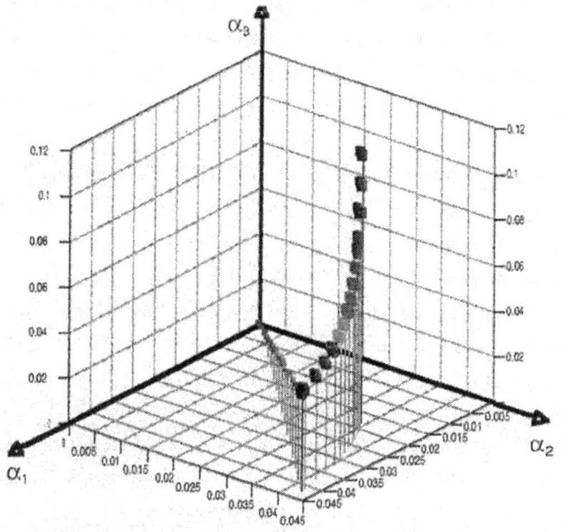

The points have a sequence of 100 GeV in energy. The last point where the "ideal" platonic straight line intercepts the theoretical prediction is at the energy of the Grand Unification. This corresponds to $E_{GU} = 10^{16.2}$ GeV. Other detailed information on the theoretical inputs: the number of fermionic families, N_F , is 3; the number of Higgs particles, N_H , is 2. The input values of the gauge couplings at the Z^0-mass is α_3 $(M_Z) = 0.118 \pm 0.008$; the other input is the ratio of weak and electromagnetic couplings also measured at the Z^0-mass value: $\sin^2 \theta_W$ $(M_Z) = 0.2334 \pm 0.0008$.

Figure 17

THE UNIFICATION OF ALL GAUGE FORCES

The Unification of all Gauge Forces and the Supersymmetry threshold with its problems are reported in Figures 18 and 19 respectively. During more than ten years (from 1979 to 1991), no one had realized that the energy threshold for the existence of the Superworld was strongly dependent on the "running" of the masses. This is now called: the EGM effect (from the initials of Evolution of Gaugino Masses).

To compute the energy threshold using only the "running" of the gauge couplings (α_1, α_2, α_3) corresponds to neglecting nearly three orders of magnitude in the energy threshold for the discovery of the first particle (the lightest) of the Superworld, as illustrated in Figure 19.

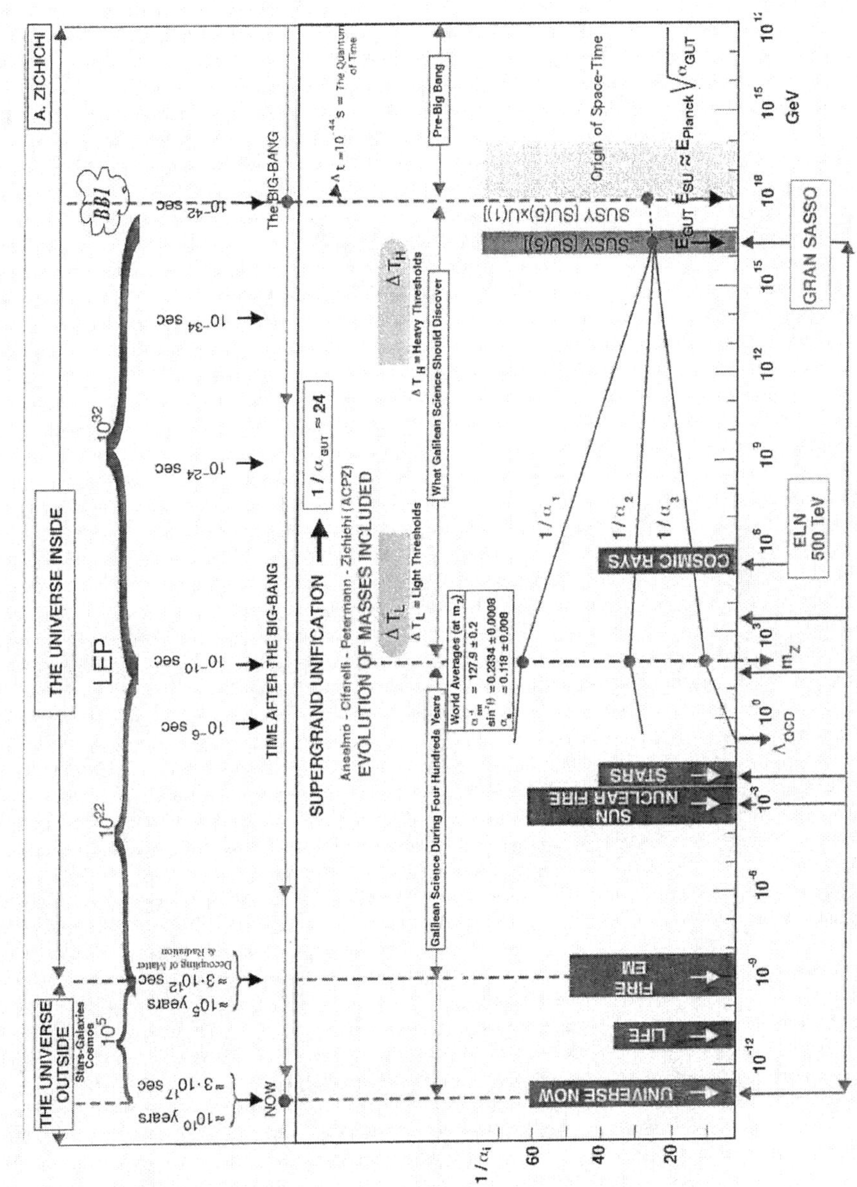
A. ZICHICHI
THE UNIVERSE OUTSIDE
THE UNIVERSE INSIDE
Stars-Galaxies Cosmos
LEP
TIME AFTER THE BIG-BANG
NOW
The BIG-BANG
SUPERGRAND UNIFICATION
Anselmo - Cifarelli - Petermann - Zichichi (ACPZ)
EVOLUTION OF MASSES INCLUDED
Galilean Science During Four Hundreds Years
What Galilean Science Should Discover
Pre-Big Bang
ΔT_L = Light Thresholds
ΔT_H = Heavy Thresholds
World Averages (at m_Z)
Origin of Space-Time
SUSY [SU(5)]
SUSY [(SU(5)xU(1)]
UNIVERSE NOW
LIFE
FIRE EM
SUN NUCLEAR FIRE
STARS
COSMIC RAYS
ELN 500 TeV
GRAN SASSO
GeV
Decoupling of Matter & Radiation

Figure 18

***

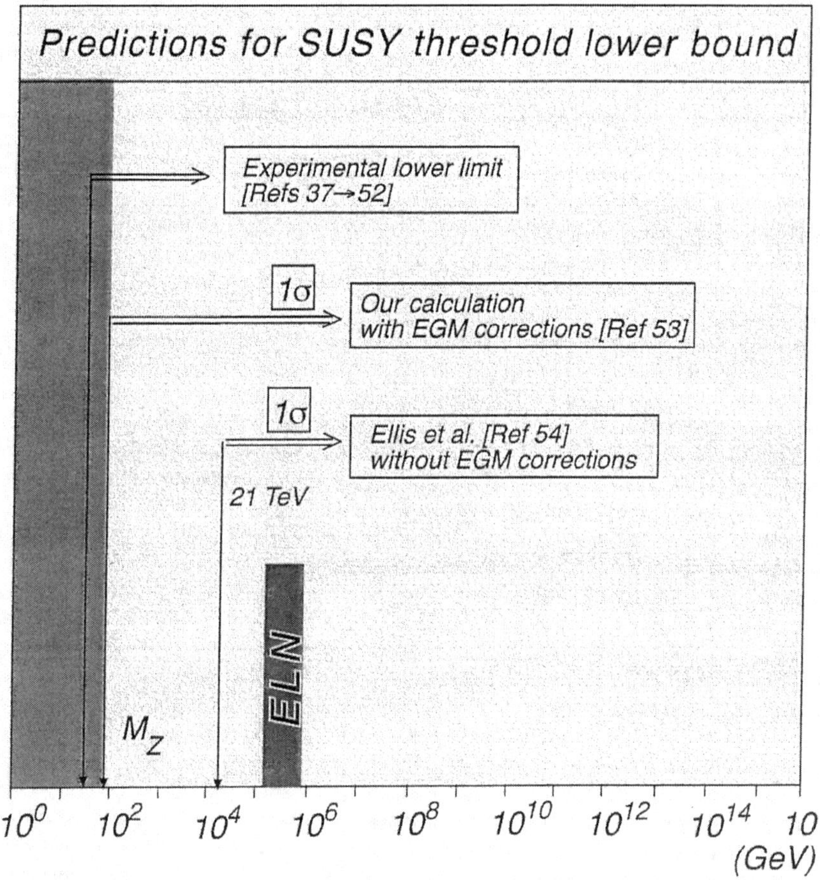

Figure 19

Talking about Supersymmetry, there is another important step: how we go from pure Theoretical Speculations to Phenomenology.

An example is given in Figure 20 where it is shown how many important properties of the physics to be described have been neglected by some authors (AdBF) whose claim was to "predict" the energy scale at which Supersymmetry is broken.

In order to attempt to give such a prediction, there are at least five "details" to be taken into account, as reported in the last five columns of Figure 20.

***

The five "details" forgotten by many authors

Authors	① Input data	② Errors	③ EC	④ M_{SUSY}	⑤ CC	⑥ UC	⑦ ΔT_L	⑧ M_X	⑨ ΔT_H	⑩ EGM
ACPZ [53, 32→34, 55→57]	WA	± 2 σ	all possible solutions (24)	Yes	physical	Yes	Yes	Yes	Yes	Yes

Authors	Input data	Errors	EC	M_{SUSY}	CC	UC	ΔT_L	M_X	ΔT_H	EGM
AdBF [58]	only one experiment	± 1 σ	only one solution	Yes	Geometrical	No	No	No	No	No

① *WA* ≡ *World Average*
② *Errors* ≡ *Uncertainty taken from all data (World Average) or from a single experiment*
③ *EC* ≡ *Evolution of Couplings*
④ M_{SUSY} ≡ *Mass Scale assumed to represent the Supersymmetry Scale breaking*
⑤ *CC* ≡ *Convergence of Couplings*
⑥ *UC* ≡ *Unification of Couplings above* E_{GUT}
⑦ ΔT_L ≡ *Low Energy threshold*
⑧ M_X ≡ *Mass Scale at the breaking of the Grand Unified Theory to the SU(3)xSU(2)xU(1)*
⑨ ΔT_H ≡ *High Energy threshold*
⑩ *EGM* ≡ *Evolution of Gaugino Masses*

Figure 20

As already mentioned these details lower the threshold energy for the discovery of the lightest supersymmetric particles by nearly three orders of magnitude.

Question: What about UEEC in other fields? For example: condensed matter.

Tony Leggett, University of Illinois, Urbana - Champaign, USA, Nobel 2003 for "Superfluidity": "*It is relatively rare in Condensed-Matter Physics to predict discoveries, it is a field where you fall over them by accident*".

3 PLATONIC SIMPLICITY

3.1 *THE PLATONIC GRAND UNIFICATION*

Let us now move towards the problem of Platonic Simplicity, taking as example the structure of a Grand Unification. The simplest way is to have one and only one basic fundamental particle, B.

This particle must obey the very simple symmetry law which puts fermions and bosons on the same basis (Figure 21).

This basic fundamental particle can therefore exist either as being a boson B_B or as being a fermion B_F.

Let us consider first

$$B_B$$

The Fundamental Forces

exist

because a

Basic Fundamental Boson

$$B_B$$

EXISTS

Figure 21

Figure 22 illustrates the simple sequence which generates all known forces of Nature.

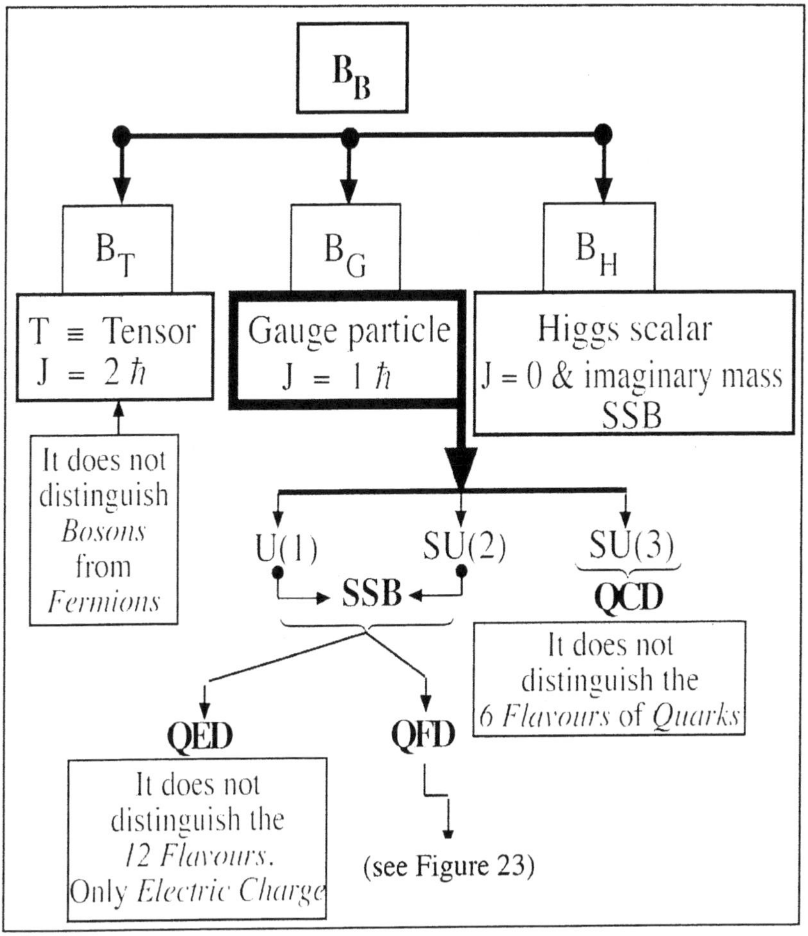

Figure 22

At the bottom of Figure 22 there is the force QFD, illustrated in Figure 23. The “Platonic” Simplicity suffers a further deviation.

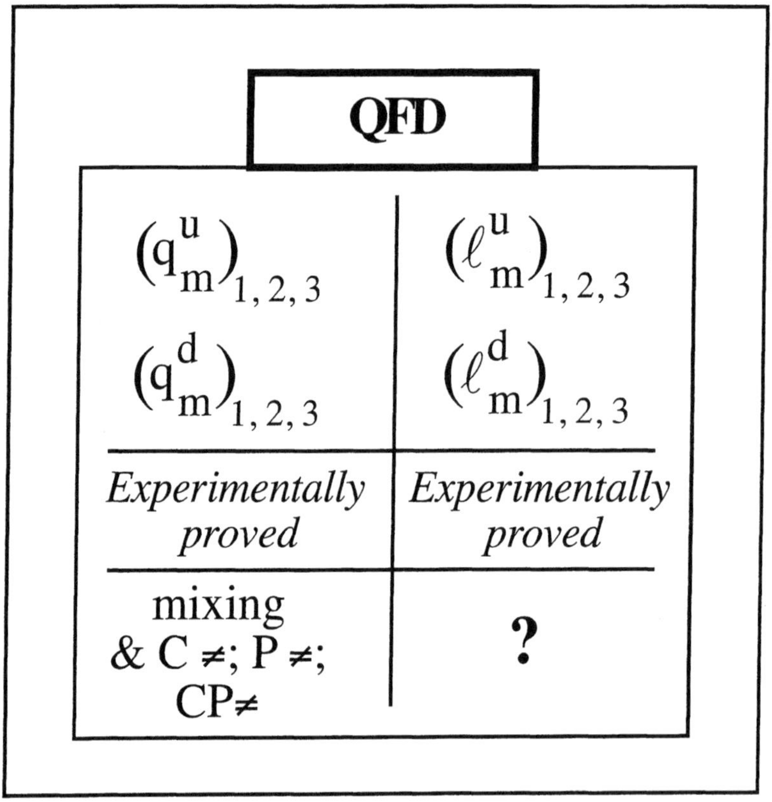

Figure 23

In fact, we need to introduce many complications. The three quarks and the three leptons are “mixed” among them, not among “quarks” and “leptons. This mixing is indicated by the index m, while the indices “u” and “d” refer to the two types of flavours

(up-type) and
(down-type)

which are present in each of the three families:

1, 2, 3.

There is a further complication.

The two mixings for the "up" and the "down" flavours **must be different**.

In the case of the quark, this mixing is experimentally measured.

In the case of the leptons, the experimental results are with nearly half a century of delay, compared with the quark case.

Mixing and violation of Symmetry Laws for charge conjugation (C), parity (P), and the product of the two (CP) are well established in the quark case.

In the leptonic sector, only future experiments will tell us if the same Symmetry Laws are violated.

There is no known reason why all these details, **mixing of states** and **Symmetry Law violations**, are needed.

They have been experimentally discovered and show how many deviations from the **simple "Platonic" structure** are needed.

So far we have developed the sequence of **Platonic deviations from *Simplicity***, starting from the basic fundamental boson B_B .

We now show in Figure 24 the **deviations needed** from the **Platonic Simplicity**, when we start from the basic fundamental fermion B_F .

It has to be with "**quark**" and "**lepton**" **flavours** and have two flavours in each class (called Family).

Total number of flavours *12*: *6 for quarks*
6 for leptons

Why so many? The answer will probably come from the Super Space with 43 Dimensions compactified into (3+1).

The quark sector interacts with two forces, QCD and QED, while the lepton sector interacts using only QED. The QFD force comes into play only after all the mixings come in.

No one knows why all these **Deviations** from the **Platonic Simplicity** are needed.

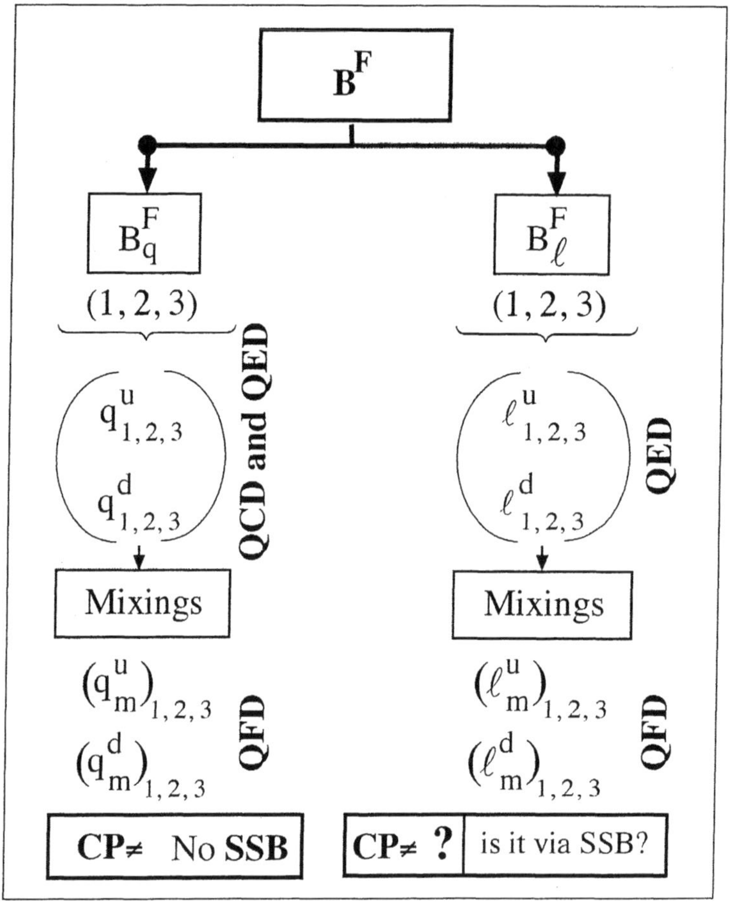

Figure 24

The bold symbols, QCD, QED in the column

$$B_q^F$$

indicate that the 6 quarks flavours interact via these two forces. In the lower part of the same column, the "mixing" indicates that the quark states are no longer "pure" states. They are "mixed"; only these mixed states

$$\left(q_m^u\right)_{1,\,2,\,3} \text{ and } \left(q_m^d\right)_{1,\,2,\,3}$$

interact via the QFD forces.

The column below

$$B_\ell^F$$

has the same structure, but the "mixings" are not the same as in the "quark" column. Furthermore, no one knows at present if the Symmetry CP is violated as it is in the quark case. This is why in the box CP ≠ there is a question mark. Another detail needs to be specified. In the quark case, the CP Symmetry breaking, CP ≠ , has been experimentally established not to be via the basic Standard Model mechanism SSB.

A further Deviation from Simplicity

In the leptonic case, we do not know if the CP Symmetry is violated. It could be it is. In this case it will be interesting to know if it follows the SSB mechanism. All these question marks are evidence of **further deviations** from the **simple Platonic descriptions** of natural phenomena.

In Figure 25 a synthesis of all deviations needed is illustrated.

THE PLATONIC GRAND UNIFICATION
AND THE DEVIATIONS NEEDED

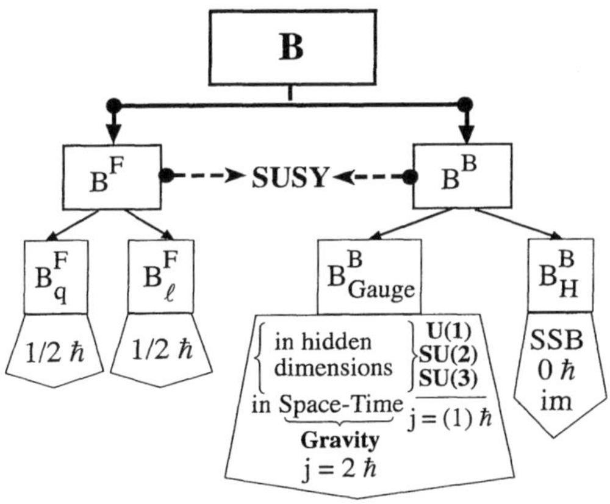

- The fundamental **forces** exist because a Basic Fundamental Boson **exists** ≡ $\mathbb{B}^B$
- The fundamental **fermions** exist because a Basic Fundamental Fermion **exists** ≡ $\mathbb{B}^F$
- The SSBs originate from the existence of a scalar with imaginary mass (im).

Figure 25

3.2 *THE PLATONIC CONCEPT OF SUPERSYMMETRY*

The Platonic Concept of Supersymmetry is schematically reported in the upper part of Figure 26, where the basic point for a Platonic Concept of Supersymmetry is given; i.e. the only fermions allowed to exist, would be the "gauginos" (with spin 1/2 $\hbar$). To go from gauginos to fermions we need the 1st deviation from Platonic Supersymmetry. The 2nd deviation is needed in order to allow fermions to be quarks and leptons. The 3rd deviation is needed in order to have not one Family but three Families. The 4th deviation is needed in order to produce mixing. The 5th deviation is needed because we need at least two mixings, one for "up" and another one for "down" type quarks. All this is illustrated in Figure 26.

THE PLATONIC CONCEPT OF SUPERSYMMETRY

The **Gauge Principle** should generate a
Gauge Force ⟶ **Gauge Bosons**
If *NATURE* was platonically *SUPERSYMMETRIC*
Supersymmetry Transformation should generate **Gauginos**

1st **DEVIATION FROM PLATONIC SIMPLICITY**
OUR FERMIONS ARE NOT THE GAUGINOS

2nd **DEVIATION FROM PLATONIC SIMPLICITY**
THE FUNDAMENTAL FERMIONS ARE OF TWO DIFFERENT CLASSES: LEPTONS AND QUARKS

3rd **DEVIATION FROM PLATONIC SIMPLICITY**
THERE IS NOT ONLY ONE BUT THREE FAMILIES OF FUNDAMENTAL FERMIONS

4th **DEVIATION FROM PLATONIC SIMPLICITY**
THE FUNDAMENTAL FERMIONS BECOME MIXED WHEN THE WEAK FORCES ARE SWITCHED ON: MIXINGS EXIST

5th **DEVIATION FROM PLATONIC SIMPLICITY**
THERE ARE DIFFERENT MIXINGS

Figure 26

SPACE AND TIME DEPARTURES FROM THE PLATONIC SIMPLICITY

A remark refers to the number of Space-Time dimensions.

In a Platonic World where Space and Time were perfectly symmetric, they should have the same number of dimensions. In real life, the number of expanded dimensions is 3 for Space and 1 for Time.

This is the first deviation from a Platonic Symmetry of Space and Time.

Let me repeat: Space and Time depart from Platonic Simplicity since

$$\boxed{N_S = N_T}$$

is not valid. In fact

$$\boxed{N_S \neq N_T .}$$

This is not all.

The nature of Space and Time is such that these two basic ingredients of our existence cannot be both real.

If Space is real, Time must be imaginary and viceversa.

This is another Deviation from the Platonic Symmetry.

To sum up: the number of Space dimensions is bigger than the number of Time dimensions, which is the minimum possible: one. They cannot be both real. One of them must be imaginary. The world appears to be very different from what could be its Platonic *Simplicity*.

3.3 *THE DEVIATION FROM THE PLATONIC SIMPLICITY AT ITS EXTREME VALUE*

In this non-Platonic world, let us see if it is possible to put on the same grounds bosonic and fermionic degrees of freedom. We will see that **Bosonic** and **Fermionic degrees of freedom** are on the same ground when the departure from Space-Time **Platonic Simplicity** is at its extreme value 9 to 1.

In our world where the departure from Platonic Space-Time Simplicity is 3 to 1, with 4 dimensions (3 Space and 1 Time), this is not the case.

For example, in the simplest case of fermions describing the neutrinos, a spinor cannot be both of Weyl and Majorana type (as illustrated in Figure 27). We recall that this difference arises when the spinor carries an additively conserved quantum number, such as the so-called "leptonic flavour".

The Weyl spinor corresponds to a neutrino with zero mass which is different from its antineutrino because the leptonic flavour is not zero.

The Majorana spinor corresponds to a neutrino with non-zero mass but zero leptonic flavour.

In **4 dimensions** a spinor **cannot be**
both **Weyl** ($m = 0$ and $q \neq \bar{q}$)
and
Majorana ($m \neq 0$ and $q = \bar{q}$)

In **10 dimensions** a spinor **can be**
both **Weyl** and **Majorana**

A **Dirac spinor** ($m \neq 0$; $q \neq \bar{q}$) in **10 dimensions**
has **32** degrees of freedom
while a Weyl ($m = 0$) or a Majorana ($q = \bar{q}$) spinor
has **16** degrees of freedom

A **Majorana-Weyl** ($m = 0$; $q = \bar{q}$) spinor
has only **8** degrees of freedom

This 8 exactly matches the number of
transverse modes of a vector in **10 dimensions**

This equality (**8** and **8**) degrees of freedom
is used to construct
SUSY theories in **10 dimensions**
where the number of **fermionic** degrees of freedom
must be equal to the
number of **bosonic** degrees of freedom

Figure 27

Let me recapitulate: **4 dimensions** a spinor **cannot be** both

Weyl $(m = 0$ and $q \neq \bar{q})$

and

Majorana $(m \neq 0$ and $q = \bar{q})$

We still have no firm basis to know the nature of our neutrinos.

It seems that they have non-zero masses and conserved leptonic flavours, i.e. $m \neq 0$ and $q \neq \bar{q}$. However massive neutrinos of the Majorana type could exist, according to some Grand Unified theories. Their mass could be very high indeed.

This short note on the neutrinos helps our understanding of the departure of the real world from the Platonic one.

As said at the beginning, the Deviation from Platonic Simplicity must be at its extreme value (9 Space and 1 Time) for the Bosonic and the Fermionic degrees of freedom to be on the same grounds.

In **10 dimensions** (9 Space + 1 Time) a spinor **can be** both **Weyl** and **Majorana**.

In the search for Space-Time dimensions where the number of degrees of freedom for fermions and bosons is the same, the result is that this number has to be 10. In fact in 10 dimensions the number of fermionic degrees of freedom is equal to the number of bosonic degrees of freedom. This number is 8 and is used to construct Supersymmetric theories in 10 dimensions, as reported in Figure 27.

In these theories, the two types of neutrinos mentioned above, the Weyl and the Majorana, have both 16 degrees of freedom.

And a neutrino with zero mass and zero leptonic flavour (that we can express in terms of a "charge" $q = 0$) has only 8 degrees of freedom.

This hypothetical neutrino would be a Majorana-Weyl type.

SUSY theories in **10 dimensions** have vector and fermions with the same degrees of freedom.

A synthesis of all this is in Figure 28. If another bosonic dimension is needed its role is to be a "gauge" coupling. The total number of bosonic dimensions is therefore 11 of which 9 for Space, 1 for Time and 1 for gauge coupling.

SUSY theories in 10 dimensions
have vector and fermions
with the same degree of freedom

A **Dirac spinor** ($m \neq 0$; $q \neq \bar{q}$) in **10 dimensions**
has **32** degrees of freedom

while a Weyl ($m = 0$) or a Majorana ($q = \bar{q}$) spinor
has **16** degrees of freedom

A **Majorana-Weyl** ($m = 0$; $q = \bar{q}$) spinor
has only **8** degrees of freedom

This 8 exactly matches the number of
transverse modes of a vector in **10 dimensions**

This equality (**8** and **8**) degrees of freedom
is used to construct
SUSY theories in **10 dimensions**
where the number of **fermionic** degrees of freedom
must be equal to the
number of **bosonic** degrees of freedom

Figure 28

3.4 *HOW MANY DEVIATIONS FROM SIMPLICITY WILL EMERGE FROM THE FUTURE TEN CHALLENGES? A NOTE ON RQST (RELATIVISTIC QUANTUM STRING THEORY)*

We have mentioned the theoretical activities based on the field of mathematics called "String Theory": more precisely Relativistic Quantum String Theory (RQST).

A very large number of theorists, the world over, is engaged in this area, whose purpose is to quantize gravity.

No one knows how long it will take.

The present difficulty is due to the fact that all these theoretical speculations produce extremely interesting results which cannot be put under experimental test.

This reminds me of the 35 years dedicated by Einstein to unify electromagnetic and gravitational forces.

We know today that his Space-Time mathematics had only bosonic dimensions and these were just 4.

The fermionic dimensions were missing.

Superspace, essential for the Grand Unification, had not been invented.

On the other hand, Einstein was convinced that all experimental basic facts were known and the difficulty with the Unification was of mathematical nature.

With Superstring Theory the story repeats itself once again; not so much because the theoretical fellows believe that the experimental information is all there, but because their theoretical speculations are in the domain of the Planck energy, while our ability to perform experiments are in the range of 17 orders of magnitude below.

Nevertheless, the **next ten years** are going to be decisive for the study of the Fundamental Logic of Nature.

What we have said in the preceding chapters ends up with a set of challenges, regardless of the success of RQST.

Even if the quantization of Space and Time will be successful, the problem of great priority is what experiments need to be implemented. And if this will produce further **deviations from Platonic Simplicity**.

If the deviations come out, they will probably arise from the ten experimental challenges reported in Figure 29.

***

THE TEN CHALLENGES

1) The Physics of Imaginary masses: SSB
2) P $\neq$, C $\neq$, CP $\neq$, T $\neq$: DSB; CPT $\neq$, and Matter-Antimatter Symmetry
3) Supersymmetry SUSY
4) Non-perturbative QCD and the Physics of deconfined colour charges
5) Anomalies and Instantons
6) Flavour mixing in the quark sector
7) Flavour mixing in the leptonic sector
8) The problem of the missing mass in the Universe
9) The problem of the hierarchy
10) The Physics at the Planck Scale, the Gap and the number of expanded dimensions

Figure 29

Let us say a few words on each of them.

1) Here the crucial point is to see if a scalar particle, the so-called Higgs (H), exists.

2) A series of experiments needs to be performed in order to understand the origin of the Direct Symmetry Breaking (DSB). The peculiar fact is that these symmetry violations are not induced by the SSB mechanism should be understood. Furthermore the problem of Matter-Antimatter Symmetry, totally disconnected from particle-antiparticle Symmetry, needs to be experimentally checked.

3) Here the point is to discover the lightest particle of the **Superworld** (The EGM effect lowers by nearly three order of magnitude the threshold for its production).

4) Here the subnuclear world where quarks and gluons are free could reveal new insights. The most promising experiments here are those of the QGCW project, illustrated in Figure 30.

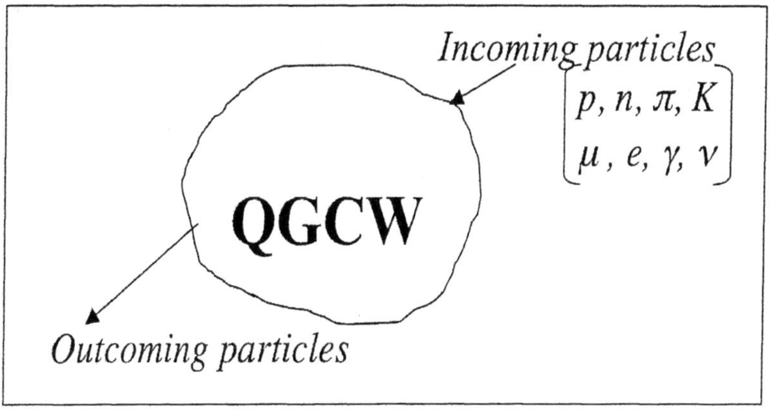

Figure 30

5) The role of the anomalies and of the Instantons in new areas of the subnuclear world could be the source of unexpected novelties.

6) So far the flavour mixings in the quark sector have been measured. What is needed is the understanding of their origin.

7) In the lepton sector the flavour mixing has been established to exist; as for the quark mixing, the problem is to understand its origin.

8) This is one of the cosmic mysteries and exists since more than half a century It should be coupled with the problem of the vacuum energy (also called "dark energy").

9) The problem is to understand why these two energy scales (the Fermi and the Planck one) exist and why they are separated by 17 orders of magnitude. How they can be kept apart is one of the biggest challenges.

10) We have seen that the Physics at the Planck scale opens new horizons in our view of the real world. The existence of the Gap (Figure 31) [1] between the two unification energy scales (the one where the three gauge couplings join together, E_{GU}, and the other predicted by the String Theory, E_{SU}) could reveal unexpected novelty. Finally, if expanded dimensions could be different from the four well known to us, the Planck scale could come down to the Fermi scale, as illustrated in Figure 32.

The GAP

[Reference 1]

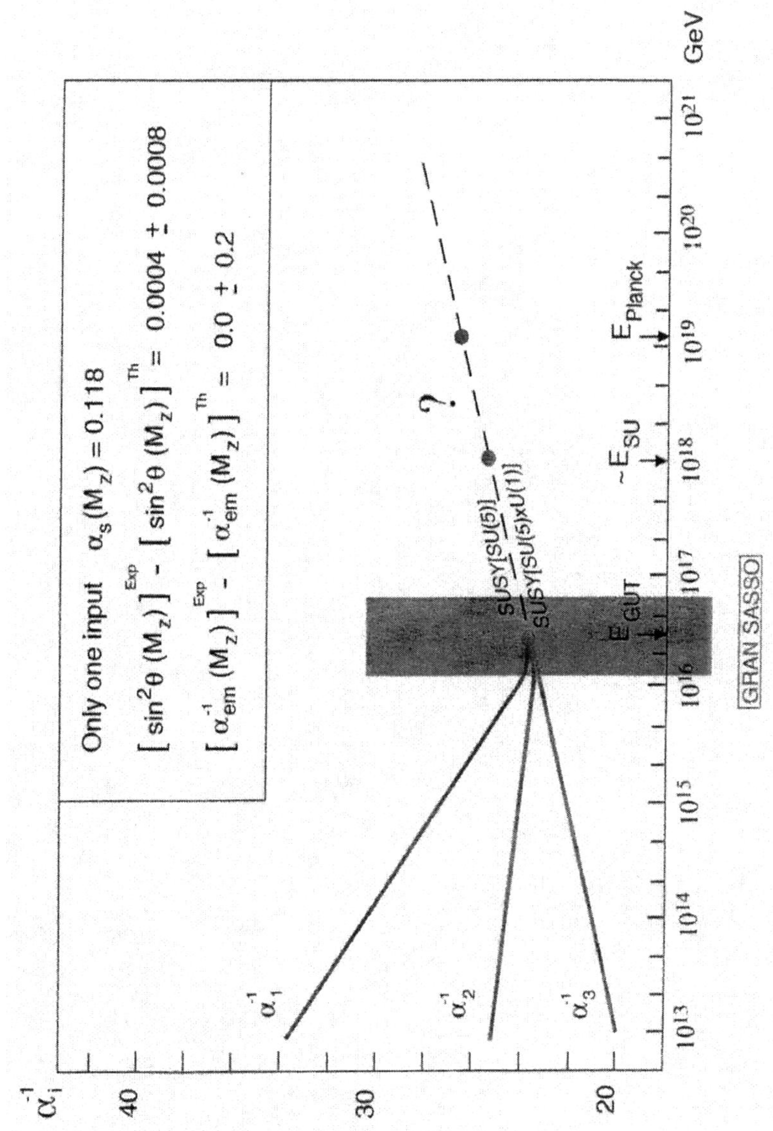

Figure 31

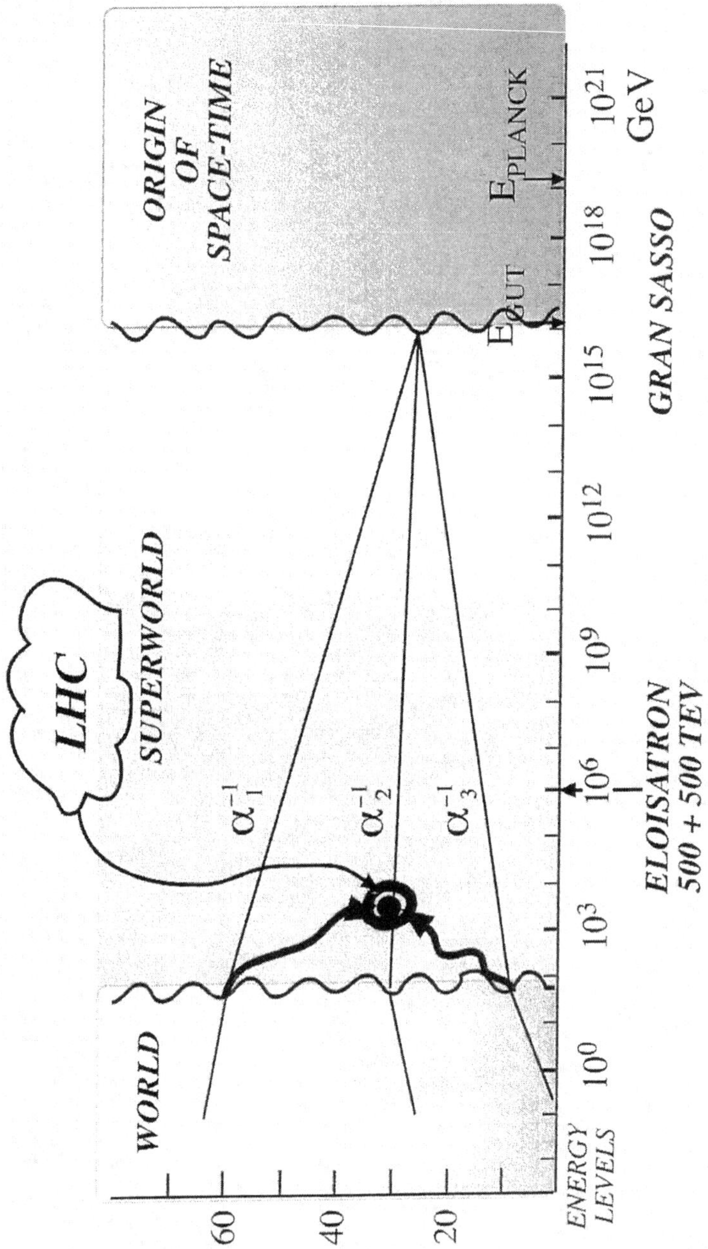
ORIGIN OF SPACE-TIME
LHC
SUPERWORLD
WORLD
α_1^{-1}
α_2^{-1}
α_3^{-1}
E_{GUT}
E_{PLANCK}
60
40
20
ENERGY LEVELS
10^0
10^3
10^6
10^9
10^{12}
10^{15}
10^{18}
10^{21}
GeV
ELOISATRON
500 + 500 TEV
GRAN SASSO

Figure 32

To study the existence of the microscopic expanded dimension is the experimental challenge which could shed new light on the meaning of the Planck scale.

When people speak of *Simplicity* in the description of the elementary phenomena, they should pay attention, not only to what has been pointed out in the previous chapters, but also to the existence of these **ten experimental challenges** and on their potential to introduce **new deviations** from the Simple Platonic view of the world.

The point is: are these challenges the source for a better understanding of the transition from "*Simplicity*" to "*Complexity*" or is this understanding going to be derived from studies which are out of the field of Galilean Science?

In Figure 33 we show the various components of our knowledge on the basis of basic quantities whose existence makes up the apparently different domains of our world.

We do not know what will be the final outcome of String Theory. What we do know is that all these attempts to understand the Logic of Nature appear to be very "*simple*" when compared with the "*Complexity*" of the world we are part of, as illustrated in the Figure 33 quoted above.

Three points need to be put in evidence:

i) so far, no new fundamental forces have been discovered when our *Simplicity* (the Standard Model and its extensions) has been expanded to include the *Complexity* of the Universe;

ii) the *Simplicity* of the Standard Model needs many deviations which are like the two basic ingredients of Complexity: the Beethoven (AFB) phenomena and the Sarajevo (UEEC) events. This means that Complexity exists at the elementary level;

iii) if we try to formulate the simplest version of our rigorous logic to describe Nature – the Platonic Simplicity – the result is that nothing in the real world looks like it, even the Standard Model.

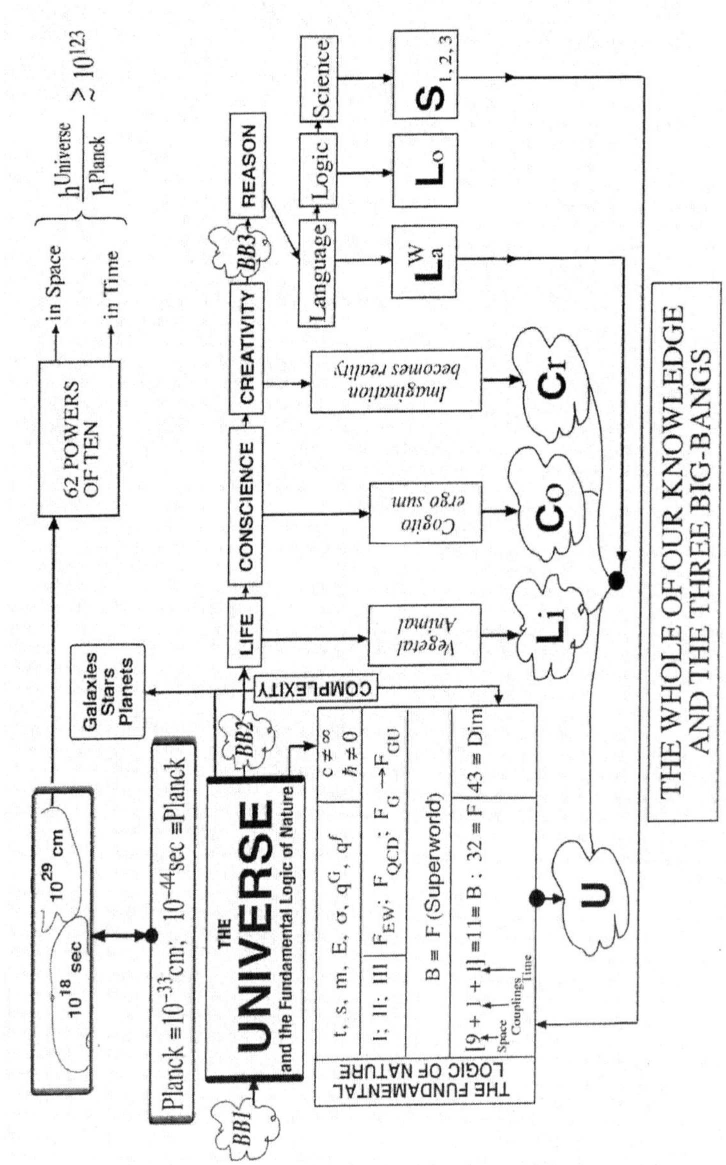
THE WHOLE OF OUR KNOWLEDGE AND THE THREE BIG-BANGS
62 POWERS OF TEN
in Space
in Time
Planck
Galaxies Stars Planets
THE UNIVERSE and the Fundamental Logic of Nature
BB1
BB2
BB3
COMPLEXITY
LIFE
CONSCIENCE
CREATIVITY
REASON
Language
Logic
Science
Vegetal Animal
Cogito ergo sum
Imagination becomes reality
Li
Co
Cr
U
THE FUNDAMENTAL LOGIC OF NATURE
B ≡ F (Superworld)
Space
Couplings
Time

Figure 33

4 CONCLUSION

I have attempted to review some of the most spectacular (UEEC) discoveries during the last decades. In case you are still not very much convinced, let me recall you that the Yukawa Centenary (2007) [59] has been for me of great value in order to draw attention to the impressive series of totally unexpected conceptual developments linked with his very simple proposal of a meson in order to explain the existence of the "nuclear glue": chirality-invariance, spontaneous symmetry breaking, symmetry breaking of fundamental invariance laws, anomalies, and 'anomaly-free condition', existence of a third family of fundamental fermions, gauge principle for non-Abelian forces, instantons and existence of a pseudoscalar particle made with the quanta of a new fundamental force of Nature acting between the constituents of the Yukawa particle.

All the pieces of the Yukawa gold mine **could not have been discovered** if the experimental technology was not at the frontier of our knowledge, as was the case for the cloud-chambers (Anderson, Neddermeyer), the photographic emulsions (Lattes, Occhialini, Powell), the high power magnetic fields (Conversi, Pancini, Piccioni) and the powerful particle accelerators and associated detectors for the discovery – the world over – of the intrinsic structure of the Yukawa particle (quarks and gluons) which has brought us to QCD and now to the QGCW Project.

The lesson needed for the future is that we must be prepared with the most advanced technology for the discovery of totally unexpected events (UEEC).

After my lecture some students asked me questions whose answers I have decided to include as Addendum 1, 2, 3, 4. For other details see the discussion following the lecture.

Addendum N° 1
THERE IS NO MIXING OF STATES AT THE FUNDAMENTAL LEVEL

The fundamental interactions are four:

1) The gravitational → flavour blind;
2) The weak → flavour mixing;
3) The electromagnetic → flavour blind;
4) The subnuclear strong → flavour blind.

The fundamental constituents, six quarks and six leptons, are all fermions with (spin ½); they are grouped into 3 families.

I family: (ud) $(\nu_e\, e^-)$ ≡ 2 quarks; 2 leptons
II family: (cs) $(\nu_\mu\, \mu^-)$ ≡ 2 quarks; 2 leptons
III family: (tb) $(\nu_\tau\, \tau^-)$ ≡ 2 quarks; 2 leptons.

The gauge coupling α_2 is not (100%) given to a quark mass eigenstate but to a mixture of "up" and "down" states. The present description attributes either to the transition

[up → "down-mixed"] the full value of α_2

or

[down → "up-mixed"] the full value of α_2.

The so called (GLCGIMKM) matrix [Gell-Mann and Levy, (GL); Cabibbo (C); Glashow, Iliopoulos e Maiani (GIM), Kobayashi and Maskawa (KM)] must be the result of the product of two matrices which describe ["up mixed"] ⇔ ["down mixed"] transitions.

The $SU(2)\times U(1) \Rightarrow SU(2)_L\times U(1)_{LR}$, electroweak force is acting between mixed states, at the fundamental level of "quarks". None of the other forces has this property.

The gravitational force is acting between eigenstates of masses, and never mixes "pure" quark states. The QED gauge coupling is fully given to one "pure" state of quark, either "up" ≡ 2/3 or "down" ≡ −1/3. The gauge couplings in the case of leptons is fully given to the charged ones (e, μ, τ) and zero to (ν_e, ν_μ, ν_τ).

The QCD force α_3 is given to a pure quark state (u, d, c, s, t, b) without any reference to the particular flavour. The reason why α_3 must be the same for all quark flavours is due to the fact that QCD is a non-Abelian force and there is a gluon-gluon interaction which fixes the value of α_3 (Figure 34).

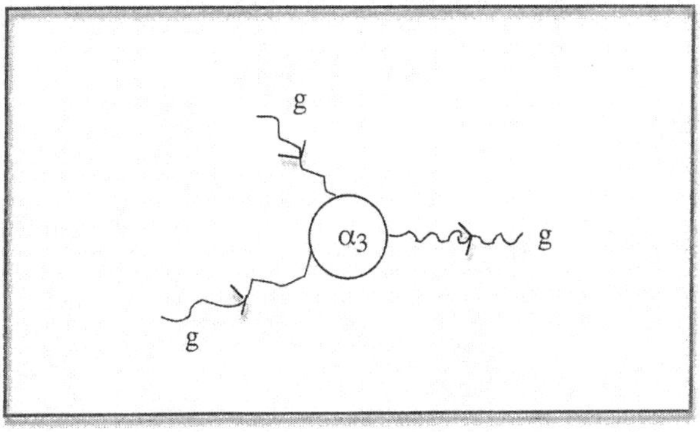

Figure 34

Each flavour has the full QCD gauge coupling: α_3.

The same value for α_3 to any quark flavour state is at the origin of the

SU(2) – isospin and of $SU(3)_{uds}$ – symmetry properties.

Let me recapitulate.

The electromagnetic gauge coupling α_1

is 100% given to the "down-type" leptonic state (e^- μ^- τ^-)

and zero to the "up-type" leptonic state (ν_e ν_μ ν_τ);

to the "down-type" quark state (d, s, b) is given −1/3;

to the "up-type" quark state (u, c, t) is given +2/3,

but never to mixed states.

The full α_2 gauge coupling is given, as said before, to a mixture of

$$\text{“up-states”} \Rightarrow u_1^m;\ u_2^m;\ u_3^m$$

and to a mixture of

$$\text{“down-states”} \Rightarrow d_1^m;\ d_2^m;\ d_3^m\ .$$

It should be noted that the mixing of “up” and the mixing of “down” quark state must be different. In order to put this in evidence we should write

$$(u_1^{m_u};\ u_2^{m_u};\ u_3^{m_u}) \ \text{ and } \ (d_1^{m_d};\ d_2^{m_d};\ d_3^{m_d}),$$

where m_u and m_d specify the different mixing needed for the up-states and down-states, respectively.

For the leptonic case we know that the gauge coupling

$$\alpha_2$$

is shared between a mixture of “top” leptonic states; in fact

$$A \left\{ \begin{array}{l} \nu_e \equiv \text{mixing of } (\nu_1\ \nu_2\ \nu_3) \\ \nu_\mu \equiv \text{mixing of } (\nu_1\ \nu_2\ \nu_3) \\ \nu_\tau \equiv \text{mixing of } (\nu_1\ \nu_2\ \nu_3). \end{array} \right.$$

Note that in (A) the three mixings must be different. If another mixing is present for the “down” leptonic states is experimentally not established. It could be that

$$B \left\{ \begin{array}{l} e^- \equiv \text{mixing of } (\text{leptonic}_1\ ;\ \text{leptonic}_2\ ;\ \text{leptonic}_3) \\ \mu^- \equiv \text{mixing of } (\text{leptonic}_1\ ;\ \text{leptonic}_2\ ;\ \text{leptonic}_3) \\ \tau^- \equiv \text{mixing of } (\text{leptonic}_1\ ;\ \text{leptonic}_2\ ;\ \text{leptonic}_3). \end{array} \right.$$

Note that in (B) the three mixings must be different.

The mixing of states at the fundamental level does exist only in the (electroweak) QFD interaction with gauge coupling α_2. All other fundamental forces act between mass eigenstates.

In my lectures at the Bologna University I discuss these problems in detail.

ADDENDUM N° 2
THE MIXINGS OF MESONIC STATES: PSEUDOSCALAR AND VECTOR

If we start with three quarks

$$\begin{cases} \text{up} & \equiv & u \\ \text{down} & \equiv & d \\ \text{strange} & \equiv & s\,, \end{cases}$$

using the charge conjugation operator C we have the correspondent antiquarks

$$\begin{cases} \text{antiup} & \equiv & \bar{u} \\ \text{antidown} & \equiv & \bar{d} \\ \text{antistrange} & \equiv & \bar{s}\,. \end{cases}$$

The only possible mesonic states would be

$$(u\bar{u}),\ (d\bar{d}),\ (s\bar{s}).$$

If states such as

$$\frac{(u\bar{u} + d\bar{d})}{\sqrt{2}} \qquad\qquad \frac{(u\bar{u} - d\bar{d})}{\sqrt{2}}$$

can be build, this is because the transitions

$$u\bar{u} \rightarrow d\bar{d} \qquad\qquad d\bar{d} \rightarrow u\bar{u}$$

are possible, and therefore, if we start with a state such as $(u\bar{u})$, due to the $u\bar{u} \rightarrow d\bar{d}$ transitions, we would end up into a mixed state

$$\frac{(u\bar{u} + d\bar{d})}{\sqrt{2}} \quad \text{or} \quad \frac{(u\bar{u} - d\bar{d})}{\sqrt{2}}. \tag{A}$$

If further transitions are possible, such as

$$u\bar{u} \rightarrow s\bar{s} \text{ and } d\bar{d} \rightarrow s\bar{s}\,,$$

then, starting with states (A), we would automatically get an $SU(3)_f$ singlet for both the pseudoscalar and the vector mesons:

$$\eta_1 \equiv \omega_1 \equiv \frac{(u\bar{u} + d\bar{d} + s\bar{s})}{\sqrt{3}} \tag{B}$$

and an 8th member of the $SU(3)_f$ octet

$$\eta_8 \equiv \omega_8 \equiv \frac{\left(u\bar{u} + d\bar{d} - 2s\bar{s}\right)}{\sqrt{6}} \qquad \text{(B')}$$

if the 3 flavours (u, d, s) obey a symmetry such as $SU(3)_f$. In this case only the mesonic states

$$(\eta_8 \ , \ \eta_1)$$
$$(\omega_8 \ , \ \omega_1)$$

would exist. In real life we have

$$(\eta \ , \ \eta')$$
and
$$(\omega \ , \ \phi).$$

Now attention: the existence of states (A), (B) and (B') means that transitions

$$u\bar{u} \to d\bar{d}, \quad u\bar{u} \to s\bar{s}, \quad d\bar{d} \to s\bar{s},$$

etc, are possible. Question: how probable are all these transitions? Answer: if the initial state is a pseudoscalar state, the transition into a final state consisting of flavours different from the initial one can be mediated by 2 gluon states:

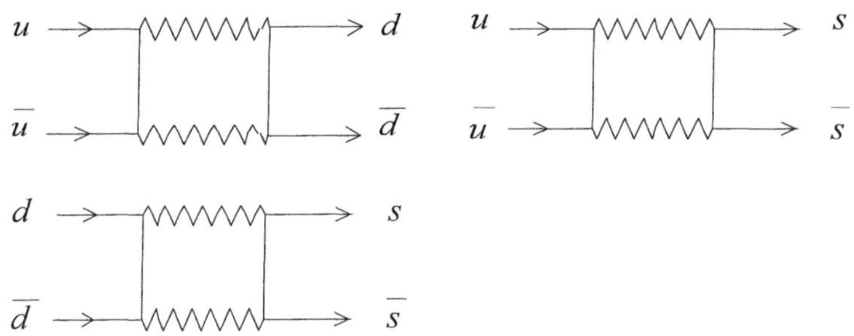

In the case of the initial state being a vector meson with $J^{PC} = 1^{--}$ the intermediate state needs to be a (3 gluon state)

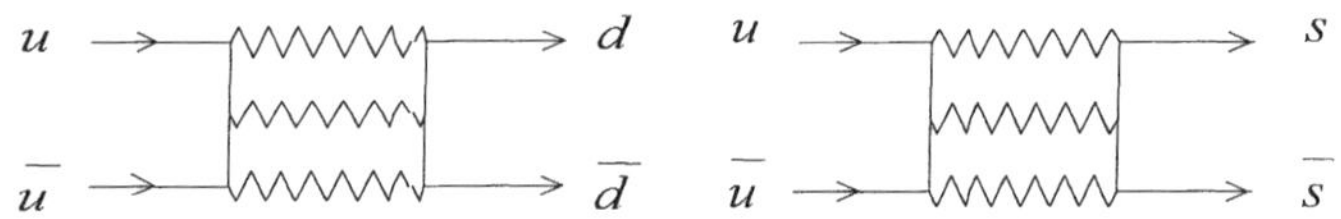

etc...

In the real world it so happens that in the vector meson sector the physical mesonic states ($u\bar{u}$, $d\bar{d}$) are almost totally separated from $(s\bar{s})$:

$$\omega \cong \frac{u\bar{u}+d\bar{d}}{\sqrt{2}}; \qquad \phi \cong (s\bar{s})$$

In the pseudoscalar sector the physical mesonic states are

$$\eta \equiv \frac{u\bar{u}+d\bar{d}-\sqrt{2}s\bar{s}}{\sqrt{4}}$$

and

$$\eta' \equiv \frac{u\bar{u}+d\bar{d}+\sqrt{2}s\bar{s}}{\sqrt{4}}$$

The wave functions correspond to $(\eta_8\,\eta_1)$ mixing with $\theta_{PS} \approx 10^\circ$, i.e.

$$\left.\begin{aligned} \eta &= \eta_8 \cos\theta - \eta_1 \sin\theta \\ \eta' &= \eta_8 \sin\theta + \eta_1 \cos\theta \end{aligned}\right\}$$

with, as said above, $\theta_{PS} \approx 10^\circ$. If θ was zero, we would have

$$\begin{aligned} \eta &\equiv \eta_8 \\ \eta' &\equiv \eta_1 . \end{aligned}$$

If $\theta_V \approx 45^\circ$ we would have

$$\begin{aligned} \eta &= (u\bar{u}+d\bar{d})\frac{1}{\sqrt{2}} \equiv\rangle\ \omega \\ \eta' &= (s\bar{s}) \equiv\rangle\ \phi \end{aligned}$$

i.e. complete separation of $(s\bar{s})$ flavour from $\left(u\bar{u}\ +\ d\bar{d}\right)$ as it is in the vector meson case. Let me recall that the $SU(3)_f$ states are (in terms of the quark composition):

$$\eta_8 = \frac{u\bar{u}+d\bar{d}-2s\bar{s}}{\sqrt{6}} \quad (8^{th}\ \text{member of}\ SU(3)_f\ \text{octet})$$

$$\eta_1 = \frac{u\bar{u}+d\bar{d}+s\bar{s}}{\sqrt{3}} \quad SU(3)_f\ \text{(singlet)}.$$

In the real world, I repeat once again, we have the physical states

$$J^{PC}=0^{-+}\begin{cases}\eta(m\cong 500\text{Mev})\equiv\dfrac{u\bar{u}+d\bar{d}-\sqrt{2}s\bar{s}}{\sqrt{4}}\\ \eta'(m\cong 950\text{Mev})\equiv\dfrac{u\bar{u}+d\bar{d}+\sqrt{2}s\bar{s}}{\sqrt{4}}\end{cases}$$

$$J^{PC}=1^{--}\begin{cases}\omega(m\cong 750\text{Mev})\equiv\dfrac{u\bar{u}+d\bar{d}}{\sqrt{2}}\\ \phi(m\cong 1020\text{Mev})\equiv s\bar{s}\end{cases}.$$

This quark mixing, when translated into the mixing of mesonic states, gives for the pseudoscalar case

$$\begin{Bmatrix}\eta=\eta_8\cos\theta_{PS}-\eta_1\sin\theta_{PS}\\ \eta'=\eta_8\sin\theta_{PS}+\eta_1\cos\theta_{PS}\end{Bmatrix}$$

$$\text{with }\theta_{PS}\approx 10°$$

and for the vector case

$$\begin{Bmatrix}\omega=\omega_8\cos\theta_V-\omega_1\sin\theta_V\\ \phi=\omega_8\sin\theta_V+\omega_1\cos\theta_V\end{Bmatrix}$$

$$\text{with }\theta_V\approx 45°.$$

Please note that all mixings discussed in this Addendum is between $(q\bar{q})$ states with total flavour equal to zero.

In Addendum 1 we have discussed mixings at the fundamental level of pure quark-states, when the α_2 is given to a mixture of quark-states with different flavours.

ADDENDUM N° 3
WHY WE NEED THE SUPERWORLD

There are fundamental reasons making the superworld a need.

1) The two energy scales, 10^{19} GeV (Planck) and 10^{2} GeV (Fermi), must be kept separate.
2) The gravitational attraction of light must be prevented from being infinite. Otherwise we could see neither the light from Stars nor our light. The "gravitino" (supergravity) allows the gravitational attraction of light to be finite.
3) Gravitational attraction is powerful but it cannot be infinite. We would be stuck to the Sun. Space would not exist between Stars and Galaxies. Cosmic expansion would not exist. In order to have a finite gravitational attraction, theories are needed in which the Euclidean concept of point is abandoned. The point is replaced by a string. No more Point-like Theories but String Theories. These theories must be supersymmetric: the already quoted supersymmetry law ($F \equiv B$) must be valid in their mathematical structure. Otherwise "tachions" would appear. This is the origin of Relativistic Quantum String Theory (RQST).
4) If we want the unification of all fundamental phenomena – the synthesis of which is provided by three "gauge couplings", α_1 α_2 α_3 – running with the energy is needed.
5) An interesting detail: "no scale-supergravity" is an infrared solution of RQST. This might allow to understand the extremely small value of the Cosmological Constant.
6) Finally: why Three Columns and Three Forces? The answer to this question should come once we will be able from the compactification of the 43-dimensional superspace to go to our present world with (3+1) Space-Time dimensions.
7) **Note**: **A big problem**. Supersymmetry does not show up at our energy scale. Hence the problem arises to compute the energy above which the ($F \equiv B$) Law starts to act. Thanks to the EGM effect, this energy level is 700 times more accessible than thought so far.

ADDENDUM N° 4
TEN PROBLEMS OPEN FOR COMPETITION

In this Addendum I recall the ten problems open for competition.

THE FIRST GROUP:
MIXINGS

- *Is there an explanation for the flavour mixing mechanisms?*
- *Why do these mechanisms produce results that differ substantially in the quark and in the lepton sectors?*
- *Why does the mixing of states not exist in any other fundamental interaction?*

THE SECOND GROUP:
ELEMENTARY AND COMPOSITE STATES

- *What is the role of instantons in the spectrum of hadrons in QCD? Where are the scalar hadronic states in QCD? Are there scalar states in the leptonic system?*
- *Why are there only fundamental particles with the minimum quark or lepton quantum numbers? Do elementary particles with higher, composite quantum numbers exist?*
- *Is there a fundamental reason why elementary fermions exist (quarks and leptons) but not elementary scalars in the same mass range?*
- *Do we really need sterile neutrinos? If yes, why? If not, why?*

THE THIRD GROUP:
SYMMETRY BREAKINGS

- *Why are the global discrete symmetries (C, P, CP, T) explicitly and not spontaneously broken, as it seems to be the case today?*
- *To what extent can we be confident that the Supersymmetry breaking threshold is not at the Planck Scale?*
  (If this were the case it would be impossible to find at LHC any evidence for Supersymmetry).
- *Are we really sure that the electro-weak symmetry breaking occurs at the Fermi scale due to a local maximum of the potential for some scalar field in the Lagrangian, also known as the "Higgs mechanism"?*
  (If this were not the case, it would not be possible to find at LHC any evidence for Higgs particles).

REFERENCES

[1] *A Study of the Various Approaches to M_{GUT} and α_{GUT}*
F. Anselmo, L. Cifarelli and A. Zichichi, *Nuovo Cimento* 105A, 1335 (1992); and
Searching for the Superworld
S. Ferrara and R.M. Mössbauer (eds), in honour of A. Zichichi on the occasion of the Sixth Centenary celebrations of the University of Turin, Italy. World Scientific Series in 20th Century Physics, Vol. 39 (2007).

[2] *Complexity Exists at the Fundamental Level*
A. Zichichi, in Proceedings of the 2004–Erice Subnuclear Physics School, Vol. 42, pages 251-327, World Scientific (2007); also presented at Desy, Hamburg, November 2005; see also "*The Ten Challenges of Subnuclear Physics*", A. Zichichi, in Proceedings of the *International Conference on "Quantum [un]speakables" in Commemoration of John S. Bell,* 10-14 November 2000, International Erwin Schrödinger Institut (ESI), Universität Wien (Austria), pages 429-477 (2002).

[3] *A Proposal to Search for Leptonic Quarks and Heavy Leptons Produced by ADONE*
M. Bernardini, D. Bollini, E. Fiorentino, F. Mainardi, T. Massam, L. Monari, F. Palmonari and A. Zichichi, *INFN/AE*-67/3, 20 March 1967.

[4] *Limits on the Electromagnetic Production of Heavy Leptons*
V. Alles-Borelli, M. Bernardini, D. Bollini, P.L. Brunini, T. Massam, L. Monari, F. Palmonari and A. Zichichi, *Lettere al Nuovo Cimento* 4, 1156 (1970);

Limits on the Mass of Heavy Leptons
M. Bernardini, D. Bollini, P.L. Brunini, E. Fiorentino, T. Massam, L. Monari, F. Palmonari, F. Rimondi and A. Zichichi, *Nuovo Cimento* 17A, 383 (1973); and Ref. 5.

[5] *The Origin of the Third Family*
C.S. Wu, T.D. Lee, N. Cabibbo, V.F. Weisskopf, S.C.C. Ting, C. Villi, M. Conversi, A. Petermann, B.H. Wiik and G. Wolf; O. Barnabei, L. Maiani, R.A. Ricci and F. Roversi Monaco Eds., a joint publication by University and Academy of Sciences of Bologna, INFN, SIF (1997); 20th Century Physics Series, Vol. 20, World Scientific (1998).

[6] *Observations of High-Energy Neutrino Reactions and the Existence of Two Kinds of Neutrinos*
G. Danby, J.-M. Gaillard, K. Goulianos, L.M. Lederman, N. Mistry, M. Schwartz and J. Steinberger, *Phys. Rev. Lett.* 9, 36 (1962).

[7] *The Discovery of Nuclear Antimatter*
Proceedings of the *Symposium to celebrate the 30th anniversary of the Discovery of Nuclear Antimatter*, L. Maiani and R.A. Ricci (eds), Conference Proceedings 53, 1, Italian Physical Society, Bologna, Italy (1995).

[8] *The Quantum Theory of the Electron*
P.A.M. Dirac, *Proc. Roy. Soc.* (London) A117, 610 (1928);
The Quantum Theory of the Electron, Part II
P.A.M. Dirac, *Proc. Roy. Soc.* (London) A118, 351 (1928).

[9] *The Positive Electron*
C.D. Anderson, *Phys. Rev.* 43, 491 (1933);
Some Photographs of the Tracks of Penetrating Radiation
P.M.S. Blackett and G.P.S. Occhialini, *Proc. Roy. Soc.* A139, 699 (1933).

[10] *Gruppentheorie und Quantenmechanik*
H. Weyl, 2nd ed., 234 (1931).

[11] *Unitary Representations of the Inhomogeneous Lorentz Group*
E.P. Wigner, *Ann. Math.*, 40, 149 (1939).

[12] *Intrinsic Parity of Elementary Particles*
G.C. Wick, E.P. Wigner, and A.S. Wightman, *Phys. Rev.* 88, 101 (1952).

[13] *Über die Operation der Zeitumkehr in der Quanten-mechanik*
E.P. Wigner, *Gött. Nach.* 546-559 (1931). Here for the first time an anti-unitary symmetry appears.

[14] E.P. Wigner, *Ann. Math.* 40, 149 (1939).

[15] J. Schwinger, *Phys. Rev.* 82, 914 (1951).

[16] *Time Reversal in Field Theory*
J.S. Bell, *Proc. Roy. Soc.* (London) A231, 479-495 (1955).

[17] To the best of my knowledge, the CPT theorem was first proved by W. Pauli in his article '*Exclusion Principle, Lorentz Group and Reflection of Space-Time and Charge*', in 'Niels Bohr and the Development of Physics' [Pergamon Press, London, page 30 (1955)], which in turn is an extension of the work of J. Schwinger [*Phys. Rev.* 82, 914 (1951); '*The Theory of Quantized Fields. II.*', *Phys. Rev.* 91, 713 (1953); '*The Theory of Quantized Fields. III.*', *Phys. Rev.* 91, 728 (1953); '*The Theory of Quantized Fields. VI.*', *Phys. Rev.* 94, 1362 (1954)] and G. Lüders, '*On the Equivalence of Invariance under Time Reversal and under Particle-Anti-particle Conjugation for Relativistic Field Theories*' [*Dansk. Mat. Fys. Medd.* 28, 5 (1954)], which referred to an unpublished remark by B. Zumino. The final contribution to the CPT theorem was given by R. Jost, in '*Eine Bemerkung zum CPT Theorem*'

[*Helv. Phys. Acta* 30, 409 (1957)], who showed that a weaker condition, called 'weak local commutativity' was sufficient for the validity of the CPT theorem.

[18] *Observation of Antiprotons*
O. Chamberlain, E. Segrè, C. Wiegand, and T. Ypsilantis, *Physical Review* 100, 947 (1955).

[19] *Anti-Neutrons Produced from Anti-Protons in Charge Exchange Collisions*
B. Cork, G.R. Lambertson, O. Piccioni, W.A. Wenzel, *Physical Review* 104, 1193 (1957).

[20] *Observation of Long-Lived Neutral V Particles*
K. Lande, E.T. Booth, J. Impeduglia, L.M. Lederman, and W. Chinowski, *Physical Review* 103, 1901 (1956).

[21] *Remarks on Possible Noninvariance under Time Reversal and Charge Conjugation*
T.D. Lee, R. Oehme, and C.N. Yang, *Physical Review* 106, 340 (1957).

[22] *Question of Parity Conservation in Weak Interactions*
T.D. Lee and C.N. Yang, *Phys. Rev.* 104, 254 (1956).

[23] *Experimental Test of Parity Conservation in Beta Decay*
C.S. Wu, E. Ambler, R.W. Hayward, D.D. Hoppes, *Phys. Rev.* 105, 1413 (1957);

Observation of the Failure of Conservation of Parity and Charge Conjugation in Meson Decays: The Magnetic Moment of the Free Muon
R. Garwin, L. Lederman, and M. Weinrich, *Phys. Rev.* 105, 1415 (1957);

Nuclear Emulsion Evidence for Parity Non-Conservation in the Decay Chain $\pi^+\mu^+e^+$
J.I. Friedman and V.L. Telegdi, *Phys. Rev.* 105, 1681 (1957).

[24] *On the Conservation Laws for Weak Interactions*
L.D. Landau, *Zh. Éksp. Teor. Fiz.* 32, 405 (1957).

[25] *Evidence for the* 2π *Decay of the* K_2^0 *Meson*
J. Christenson, J.W. Cronin, V.L. Fitch, and R. Turlay, *Physical Review Letters* 113, 138 (1964).

[26] *Experimental Observation of Antideuteron Production*
T. Massam, Th. Muller, B. Righini, M. Schneegans, and A. Zichichi, *Nuovo Cimento* 39, 10 (1965).

[27] The first report on 'scaling' was presented by J.I. Friedman at the 14th International Conference on *High Energy Physics* in Vienna, 28 August-5 September 1968. The report was presented as paper n. 563 but not published in the Conference Proceedings. It was published as a SLAC preprint. The

SLAC data on scaling were included in the Panofsky general report to the Conference where he says «... the apparent success of the parametrization of the cross-sections in the variable ν / q^2 in addition to the large cross-section itself is at least indicative that point-like interactions are becoming involved». '*Low q^2 Electrodynamics, Elastic and Inelastic Electron (and Muon) Scattering*', W.K.H. Panofsky in Proceedings of 14th International Conference on *High Energy Physics* in Vienna 1968, J. Prentki and J. Steinberger (eds), page 23, published by CERN (1968). The following physicists participated in the inelastic electron scattering experiments: W.B. Atwood, E. Bloom, A. Bodek, M. Breidenbach, G. Buschhorn, R. Cottrell, D. Coward, H. DeStaebler, R. Ditzler, J. Drees, J. Elias, G. Hartmann, C. Jordan, M. Mestayer, G. Miller, L. Mo, H. Piel, J. Poucher, C. Prescott, M. Riordan, L. Rochester, D. Sherden, M. Sogard, S. Stein, D. Trines, and R. Verdier. For additional acknowledgements see J.I. Friedman, H.W. Kendall and R.E. Taylor, '*Deep Inelastic Scattering: Acknowledgements*', *Les Prix Nobel 1990*, (Almqvist and Wiksell, Stockholm/Uppsala 1991), also *Rev. Mod. Phys.* 63, 629 (1991). For a detailed reconstruction of the events see J.I. Friedman '*Deep Inelastic Scattering Evidence for the Reality of Quarks*' in '*History of Original Ideas and Basic Discoveries in Particle Physics*', H.B. Newman and T. Ypsilantis (eds), Plenum Press, New York and London, 725 (1994).

[28] *Quark Search at the ISR*
T. Massam and A. Zichichi, *CERN preprint*, June 1968;

Search for Fractionally Charged Particles Produced in Proton-Proton Collisions at the Highest ISR Energy
M. Basile, G. Cara Romeo, L. Cifarelli, P. Giusti, T. Massam, F. Palmonari, G. Valenti and A. Zichichi, *Nuovo Cimento* 40A, 41 (1977); and

A Search for quarks in the CERN SPS Neutrino Beam
M. Basile, G. Cara Romeo, L. Cifarelli, A. Contin, G. D'Alì, P. Giusti, T. Massam, F. Palmonari, G. Sartorelli, G. Valenti and A. Zichichi, *Nuovo Cimento* 45A, 281 (1978).

[29] *Subnuclear Physics - The First Fifty Years: Highlights from Erice to ELN*
A. Zichichi; O. Barnabei, P. Pupillo and F. Roversi Monaco (eds), Academy of Sciences and University of Bologna, published by World Scientific, Series in 20th Century Physics, Vol. 24 (2000).

[30] *New Developments in Elementary Particle Physics*
A. Zichichi, *Rivista del Nuovo Cimento* 2, n. 14, 1 (1979). The statement on page 2 of this paper, «*Unification of all forces needs first a Supersymmetry. This can be broken later, thus generating the sequence of the various forces of nature as we observe them*», was based on a work by A. Petermann and A. Zichichi in which the renormalization group running of the couplings using supersymmetry was studied with the result that the convergence of the three

couplings improved. This work was not published, but perhaps known to a few. The statement quoted is the first instance in which it was pointed out that supersymmetry might play an important role in the convergence of the gauge couplings. In fact, the convergence of three straight lines ($\alpha_1^{-1}\alpha_2^{-1}\alpha_3^{-1}$) with a change in slope is guaranteed by the Euclidean geometry, as long as the point where the slope changes is tuned appropriately. What is incorrect about the convergence of the couplings is that, with the initial conditions given by the LEP results, the change in slope needs to be at $M_{SUSY} \sim 1$ TeV as claimed by some authors not aware in 1991 of what was known in 1979 to A. Petermann and A. Zichichi.

[31] *The Creation of Quantum ChromoDynamics and the Effective Energy*
V.N. Gribov, G. 't Hooft, G. Veneziano and V.F. Weisskopf; L.N. Lipatov (ed), a joint publication by the University and the Academy of Sciences of Bologna, Italy (1998); World Scientific Series in 20th Century Physics, Vol. 25 (2000).

[32] *The Effective Experimental Constraints on M_{SUSY} and M_{GUT}*
F. Anselmo, L. Cifarelli, A. Petermann and A. Zichichi, *Nuovo Cimento* 104A, 1817 (1991).

[33] *The Simultaneous Evolution of Masses and Couplings: Consequences on Supersymmetry Spectra and Thresholds*
F. Anselmo, L. Cifarelli, A. Petermann and A. Zichichi, *Nuovo Cimento* 105A, 1179 (1992).

[34] *A Study of the Various Approaches to M_{GUT} and α_{GUT}*
F. Anselmo, L. Cifarelli and A. Zichichi, *Nuovo Cimento* 105A, 1335 (1992). This is the paper on the study of the Gap given as Reference [1].

[35] *Are Matter and Antimatter Symmetric?*
T.D. Lee, in Proceedings of the '*Symposium to celebrate the 30th anniversary of the Discovery of Nuclear Antimatter*', L. Maiani and R.A. Ricci (eds), Conference Proceedings 53, page 1, Italian Physical Society, Bologna, Italy (1995).

[36] *String Theory: the Basic Ideas*
B. Greene, Erice Lectures - Discussion 1999 in '*Basics and Highlights in Fundamental Physics*', A. Zichichi (ed), World Scientific (2001).

[37] *Search for Supersymmetric Particles using Acoplanar Charged Particle Pairs from Z^0 decays*
ALEPH Collab., D. Decamp et al., *Phys. Lett.* B236, 86 (1990).

[38] *Search for Neutral Higgs Bosons from Supersymmetry in Z decays*
ALEPH Collab., D. Decamp et al., *Phys. Lett.* B237, 291 (1990).

[39] *Search for Neutralino Production in Z decays*
ALEPH Collab., D. Decamp et al., *Phys. Lett.* B244, 541 (1990).

[40] *Search for the Neutral Higgs Bosons of the MSSM and other two Doublet Models*
ALEPH Collab., D. Decamp et al., *Phys. Lett.* B265, 475 (1991).

[41] *Search for Heavy Charged Scalars in Z^0 decays*
DELPHI Collab., P. Abreu et al., *Phys. Lett.* B241, 449 (1990).

[42] *Search for Pair Production of Neutral Higgs Bosons in Z^0 decays*
DELPHI Collab., P. Abreu et al., *Phys. Lett.* B245, 276 (1990).

[43] *Search for Scalar Quarks in Z^0 decays*
DELPHI Collab., P. Abreu et al., *Phys. Lett.* B247, 148 (1990).

[44] *A Search for Sleptons and Gauginos in Z^0 Decays*
DELPHI Collab., P. Abreu et al., *Phys. Lett.* B247, 157 (1990).

[45] *Mass Limits for Scalar Muons, Scalar Electrons and Winos from e^+e^- Collisions near S**(1/2) = 91–GeV*
L3 Collab., B. Adeva et al., *Phys. Lett.* B233, 530 (1989).

[46] *Search for the Neutral Higgs Bosons of the Minimal Supersymmetric Standard Model from Z^0 Decays*
L3 Collab., B. Adeva et al., *Phys. Lett.* B251, 311 (1990).

[47] *Search for the Charged Higgs Boson in Z^0 decay*
L3 Collab., B. Adeva et al., *Phys. Lett.* B252, 511 (1990).

[48] *A Search for Acoplanar Pairs of Leptons or Jets in Z^0 decays: Mass Limits on Supersymmetric Particles*
OPAL Collab., M.Z. Akrawy et al., *Phys. Lett.* B240, 261 (1990).

[49] *A Search for Technipions and Charged Higgs Bosons at LEP*
OPAL Collab., M.Z. Akrawy et al., *Phys. Lett.* B242, 299 (1990).

[50] *A Direct Search for Neutralino Production at LEP*
OPAL Collab., M.Z. Akrawy et al., *Phys. Lett.* B248, 211 (1990); P.D. Acton et al., preprint CERN-PPE/91-115, 22 July 1991.

[51] *Searches for Supersymmetric Particles Produced in Z Boson decay*
MARK II Collab., T. Barklow et al., *Phys. Rev. Lett.* 64, 2984 (1990).

[52] *Searches for New Particles at LEP*
M. Davier, LP-HEP 91 Conference, Geneva, CH, Preprint LAL 91-48, December 1991.

[53] *The Evolution of Gaugino Masses and the SUSY Threshold*
F. Anselmo, L. Cifarelli, A. Peterman and A. Zichichi, *Nuovo Cimento* 105A, 581 (1992).

[54] *A Detailed Comparison of LEP Data with the Predictions of the Minimal Supersymmetric SU(5) GUT*
J.R. Ellis, S. Kelley, D.V. Nanopoulos, preprint CERN-TH/6140-91, *Nucl. Phys.* B373, 55 (1992).

[55] *The Convergence of the Gauge Couplings at* E_{GUT} *and above: Consequences for* $\alpha_3(M_Z)$ *and SUSY Breaking*
F. Anselmo, L. Cifarelli, A. Peterman and A. Zichichi, *Nuovo Cimento* 105A, 1025 (1992).

[56] *Analytic Study of the Supersymmetry-Breaking Scale at Two Loops*
F. Anselmo, L. Cifarelli, A. Peterman and A. Zichichi, *Nuovo Cimento* 105A, 1201 (1992).

[57] *A* χ^2*-Test to Study the* α_1, α_2, α_3 *Convergence for High-Precision LEP Data, Having in Mind the SUSY Threshold*
F. Anselmo, L. Cifarelli and A. Zichichi, *Nuovo Cimento* 105A, 1357 (1992).

[58] U. Amaldi, W. de Boer and H. Furstenau, *Phys. Lett.* B260, 447 (1991).

[59] *From the Yukawa Particle to the QGCW*
A. Zichichi, in Proceedings of the *Symposium for the Centennial Celebration of Hideki Yukawa*, International Nuclear Physics Conference, Tokyo, Japan, June 3-8 (2007).

CHAIRMAN: A. ZICHICHI

Scientific Secretaries: G. Inguglia, S. Yasnopolsky

DISCUSSION

-'t Hooft:
You've said that states mix only in the electroweak sector. But there's mixing of states everywhere.

- Zichichi:
There's no transition between quarks, QCD is flavor-blind, which means zero transition.

- 't Hooft:
What about the mesonic states?

- Zichichi:
"Mesons" means quark-antiquark. The fundamental objects are the quarks not the mesons.

- 't Hooft:
Yes, but there the states mix, as soon as quantum numbers allow it.

- Zichichi:
These are meson mixings, not quark mixings; at the fundamental level, QCD does not allow any mixing.

-'t Hooft:
If these quarks are massless, then everything is allowed. The mixing of the mass of the quarks comes with the diagonalization of the Kobayashi-Maskawa matrix.

- Zichichi:
The true matrix is the square root of the Kobayashi-Maskawa matrix; there are two matrices in fact that mix together and give you the Kobayashi-Maskawa mixing. QCD is flavor-blind while QFD is not; that's what I mean.

- 't Hooft:
Then you spoke about some very special kind of mixing. Also mixing particles at different positions in physics space or in momentum space is mixing. Quantum mechanics is the theory of mixing states.

- Zichichi:
This is not the fundamental mixing. At the fundamental level, the mixing does not exist for QCD, because QCD is flavor-blind. In other words, QCD could exist with only one flavor. No one knows why there are six flavors. QCD is replicated for each flavor. So, at the fundamental level of QCD you can never go from up to down, from any flavor to another flavor, which would be what I call mixing.

- 't Hooft:
Think of the kind of experiments where you have spin up, spin sideways and spin down; those are mixings of states.

- Zichichi:
But these are mixings of the same state. At the fundamental level of our existence there are twelve fermions: six quarks and six leptons. The electric charge is never shared. One particle has the full

charge, while the weak charge is shared between states, and the mixing of these states is expressed by KM matrix, whose origin is unknown. This is what I mean by mixing.

- *Ferrara*:
What we have in the meson mixing is a state mixing, it is not flavor mixing. For example, the rho-zero ω-mixing is just a state mixing, between states having a total of zero flavor.

- *Zichichi*:
At CERN, we measured the mixing between "pseudoscalar" mesons (η, η') and the mixing between "vector" mesons (ω, φ). The mesons are mixtures of q-q bar states and therefore are not basic.

- *Ferrara*:
The existence of three families is not explained. The presence of one family is due to anomaly cancellation, but there's no reason for having three of them. It's not needed by anything, simply physics happens to be like that. However there is an interesting connection between asymptotic freedom and number of states.

- *Aschieri*:
You said that physics is the "queen of sciences". You don't have to convince us here, but this statement has been challenged. For example, some people say that it was the queen of sciences in the past, but that it is not any more, and that now it is biology or genomics or something like that. Can you provide us with an explicit motivation for your statement?

- *Zichichi*:
At the fundamental level there is no reason why QED has been discovered by Maxwell and why two hundred years of experiments with totally unexpected results in electricity, magnetism and optics (each one though to be in the 19th Century a titally independent field of physics) have been synthesized by the Maxwell equations, which then produced QED (quantum electrodynamics). Suppose we knew nothing about QED and we picked up living matter and study biology. At the end of all these studies we would rediscover QED. This is out of question, and it is why physics is the queen of sciences. But if we ourselves don't say it, nobody else will. So we need to say this. If we don't know how to defend our activity, we will disappear. However what you say is correct. Now biological studies are presented as the new frontier of science, but they're not the new frontier of basic science. Two years ago here in Erice we held a conference on the so-called "minimal life" problem. The minimal life problem corresponds to find out how many pieces of non living matter you need to build the most elementary example of a living object, namely a cell. This problem is still open, and I would like to have another conference about it. So, the basic problem of living matter science, i.e. biology, is to understand the transition from the non-living matter, to living matter. At the end of their studies, biologists will be able to solve this problem, but at present they are not. This is why physics remains the basic science. Moreover, at the end of the story, they will discover QED.

- *Burda*:
During your lecture you spoke about the complexity of the Standard Model and its extensions. I wonder whether maybe this complexity is just due to the fact that we study the "effective" theory, in the low-energy scale. Perhaps, if we discover the "master" theory which we could apply also to very high energies, everything will be platonically simple. Maybe we'll simply find prime numbers at the origin of this theory. Maybe, at the very beginning, nature is very simple, and complexity is just an effective phenomenon.

- *Zichichi*:
Very interesting. In fact I'm the author of the "Platonic Grand Unification". The point is that the experimentally observable quantities needed to conclude that we have complexity, are two, as I

have been able to understand from many talks about complexity in many fields: the AFB (Anderson-Feynman-Beethoven) phenomena, and the Sarajevo-type events, the so-called UEEC (Unexpected Events with Enormous Consequences). These are the two experimentally observable quantities. So, if you are in a field like social science, biology or anything else, and you have these two quantities, you can conclude that it is a complex field, a field dominated by complexity. In fact, there are 70 definitions of complexity, of which I have selected 17. One of them is due to Professor 't Hooft. Once you carry on this analysis in physics, you can conclude that our field is a complex one, because it is dominated by UEEC events. There are many examples of unexpected events, as well as examples of the Anderson-Feynman-Beethoven phenomenology. The most famous of the latter is the behavior of protons and neutrons: they are completely blind to the fact that they are made of quarks and gluons, they behave like two single individual particles. Our fathers were studying nuclear physics because these particles behave as if nuclear physics was a fundamental law of nature. The European Centre for Nuclear Research (CERN) is called this way because when it was founded nobody could imagine that the fundamental force was QCD, from which nuclear forces are generated. But the transition between the two is still an open issue.

- *Vikman*:
Would you agree that our field is infinitely complex due to the Gödel's incompleteness theorem?

- *Zichichi*:
This is not what I can say. Professor Douglas and many other people are engaged in this field. Two years ago there was a lecture by Professor Douglas and Denef, who analyzed the most famous model of Busso and Polchinski and proved that it was NP-complete. It is a very hot topic which has being investigated, in terms of whether what we call theoretical physics is NP-complete or not. These are not new issues. Professor Gell-Mann himself realized this much earlier than anyone else, and for this reason he decided to study complexity. He said that if this is the case, then we are lost. Wait and see.

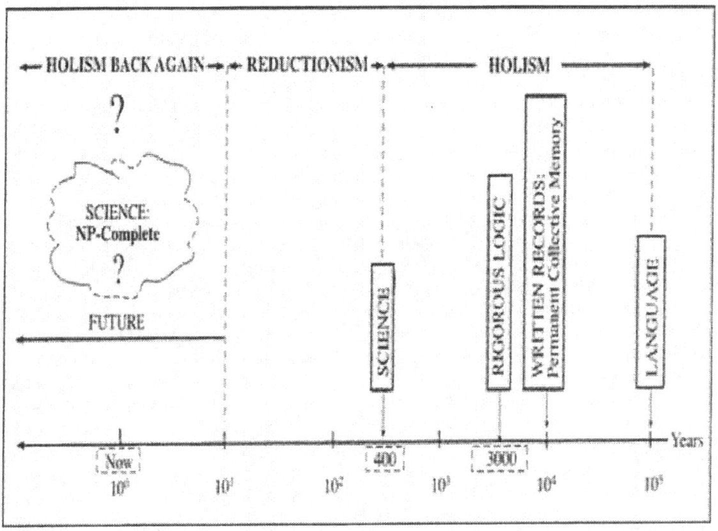

This diagram depicts the battle that I've once undertaken with philosophers and all the people who drew the attention of decision-making leaders to the fact that our field is now over. What they say is that science is a reductionistic approach to the study of nature. Namely, take one problem, solve it, but don't put many problems together. They claim that we should stop doing it, that we need holism. In this way they forget that for ten thousand years mankind has been warped in holism, without understanding anything. Science has first to state that the reductionistic approach is the

most incredible success of human intellect. And the question that is now open for competition is whether science, i.e. physics, is NP-complete.
A theory is NP-complete when it has no predictions, when it is non-polynomial. For example, today we learned that elementary particles could be black holes; this is a fantastic new point. We derive from scale invariance, but we live in a non-scale-invariant world. These are fundamental problems which are at the centre of our attention, and these are problems that should be understood by the public. We should not forget that we need public support. Besides, it is not just Professor Douglas who's studying this. Professors Denef and Douglas are working on this open problem, they are the two people who proved that the Busso-Polchinski model is NP-complete.

- *Tapia Takaki*:
One of the motivations in heavy-ion physics is to try to understand the restoration of chiral symmetry, the deconfinement, and how low-mass quarks get their masses. I remember that during the lecture you gave last Sunday you mentioned that there are still things that are not clear to you concerning the Higgs mechanisms. I'm talking about this because qualitatively this chiral symmetry restoration is quite similar to the Higgs mechanism. Do you consider this to be a fundamental problem?

- *Zichichi*:
The specialist is Professor 't Hooft. He used this mechanism to solve a big problem. My personal feeling is that the Higgs field all over the world cannot be there; this does noit mean that the Higgs particledoes not exist. The Higgs field appears to me like being the Dirac Sea. Where is the Dirac Sea? In matter physics, not in fundamental physics.

- *'t Hooft:*
Except for that the Higgs particle is a boson of course. I think most of us believe that there exists a particle of that sort. A scalar particle, or more than one scalar particle, and it doesn't have to be a fundamental particle, it could be a composite object.

- *Zichichi*:
If there are more than one scalar particle, I would be delighted.

- *'t Hooft:*
It doesn't have to be a fundamental particle, it could be a composite object.

- *Zichichi*:
While listening to your lecture I received a paper which I did not have time to read, where it was announced that FermiLab defeats about the Higgs particle. I could not read it since I wanted to carefully listen to your very exciting new theoretical approach to understand nature which starts from scale invariance and allows us to live in a non-scale-invariant world.

Highlights from the Relativistic Heavy Ion Collider

Paul Richard Sorensen

Brookhaven National Laboratory, Upton, New York 11973, USA

Abstract

At the Relativistic Heavy Ion Collider (RHIC) at Brookhaven National Laboratory, heavy nuclei are collided at high energies in order to create matter that is hot and dense enough to dissolve hadrons into a quark-gluon-plasma (QGP). In the two lectures summarized in these proceedings, I presented a basic introduction to heavy-ion collisions and presented the highlights from the first decade of RHIC results.

1 Introduction

The motivation for the construction of the Relativistic Heavy Ion Collider at Brookhaven National Laboratory was to collide together nuclei with large atomic numbers in order to form a state of matter called the Quark Gluon Plasma(QGP) [1, 2]. These collisions can deposit relatively large amounts of energy into a region the size of a nucleus. The matter left behind in that region is so hot and dense that hadronic matter undergoes a phase transition into a form of matter where quarks and gluons are the relevant degrees of freedom, not hadrons [3]. This is the state of matter that existed when the universe was young (less than a microsecond old) and very hot.

The first clues to the existence of a transition from hadronic matter to a different state of matter predate the discovery of quarks and the advent of Quantum Chromodynamics. Even before the quark model was proposed, Hagedorn noticed that the spectrum of hadronic states (shown in figure 1) grows exponentially with mass; that is, the number of hadronic states with a given mass, increases exponentially as the mass goes up. Hagedorn realized that this feature of the hadronic mass spectrum implies that there is a maximum temperature in a hadronic gas [4]. As we put more energy into the hadron gas, the energy can go into creating higher mass states instead of increasing the temperature. Based on this observation, one could wonder, what existed in the universe when it was hotter than the limiting hadronic temperature? The answer to that question is provided by the theory of Quantum Chromodynamics.

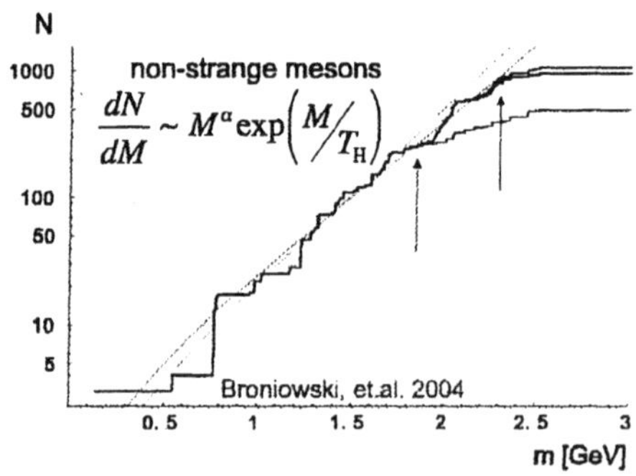

Figure 1: The hadronic mass spectrum; the number of hadronic states increases exponentially with mass. This exponential increase implies a maximum temperature for hadronic matter called the Hagedorn temperature. Figure from Ref. [5]

With the discovery of quarks and gluons, the constituents of hadrons and the development of QCD, the theory that describes the interaction of quarks and gluons, it becomes clear what happens above Hagedorn's limiting temperature for hadronic matter. That matter undergoes a phase transition into a plasma of color charged quarks and gluons called the quark gluon plasma (QGP) [6]. The theory of QCD has been tested and verified in high energy, short wavelength, perturbative limit but the QGP is dominated by non-perturbative interactions. In this case, we rely on computationally intense Lattice QCD calculations to theoretically study the transition from hadronic matter to QGP [3]. Figure 2 shows a Lattice QCD calculation of energy density scaled by the fourth power of temperature versus the temperature. This quantity is proportional to the number of degrees of freedom. As expected, as the temperature is increased, the number of degrees of freedom begins to rapidly increase. This is the effect expected from the rise in the number of more massive hadronic states that become accessible at higher energies. Then, as the energy continues to grow, the number of degrees of freedom levels off and nearly saturates. That saturation is a clear demonstration of a transition into a QGP where the number of degrees of freedom is determined by the number of quarks and gluons, not by the ever increasing number of hadronic states. These calculations demonstrate that QGP is a state of matter described by QCD. The challenge then becomes to create that state and to study it's properties. That is the task undertaken by RHIC and other heavy ion experimental facilities [7].

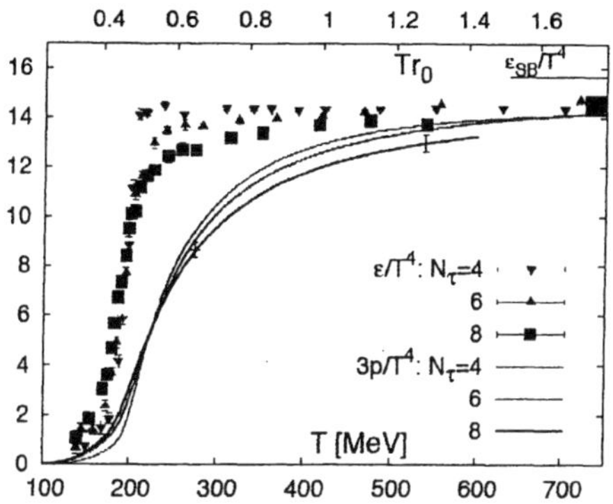

Figure 2: Lattice calculations of the energy density scaled by T^4 showing the transition to QGP [3]. The energy density of a hadron gas would grow to infinity as the temperature approached the Hagedorn temperature but instead, the hadronic matter dissolves into QGP.

2 RHIC Collisions

The creation of a QGP at RHIC is typically initiated by colliding together Gold nuclei at $\sqrt{s_{nn}} = 200$ GeV, the maximum center of mass energy possible for those species at that facility. The RHIC facility is flexible however and can collide a range of nuclei from single protons up to even heavier elements like Uranium 238 and with energy ranges from approximately 6 GeV to 200 GeV (or even up to 500 GeV for protons). With this broad flexibility and with two large detector systems (the STAR and PHENIX detectors), RHIC is the most ideal facility in the world to study the transition from hadronic matter to QGP.

Figure 3 (left) shows a Monte-Carlo Model of the collision of two highly Lorentz contracted Gold nuclei [8]. For collisions at 200 GeV, the Gold nuclei are contracted along their direction of motion by a factor of 100. The illustration shows the protons and neutrons distributed within each nucleus — those that are

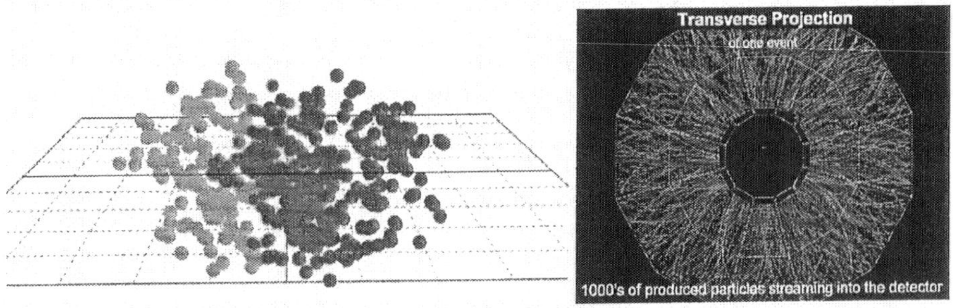

Figure 3: Transverse view of the initial overlap region and a transverse view of the particles streaming from the collision into the detector.

expected to participate in the collision are shown in red. On the right, is an image from the STAR detector [9] showing the tracks left behind by thousands of charged particles produced in a 200 GeV Au+Au collision. From these tracks, we attempt to infer the properties of the tiny speck of matter that briefly existed after the collision. The size of the matter is approximately the size of a Au nucleus (10^{-14} meters or 10 fm), and it exists for $t \approx 20 fm/c$ ($\approx 10^-22$ seconds) before the interactions between the fireball's constituents cease and hadrons freely stream to the detectors. Since it is impossible to directly observe the matter that is formed, we look at the abundances, and distributions of various particles species, and the patterns with which they are emitted to infer the properties of the matter.

3 RHIC Highlights

3.1 Particle Ratios and Chemical Freeze-out

One of the firsts questions asked is whether regions of the fireball were locally able to achieve equilibrium before the system blasted apart. We specify *local* because causality dictates that the system cannot be completely equilibrated until it has lived for $c_s \tau = R$, where c_s is the speed of sound in the medium and R is the size of the system. Evidence for local thermal equilibration can be found by investigating the ratios of various particle species. A thermal model can be constructed to predict these relative abundances based on simple thermodynamic arguments [11]. For a thermally equilibrated system, the relative abundances of particles should only depend on the temperature of the system, the chemical potentials of the quarks and the masses of the particles. Figure 4 shows a fit of such

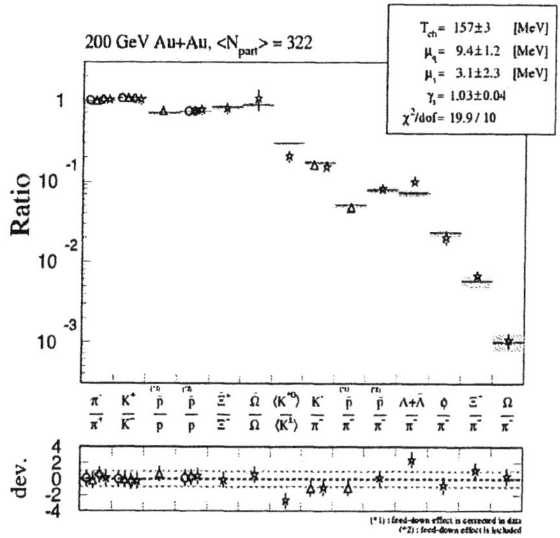

Figure 4: The ratios of various particle species produced in heavy-ion collisions compared to expectations of thermal equilibrium [10].

a model to the particle ratios measured at RHIC. The fit demonstrates that particle production from 200 GeV Au+Au collisions is consistent with that expected from an equilibrated medium, thus satisfying a basic requirement for the formation of a state of matter.

3.2 Anisotropy from Coordinate to Momentum Space

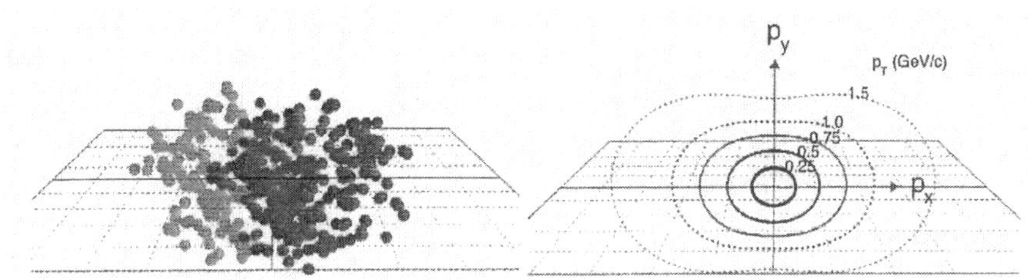

Figure 5: Conversion from coordinate space into momentum space.

If interactions in the fireball are strong enough to create a locally thermalized medium, then we may expect the interactions to influence the patterns with which particles are emitted. Typically, the region where the two nuclei overlap is not isotropic; for non-central collisions, it possesses an elliptic shape. If particles free stream from that region, the elliptic shape will not be transferred into momentum space, and the particles will be emitted isotropically. Interactions on the other hand will lead to anisotropies in the momentum space distributions of particles. In particular, if we consider the possibility of pressure gradients in the system, we see that the largest pressure gradient will be along the short axis of the elliptic overlap region (see figure 5). That is the axis where the density goes most abruptly from the maximum at the center to the minimum at the edge. If pressure gradients are acting on the emitted particles, that anisotropy in the pressure gradient will lead to an elliptic anisotropy in the momentum space distribution of emitted particles (as shown in the right panel of figure 5) [12]. That elliptic anisotropy is quantified by v_2, the second term in a Fourier Series expansion of the angular dependence of particle emission in the transverse plane, where the azimuthal angles are taken relative to the axis connecting the two colliding nuclei. The right panel of figure 5 illustrates how anisotropic the distributions are for particles emitted with different transverse momenta. In semi-central Au+Au collisions at 200 GeV, there are nearly twice as many 1.5 GeV particles emitted in the direction of the short axis of the overlap region as there are in the direction of the long axis: note this ratio will go as $\frac{dN_{short}}{dN_{long}} = \frac{1+2v_2}{1-2v_2} \approx 1+4v_2$ so that $v_2 = 0.25$ corresponds to $\frac{dN_{short}}{dN_{long}} \approx 2$. This is a remarkably large deviation from expectations of free-streaming that indicates the fireball maintains a memory it's original geometry through interactions.

3.3 Elliptic Flow and Quark Number Scaling

Measurements of v_2 for different particle species at different transverse momenta are shown in figure 6 [12]. In the region below $p_T \sim 2$ GeV/c, v_2 follows mass ordering with heavier particles having smaller v_2 at a given p_T. Above this range, the mass ordering is broken and the heavier baryons take on larger v_2 values. A hydrodynamic model for $v_2(p_T)$ is also shown which describes the v_2 in the lower p_T region well [13]. The hydrodynamic model assumes local thermal equilibrium and zero mean-free-path for interactions. The mass ordering in the model and data are a feature expected for particle emission from a boosted source. In the case that particles move with a collective velocity, more massive particles will receive a larger p_T kick. As the particles are shifted to higher p_T, the lower momentum regions become depopulated with a larger reduction in the direction with

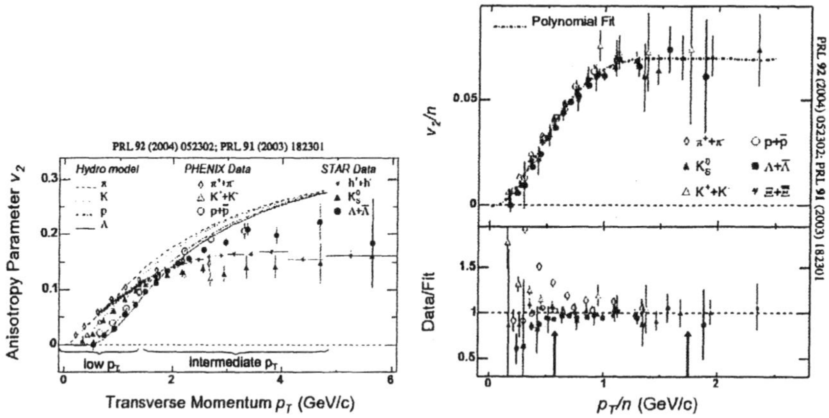

Figure 6: The momentum-space anisotropy and the same scaled by the number of quarks in each hadron.

the largest boost (in-plane). This reduction reduces v_2 at a given p_T, with the reduction largest for more massive particles, as observed in the data. This is clear evidence for a fireball expanding under the influence of anisotropic pressure gradients. The successful description of the v_2 data in the momentum range where 99% of the particles are produced with a ideal hydrodynamic models led physicists at RHIC to dub the matter a perfect liquid [14]. In fact preliminary estimates of the viscosity of the QGP created at RHIC indicate that it is lower than any known substance and consistent with the lowest value possible [15].

At higher p_T, v_2 no longer rises with p_T and the mass ordering is broken. Above $p_T \sim 2$ GeV/c the more massive baryons exhibit a larger v_2 than the mesons. While the pion and kaon v_2 reach a similar maximum of $v_2 \approx 0.14$ at $p_T \approx 2.5\ GeV/c$, the baryon v_2 continues to rise until it reaches a maximum of $v_2 \approx 0.20$ at $p_T \approx 4.0$ GeV/c. For still larger p_T, the v_2 values exhibit a gradual decline until v_2 for all particles is consistent with $v_2 \approx 0.10$ at $p_T \approx 7$ GeV/c (not shown in this figure). At these higher p_T values one expects that the dominant process giving rise to v_2 is jet-quenching [16] where hadron suppression is larger along the long axis of the overlap region than along the short axis [17, 18, 19]. Jet quenching occurs as fast moving partons radiate energy in the expanding medium. This leads to a dramatic suppression of high momentum particles which will be discussed later. For very large energy loss, the value of v_2 should be dominated by the geometry of the collision region.

At an intermediate p_T range, it was noted that mesons and baryons spread out into two bands with the generally heavier baryons developing a larger v_2, contrary to the behavior in the lower momentum region. It was also noticed that if the v_2 and p_T of the hadron were scaled by the number of constituent quarks in the hadron, the v_2/n vs p_T/n of various species all lined up with one another [20]. This scaling is shown in the right panel of figure 6. Figure 7 shows that this grouping of mesons and baryons extends to the heavier, multiply strange hadrons, the ϕ-meson and Ω-baryon. The data in figure 7 is important because it shows that although the ϕ-meson is as heavy as the proton, it behaves like other mesons. Indeed, the number of constituent quarks seems to be more relevant than the mass of the hadron.

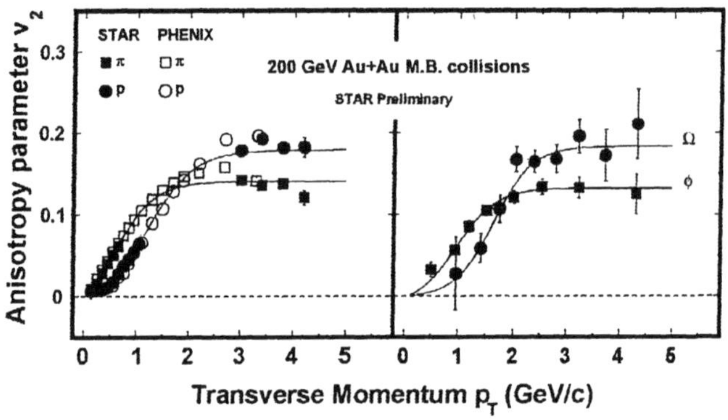

Figure 7: Elliptic flow of strange and multiply strange hadrons showing that the quark number scaling also holds for particles with larger masses and that the ϕ-meson groups with other particles having the same number of constituent quarks, rather than with particles having similar masses like protons.

3.4 The Baryon Enhancement

The observation of the quark-number dependence of v_2 at intermediate p_T and the constituent quark number scaling led to speculation that hadron formation was happening via the coalescence of dressed quarks at the hadronization phase boundary, leading to an amplification of v_2 with baryons getting amplified by a factor of 3 while mesons were amplified by a factor of 2[21, 22, 23, 20, 24, 25, 26, 27, 28, 29, 30, 20]. This picture was subsequently strengthened by the observation that a similar quark-number dependence arises in R_{CP}[31, 32]: the ratio of the single particle spectra in central collisions to that in peripheral collisions. At intermediate p_T the R_{CP} values for various particle species are also grouped by the number of constituent quarks, with baryons having a larger R_{CP}. The larger R_{CP} for baryons signifies that baryon production increases with collision centrality faster than meson production; an observation consistent with the speculation that hadrons from Au+Au collisions are formed by coalescence such that baryon production becomes easier as the density of the system increases. The more general and less model dependent statement is that the baryon versus meson dependence arises from high density and therefore most likely from multi-quark or gluon effects or sometimes called "higher twist" effects. The combination of large baryon v_2 and large baryon R_{CP} also immediately eliminates a class of explanations attempting to describe one or the other observation: *e.g.* originally it was speculated that the larger R_{CP} for baryons might be related to a smaller jet-quenching for jets that fragment to baryons than for jets that fragment to mesons. This explanation would lead to a smaller baryon v_2 and is therefore ruled out by the larger v_2 for baryons. The same can be said for color transparency models[33] which would account for the larger baryon R_{CP} in this p_T region but would predict a smaller baryon v_2. Color transparency may still be relevant to the particle type dependencies at $p_T > 5$ where R_{CP} for protons is slightly larger than R_{CP} for pions[34] and the v_2 measurements are not yet precise enough to conclude whether the baryon v_2 is also smaller than the meson v_2. This is a topic that needs to be studied further. Figure 8 shows baryon to meson ratios for p+p collisions, mid-central Au+Au collisions, and central Au+Au collisions. At intermediate p_T values, the number of baryons produced is similar to the number of mesons. v_2 measurements also demonstrate that baryons are more abundantly produce along the short axis of the overlap region rather than perpendicular to it. The two observations are directly related to each other.

3.5 Our Standard Picture

The data presented above along with a large amount of other observations has led to a standard picture of heavy-ion collisions which is summarized in the diagram in figure 9. Energy is deposited in the overlap region of the two colliding nuclei. That matter reaches local thermal equilibrium in a surprisingly short time (about 1 fm/c) and the quark-gluon-plasma phase of the expansion begins. The QGP expands and cools with cooler regions passing through the phase transition to hadronic matter. Eventually, all of the QGP has converted to hadronic matter. At some point, inelastic interactions cease and the particle ratios are fixed. This is

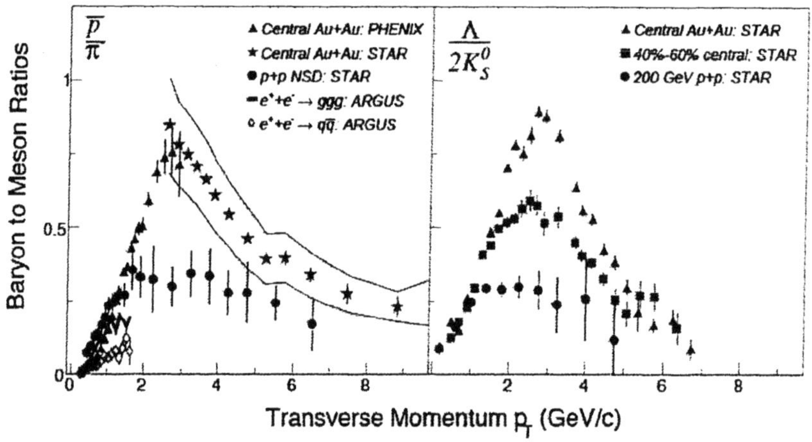

Figure 8: Baryon to meson ratios in 200 GeV Au+Au and p+p collisions. The data shows that the denser Au+Au collisions produce more baryons relative to mesons than the less central Au+Au collisions or p+p collisions.

called chemical freeze-out. Finally the elastic interactions cease and the momentum of the particles cease to change. This is called kinetic freeze-out. After this, the particles stream freely to the detector.

3.6 Correlations and Fluctuations

It is apparent from the v_2 measurements that the fireball converts features of the initial overlap geometry into momentum space. The initial overlap region, however, is neither a perfect sphere nor a perfect ellipse. It is expected to contain regions of high and low energy density. One can ask if the fireball transfers those features into the final momentum space distributions. If so, then they should show up as correlations between produced particles. There are many conjectured processes that can lead to correlations between the produced particles; resonance decays, fragmentation of jets, bubble nucleation or spinodal decomposition at a phase boundary, and quantum interference are some of the sources of correlations frequently considered. Of all these correlations, only correlations from the very early stages of the evolution can spread out very far in the longitudinal direction. The reason for this is illustrated in figure 10. The left of the figure shows the space time picture of a heavy ion collision. The trajectories of the incoming nuclei and

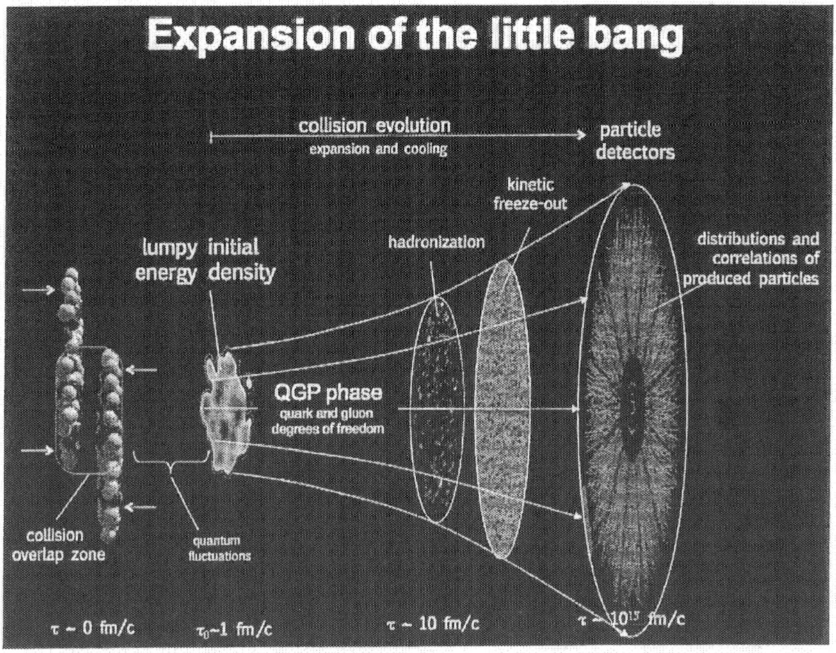

Figure 9: A sketch of the expansion of a heavy ion collision.

the outgoing nucleons moving near the speed of light are shown as thick grey lines. The horizontal axis shows the longitudinal rapidity y and the vertical axis time. Causality dictates that correlations at later times, cannot spread out far in rapidity [35]. Only correlations from early in the evolution can stretch far in rapidity. One way to look for correlations left over form the lumpiness in the initial overlap region, is to look for long range correlations.

The left panel of figure 11 shows measurements of two particle correlations where particles are required to have momentum above 4 and 2 GeV/c [37]. These selections are usually made to search for correlations arising from jets. A jet will lead to correlations that are narrow in the azimuthal and longitudinal direction. Such a correlation is apparent in the peak at $\Delta\eta$ and $\Delta\phi$ near zero where η is the longitudinal pseudo-rapidity variable. In addition though, one notices a correlation which extends in the $\Delta\eta$ across the entire acceptance of the detector. This ridge-like correlation structure has the characteristics one would expect from lumpiness in the initial overlap zone.

The middle panel shows two particle $p_T p_T$ correlations between all observed

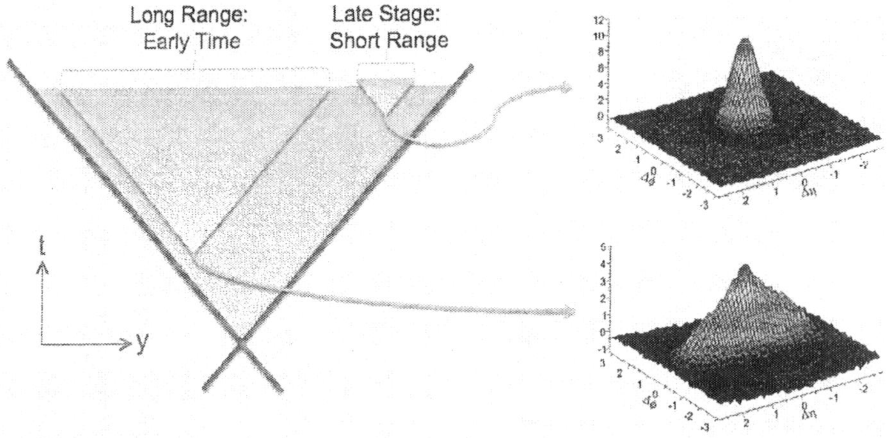

Figure 10: The relationship of long and short range correlations to early and late times [36].

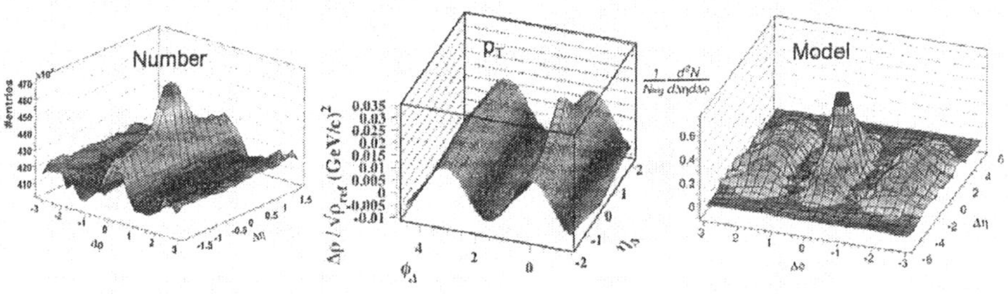

Figure 11: Measured and calculated long range correlations.

hadrons (with no p_T selection) [38]. This data shows the correlations expected from v_2 (a $\cos 2\Delta\phi$ modulation) with an additional peak centered at $\Delta\eta$ and $\Delta\phi = 0$ which is also broadened longitudinally. The model calculations shown in the right panel are from a hydrodynamic model with a lumpy initial overlap density as the initial conditions [39]. The granular nature of the initial conditions clearly show up after the hydrodynamic expansion as correlations similar to those seen in the data. This comparison therefore confirms the standard picture of heavy-ion collisions described above.

Correlations and fluctuations can also develop at the phase boundary between QGP and hadronic matter. If the phase transition is second or first order we would expect the correlations to be larger for larger systems. These correlations and fluc-

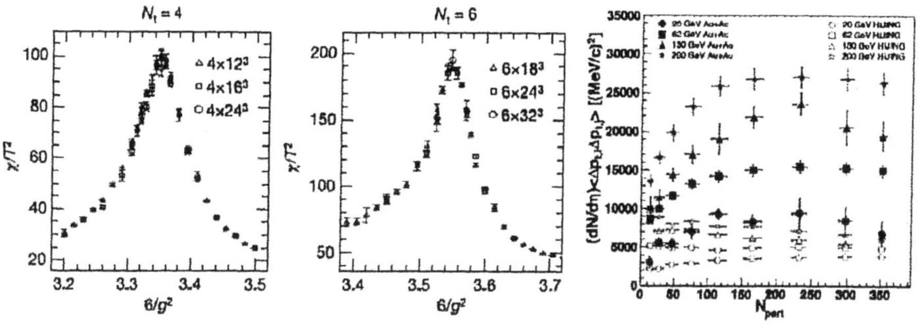

Figure 12: The size dependence of susceptibilities in lattice calculations [40] and the measured pt fluctuations in data [41].

tuations are related to the susceptibilities in the system. Lattice QCD calculations show that the magnitude of the peak in the susceptibilities at the phase boundary does not change when the size of the lattice changes. This indicates that the phase transition in QCD at zero baryon density is a smooth cross-over. In data, as discussed in the previous paragraph, the measured correlations and fluctuations seem to be dominated by the initial conditions rather than by correlations and fluctuations from the phase boundary. The right panel of figure 12 shows $p_T p_T$ fluctuations for several different energies vs an estimate of the number of nucleons participating in the collision. Above about 100 participants, the fluctuations are independent of the system size. This is consistent with the smooth crossover from QGP to hadron gas as expected from Lattice QCD calculations.

3.7 The RHIC Critical Point Search

Although Lattice calculations show that the QGP to hadron phase transition is a smooth crossover at zero baryon density, it is thought to be a first-order phase transition at higher baryon densities [42]. Lattice calculations suffer from the sign problem at non-zero baryon chemical potential so it is not easy to infer from QCD where the transition from a smooth crossover to a first order phase transition occurs. The point where the smooth crossover and first order phase transition meet is called a critical point and represents an important landmark on the phase diagram of nuclear matter. Figure 13 shows a schematic diagram of the phases of QCD. Full energy heavy-ion collisions at RHIC and those that will be upcoming at the LHC, are nearly net baryon free and probe along the left axis of the dia-

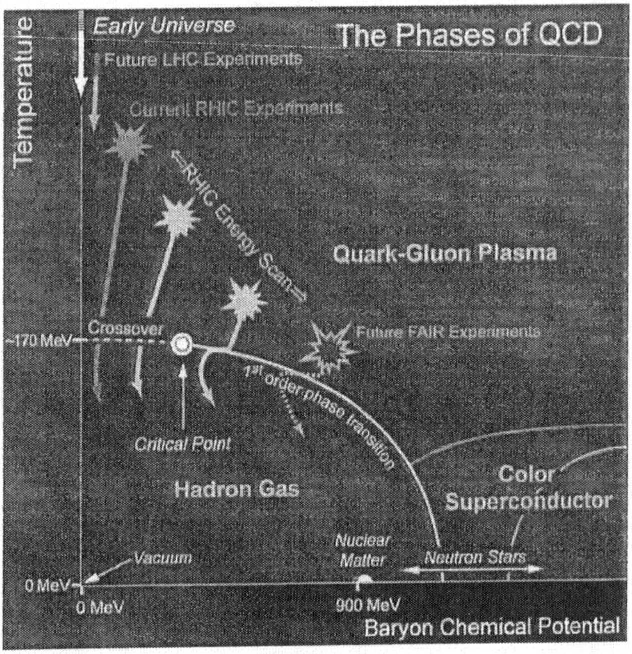

Figure 13: The phase diagram of nuclear matter.

gram. The flexibility of RHIC makes it possible to probe deeply into the non-zero baryon chemical potential axis. By lowering the energy of the colliding beams, one initiates a system with larger net baryon density. By scanning down to lower and lower energies, the RHIC experiments can search for evidence of a critical point or of a first order phase transition. This program is already underway at RHIC [43].

3.8 The Temperature of the Plasma

In order to map out the regions of the phase diagram probed by heavy-ion collisions, it's important to estimate the temperature of the QGP when it is first formed. There are several ways to do this including by attempting to measure the color of the photons emitted from the source. This method requires disentangling various sources of photons and then inferring the temperature from the slope of the p_T spectra of the thermal photons. Recent studies indicate an inverse slope consistent with T=220 MeV [44], which is just above the expected transition temperature of

165 MeV. The photon data is also consistent with hydrodynamic models which start with temperatures in the range from 300 MeV to 600 MeV. This suggests again, that indeed we are probing above the phase transition.

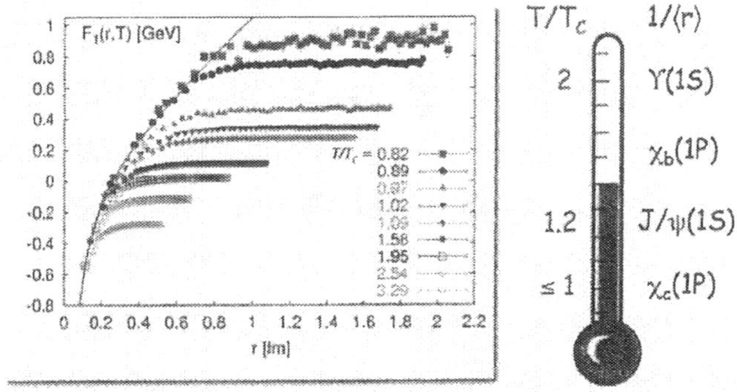

Figure 14: Quarkonium as a tool to measure the temperature. The left panel shows screening of the free energy for heavy quark and anti-quark pairs at finite temperature. The sequential melting of states can provide a thermometer for the QGP [50, 51].

Another valuable tool for determining the temperature of the produced matter is the Debye screening of heavy quark pairs (Quarkonium). The idea is that in a hot and dense deconfined QGP, the abundant light quarks can screen the force acting to bind the heavy quark pairs in a quarkonium state [45]. The expectation of screening has been confirmed in lattice QCD calculations [46, 47]. Figure 14 shows the free-energy between two heavy quarks placed inside a bath of light quarks. The zero temperature free energy is shown as a black curve. As the temperature is increased the free energy drops further and further from the zero temperature case. At first the deviation is only at large distance, but as the temperature is increased, the screening sets in at smaller and smaller distances. This means that the large, loosely bound, quarkonium states will melt at low temperatures while the more tightly bound states will survive until hotter temperatures are reached. This sequential screening provides and effective thermometer for determining the temperature of the QGP by observing which Quarkonium states survive and which were unable to form in the QGP [48, 49]. Measurements show that the J/ψ is suppressed at RHIC [52]. Measurements of higher excited states and the Upsilon still await more precise measurements.

3.9 The Opacity of the Plasma

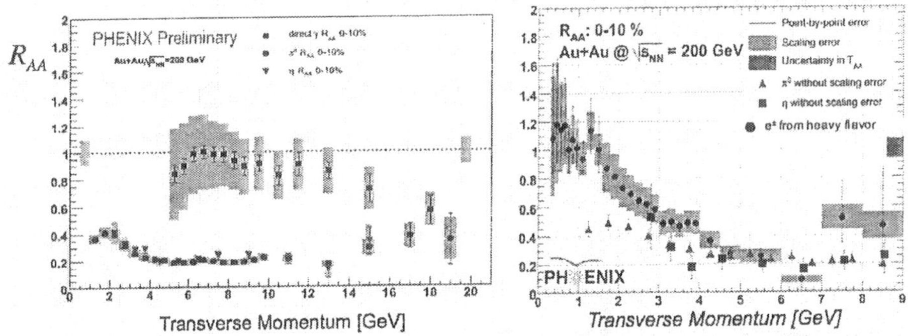

Figure 15: Suppression of high momentum particles do to the opacity of the QGP. The left panel shows that while high momentum pions and η-mesons are suppressed by a factor of 5, the color neutral photons are not.

Another consequence of the formation of a hot and dense medium is jet quenching [17, 18, 19]. In a heavy ion collision, the hard scattered partons that form jets must traverse the hot and dense matter before they escape and hadronize in the vacuum. The time-scales require that hard processes occur early in the collision and the soft fragmentation process occurs late. As the fast moving parton traverses the dense medium, it can lose energy to the medium and be quenched. Since the number of high momentum particles falls quickly, even a modest loss of energy can result in a large suppression of high momentum hadrons observed in a particular bin. Figure 15 shows the ratio of particles observed in heavy-ion collisions compared to what is expected based on scaling of p+p collisions (R_{AA}) [53, 54, 55, 56]. The left panel shows that the number of pions and η-mesons with momentum above 5 GeV is suppressed by nearly a factor of 5. The figure also shows that photons in the same region are not suppressed, as is expected since they are color neutral and do not lose energy in the QGP.

It is also of quite some interests to note that, as seen in the right panel of figure 15, even electrons coming from the decay of heavy flavor hadrons are suppressed [59]. This indicates that the QGP is able to quench even high momentum heavy-quark jets. Heavy quarks (*e.g.* charm and bottom quarks) where expected to lose less energy to the QGP matter[60, 61]. It's not a priori obvious that heavy quarks will couple significantly to the medium and be influenced by its apparent expansion. The extent to which they do couple to the medium should be reflected in how large v_2 for heavy flavor hadrons becomes and how much the nuclear modification (R_{AA}) deviates from unity. Precision measurements of Heavy Flavor mesons or baryons are not yet available from the RHIC experiments. As a proxy for identifying D-mesons, the STAR and PHENIX experiments have measured non-photonic electrons[57, 58]. Non-photonic electrons are generated from the weak-decays of heavy flavor hadrons and after various backgrounds have been accounted for can, with some caveats[62], be used to infer the R_{AA} and v_2 of D-mesons.

The right panel of figure 15 shows R_{AA} for non-photonic electrons[63, 59]. Prior to the measurement of non-photonic electron R_{AA}, it was expected that heavy-flavor hadrons would be significantly less suppressed than light flavor hadrons. These expectations based on a decrease in the coupling of charm quarks to the medium because of the dead-cone effect[60], are contradicted by the data; At $p_T \approx 5$ GeV/c, non-photonic electrons are as suppressed as pions. This suppression suggests a stronger than expected coupling of charm quarks to the medium. This coupling apparently also leads to significant v_2 for non-photonic electrons. The QGP created at RHIC turns out to be exceptionally opaque, even to heavy-quark jets.

Evidence of the opacity of the QGP is also evident in the correlations between high momentum particles [64, 65]. These measurements serve as a stand-in for jet studies which are very difficult to carry out in heavy-ion collisions and are just not coming to fruition. The two particle correlations however, already demonstrate the opacity of the QGP to jets. The figure shows the angular difference between an 8 GeV hadron and a lower momentum associated hadron which satisfies the p_T cut specified in the figure. The left panels of figure 16 show the results for d+Au collisions while the center and right panels show the results for mid-central and central Au+Au respectively. The peak near zero, is caused by the correlation between two particles coming from the same parton. The peak at π is caused by the correlation between a hadron from a jet and a hadron from the away-side jet created in the original scattering process. For d+Au collisions, the away-side peak contains more pairs than the near-side peak. For central Au+Au collisions the away-side peak is suppressed. This suppression is consistent with an opaque medium. When looking for high momentum particles, we have a preference for finding particles originating from jets near the surface of the medium. When we look at the away-side, we see the correlations from the jet that needed to traverse the medium before escaping. The away-side jet is shown to be quenched in the opaque medium.

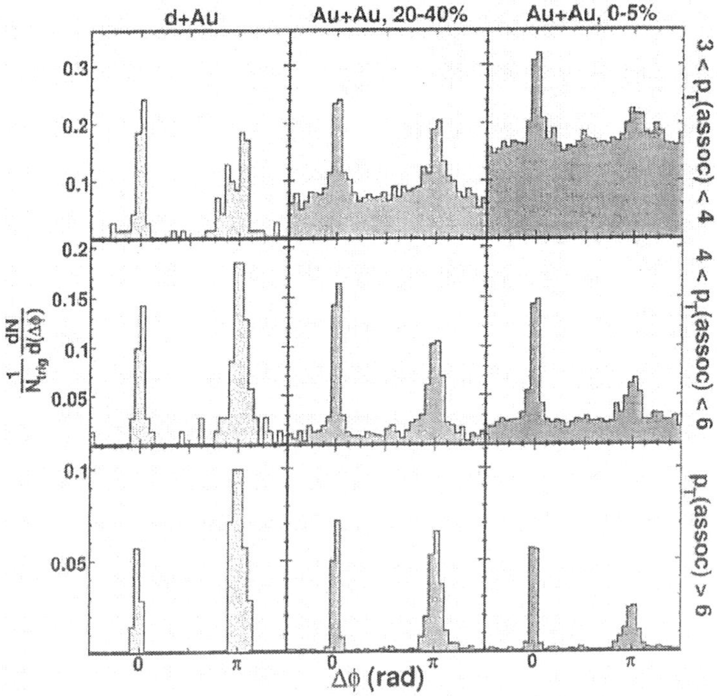

Figure 16: Jet-like correlations between the remaining high momentum hadrons.

4 Conclusion

Studies at RHIC have revealed that the collision of Au+Au nuclei at ultra-relativistic energy creates a hot and dense QGP that is opaque to fast moving color charges, acts like a low viscosity liquid, and transforms into hadronic matter through a smooth cross-over. Future studies will search for a first order phase transition at higher baryon chemical potentials, and attempt to better determine the properties of the medium. RHIC is ideally suited for these studies and continues to provide fascinating insights on the nature of QCD matter.

References

[1] W. Reisdorf and H. G. Ritter, Ann. Rev. Nucl. Part. Sci. **47**, 663 (1997).

[2] N. Herrmann, J. P. Wessels and T. Wienold, Ann. Rev. Nucl. Part. Sci. **49**, 581 (1999).

[3] C. Schmidt, PoS C **POD2006** (2006) 002 [arXiv:hep-lat/0701019]; F. Karsch, PoS C **POD07** (2007) 026 [arXiv:0711.0656 [hep-lat]]; PoS **LAT2007** (2007) 015 [arXiv:0711.0661 [hep-lat]].

[4] R. Hagedorn, Statistical Thermodynamics of Strong Interactions at High Energies, Nuovo Cim. Suppl. 3, 147186, 1965.

[5] W. Broniowski, W. Florkowski and L. Y. Glozman, Phys. Rev. D **70**, 117503 (2004) [arXiv:hep-ph/0407290].

[6] E. V. Shuryak, Phys. Rept. **61**, 71 (1980).

[7] I. Arsene *et al.* [BRAHMS Collaboration], Nucl. Phys. A **757**, 1 (2005); B. B. Back *et al.* [PHOBOS Collaboration], Nucl. Phys. A **757**, 28 (2005); J. Adams *et al.* [STAR Collaboration], Nucl. Phys. A **757**, 102 (2005); K. Adcox *et al.* [PHENIX Collaboration], Nucl. Phys. A **757**, 184 (2005).

[8] M. L. Miller, K. Reygers, S. J. Sanders and P. Steinberg, Ann. Rev. Nucl. Part. Sci. **57**, 205 (2007) [arXiv:nucl-ex/0701025].

[9] STAR Collaboration, C. Adler *et al.*, Nucl. Instr. Meth. A **499**, 624 (2003).

[10] O. Y. Barannikova [STAR Collaboration], arXiv:nucl-ex/0403014.

[11] P. Braun-Munzinger, K. Redlich and J. Stachel, arXiv:nucl-th/0304013.

[12] K. H. Ackermann *et al.* [STAR Collaboration], Phys. Rev. Lett. **86**, 402 (2001); C. Adler *et al.* [STAR Collaboration], Phys. Rev. Lett. **89**, 132301 (2002); J. Adams *et al.* [STAR Collaboration], Phys. Rev. Lett. **92**, 052302 (2004); J. Adams *et al.* [STAR Collaboration], Phys. Rev. C **72**, 014904 (2005); J. Adams *et al.* [STAR Collaboration], Phys. Rev. Lett. **95**, 122301 (2005).

[13] P. Huovinen, P. F. Kolb, U. W. Heinz, P. V. Ruuskanen and S. A. Voloshin, Phys. Lett. B **503**, 58 (2001); D. Teaney, Phys. Rev. C **68**, 034913 (2003).

[14] B. Muller, Acta Phys. Polon. B **38**, 3705 (2007) [arXiv:0710.3366 [nucl-th]]; W. A. Zajc, Nucl. Phys. A **805**, 283 (2008) [arXiv:0802.3552 [nucl-ex]].

[15] P. Danielewicz and M. Gyulassy, Phys. Rev. D **31**, 53 (1985); P. Kovtun, D. T. Son and A. O. Starinets, Phys. Rev. Lett. **94**, 111601 (2005).

[16] X. N. Wang and M. Gyulassy, Phys. Rev. Lett. **68**, 1480 (1992).

[17] X. N. Wang, Phys. Rev. C **63**, 054902 (2001) [arXiv:nucl-th/0009019].

[18] M. Gyulassy, I. Vitev, X. N. Wang and P. Huovinen, Phys. Lett. B **526**, 301 (2002) [arXiv:nucl-th/0109063].

[19] M. Gyulassy, I. Vitev and X. N. Wang, Phys. Rev. Lett. **86**, 2537 (2001) [arXiv:nucl-th/0012092].

[20] R. J. Fries, V. Greco and P. Sorensen, arXiv:0807.4939 [nucl-th].

[21] S. A. Voloshin, Nucl. Phys. A **715**, 379 (2003) [arXiv:nucl-ex/0210014].

[22] Z. w. Lin and C. M. Ko, Phys. Rev. C **65**, 034904 (2002) [arXiv:nucl-th/0108039].

[23] D. Molnar and S. A. Voloshin, Phys. Rev. Lett. **91**, 092301 (2003) [arXiv:nucl-th/0302014].

[24] R. C. Hwa and C. B. Yang, Phys. Rev. C **67**, 064902 (2003).

[25] R. J. Fries, B. Muller, C. Nonaka and S. A. Bass, Phys. Rev. C **68**, 044902 (2003).

[26] V. Greco, C. M. Ko and P. Levai, Phys. Rev. C **68**, 034904 (2003).

[27] V. Greco, C. M. Ko and P. Levai, Phys. Rev. Lett. **90**, 202302 (2003).

[28] R. J. Fries, B. Muller, C. Nonaka and S. A. Bass, Phys. Rev. Lett. **90**, 202303 (2003).

[29] S. Pratt and S. Pal, Nucl. Phys. A **749**, 268 (2005) [Phys. Rev. C **71**, 014905 (2005)] [arXiv:nucl-th/0409038].

[30] L. Ravagli, H. van Hees and R. Rapp, arXiv:0806.2055 [hep-ph].

[31] J. Adams *et al.* [STAR Collaboration], Phys. Rev. Lett. **92**, 052302 (2004) [arXiv:nucl-ex/0306007].

[32] S. S. Adler *et al.* [PHENIX Collaboration], Phys. Rev. Lett. **91**, 172301 (2003) [arXiv:nucl-ex/0305036].

[33] S. J. Brodsky and A. Sickles, Phys. Lett. B **668**, 111 (2008) [arXiv:0804.4608 [hep-ph]].

[34] B. I. Abelev *et al.* [STAR Collaboration], Phys. Lett. B **655**, 104 (2007) [arXiv:nucl-ex/0703040].

[35] A. Dumitru, F. Gelis, L. McLerran and R. Venugopalan, Nucl. Phys. A **810**, 91 (2008) [arXiv:0804.3858 [hep-ph]].

[36] A. Mocsy and P. Sorensen, Phys. Lett. B **690**, 135 (2010) [arXiv:0908.3983 [nucl-th]].

[37] J. Adams et al. [STAR Collaboration] Phys. Rev. Lett. 95:152301, (2005); F. Wang [STAR Collaboration], J. Phys. G **30** (2004) S1299; J. Adams *et al.* [STAR Collaboration], Phys. Rev. C **73** (2006) 064907; J. Putschke, J. Phys. G **34** (2007) S679; J. Adams *et al.* [Star Collaboration], Phys. Rev. C **75** (2007) 034901; Brijesh Srivastava for the STAR Collaboration, Int. J. Mod. Phys. E16, 3371 (2008); A. Adare *et al.* [PHENIX Collaboration], Phys. Rev. C **78** (2008) 014901; B. Alver *et al.* [PHOBOS Collaboration], J. Phys. G **35**, 104080 (2008).

[38] J. Adams *et al.* [STAR Collaboration], J. Phys. G **32**, L37 (2006).

[39] J. Takahashi, B. M. Tavares, W. L. Qian, F. Grassi, Y. Hama, T. Kodama and N. Xu, arXiv:0902.4870 [nucl-th].

[40] Y. Aoki, G. Endrodi, Z. Fodor, S. D. Katz and K. K. Szabo, Nature **443**, 675 (2006) [arXiv:hep-lat/0611014].

[41] J. Adams *et al.* [STAR Collaboration], Phys. Rev. C **72**, 044902 (2005) [arXiv:nucl-ex/0504031].

[42] J. Cleymans, H. Oeschler, K. Redlich and S. Wheaton, Phys. Rev. C **73**, 034905 (2006) [arXiv:hep-ph/0511094]; F. R. Brown *et al.*, Phys. Rev. Lett. **65**, 2491 (1990); A. Mocsy, F. Sannino and K. Tuominen, Phys. Rev. Lett. **92**, 182302 (2004) [arXiv:hep-ph/0308135]; F. Karsch and M. Lutgemeier, "Deconfinement and chiral symmetry restoration in an SU(3) gauge theory Nucl. Phys. B **550**, 449 (1999) [arXiv:hep-lat/9812023]; M. Asakawa and

K. Yazaki, Nucl. Phys. A **504**, 668 (1989); A. Barducci, R. Casalbuoni, G. Pettini and R. Gatto, Phys. Rev. D **49**, 426 (1994); J. Berges and K. Rajagopal, "Color superconductivity and chiral symmetry restoration at nonzero baryon Nucl. Phys. B **538**, 215 (1999) [arXiv:hep-ph/9804233]; A. M. Halasz, A. D. Jackson, R. E. Shrock, M. A. Stephanov and J. J. M. Verbaarschot, Phys. Rev. D **58**, 096007 (1998) [arXiv:hep-ph/9804290]; O. Scavenius, A. Mocsy, I. N. Mishustin and D. H. Rischke, Phys. Rev. C **64**, 045202 (2001) [arXiv:nucl-th/0007030]; N. G. Antoniou and A. S. Kapoyannis, Phys. Lett. B **563**, 165 (2003) [arXiv:hep-ph/0211392]; Y. Hatta and T. Ikeda, Phys. Rev. D **67**, 014028 (2003) [arXiv:hep-ph/0210284]; M. A. Stephanov, Prog. Theor. Phys. Suppl. **153**, 139 (2004) [Int. J. Mod. Phys. A **20**, 4387 (2005)] [arXiv:hep-ph/0402115]; M. A. Stephanov, K. Rajagopal and E. V. Shuryak, Phys. Rev. Lett. **81**, 4816 (1998) [arXiv:hep-ph/9806219].

[43] P. Sorensen [STAR Collaboration], PoS **CPOD2006** (2006) 019 [arXiv:nucl-ex/0701028]; T. Satogata, *et.al.*, PoS **CPOD2007** (2007) 051.

[44] A. Adare *et al.* [PHENIX Collaboration], Phys. Rev. Lett. **104**, 132301 (2010) [arXiv:0804.4168 [nucl-ex]].

[45] T. Matsui and H. Satz, Phys. Lett. B **178**, 416 (1986).

[46] O. Kaczmarek, F. Karsch, F. Zantow and P. Petreczky, "Static quark antiquark free energy and the running coupling at finite Phys. Rev. D **70**, 074505 (2004) [Erratum-ibid. D **72**, 059903 (2005)] [arXiv:hep-lat/0406036].

[47] K. Petrov [RBC-Bielefeld Collaboration], PoS **LAT2006**, 144 (2006) [arXiv:hep-lat/0610041].

[48] F. Karsch, M. T. Mehr and H. Satz, Z. Phys. C **37**, 617 (1988).

[49] S. Digal, P. Petreczky and H. Satz, Phys. Rev. D **64**, 094015 (2001) [arXiv:hep-ph/0106017].

[50] A. Mocsy and P. Petreczky, Phys. Rev. D **73**, 074007 (2006) [arXiv:hep-ph/0512156].

[51] A. Mocsy and P. Petreczky, Phys. Rev. Lett. **99**, 211602 (2007) [arXiv:0706.2183 [hep-ph]].

[52] A. Adare *et al.* [PHENIX Collaboration], Phys. Rev. Lett. **98**, 232301 (2007) [arXiv:nucl-ex/0611020].

[53] S. S. Adler *et al.* [PHENIX Collaboration], Phys. Rev. Lett. **91**, 072301 (2003) [arXiv:nucl-ex/0304022].

[54] J. Adams *et al.* [STAR Collaboration], Phys. Rev. Lett. **91**, 172302 (2003) [arXiv:nucl-ex/0305015].

[55] S. S. Adler *et al.* [PHENIX Collaboration], Phys. Rev. Lett. **94**, 232301 (2005) [arXiv:nucl-ex/0503003].

[56] S. S. Adler *et al.* [PHENIX Collaboration], Phys. Rev. C **75**, 024909 (2007) [arXiv:nucl-ex/0611006].

[57] S. S. Adler *et al.* [PHENIX Collaboration], Phys. Rev. Lett. **96**, 032301 (2006) [arXiv:nucl-ex/0510047].

[58] B. I. Abelev *et al.* [STAR Collaboration], Phys. Rev. Lett. **98**, 192301 (2007) [arXiv:nucl-ex/0607012].

[59] A. Adare *et al.* [PHENIX Collaboration], Phys. Rev. Lett. **98**, 172301 (2007) [arXiv:nucl-ex/0611018].

[60] Y. L. Dokshitzer and D. E. Kharzeev, Phys. Lett. B **519**, 199 (2001) [arXiv:hep-ph/0106202].

[61] M. Djordjevic and M. Gyulassy, Nucl. Phys. A **733**, 265 (2004) [arXiv:nucl-th/0310076].

[62] P. R. Sorensen and X. Dong, Phys. Rev. C **74**, 024902 (2006) [arXiv:nucl-th/0512042].

[63] S. S. Adler *et al.* [PHENIX Collaboration], Phys. Rev. C **72**, 024901 (2005) [arXiv:nucl-ex/0502009].

[64] C. Adler *et al.* [STAR Collaboration], Phys. Rev. Lett. **90**, 082302 (2003) [arXiv:nucl-ex/0210033].

[65] J. Adams *et al.* [STAR Collaboration], Phys. Rev. Lett. **97**, 162301 (2006) [arXiv:nucl-ex/0604018].

CHAIRMAN: P. SORENSEN

Scientific Secretaries: R. Ichou, H. Perrey

DISCUSSION

- *Preghenella:*

What do you think about trying to see if elliptic flow comes from p-p collisions at the LHC ? Is that something weird?

- *Sorensen:*

No, I don't think it is completely weird. I'm not sure exactly how to do it, it is a little bit difficult. We use p-p as a reference. You can just measure the correlations and then compare that to Au-Au collisions. I wish I had a plot to show you, but maybe I can just draw it. You can draw something, which goes like multiplicity times v_2, as a function of p_T times the pt integrated v2. This is something easy to measure in both p-p, d-Au and Au-Au collisions. What you see in p-p collisions, is something that goes like this, in d-Au like this, and in Au-Au, like that. We assume that the p+p correlations here are due to things like jets and other things that are not necessarily associated with the reaction planes. So I think, to do what you are talking about, one would have to look in very close in details at the shape of these correlations and try to confirm if there is something beyond correlations like jets. I don't know how to define a reaction plane in p-p, beyond just defining your jet axis.

- *Preghenella:*

That's what I was wondering.

- *Sorensen:*

Yes, so I don't have any answer. That is something we should keep thinking about.

- *Noferini:*

I have a question about slide 25. Did you try a similar plot scaling the x-axis for the number of constituent quarks? What do you expect in that case? (Slide 25 shows a plot of Rcp vs pt with a variety of particles plotted.)

- *Sorensen:*

Yes, we checked for example when you look at baryons here and mesons here, if they are following some kind of self-similar behavior. So if you do divide the pt by the number of constituent quarks, then the baryons move over, and the vertical scale is not the same but their p_T dependence is the same. So you do see the same behavior, with respect to the R_{CP}, compared to the v_2. That was one of the first things we checked. So if what you say is that you seem to see the same kind of shapes here, it makes sense that if you divide the pt axis by n, the shapes will come out the same. In the y axis I don't know what to scale by.

- *Noferini:*

But your axis is a ratio between two things pretty different, so it doesn't change if you scale, in fact you don't scale.

- *Sorensen:*

That's right, you don't scale. I don't know how to scale Rcp in this respect.

- *Noferini:*

So it should be able to become similar.

- *Sorensen:*

Yes, when you scale, the curves will lie on the top of each other, but with a different amplitude.

- *Albrecht:*

Can you comment on the possibility to measure P or CP violation? Does this address the strong CP problem?

- *Sorensen:*

No, because this would be local. The strong CP problem refers to the fact that you find that you even though Parity violation is allowed in QCD and expected to be of order 1, it is measured to be less than 10^{-10}. What is talked about in heavy ions collisions, is the possibility that you can have local fluctuations, that happen at short range and for short time, and in heavy ion collisions, you have the possibility of observing these local parity violations. What you have in the collision, as the two nuclei pass each other, you have a lot of charge flowing by You end up with a very strong magnetic field which has been calculated. And for very short time, you have a magnetic field, which is larger than any other magnetic field in the universe. And then it decays very quickly. So the idea is that, if you have something like this, at the time when you have that strong magnetic field, you can get charge separation. You have parity violating regions and in those regions you can get charge separation. This is something you can measure. The problem is that this charge separation is random, from event-to-event, and also from region to region, so you can't average over many events and say for example " I look always only in the top side, and always in the bottom side", because the separation could flip from one to the other, and so you have to define a correlation to do that. The correlation, which is defined in order to measure the parity violation effects, is P even. So it doesn't directly prove that you have seen a P violation, when you measure a signal. What is expected for P violation and C separation is shown in this figure and depends on what you are correlating, positive particles with negative particles, or positive particles with positive particles. So you will expect a different signal, like the ones I show here. In that measurement we have made, I think four years ago the first time, the C sign didn't fit with what people expected in terms of this simple picture of C separation. Later there was a discussion about what the cause might be, and there is a paper that explains what might happen. The authors explanation is that you are going to have these topological bubbles but we talked a lot about the medium opacity, jet quenching etc, so you can't just take and separate C across the whole system, you might have an effect where the charge here might get quenched in the system or might get diffused away. The challenge from there is to try to prove that what you see here is due to P violation, even though it is a P- even signal. And so that requires a lot of work in terms of using different models that don't include P violation to see if you can mach up this signal with some reasonable physics models. So far, you don't find any model, which seems to predict the signal that is seen and so for that reason, it is considered possible evidence for P violation. So this is a new definition, which is not quite fitting with the 5sigma versus 2sigma rules. But since its a P-even observable, the challenge is to eliminate all the reasonable scenarios, and then once you have eliminated a lot of reasonable scenarios, then you can consider P violation as perhaps the leading explanation. And that is the state we are in it. The data will be published very soon, or at least submitted in the journal very soon. I think this is done already by the STAR collaboration, I think that the people in the STAR collaboration are interested to see if somebody can explain this with some other reasonable models. Because if you have a way to explain the signal that is seen, which doesn't involve P violation, and which is a reasonable model, you can kill the signal, but otherwise, P violation, which is a standard part of QCD, seems to show up in heavy ion collisions, through this effect.

- *Cifarelli:*

Concerning the azimuthal oscillations, does the Fourier Transform interpretation rule out the Mach cone effect ?

- *Sorensen:*

No, it is a different interpretation of the data, because the data is the same, actually the Fourier Transform is just a mathematical re-representation of that same correlation. The question is what physics do you interpret, based on that ? I believe those correlation functions. There is a lot of talk about what happens exactly in the v_2 subtraction and things like that. But I do believe that you see these funny correlations. And those shapes are real. But the question is how do you interpret them. I was trying to give another alternative interpretation besides the Mach cone.

- *Korthals-Altes:*

Coming back to the question of P or CP you mentioned – except for the "chiral" effect – there is the idea that there are bubbles inside the plasma defined by a CP odd order parameter that would decay into the thermal plasma ground state and give CP violating effects, again on an event by event basis. In fact Stephene Broneff and myself proposed in 1998 a model that went along similar lines. Have experimentalist looked into this?

- *Sorensen:*

With respect to this theoretical detail I cannot say what the difference is between the mentioned model and the models being discussed by experimentalists.

- *Kovetz:*

Can you explain the relation between WMAP and the superhorizon fluctuations in your work?

- *Sorensen:*

Slide 18 of my presentation shows an analogy between WMAP and the correlation structures that we see. You can take a power spectrum of e.g. the pt,pt-correlations the same way you take the power spectrum of the temperature fluctuations in WMAP. Of course in heavy ion collisions there is no inflation, but nevertheless fluctuations in the initial state do get translated into what we measure. I draw this analogy to tie the correlations we see to something people are familiar with so you do not think of the correlation structure as just jets.

- *t' Hooft:*

You have mentioned the speed of sound several times during your lecture. What would you think might be the most elegant, accurate, and direct way to measure this quantity?

- *Sorensen:*

This is the equation of state which is the "holy grail" we are looking for. But one idea people have discussed to measure the speed of sound is to look at the Mach-cone signal. If in gold-gold collisions there actually is a Mach-cone set off by a jet traversing the matter then you can extract the speed of sound from the resulting correlations. The "hump"-structure in the correlations around pi, shown in the right plot on slide 12 compared to p-Au data, has been interpreted as such a jet setting off a Mach-cone.Another way to determine the equation of state is to look at the final spatial eccentricity of the system in comparison to the initial spatial eccentricity. If the speed of sound drops, the final eccentricity will drop as well. In that way you may learn about whether the equation of state has a soft point.

- *Agols:*

Do you believe that the Mach-cone model is correct?

- *Sorensen:*

No, I don't. The current theoretical calculations of where you actually would see the Mach-cones do not really match the data we measure.

- *Dixon:*

Can you look for "dual Schwinger pair production" of magnetic monopole pairs in the large magnetic fields set up in the collision, via heavily ionizing tracks?

- *Sorensen:*

That would be interesting. It sounds like it would be a clear signal in the detector that should be easy to pick up.

THE LHC AND BEYOND

The Energy Frontier

R.-D. Heuer, CERN, CH-1211 Geneva 23 Switzerland

Abstract

Following an outline of the Large Hadron Collider, this paper will analyze CERN's scientific plans for high-energy colliders at the energy frontier for the years to come. The immediate plans include the upgrades to the Large Hadron Collider and its injectors. This may be followed by a linear electron-positron collider, based on the technology being developed by the Compact Linear Collider or by the International Linear Collider. This paper describes the design of these colliders, all of which have a unique value to add to experimental particle physics, and concludes by outlining key messages for the way forward.

FEATURES OF PARTICLE PHYSICS

Furthering our understanding of key questions in particle physics has relied on the interplay and synergy between various tools, such as accelerators, cosmic-ray detectors and reactors, and at different facilities such as a lepton collider, hadron collider and lepton-hadron collider, at both the energy frontier and the intensity frontier. Of much topical interest presently is the synergy between lepton and hadron colliders in the elaboration of the origin of mass and the search for the Higgs boson. This interplay will continue to be central in extending our knowledge in other areas, such as the unification of forces, the fundamental symmetry of forces and matter, the number of space-time dimensions and the understanding of dark matter and energy. Moreover, the development of particle accelerators and particle detectors needs to proceed together with advanced theoretical calculations.

THE LARGE HADRON COLLIDER

The Large Hadron Collider (LHC), shown in Figure 1, is primarily a proton-proton collider with a design centre-of-mass energy of 14 TeV and nominal luminosity of 10^{34} $cm^{-2}s^{-1}$ [1], but will also be operated in heavy-ion mode. The high 40 MHz collision rate and the tens of interactions per crossing result in an enormous challenge for the detectors and for the collection, storage and analysis of the data.

By colliding unparalleled high-energy and high-intensity beams, the LHC will open up previously unexplored territory at the TeV scale in great detail, allowing the experiments to probe deeper inside matter and providing an understanding of processes that occurred very early in the history of the Universe.

Of central importance for the LHC is to elucidate the nature of electroweak symmetry breaking, for which the Higgs mechanism and the accompanying Higgs boson(s) are presumed to be responsible. In order to make significant inroads into the Standard Model Higgs search, sizeable integrated luminosities of several fb^{-1} are needed. However, even with 1 fb^{-1} per experiment, discovery of the Standard Model Higgs Boson is still possible in certain mass regions.

In addition, the potential for discovery of particles predicted in Supersymmetry (SUSY) theory is sizeable already at LHC start-up. Due to their high production cross-sections, squarks and gluinos can be produced in significant numbers even at modest luminosities. This would enable the LHC to probe the nature of dark matter.

The LHC will also provide information on the unification of forces, the number of space-time dimensions and on matter-antimatter asymmetry. With the heavy-ion collision mode, the LHC will probe the formation of the quark-plasma at the origin of the Universe.

The start-up of the LHC on 10 September 2008 was a great success for both the accelerator and the experiments. Circulating beams were established rapidly and the beams were captured by the RF system with optimum injection phasing and with the correct reference. The incident of 19 September 2008, caused by a faulty inter-magnet bus-bar splice, resulted in significant damage in Sector 3-4 of the accelerator. Actions were taken immediately to repair the damage and to introduce measures to avoid any re-occurrence. The damaged thirty-nine main dipole magnets and fourteen quadrupole magnets were removed and replaced. Fast pressure release valves were added on the main magnets, an improved anchoring on the vacuum barriers was introduced around the ring, and an enhanced quench protection system was implemented. This had been a major amount of work and any remaining risks to the LHC are minimized by limiting the top beam energy in the first years of LHC operation. In view of the excellent progress in the above repair and consolidation work, the LHC physics run is expected to start soon, with the re-start of the machine with proton beams scheduled for November 2009.

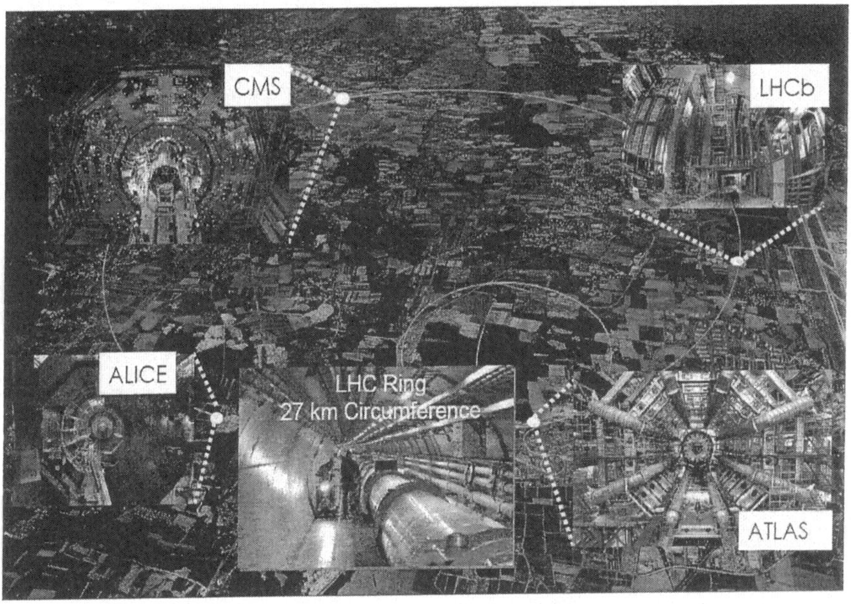

Figure 1: The LHC accelerator and experiments.

THE EUROPEAN STRATEGY FOR PARTICLE PHYSICS

European particle physics is founded on strong national institutes, universities and laboratories, working in conjunction with CERN. The increased globalization, concentration and scale of particle physics require a well-coordinated European strategy. This process started with the establishment of the CERN Council Strategy Group [2], which organized an open symposium in Orsay in 2006, a final workshop in Zeuthen in May 2006 and with the strategy document being signed unanimously by Council in July 2006 in Lisbon. CERN considers that experiments at the high-energy frontier to be the premier physics priority for the coming years. This direction for future colliders at CERN follows the priorities set in 2006.

The European Strategy for Particle Physics includes four key areas. The highest priority rests with exploiting fully the physics potential of the LHC at its design performance. In parallel, motivated by the operation experience and findings at the LHC, a vigorous R&D programme for the LHC accelerator and detectors needs to be pursued and centrally-organized with the aim of increasing the LHC luminosity by up to a factor of ten from the nominal design value of 10^{34} cm^{-2} s^{-1}. To this end, funding during the coming years will be used by CERN to gradually increase the performance of the LHC towards these specifications. This effort will be undertaken in stages. The first-stage luminosity upgrade of the LHC is through the construction of LINAC4 [3], which will reduce the risk to LHC operation by replacing the ageing LINAC2 which came into operation in 1978. In parallel, the effort includes the R&D and subsequent implementation necessary for a significant luminosity increase of the LHC, if necessitated by the physics and/or running experience, in particular concerning the focusing elements in the interaction regions and the upgrades of the injector chain.

The LHC will provide a first indication of any new physics at the TeV scale. Many of the open questions left by the LHC and its upgrades may be addressed best by an electron-positron collider, based on technology developed by the Compact Linear Collider (CLIC) and International Linear Collider collaborations.

FUTURE HIGH ENERGY LINEAR COLLIDERS

The Compact Linear Collider (CLIC)

The conceptual lay-out of the CLIC [4] is shown in Figure 2. The collider is based on using low-energy electron beams to drive high-energy beams. The fundamental principle is that of a conventional AC transformer. The low-energy drive beam serves as an RF source that accelerates the high-energy main beam with a high-accelerating gradient. The nominal centre-of-mass energy is 3 TeV, the luminosity exceeds 10^{34} $cm^{-2}s^{-1}$, the main linear accelerator frequency is 12 GHz, the accelerating gradient is 100 MeV/m and the total length of the main linear accelerators is 48.3 km.

CLIC requires more R&D. In particular, the target accelerating gradient is considerably high and requires very aggressive performance from the accelerating structures. The nominal CLIC accelerating gradient has been exceeded in an unloaded structure with a very low breakdown probability of less than 3×10^{-7} per metre after RF conditioning for 1200 hours.

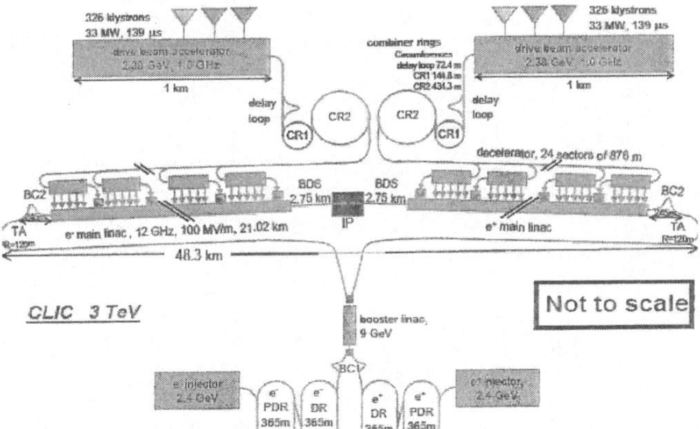

Figure 2: The conceptual lay-out of CLIC.

The mandate of the CLIC team is to demonstrate the feasibility of the CLIC concept by the end of 2010 in a Conceptual Design Report. If this effort is successful, and if the new physics revealed by the LHC warrants, the next phase of R&D on engineering and cost issues will be launched. This would serve as the basis for a Technical Design Report and a request for project approval.

The International Linear Collider (ILC)

The ILC [5], which is an option for a linear electron-positron collider at lower energies than CLIC, is based on a more conventional design for acceleration using superconducting standing wave cavities with a nominal accelerating field of 31.5 MeV/m and a total length of 31 km. at 500 GeV centre-of-mass energy. A two-stage technical design phase during 2010-2012 is presently underway. A major contribution from Europe and from DESY to the ILC Global Design Effort is the European X-ray Laser Project XFEL at DESY. The purpose of the facility is to generate extremely brilliant and ultra-short pulses of spatially-coherent X-rays. The electron energy is brought up to 20 GeV through a superconducting linear accelerator, of length one-tenth that of the ILC superconducting linear accelerator, and conveyed to long undulators where the X-rays are generated and delivered to the experimental stations. The XFEL technical design is ready and construction of the XFEL will commence soon with the aim of starting operation in 2013.

The strategy to address key issues common to both linear colliders involves close collaboration between ILC and CLIC. Recent progress has been encouraging in this respect and a first common meeting between ILC and CLIC was held in February 2008.

Detector Challenges

R&D on key components of the detector for a linear collider is mandatory and is also well underway. High-precision measurements demand a new approach to the reconstruction. Particle flow, namely reconstruction of all particles, is thus proposed requiring unprecedented granularity in three dimensions of the detection channels.

KEY MESSAGES

Great opportunities are in store at the TeV scale and a fuller understanding of Nature will come about through a clearer insight at this energy level. As in the past, there is a synergy between collider types – proton-proton and electron-positron. An upgrade to the LHC will not only provide an increase in luminosity delivered to the experiments, but will also provide the occasion to renew the CERN accelerator complex. The ILC could be constructed now whereas further R&D is needed for CLIC. There is a drive to converge towards a single electron-positron linear collider project. The above effort on accelerators should advance in parallel with the necessary detector R&D. First results from the LHC will be decisive in indicating the direction that particle physics will take in the future.

Particle physics will need to adapt to the evolving situation. Facilities for high-energy physics (as for other branches of science) are becoming larger and more expensive. Funding for the field is not increasing and the timescale for projects is becoming longer, both factors resulting in fewer facilities being realized. Moreover, several high-energy physics laboratories are changing their missions.

All this leads to the need for more co-ordination and more collaboration on a global scale. Expertise in particle physics needs to be maintained in all regions, ensuring the long-term stability and support through-out. It will be necessary to engage all countries with particle physics communities and to integrate the communities in the developing countries. The funding agencies should in their turn provide a global view and synergies between various domains of research, such as particle physics and astroparticle physics, should be encouraged.

Particle physics is now entering a new and exciting era. The start-up of the LHC allows particle physics experiments at the highest collision energies. The expectations from the LHC are great, as it would provide revolutionary advances in the understanding of the microcosm and might change our view of the early Universe. Due to the location of the LHC, CERN is in a unique position to contribute to further understanding of the microcosm in the long term.

Results from the LHC will guide the way in particle physics for many years. It is expected that the period of decision-making concerning the energy frontier will be in the next few years. Particle physics is now in an exciting period of accelerator planning, design, construction and running and it is now the time for intensified efforts in R&D and technical design work to enable the decisions for the future course and global collaboration coupled with stability of support over long time scales.

The particle physics community needs to define now the most appropriate organizational form and needs to be open and inventive in doing so, and it should be a dialogue between the scientists, funding agencies and politicians. It is mandatory to have accelerator laboratories in all regions working as partners in accelerator development, construction, commissioning and exploitation. Furthermore, planning and execution of high-energy physics projects today require world-wide partnerships for global, regional and national projects, i.e. for the whole particle physics programme. The exciting times ahead should be used to establish such partnerships.

CONCLUSIONS

With the start-up of the LHC, particle physics is entering a new phase of research and discovery, and results from this physics programme will provide directions for the future of particle physics. In the coming years, the ordered priorities are the full exploitation of the LHC, together with preparation for a possible luminosity upgrade and the consolidation and optimization of the CERN infrastructure and the LHC injectors. This may be followed by a linear electron-positron collider based on the technology being developed by CLIC or the ILC. All these future projects need to be carried-out through an enhanced collaboration on a global scale.

ACKNOWLEDGEMENTS

I would like to thank the organizers for the invitation to make this contribution and for the excellent organization of the very interesting workshop. I would also like to thank Emmanuel Tsesmelis for his help in preparing this contribution.

REFERENCES

[1] LHC Design Report, Volumes I, II and III, http://lhc.web.cern.ch/lhc/LHC-DesignReport.html

[2] The European Strategy for Particle Physics, http://council-strategygroup.web.cern.ch/council-strategygroup/Strategy_Statement.pdf

[3] http://linac4.web.cern.ch/linac4/

[4] http://clic-study.web.cern.ch/CLIC-Study/

[5] http://www.linearcollider.org/cms

CHAIRMAN: R. D. HEUER

Scientific Secretaries: N. van der Kolk, K. Perez

DISCUSSION

- *Preghenella:*

What is the centre-of-mass energy foreseen for Pb-Pb collisions in 2010?

- *Heuer:*

We have to see how the machine runs with p-p, and if the machine runs with p-p up to 5 TeV, then I don't see any problem with Pb-Pb collisions at a similar energy. But I would not go beyond a current in the magnet that is needed for 5 TeV collisions.

- *Preghenella:*

What is still to be done to safely run the LHC at 7 TeV beam energy?

- *Heuer:*

There are several issues. One issue is to install the additional safety valves, the so-called DN-200, that is 200 mm diameter valves, for the release of gaseous helium. We have equipped the whole ring with 100 mm diameter valves, which could even be done in the cold sectors. In the cold sectors we could not do these 200 mm valves because we have to drill openings into the magnet, and this we cannot do once they are cold. In the configuration at the moment I would not go above 5 TeV beam energy, because if such an incident happens again, then up to 5 TeV, the 100 mm valves are sufficient. Beyond 5 TeV the 200 mm valves are needed in the worst-case incident. So that is the kind of mechanical work that we have to do in any case. The bigger problem is what to do with the connections. It is not so easy to measure resistances with high accuracy in a system like the LHC. There is a lot of electronic noise which you have to filter out, and you have to measure all the resistances in order to be sure that all these splices are below nanoohm in order to be able to support a current corresponding to 7 TeV in energy. That is one thing. Secondly, I showed you these copper casings. They need to have about 10 microohms resistance for the dipoles and about 16 microohms for the quadrupoles. These are even more difficult to measure, especially once you are not at room temperature, because we measure not a single copper casing, but a series of copper casings, since we have only access at certain points. And then you have to, again, filter out the noise, although we have measured now a distribution which is a very nice Gaussian centered around zero after subtraction of the design values. The measurement resolution therefore determines the limit for the outliers. So the energy value of 3.5 TeV comes from the fact that we want to restrict ourselves to a value of the resistance which is safe for running.

We have to learn, we have to measure more, and then we have to see: do we have to change some of the connections? Do we have to change all of the connections? We don't know, and I don't want to pre-empt what we have to do. So we need more experience in order to do more measurements, in order to digest the measurements, and then we see what we have to do.

Then there are other things. At the moment we don't have enough storage for the helium, so if we warm up the machine, we can only warm up some sectors. To warm up other sectors we have to move the helium into other sectors which are not warm. We are buying new storage tanks, but to my surprise there are very few firms in the world that can build such tanks.

So these are a few of the major things which we have to do, which means that the next shutdown must be a pretty long shutdown, in particular in the case we must redo some of the connections or all of the connections.

- *Klopot:*

Is the data that will be collected during the first year of the LHC sufficient for the solving of the Higgs quest? How long does it take to collect a sufficient amount of data?

- *Heuer:*

That question is easy to answer: it is not. I mean, with 250 inverse picobarns there is no chance to discover it in the whole mass region still allowed.

- *Klopot:*

How long does it take to collect a sufficient amount of data?

- *Heuer:*

If you give me a crystal ball, then I could tell you, but I don't know. As I told you in my talk, it will take time, and it depends on the mass of the Higgs. And it depends on the way we get the LHC running. We will start slowly, and then we go up with the current, and with the luminosity...it's very difficult to say. I would say that if there is a low-lying supersymmetric state, then the chance to find it before the Higgs is much higher.

- *Tapia Takaki:*

What are the CERN plans to integrate more deeply non-member states to its projects? Does the European Union consider CERN as the European Laboratory of sub-nuclear physics?

- *Heuer:*

For me it is important to grow the membership of CERN more by offering different kinds of memberships to partners. That is for particle physics developed countries. Next week, on Tuesday and Wednesday, we have representatives from eight countries discussing with us how an enlargement could globally take place. This is the U.S., Canada, Brazil, Japan, China, Korea, India, and Russia. We will see what they would like to propose. CERN has at the moment the possibility of associating membership for non-European countries, but nobody is an associate member because there is not the right balance between obligations and rights. If you have no balance between obligations and rights, why should I become an associate member? So you have to make a balance between obligations and rights. That is one thing. Then another part of this is the particle physics developing countries. This should be in the area of education and training: teaching programs, school programs, innovation, etcetera. Considering the European Union, there is a process in Europe, of course, at the moment, in science, that's called the ESFRI process. That is, projects that are larger than can be realized on a national basis. Now, scientists and governments are giving their input into this process. Particle physics has in 2006 developed its European strategy, which is supposed to be updated every five years. I don't think it will be updated in 2011 because there will not yet be results from the LHC, and it doesn't make sense to update a year before you have new results.

The European Union accepts that CERN Council speaks for European particle physics. It's not CERN; it is the CERN Council. And we have signed a new agreement with the European Commission that clearly states that in all aspects of particle physics it is the CERN Council that speaks internationally for European particle physics. But we want to have a very close connection between the Commission and between CERN and the CERN Council. And again this is not only in particle physics, but it is again in education, in outreach, in technology transfer, on knowledge transfer, on many different topics, so that everyone can profit very much from that. And again next week, there are meetings on how to go on after this signature.

- *Iglovikov:*

The quality of some predictions of the LHC depends on the volume of experimental data gained during the life of the LHC. How has the incident decreased this volume?

- *Heuer:*

It only shifts it by one year.

- *Iglovikov:*

But some parts were in use during this year, and may influence the age of the LHC.

- *Heuer:*

Concerning the LHC complex, nothing was used. Do you mean the injector complex?

What counts for use is the integrated luminosity and the energy, of course. The integrated luminosity is a function of the instantaneous luminosity, the luminosity lifetime, and the amount of time you are running. The fact that we have lost a year will be made partially up by the decision we have made earlier this year of not going into a shutdown over the winter. So we run much longer than was originally foreseen for the first months. Normally a run at CERN goes from April until November, but now we have 12 months, from November until November. Concerning the energy, we start at a lower energy and by comparison, I think that once you buy a new car, you don't immediately go up, at the very beginning, to the full speed. You first start the car at a lower speed. And LHC is a very complicated car.

- *Ahmad:*

In the LHC mass is converted into energy. How can we control this energy? How is this different from the CBM experiment?

- *Heuer:*

The difference with a fixed target experiment is that in the center of mass of the proton the energy is much lower. You can compare this to two cars driving at 100 km/h onto each other. The damage is much larger than when one car hits a brick wall at the same speed.

- *Ahmad:*

Is there any danger to the earth?

- *Heuer:*

There is no danger at the LHC. Unlike a nuclear reactor, at the LHC when you switch off the machine the particles are gone.

- *Tannenbaum:*

How long do you anticipate it will take to switch from protons to ions at the same magnet current the first time?

- *Heuer:*

That is difficult to predict. What we are doing is to prepare the injector such that the ions are going all the way to the beam dump before entering the LHC. How quickly we can come up to the high energy with the acceleration in the LHC I do not know, but it should be reasonably fast. That is the prediction.

It should be relatively easy after all the experience with the proton-proton collisions. The aim is, of course, to have a decent run with heavy ions.

- *Tannenbaum:*

One of the problems with the CERN fixed target heavy ion program was the lack of p-p (and p-A) comparison data at the same nucleon-nucleon CM energy. How will you overcome this problem at the LHC? Will you make p-p comparison runs at the same CM energy?

- *Heuer:*

Next year we want to concentrate on the highest possible proton-proton energy we can do. And then we are going to the highest possible lead-lead energy that we can do. And then we have to decide how long the shutdown will be and what we will do as a next step.

These questions about the future I fully understand. However, my attitude has to be: "Let's get this machine on. Let's run it, learn how to run it". Only then can we make these decisions. Therefore I do not want to take any decisions now concerning what happens after one year of running.

In January we have again a Chamonix meeting. I have resumed these Chamonix meetings which were very successful during the LEP time. The machine and the experiments were talking for one week with each other and developing the program. It was very good for this year and will again be very good for next year, with the first experience of running and then see how to continue.

- *Parke:*

What can you tell us about some of the magnets losing their training?

- *Heuer:*

Yes, some of the magnets have Alzheimer.It is not clear why they need more training again after some time. It seems to be connected to one of the three firms. But it is not connected to the performance of the magnets, they just need the training again. We have to see how it is after they have been in the tunnel for some time. These magnets have been stored at the surface for a long time, I do not know what could have happened to them there, but the performance is by no means influenced.

Again, I would like to see how they behave now when we go to 3.5 and later to 5 TeV. Then when we go beyond 5 TeV we have to see again how they behave and how quickly we can go up to 7 TeV.

They should not go through too many warm-cold cycles. When they are kept cold it's much better. At the moment I'm not afraid of this problem.

- *Dixon:*

What kind of milestones, such as ability to abort the beam, do you need to pass before increasing the beam current and hence the LHC luminosity?

- *Heuer:*

There are two main milestones: we have to be sure that the beam orbit monitors are all working correctly, as well as the fastkickers, which abort the beam.

- *Tapia Takaki:*

In your opinion, what are the outcomes from the recent workshop on new physics initiatives at CERN?

- *Heuer:*

here is no final outcome yet. At the beginning of October there will be a second workshop which will concentrate on neutrino physics. The first workshop concentrated on other fixed target experiments. There are discussions about new ideas, and these are encouraged. At the moment there is no funding, but if there is a good idea I'm pretty sure the money will be found to realize it.

THE GRAN SASSO UNDERGROUND LABORATORY

Eugenio COCCIA
Department of Physics, University of Rome "Tor Vergata" and INFN, Italy
E-mail: coccia@roma2.infn.it

ABSTRACT

Thirty years have passed since, thanks to Antonino Zichichi, the project for the largest underground laboratory in the world was conceived and brought to the attention of Italian authorities. The Gran Sasso National Laboratories of INFN have become a scientific reality of worldwide pre-eminence, in an expanding area of research where elementary particle physics, astrophysics and cosmology overlap. I briefly present here the main scientific challenges of underground laboratories and the activity and future perspectives of the INFN Gran Sasso Laboratory.

1. Underground physics

400 years ago, Galileo Galilei observed with his telescope for the first time the Jupiter moons, realizing that Jupiter was the center of a small celestial system in its own. The discovery of bodies orbiting something other than the Earth dealt a blow to the then-accepted Ptolemaic world model and provided evidence in support of the Copernican system. Underground scientists are observing today the cosmos with new telescopes, very different from Galilei's one, but with the same aim and enthusiasm for discovering our place in the Universe.

Underground laboratories, shielded by the Earth's crust from the particles that rain down on the surface in the form of cosmic rays, provide the low radioactive background environment necessary to host key experiments in the field of particle and astroparticle physics, nuclear astrophysics and other disciplines that can profit of their characteristics and of their infrastructures. The "cosmic silence" condition existing in these laboratories allows the search for extremely rare phenomena and the exploration of the highest energy scales that cannot be reached with accelerators.

We go deep underground to respond to vital scientific questions: What is the universe made of? What is the nature of dark matter? What happened to the antimatter? What are neutrinos telling us about the fundamental symmetries of nature, and about the stars and the Earth?

The realization of the largest underground laboratory in the world was proposed in 1979, thirty years ago, by Zichichi, then the President of INFN, and approved by the Italian Parliament in 1982. Since the original project the orientation of the halls was towards CERN, in order to host detectors to study neutrino physics on a beam produced at that laboratory.

Civil engineering works, under the responsibility of ANAS, the Italian Road Department, started in autumn 1982 and were completed by 1987.

I review here the present activity and future perspectives of the INFN Laboratori Nazionali del Gran Sasso (LNGS) and conclude with some notes on the L'Aquila earthquake of Aril 6, 2009. More information can be found on the WEB-site http://www.lngs.infn.it.

2. The Gran Sasso National Laboratory

The INFN Laboratori Nazionali del Gran Sasso (LNGS) is located besides a freeway tunnel under the Apennines, in central Italy near the town of L'Aquila, 120 km from Rome.

The mission of the Laboratory is to host experiments that require a low background environment in the field of particle and astroparticle physics and nuclear astrophysics and other disciplines that can profit of its characteristics and of its infrastructures.

The geographical location (inside the National Park of Gran Sasso - Monti della Laga) and the special operating conditions (underground, near a highway tunnel and in close proximity to a large water basin) demand that special attention is paid to the safety and environmental aspects of the activities.

The horizontal access allows easy transportation and installation of large pieces of apparatus. The underground facilities consist of three experimental halls and a set of connecting tunnels and service areas, for a total surface area of 18000 m^2. The three halls are approximately 100x18x18 m^3. An almost angle-independent 1400m rock overburden provides a μ flux attenuation of 10^{-6}. The neutron flux is $= 3.7 \pm 0.3 \times 10^{-2} m^{-2} s^{-1}$.

Figure 1: The underground facilities and the Gran Sasso massif geology

The infrastructures of the laboratory are completed by the surface site in Assergi, nearby the western entrance of the tunnel, hosting the directorate, offices, laboratories, workshops, library, canteen, etc. General services include administrative and user support offices, the prevention, safety and environmental protection services, maintenance and, in direct support to the experiments, low activity techniques (including very low background facilities with Ge detectors underground), cryogenics, mechanics, chemistry, electronics, computing and networking.

Last but not least, the laboratory has an important activity in the outreach, in the diffusion of the scientific culture and in the training of young people, with the active support of the Abruzzo Region.

The permanent staff is composed by 77 persons.

Presently, the user are about 800 in number, from 25 different countries.

Main research topics of the present scientific programme are: neutrino physics with neutrinos naturally produced in the Sun and in Supernova explosion and neutrino oscillations with a beam from CERN (CNGS program), search for neutrino mass in neutrinoless double beta decays, dark matter search, studies of nuclear reactions of astrophysical interest.

Let us summarise the recent activity of the main research lines.

3. Neutrinos and stellar physics

Solar neutrino physics is one of the traditional research sectors of the laboratory. After the glorious life of GALLEX and GNO, the focus is now on Borexino, which is dedicated mainly to the measurement of the Be line component of the solar neutrino spectrum. The data taking started in May 2007.

The excellent radio-purity reached by the Borexino experiment, in some cases two orders of magnitude higher than what expected, open a new window in the solar neutrino physics. In particular, recent results published by the Borexino Collaboration on the measurement of the solar neutrino flux from ^7Be, demonstrates that, with enough statistics, systematic errors will dominate over the statistical ones. The actual statistical error of 10% will be reduced at less than 1% in 1.5 years. Such precise measurement would represent a strong constrain on the actual Solar Model, providing a double test of our knowledge of the energy production mechanisms in the Sun and of the LMA-SMW flavor oscillation model.

The solar models are based on data and extrapolations; in particular the thermonuclear cross sections of the involved reactions are not measured in the relevant energy range but rather extrapolated from higher energies.

The direct measurements are made very difficult by the very low values of the cross sections. Using the new 400 kV accelerator, LUNA is continuing his activity for the measurements of the cross section of nuclear reaction of astrophysical interest.

Stars substantially more massive than the Sun end their life through spectacular stellar explosions known as core collapse supernovae. Supernovae are among the most violent phenomena in the universe. The huge amount of the binding energy of the collapsed core is emitted in the form of neutrinos and anti-neutrinos of all flavours.

The observation of neutrinos released in the various stages of the explosion will provide direct information about the dynamics at the centre of the Supernova and therefore of the explosion mechanism. Neutrinos are in fact a unique tool to study the central region of the star, and hence the nuclear matter in the extreme state of temperature and density encountered in the Supernova core.

Neutrinos originated from Galactic supernovae are continuously monitored, with 100% live time, by the LVD detector, constituted by 1000 tons of liquid scintillator.

LVD is part of the SNEWS (Supernovae Early Warning System), which involves an international collaboration of experimenters representing several of the current supernova neutrino detectors in the world. The goal of SNEWS is to provide the astronomical community with a prompt alert of the occurrence of a Galactic core collapse event and enable optical observation of the early phases of the Supernova. LVD is also monitoring the neutrino beam from CERN.

4. Neutrino mass and nature

Elementary particles and their antiparticles are characterized by opposite signs of all charge quantum numbers. The standard model assumes that neutrinos have only one charge, the lepton number. But, if this charge is not conserved, neutrinos and antineutrinos can be two states of the same particle. In this case well-specified nuclides would decay through the neutrino-less double beta channel. The Laboratory hosts today experiments searching for these very rare decays, employing different and complementary techniques, and is preparing new important activities.

The Heidelberg-Moscow experiment, which with a sensitive mass of 11 kg of enriched ^{76}Ge was the most sensitive experiment in the world for several years and accumulated data for a 75 kg y exposure, was terminated in 2003, claiming a positive detection. In the same year CUORICINO, which employs TeO_2 bolometers for a total of 42 kg, started taking data and is the most sensitive running experiment today.

The Laboratory approved two new experiments: CUORE, which brings the CUORICINO technique to a mass of more than 400 kg of TeO_2 bolometers, and GERDA planning to employ 500 kg of enriched ^{76}Ge. These activities represent the state of the art of the field. They will start taking data, respectively, in 2010 and in 2011.

One of the major commitments of the Gran Sasso laboratory is the the CNGS project (CERN Neutrinos to Gran Sasso), aiming at the search for the appearance of tau neutrinos in the artificial muon neutrino beam built at CERN.

A beam of protons, with an energy of 400 GeV, produced in the SPS accelerator of CERN, collides with a target of graphite to produce a secondary beam consisting partially of pions (π) and kaons (K), charged particles with short lifetimes. After having been focused and directed towards the Gran Sasso, the pi+ and K+ decay, giving origin to neutrinos that maintain practically the same direction of the particles that generated them. The neutrino beam contains almost exclusively muon neutrinos with an average energy of 17.4 GeV and a contamination on the level of a few percent of anti-muon neutrinos, electron and anti-electron neutrinos.

The beam is directed through the Earth crust to Gran Sasso at 732 km distance and reaches the Gran Sasso Laboratory with a flux of thousands of billions per day. The first neutrinos arrived in august 2006, and the beams is working regularly since 2008.

The OPERA experiment is designed for the direct observation of tau neutrinos resulting from oscillations of the muon neutrino of the beam. This search requires both micrometer scale resolution, obtained with modern emulsion techniques and large sensitive mass (1800 t) obtained with Pb sheets interleaved by emulsion layers.

The experimental apparatus is made up of two *supermodules*, each formed by a target followed by a muon spectrometer.

The target consists of 150,000 bricks, where the neutrino interactions occur. The brick walls are interleaved with layers of Target Trackers, detectors that electronically track the event, permitting the determination in real time of the position. The measurement of the position of the neutrino interaction allows the identification of the brick in which the interaction occured. That brick can be extracted and, after developing the photographic emulsions, can be analyzed using automated microscopes. While the events induced by muon neutrinos are relatively frequent (about 50 per day), only a few tau neutrino events are predicted per year year.

The production and installation of the 150,000 bricks containing the Pb sheets and the emulsion layers is terminated and the experiment is in full data taking.

ICARUS is a general-purpose detector, with a broad physics programme, based on the novel concept of the liquid Argon time-projection chamber. After the development of the "definitive project" and of the risk analysis, completed in 2003, the 600 ton module was transported from Pavia to the underground Laboratory in early 2005 and is now at the end of the phase of installation and completion. The experiment should be ready for the beginning of 2010.

5. Dark Matter

Let me report now on the Laboratory intense activity on Dark Matter.

There are many evidences today that most of the matter in the Universe does not significantly emit or absorb electromagnetic radiation, but manifests itself only through its gravitational interaction with luminous matter (stars and interstellar gas).

The conclusion from cosmological background observations, in particular the ones of WMAP, and nucleosynthesis arguments is that within an apparently flat universe at least 85% of the matter in the universe is in the form of non-baryonic, non-luminous matter.

Models of galaxy formation and simulations indicate that most of this dark matter component must be *cold*, i.e. consisting of particles that were non-relativistic at the time the universe became matter dominated and that are weakly interacting. Hence the denomination for the elusive particles of this dark matter as Weakly Interacting Massive Particles, or WIMPs. Galaxy rotation curves measurements lead to consider a WIMP halo of local density of 0.3 GeV/cm, a maxwellian velocity distribution with rms value of 220 km/s, and a relative velocity of the sun within the halo of 230 km/s.

One of the most natural WIMP candidates is provided by supersymmetric theories: If R-parity is conserved, the lightest supersymmetric particle is stable and leads to a relic density compatible with the missing mass component when the interaction cross-section is of the order of the weak interaction cross-section. A natural candidate is the neutralino. Its predicted mass range is bound at low masses by LEP to 45 GeV, and at high masses to a few TeV, which gives it a non-negligible chance of being produced at LHC in the coming years.

An indirect method to detect such particles is looking for products of WIMP annihilations in the Universe.

The direct method is looking directly for WIMP interactions within a target material.

The signal expected from a WIMP contribution have several distinctive signatures.

WIMPs, like neutrons, scatter elastically off a nucleus in the target material, while gammas and betas scatter off the electronic cloud. If one can discriminate in the data a nuclear recoil from an electronic recoil, a large part of the background will have thus been eliminated.

Moreover, for given WIMP characteristics (mass and cross-section) and for a given target material, a recoil energy spectrum is predicted, which can ideally be identified to the data, or at least allow to set an upper limit on the mass and cross-section.

Last but not least, due to the relative speed distributions between this halo and an observer on Earth, both an annual and diurnal modulation of the WIMP signal flux and directionality, respectively, are expected. The amplitude of the annual modulation amount at best to a few percent of the total WIMP signal.

The principles behind direct detection boils down to measuring the energy deposited by a WIMP interaction, which is expected to be in the keV to tens of keV range. There are three ways of measuring this energy, each choice leading in turn to a choice of target material, and which can eventually be combined:

- light signal, produced by a scintillation process, in pure crystals or in a liquid noble gas (fast).
- ionization signal from semi-conductor materials like Si, Ge, or CdTe (slower but very efficient). It is present also in liquefied noble gases.

- heat signal from a target material cooled at low temperature (slow but also very efficient).

Because of the very small interaction cross-section of WIMPs with ordinary matter, an extremely low event rate of events is expected typically, less than a few events per kg of target material and per year for preferred SUSY models. Background suppression is therefore a vital issue and operation deep underground is mandatory. It is also very important that the level of radio-purity of each component is kept to a minimum, in particular the parts closer to the detectors.

Taking into account all these considerations has led to a number of experiments set in various underground sites all over the world and to the design of a new generation of large mass detectors, up to one ton.

At Gran Sasso several experiments, using different techniques, are active and new experiments have been approved and started operating in the last years.

DAMA/LIBRA employs NaI crystals to detect the WIMPs by means of the flash of light produced in the detector by an Iodine nucleus recoiling after having been hit by a WIMP. To distinguish these events from the background, DAMA searches for an annual modulation of the rate, a behaviour that has several aspects that are specific to the searched effect and not to the main backgrounds. With about 100 kg sensitive mass DAMA was the only experiment world wide sensitive to the annual modulation signature. After the conclusion of the experiment, results were published showing a signal of annual modulation.

The larger experiment LIBRA, with 250 kg sensitive mass, published in 2009 results from 4 years of data, confirming the annual modulation signal to better than 8 sigma.

CRESST searches for WIMPs with a cryogenic technique, looking for a very tiny temperature increase in the detector, due to the energy deposited by nuclei hit by the WIMPs. After the final design, the CRESST2 $CaWO_4$ detector has been installed and is operating.

Two new dark matter experiments, WARP and XENON have started their operation during 2006. They are based on the simultaneous detection of the ionization and scintillation signals in, respectively, liquid argon and liquid xenon. WARP performed preliminary measurements underground with a small (2.3 l) prototype and XENON took data with a 10 l apparatus. Both experiments have realized their 100 liter size detectors and will provide data in the next year.

6. More on Gran Sasso Science

The main activity of the theory group, staff and visitor scientists, has been focused on astroparticle physics, including solar and Supernova neutrinos, massive neutrinos, ultra high energy cosmic rays, topological defects and relativistic astrophysics. Important activity took place also in particle phenomenology and computer simulations of Lattice Field Theories.

In 2008 an Astroparticle Physics Center (in Italian Centro di Fisica Astroparticellare), has been created at LNGS. CFA is a network of Italian research centres around the main node constituted by the Gran Sasso National Laboratory, and has the aim of promoting the research in the field of astroparticle physics through actions like visiting fellowships and post doc positions at Gran Sasso Lab. It is funded by Abruzzo Region and EC through the Piano Operativo Regionale 2007-2013, with a MoU signed by INFN and Regione Abruzzo.

Housed at the Gran Sasso National Laboratory and led by Veniam Berezinsky, CFA includes at present other two nodes: the University of L'Aquila and the University of Rome Tor Vergata. Among its main research areas there are neutrino physics, dark-matter physics, high energy astrophysics and gravitational waves. For more info see http://www.lngs.infn.it .

There are other fundamental questions that make the interest in underground science increase. Questions araising from life sciences: What makes life successful at extreme depth? And other questions araising from Earth sciences: Can we understand slippage mechanisms in high-stress environments, in conditions as close as possible to tectonic faults/earthquakes? There are more questions on rock mechanics at large scales and on geo-hydrology. Some activities in these fields started and are going on at Gran Sasso.

7. LNGS and Europe

The directors of the four Western Europe underground laboratories (Gran Sasso, Modane, Canfranc and Boulby) initiated actions to co-ordinate activities and to optimise the use of the available resources, taking into account the different characteristics of the infrastructures. In the 6th "framework programme" the ILIAS project (Integrated Large Infrastructures for Astroparticle Science, putting together the underground and the gravitational wave communities) was submitted successfully to European Commission (EC) by ApPEC, the European inter-agency committee for astroparticle physics.

ILIAS was launched on April 2004 as a 5 years project, and got a total EC financial contribution of 7,5 Million. ILIAS contained several elements: the co-ordination of a number of activities of the laboratories, such as environmental background measurement and control, safety procedures, outreach activities, and the funding for access of new users to the labs. The continuation of this program (a new proposal called ILIAS-Next has been submitted to EC) and the continuation of the coordination among the four underground labs (a new agreement called EuLAbs is being signed by the agencies Presidents) has been favored by ASPERA. Created by ApPEC and funded by the EC through the 7th Framework Programme, ASPERA is an European Union ERANET project for the coordination of national research efforts in Astroparticle Physics in Europe. See http://www.aspera-eu.org/ . In the same Framework Programme, EC also funded a Design Study project (LAGUNA) focusing on the design of an European infrastructure able to host underground new large-volume instruments with a fiducial mass from several tens of ktons up to 1 Mton for low-energy neutrino astronomy and proton decay search.

8. Outreach

New initiatives have been added to the interesting projects already afoot (like the "open days", the competition "Anch'io scienziato" (I am scientist too) and the other initiatives with the schools in collaboration with AIF (the Italian association for physics teaching)).

8.1. The Gran Sasso-Princeton Summer School.

Thanks to colleagues in Princeton, it has been possible since 2004 to organize an annual summer school at the Lab for 20 young people of the Abruzzi - chosen by means of a written examination among students of the fourth and fifth year of the lyceum (high school). The programme of this school includes spending three weeks in the Princeton University campus, making use of the University rooms and labs, where students attend Physics and English classes, understand the deep interest of our researches and where, during the weekends, it is possible for them to make trips to New York, Washington and Philadelphia. The initiative has been fully supported by the Presidents of the Abruzzo Region and by INFN. And, what is more important, it has been adhered to enthusiastically by hundreds of boys and girls who took part in the selection process and played the crucial role, also within their own schools and their own families, of exceptional ambassadors of the scientific activities of the Lab.

Since 2008 the School has attracted the attention of South Dakota, home of the legendary underground Homestake Lab, which has hosted the experiment on solar neutrinos by Ray Davis who was awarded the 2002 Nobel Prize in Physics for that. But, above all, it is the designated site for the future large underground Lab DUSEL. The Governor of the State, following the advice of Frank Calaprice, has adhered to the spirit of the School and has financed the participation of students from South Dakota who, first visit the Gran Sasso Labs and then follow the School in Princeton together with our students of the Abruzzi. Since 2009 the initiative has been called Gran Sasso-South Dakota-Princeton Summer School.

8.2. Teramo Museum.

Created by the same 1990 law approving an enlargement of the underground Lab and the realisation of the third tunnel, the Physics and Astrophysics Museum was officially opened in Teramo in 2008. According to the establishment law, the Museum is to be run by INFN and, therefore, ultimately by the Labs which have closely collaborated with the Municipality of Teramo towards the long-awaited opening.

The potential of this structure is great, and thanks should be given to those who imagined it twenty years ago. It is made up of two buildings: one of which hosts an exhibition on radioactivity, whereas the other is awaiting to host part of the beautiful exhibition "Astri e Particelle", which has already been displayed on a large scale, in Rome. The buildings are part of a complex known as "Parco della Scienza" (Science Park), which includes a multimedia centre with a 400-seat conference hall. The Director of the Museum is the LNGS senior scientist Aurelio Grillo.

8.3. The EEE Project.

Conceived on the iniatitive and under the guidance of Antonino Zichichi, the main aim of the Extreme Energy Events project is to understand where, when and how cosmic rays originate, studying the characteristics of extensive showers of secondary particles produced by the impact of the highest-energy particles with the Earth atmosphere. At the same time, since cosmic rays play an emblematic and interdisciplinary role in scientific culture, studying them may be considered "the ideal instrument for bringing Science into the heart of young people" (see the site http://www.centrofermi.it/eee.html).

The EEE Project is structured on a modular basis. In numerous Italian Schools (Lyceums or Technical Institutes) a "telescope" is installed, made up of sensors called MRPCs (Multigap Resistive Plate Chambers), which allow the observation and the measurement of cosmic muons. In every School the students take part in the setting up and the start-up of the telescope.

Thus a direct interest in scientific questions is stimulated in the students. They feel they are becoming part of the construction of an instrument and of the elaboration of data which are at the cutting edge of contemporary research. Naturally "our" Schools (two of them) in L'Aquila are at the forefront in this initiative, also as regards the attainment of interesting results. Teramo is ready to follow the lead.

8.4. The Visitors Centre.

It is not always possible to visit the underground laboratories, and yet it is always important to explain our activities. Therefore the decision has been taken to realize a Visitor Hall, the realizaation of which is under way and which has benefited from the generous contribution of the "Fondazione della Cassa di Risparmio de L'Aquila".

9. Training and higher education

The desire to enhance and to make the transfer of technological and scientific knowledge that our lab carries out towards the surrounding territory valuable and visible, has led to new initiatives in the field of higher education. The principal way to realize our intent has been to present the Lab to the Abruzzo Region as a training and higher education centre, suggesting an innovative programme based on the features of an INFN Lab, which could be financed by the EU funds from the European Social Fund granted to the Region. The aim should be to place this activity in the wider context of leading the LNGS, as a research centre in the Abruzzo Region, into the perspective of a European-wide strategy for technical and scientific training, and for the transfer of the results from research centres to the local production systems.

It should be pointed out, to those who are not acquainted with our activities, that the technological research of INFN, and therefore also of LNGS, is driven primarily by the need to develop new ways of detecting and studying particles, new methods for the acquisition and the analysis of data. These methods represent a unique source of innovative technology in fields as electronics, superconductivity, precision mechanics, computing and high-performance networks, diagnostics imaging, adrotherapy, techniques applicable to the conservation of cultural heritage, and many others. It is natural for INFN to transfer the knowledge it has acquired in the course of its research activities to production companies.

The impact of the interaction with INFN on these companies may be estimated in terms of the capacity to widen their range of products, and in terms of the growth of their planning and output capacity. It may be useful to underline the significant territorial impact of this interaction, which is particularly incisive as regards small and medium-sized companies.

The LNGS are playing a role as stimulus for the innovation in the Abruzzo territory. The formal steps consisted in signing the memorandums of understanding between INFN and the Abruzzo Region, including the programme to be carried out in the POR (Regional Operational Plan) for Abruzzo: the first in 2006, with the participation of the GARR Consortium, the second in 2008 with the setting up of POR Abruzzo for 2007-2013.

We have used our infrastructure to train, by means of scholarships and post-doc positions, dozens of students of Abruzzo, and to offer higher education opportunity to all those involved in this initiative. Among these we must include all those who, thanks to this initiative, have attended our Labs, those who have had the chance of carrying out part of their activities working for companies associated with this project and those who have followed, either personally or in e-learning modality, the courses that have been on offer, often as an innovative packaging.

Among the latest financed projects I would like to highlight the following: the station for chemical-physical analysis, the activities for university vocational guidance for secondary-school students, and the creation of the centre for astroparticle physics, already mentioned in section 2 above.

10. The L'Aquila earthquake

At 3:32 on April 6, 2009 I was not in my flat in the L'Aquila centre. I was in Pisa for a meeting. Already in the first hours of the morning I could realize that, luckily, the Lab people had not suffered losses and that the Lab structures, both on the surface and underground, had well withstood. Arriving at the Lab, confronted with what I had seen crossing the territory, I thought that the very same shock which had caused dead and wounded people and destruction, should have also drawn, from inside ourselves, all the positive energies we were able of and even those unknown to ourselves.

Actually the Lab did not suffer serious damages: the outer premises had just few small cracks in the plasters of a couple of buildings; the underground Lab, where all the experiments had been set up according to severe anti-seismic criteria, had remained undamaged. But there was a great problem: by far the majority of the employees had lost their houses. Some of them had been trapped under the ruins and had miraculously remained uninjured. Several had been evacuated in hotels along the cost, others had found a shelter under the tents, or were guests of relatives or friends. In all their eyes tears, disbelief and anxiety for what was happening. Everyday the land went on shaking and seemed not to give rest to the already heavily tried nerves. The watchmen and the fire-fighting technicians, who safeguard the underground Lab 24 hours a day and whose presence is mandatory (by law too), underwent a particular psychological stress. For a moment let us put ourselves in their shoes: you are homeless, your relatives are evacuated or in a tent, they need you and are shocked by (irresponsible) rumors of a new, imminent very strong earthquake, and you must tell them that you go for your 8 hours work shift to an underground place 5 km far from the nearest way out, coming out on top of their invocations not to go. This is not an easy situation. Nevertheless I can directly witness, that the shifts have gone on non-stop. I can as well witness the sense of duty and the loyalty to the Lab of so many employees. Those who already were underground on April 6 at 4 a.m. checking the equipments; those who, without asking anything, were immediately willing to check up on the situation of all their colleagues; those who came back to work also to show that the life of their Lab went on. I feel a great admiration for all of them who have set such a good example.

11. Conclusions

The activity of the Gran Sasso Laboratory gives an idea of how exciting are the perspectives for underground physics in the next decade. The main scientific goals of the underground research in the next decade are:

- The direct detection of dark matter particles, searching down to the lowest cross-section limits (10^{46} cm^2) with ton scale detectors.
- The discovery of the nature of neutrino and the resolution of the neutrino mass spectrum with ton scale neutrinoless $\beta\beta$ detectors.
- A realistic project of a Megaton-scale detector (and the definition of a site and of an international collaboration) for the search for proton decay, for neutrino astrophysics and for the investigation of neutrino properties and CP violation in the leptonic sector.

Both in the search of dark matter and in $\beta\beta$ decay the sensitive masses of the experiments need to reach the ton to multi-ton size. Experiments starting to produce data now, as those based on noble fluids (liquids and gases) with 100 kg mass, appear to have good chances to be scalable at larger masses at not prohibitive costs. This will be useless, however, if the present levels of radioactive backgrounds will not be reduced by as many orders of magnitude below the present status of the art. Only a limited fraction of these backgrounds can be reduced going at larger depths. To reduce the largest fraction of the background, a large investment in *R&D* will be necessary. We have to consider that, because of the increase of the thickness of the shielding structures, experimental halls having heights and diameters in the 20 m range will be required; and because of the limited number of producers of the key materials used in these experiments, their procurement in large amount could be difficult and exceedingly expansive. So innovation will be of paramount importance, and young scientists should be encouraged to think differently and to risk. We do not know today which will be the successful approach. Consequently, premature choices should be avoided, as, for example, between liquid Xe and liquid Ar in Dark Matter, or between one or the other isotope in $\beta\beta$. On the other hand, duplications of the same, or almost so, technique should be avoided.

On the Megaton scale project, there is an increasing interest in realizing a large particle detector based on liquid Argon technology to afford exciting searches on neutrino physics with artificial sources, investigating CP violation in the leptonic sector, and preparing the technology for a proton decay experiment. Such perspectives could drive the realization of a new shallow depth laboratory in the Gran Sasso area and exploit again a dedicated CERN to Gran Sasso neutrino beam.

Finally, I think we have to favor the increasing interest for underground facilities, including Gran Sasso, from other scientific communities. It is certainly a benefit to widen the underground frontier and look to underground labs from a global multidisciplinary perspective.

Information on the experiments and their results can be found on the WEB-site http://www.lngs.infn.it.

CHAIRMAN: E. COCCIA

Scientific Secretaries: A. Achilli, B. Guerzoni

DISCUSSION

- *Tapia Takaki:*

How do you identify the brick at which the neutrino interaction occurs?

- *Coccia:*

From the interaction of a neutrino with the brick, charged particles emerge. They are measured by the scintillator detector, so you can reconstruct back the vertex of the interaction and individuate the exact brick in which the vertex is. This brick is taken and analyzed for looking exactly to the features of the event.

- *Tapia Takaki:*

What is the detector resolution and how do you calibrate the bricks?

- *Coccia:*

Detector resolution is 1 micron. To calibrate we use cosmic rays.

- *Tapia Takaki:*

But do you need to recalibrate after the interaction?

- *Coccia:*

No

- *Caliva:*

This morning you talked about an annihilation process between neutrino emitted in the first single beta decay and the antineutrino emitted in the second one. I think this is not correct because if it was an annihilation process other particles should come out from that point otherwise the energy would not be conserved. The correct process should be the exchange of a virtual neutrino and not an annihilation process.

- *Coccia:*

It is correct to say that the event is not the result of real annihilation process but of a virtual one. I talked simply of an annihilation process just to abbreviate.

- *Baldini:*

Can you compare the 2 new experiments looking for dark matter, Liquid Argon (WARP) and Liquid Xenon (XENON), to DAMA and LIBRA?

- *Coccia:*

The dark matter signature in DAMA LIBRA is the annual modulation of the signals. The aims of these new noble liquid experiments is to have the power of discriminating event by event the interaction due to an impinging massive neutral particle and to work with zero background. Let me also add that Liquid Argon and Xenon in this game have important complementarities. One is sensitive to spin dependent interaction of the WIMPS, the other also to spin independent one. Argon is cheaper than Xenon but the background is not as good as that of Xenon. There are many reasons to perform both experiments. For liquid Argon one potential problem is the presence of an isotope, the Argon-39, which is radioactive and constitutes an important background. Since a few years it has been discovered an efficient way for extracting Argon gas not from the atmosphere but from the underground directly. Underground gas is depleted respect to this Argon-39, and this makes Argon extremely competitive.

- *Baldini:*

So after a while we should know if DAMA and LIBRA results are right or no.

- *Coccia*

A few years of data taking from WARP and XENON will contribute to the overall picture.

- *Bettini:*

To test the DAMA results one needs to reproduce the DAMA experimental conditions as close as possible. In these experiments there are two basic backgrounds: those due to gamma rays and those due to neutrons. All the experiments, but DAMA strongly, reduce the gamma-ray background by pulse-shape analysis or other techniques. If there is a dark matter component that produces only an electromagnetic signal it would escape detection by all but DAMA. For this reason there are two projects to check DAMA results: the KIMS experiment in Korea, presently taking data with more than 100 Kg of CsI and ANAIS at LSN planning a 100 Kg of NaI detector.

- *Coccia*

It will not only be interesting to run this new experiments but also very nice to repeat the DAMA experiment on the other hemisphere where many sources of modulation have opposite phase, including cosmic rays, temperature and pressure and so on. But I don't know about the existence of underground labs in the other hemisphere or groups wanting to perform there this kind of experiment.

- *Badziak*

Is it possible that any dark matter experiment performed in Gran Sasso Laboratories will confirm or exclude the PAMELA satellite signal? Can these experiments probe the same mass range of dark matter that PAMELA reported?

- *Coccia:*

I don't think that PAMELA results are a convincing indication of dark matter. I think there are explanations which are much easier like supernovae or pulsars. The direct detection underground with present and future experiments should say something in a definite way about that.

- *Parke:*

Will the CNGS beam reach design intensity of 4.5 10^{19} POTS next year?

- *Coccia:*

This year we should reach 3.5 10^{19} POTS, so I am optimist for the next year.

- *Parke:*

Will the total number of POTS be 2.2 10^{21} after 5 years from the beam started (August 2006)?

- *Coccia:*

We expect a total of 2.2 10^{21}. So we cannot count 5 years from when the beam started, August 2006, because in the last years the provided beam was not at all the nominal one. So I think you have to count few more years to get the promised total amount of proton on target.

Highlights from Fermilab

Stephen J. Parke

Theoretical Physics Department,
Fermi National Accelerator Laboratory
P.O.Box 500, Batavia, IL 60510, USA
parke@fnal.gov

ABSTRACT

In these two lectures I will chose some highlights from the Tevatron experiments (CDF/D0) and the Neutrino experiments and then discuss the future direction of physics at Fermilab after the Tevatron collider era.

1. Tevatron Highlights:

Although there are many highlights of the Tevatron program, I have chosen to speak about those that are relevant for the Tevatron Higgs search: W-boson mass, Top quark mass, single Top production, di-Vector boson production and the Higgs search. But first I would like to say a little about the current status of the machine and the prospects for the future. Currently, the Tevatron has delivered ~ 7 fb^{-1} of data to each experiment with a peak luminosity of 3.5×10^{32} cm^{-2}sec^{-1}. The expectation is that by the end of fiscal 2011 (9/2012) each of the Tevatron experiments will have ~ 12 fb^{-1} of data of tape, essentially doubling the current amount of data collected.

At the peak instantaneous luminosity the number of anti-protons burnt by collisions is 7.0×10^7 sec^{-1} ($2 \times 3.5 \times 10^{32}$ cm^{-2} sec^{-1} $\times$ 100 mb $\approx 7.0 \times 10^7$ sec^{-1}) which is very close to the peak anti-proton collection rate of 7×10^7 sec^{-1}. Of course this a statement about the peak luminosity and peak collection rates but I think it illustrates the issue very effectively and thus, only modest increases in the luminosity of the Tevatron can be envisaged over the remaining life of the machine without significant improvements in the collection and handling of anti-proton.

In the Standard Model (SM) the Higgs mass is related to the mass of the W-boson and Top quark mass through radiative corrections. Thus, a precise determination of the W-boson mass and the Top quark mass will tell us the mass of the SM Higgs boson. However, the precision required is extreme, especially for the W-boson where a 6 MeV uncertainty is equivalent to a 1 GeV uncertainty on the Top quark mass on the range of allowed Higg masses.

At the Tevatron the cross section for producing a Higgs boson is at the pb level so that one needs to observe a number of background processes which have larger cross sections such as single top quark production and di-Vector boson production. Both of these two cross sections will be discussed before addressing the Tevatron Higgs searches. Many of the results reported in this section for the Tevatron experiments can be found the slides and proceedings of Lepton-Photon 2009[1]).

1.1. W-boson Mass

Determination of the W-boson mass at the Tevatron requires a detailed understanding of the detector response to both the charge lepton and jet P_T. The Tevatron Average (CDF+D0) for W-boson mass is

$$m_W = 80420 \pm 31 \text{ MeV}. \tag{1}$$

See Fig. 1. This result has a smaller uncertainty than the LEP2 average! When combine with LEP2 gives the World Average W-boson mass is

$$m_W = 80399 \pm 23 \text{ MeV}. \tag{2}$$

As will be seen shortly the uncertainty in the W-boson mass dominates our uncertainty on the expected mass of the SM Higgs boson.

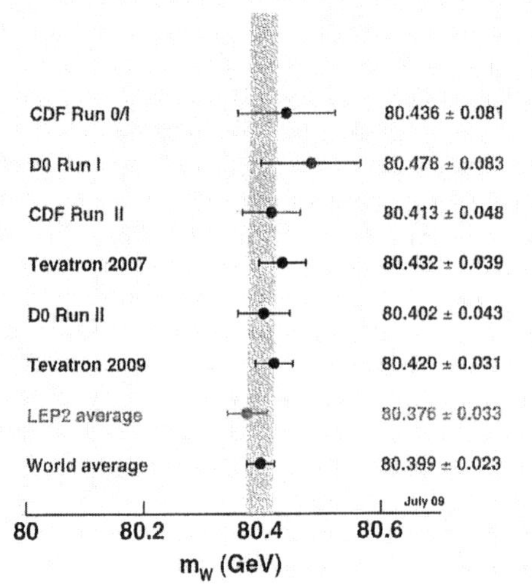

Figure 1: Summary of W-boson mass measurements from the various experiments.

1.2. Top Quark Mass

The mass of the Top quark has been measured in numerous channels and with a variety of techniques, see Fig. 2, and the Tevatron Average Top quark mass is given by

$$m_t = 173.1 \pm 1.3 \text{ GeV}. \tag{3}$$

D0 has also measured the difference in the mass of the Top and Anti-Top quarks and finds that the difference is given by

$$\Delta m_t = 3.8 \pm 3.7 \text{ GeV}. \tag{4}$$

The symmetry CPT implies that the mass difference between the Top and the anti-Top quarks should be strictly zero, hence, there is no evidence for CPT violation in the mass of the Top and anti-Top quarks.

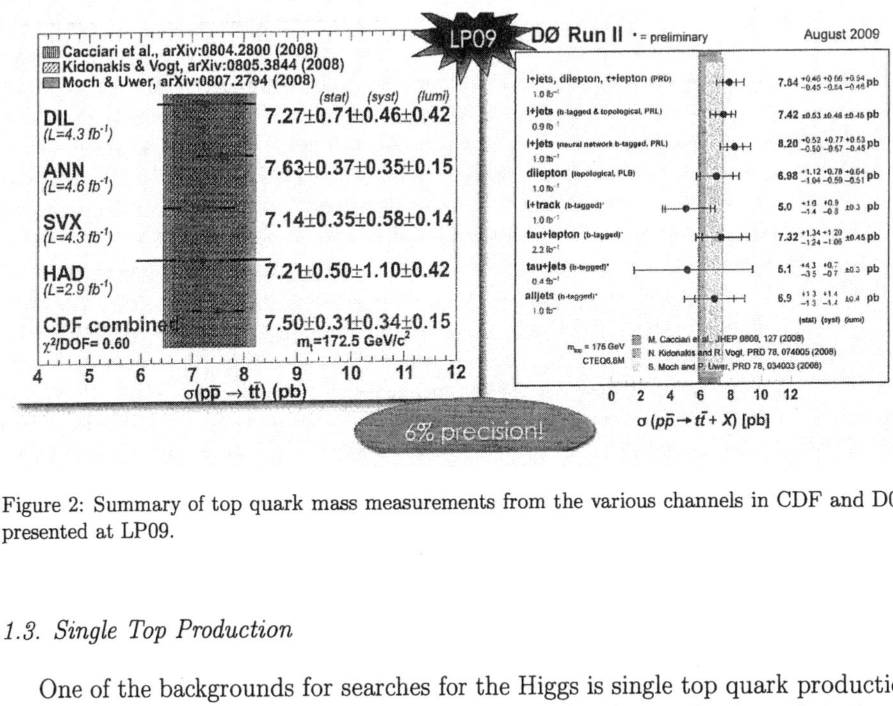

Figure 2: Summary of top quark mass measurements from the various channels in CDF and D0 as presented at LP09.

1.3. Single Top Production

One of the backgrounds for searches for the Higgs is single top quark production. Both CDF and D0 have better than 5σ evidence for the production of single top quarks. The combined CDF and D0 result for the sum of the s and t channel cross sections is given as

$$\sigma_t = 2.76^{+0.58}_{-0.47} \text{ pb}. \tag{5}$$

See Fig. 3. Understanding how to extract this signal is a very important milestone on the way to discovering the Higgs boson.

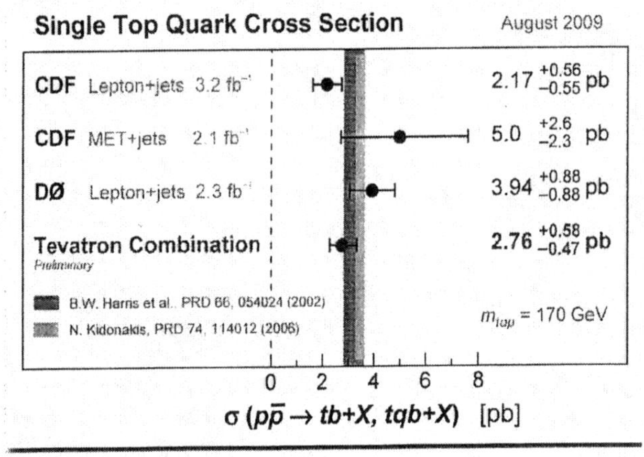

Figure 3: The single top cross section for CDF and D0 for the combined s and t channel processes.

1.4. Cross Sections for VV

Another milestone on the way to discovery of the Higgs boson is the observation of $WW/WZ/ZZ$. Both CDF and D0 have observed ZZ in the four charged lepton channel with a significance greater than 5σ. The cross sections are consistent with the SM cross section of 1.4 ± 0.1 pb. ZZ to dilepton plus dijets as well as ZZ dijets plus neutrinos have not been observed so far.

For the $WW/WZ/ZZ$ channels, CDF has observed then in the mode where one of the vector bosons decays hadronically recoiling against the leptonic decay of the other vector boson, see Fig. 4. The measured cross section is

$$\sigma(WW/WZ/ZZ) = 18.0 \pm 2.8 \pm 2.6 \text{ pb}. \tag{6}$$

The next milestone in Higgs search is to look for WZ and ZZ with $Z \to b\bar{b}$ the last benchmark before observation of the low mass Higgs in WH and ZH channels.

1.5. Search for the Higgs Boson

For the constraint coming from the mass of W boson and top quark see Fig. 5. For masses of the Higgs boson below 135 GeV, the dominant decay mode is $H \to b\bar{b}$ whereas for masses above 135 GeV the dominant decay mode is WW^*. For the high mass Higgs one can use the events from gluon fusion $gg \to H$ whereas for low mass Higgs one uses $q\bar{q} \to H + W$ or Z with the W or Z decaying leptonically. See Fig. 6. The combined current Tevatron limit is given in Fig. 7.

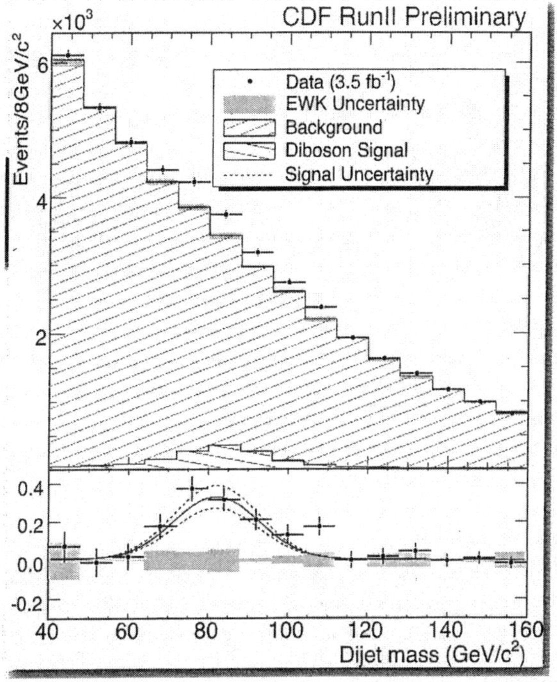

Figure 4: The dijet invariant mass distribution showing the contribution from W and Z bosons..

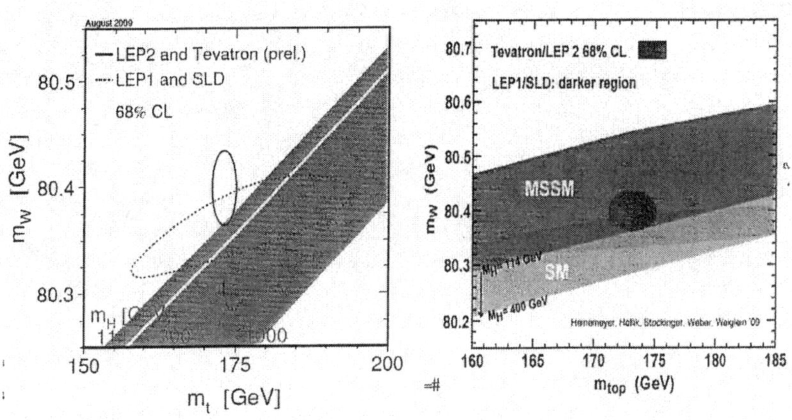

Figure 5: The constraint on the Higgs mass coming from the mass of the W-boson and the top quark. Left panel is for the Standard Model only whereas the right panel includes the allowed region for SUSY models. In the Standard Model the mass data favors light Higgs masses.

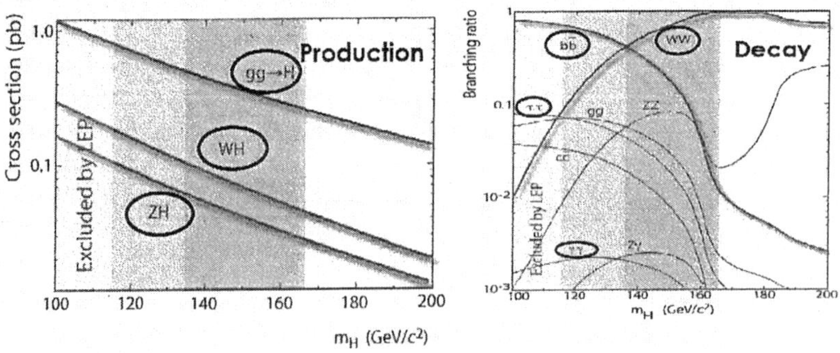

Figure 6: The production cross sections for the Higgs boson at the Tevatron along with the Higgs branching fraction to various decay modes.

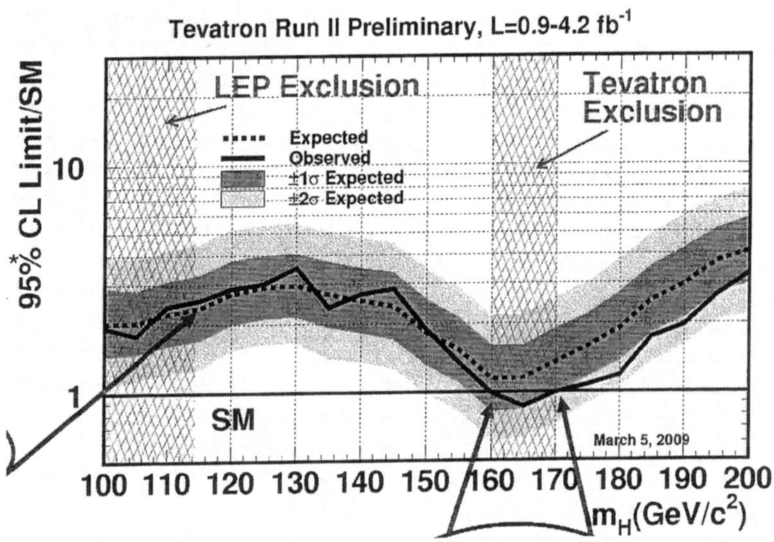

Figure 7: The combined CDF and D0 exclusion limits for the Higgs boson. At the 95% C.L. the SM Higgs is excluded between 160 and 170 GeV.

2. Fermilab Neutrinos, Future Program & Long Term Strategy:

2.1. Current Neutrino Program

Currently there are two operating neutrino beamlines at Fermilab with another being proposed. These are

- The Booster Neutrino Beamline which uses 8 GeV protons from the Booster. Currently only miniBooNE is taking data in this beamline since SciBooNE has completed it's data taking and mircoBooNE is still in the approval process.
- NuMI (Neutrinos from the Main Injector) which uses 120 GeV protons from the main injector. This beamline sends neutrinos to the MINOS near and far detectors. Also in the near detector hall is ArgoNut and MINERVA. ArgoNut is taking data and MINERVA will start data taking in 2010. The NOνA near and far detectors will also use neutrinos from this beamline.
- In addition there is a proposal to build a third neutrino beamline to send a neutrino beam from Fermilab to DUSEL in the Homestake mine. This beamline will initially be powered by protons from the main injector but will eventually be powered by a very high intensity proton beam (>2MW) from Project X.

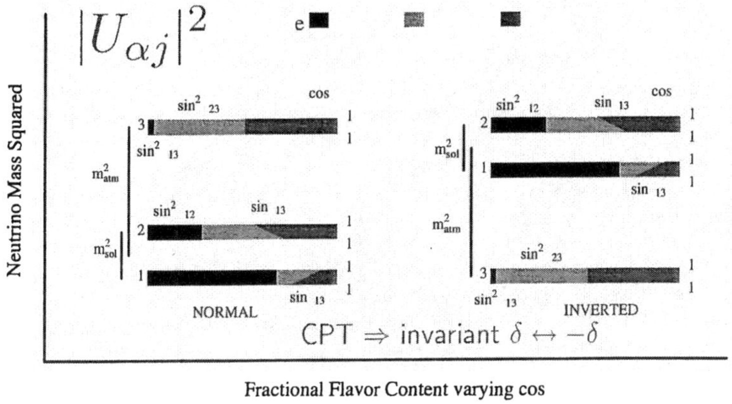

Figure 8: The flavor content of the neutrino mass eigenstates[3)]. The width of the lines is used to show how these fractions change as $\cos \delta_{CP}$ varies from -1 to +1. Of course, this figure must be the same for neutrinos and anti-neutrinos if CPT is conserved.

MINOS uses 120 GeV protons from the main injector with a nominal beam power of 300kW and currently has the world's best measurement of $|\Delta m^2_{32}|$ and is currently running in anti-neutrino mode to check whether or not CPT is violated in the neutrino sector. MiniBooNE and SciBooNE are measuring neutrino cross sections relevant for future neutrino experiments. Also, MiniBooNE is also looking for deviations from the three active neutrino Standard Model that has been assembled in the last ten years and in particular to confirming or refuting the LSND anomaly.

2.2. Neutrino Standard Model

The flavor content of the three neutrino mass eigenstates is given in Fig. 8. The current questions in this model are what is the size of

$$\sin^2\theta_{13}, \quad (\sin^2\theta_{23} - \frac{1}{2}), \quad (\sin^2\theta_{12} - \frac{1}{3}), \quad \mathrm{sign}(\delta m^2_{31}) \quad \text{and} \quad \sin\delta_{CP} \tag{7}$$

It is possible that the first three of these are related to some small parameter like $\delta m^2_{21}/\delta m^2_{31} \approx 0.03$ raised to some power, n. From current experiments the size of this power is constrained to be larger than one, $n > 1$, except for $(\sin^2\theta_{23} - \frac{1}{2})$ where the lower bound is weaker, approximately $n > 1/2$.

This could be due to some broken Tri-Bi-Maximal symmetry which would be very interesting and exciting. Another possibility is that it is coincidental which would be much less interesting from a theorists perspective.

2.3. NOνA

The NOνA project consists of increasing the beam power in NuMI (Neutrinos from the Main Injector) from 300 kW to 700 kW and to build the 15 kton liquid scintillator NOνA detector in Ash River, 810 km from Fermilab, so as to explore both $P(\nu_\mu \to \nu_e)$ and $P(\bar{\nu}_\mu \to \bar{\nu}_e)$ at the sub 1% level. This will put a limit on $\sin^2 2\theta_{13}$ at the 0.01 level, see Fig. 9. This experiment will also give us the first glimpse of the neutrino mass hierarchy and/or the first restrictions of the range of the CP violating phase δ_{CP}, see Fig. 10.

Also the MINERvA experiment operating in the NuMI near detector hall will perform precision measurements of neutrino cross sections above 1 GeV on various nuclear targets. The information obtained from this experiment will be invaluable for future very long baseline neutrino oscillations.

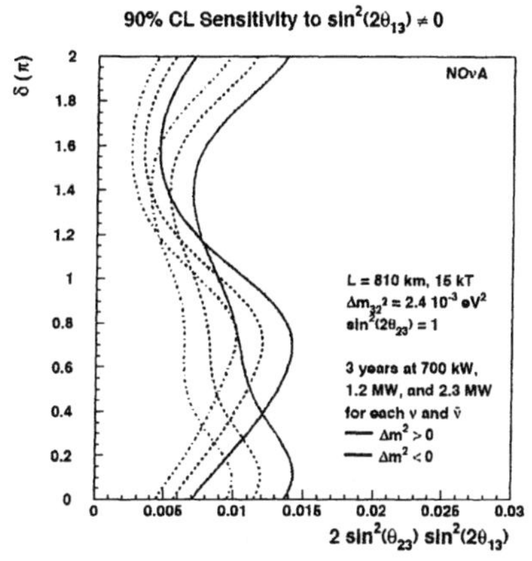

Figure 9: The 90% sensitivity to $\sin^2 2\theta_{13}$ for NOνA [2] assuming equal running time for neutrinos and anti-neutrinos. The blue (red) curves is for the normal (inverted) hierarchy. The three lines from right to left are for 0.7, 1,2 and 2.3 MW of protons on target respectively. These curves correspond to $P(\nu_\mu \to \nu_e)$ and $P(\bar{\nu}_\mu \to \bar{\nu}_e)$ at the sub 1% level.

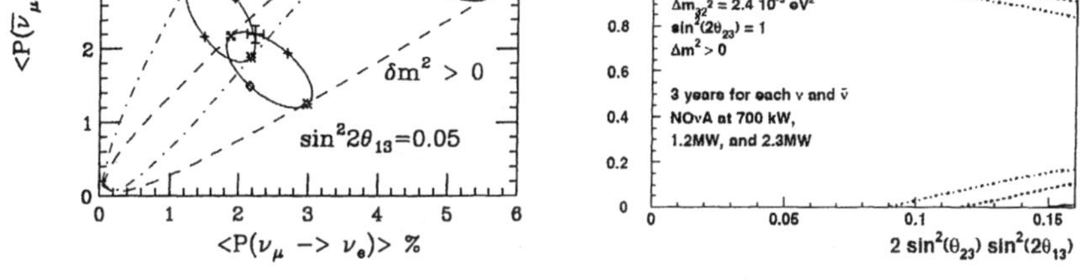

Figure 10: The left panel is the bi-probability plot, $\nu_\mu \to \nu_e$ verses $\bar{\nu}_\mu \to \bar{\nu}_e$ for the NOνA experiment showing the critical value of θ_{13} ($\sin^2 2\theta_{13} = 0.11$) as well as two (out of four) of the ellipses which pass through the data point (large cross) with $\sin^2 2\theta_{13} = 0.05$ [3] . The right panel shows the parameters in $\sin^2 2\theta_{13}$ v δ plane that NOνA [2] determines the hierarchy assuming it is normal.

2.4. The Physics of Long Baseline: $\boldsymbol{\nu_\mu \longrightarrow \nu_e}$

The amplitude for $\nu_\mu \to \nu_e$ can be simple written a sum of three amplitudes, one associated with each neutrino mass eigenstate,

$$U^*_{\mu 1} e^{-im_1^2 L/2E} U_{e1} + U^*_{\mu 2} e^{-im_2^2 L/2E} U_{e2} + U^*_{\mu 3} e^{-im_3^2 L/2E} U_{e3}.$$

The first term can be eliminated using the unitarity of the MNS matrix and thus the appearance probability can be written as follows[5)]

$$P(\nu_\mu \to \nu_e) \approx \left| \sqrt{P_{atm}} e^{-i(\Delta_{32}+\delta)} + \sqrt{P_{sol}} \right|^2. \tag{8}$$

Δ_{jk} is used as a shorthand for the the kinematic phase, $\delta m^2_{jk} L/4E$. As the notation suggests the amplitude $\sqrt{P_{atm}}$ only depends on δm^2_{31} and $\sqrt{P_{sol}}$ only depends on δm^2_{21}. For propagation in the matter, these amplitudes are simple given by

$$\begin{aligned} \sqrt{P_{atm}} &= \sin\theta_{23} \, \sin 2\theta_{13} \, \frac{\sin(\Delta_{31} - aL)}{(\Delta_{31} - aL)} \, \Delta_{31} \\ \sqrt{P_{sol}} &= \cos\theta_{23} \, \sin 2\theta_{12} \, \frac{\sin(aL)}{(aL)} \, \Delta_{21}. \end{aligned} \tag{9}$$

The matter potential is given by $a = G_F N_e/\sqrt{2} \approx (4000\ km)^{-1}$ and the sign of Δ_{31} (and Δ_{32}) determines the hierarchy; normal $\Delta_{31} > 0$ whereas inverted $\Delta_{31} < 0$. When a is set to zero one recovers the vacuum result. See Fig.11 [6)].

For anti-neutrinos $a \to -a$ and $\delta \to -\delta$. Thus the phase between $\sqrt{P_{atm}}$ and $\sqrt{P_{sol}}$ changes from $(\Delta_{32} + \delta)$ to $(\Delta_{32} - \delta)$. This changes the interference term from

$$2\sqrt{P_{atm}}\sqrt{P_{sol}} \cos(\Delta_{32} + \delta) \;\Rightarrow\; 2\sqrt{P_{atm}}\sqrt{P_{sol}} \cos(\Delta_{32} - \delta). \tag{10}$$

Expanding $\cos(\Delta_{32} \pm \delta)$, one has a CP conserving part $2\sqrt{P_{atm}}\sqrt{P_{sol}} \cos\Delta_{32} \cos\delta$ and the CP violating part

$$\mp 2\sqrt{P_{atm}}\sqrt{P_{sol}} \sin\Delta_{32} \sin\delta. \tag{11}$$

Therefore CP violation is maximum when $\Delta_{32} = (2n+1)\frac{\pi}{2}$ and grows as n grows. Notice also, that for this term to be non-zero the kinematical phase Δ_{32} cannot be $n\pi$. The first person who emails me having noticed this statement will get a prize. I'm checking to see if anybody reads these things. This is the neutrino counter part to the non-zero strong phase requirement for CP violation in the quark sector.

The asymmetry between $P(\nu_\mu \to \nu_e)$ and $P(\bar{\nu}_\mu \to \bar{\nu}_e)$ is a maximum when $\sqrt{P_{atm}} = \sqrt{P_{sol}}$. At the first oscillation maximum, $\Delta_{31} = \pi/2$, this occurs when $\sin^2 2\theta_{13} = 0.002$ in vacuum. For values of $\sin^2 2\theta_{13} < 0.002$ the oscillation probabilities are dominated by P_{sol} and thus observing the effects of non-zero $\sin^2 2\theta_{13}$ become increasing more challenging.

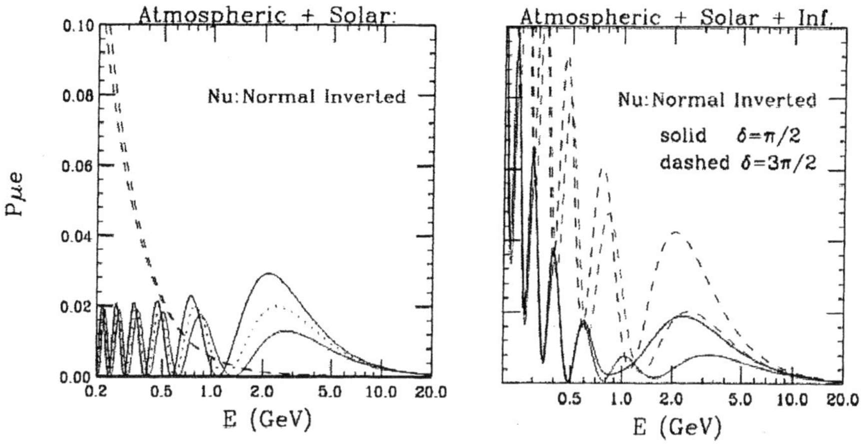

Figure 11: The left panel shows the two components P_{atm} and P_{sol} in matter for the normal and inverted hierarchies for $\sin^2 2\theta_{13} = 0.04$ and a baseline of 1200 km. The right panel shows the total probability including the interference term between the two components for various values of the CP phase δ for the neutrino. Notice that the coherent sum of two amplitudes shows a rich structure depending on the hierarchy and value of CP phase. These curves can also be interpreted as anti-neutrino probabilities if one interchanges the hierarchy AND the values of the CP phase.

2.5. DUSEL and LBNE

Even if $\sin^2 2\theta_{13} > 0.01$, a further experiment will be needed to determine the precise value of the CP violating phase, δ_{CP}. Fermilab has proposed to build a new neutrino beamline to DUSEL and to build a large neutrino detector there. This experiment is called the Long Baseline Neutrino Experiment (LBNE). The detector would consist of 300 ktons of water Cerenkov detectors or 50 ktons of Liquid Argon detectors or some combination of both.

The new neutrino beamline would be initially powered by the main injector but then by Project X which would deliver 2.2 MW of protons at 50-120 GeV. Project X in addition could provide up to 2 MW of protons at 2-3 GeV for Kaon, Muon and other experiments.

The detector technology would be either water cerenkov similar to SuperK or a LAr TPC or a combination of both. The water cerenkov detector could be built with little R&D but would have to be larger than LAr due to it's lower efficiency. Whereas substantial R&D is required for a large LAr TPC but this technology has a higher efficiency than water cervenkov due to it's better discrimination of electron and gamma (π^0) events. LAr also has a enhanced sensitivity to proton decay in the $K^+\nu$ channel over water Cerenkov which also makes it an attractive alternative assuming a successful R&D program. For both detector technology a modular design is probably necessary to get the very large fiducial volumes required. If affordable, a combination of both detectors would be very powerful. For a possible evolution of this R&D see Fig.12.

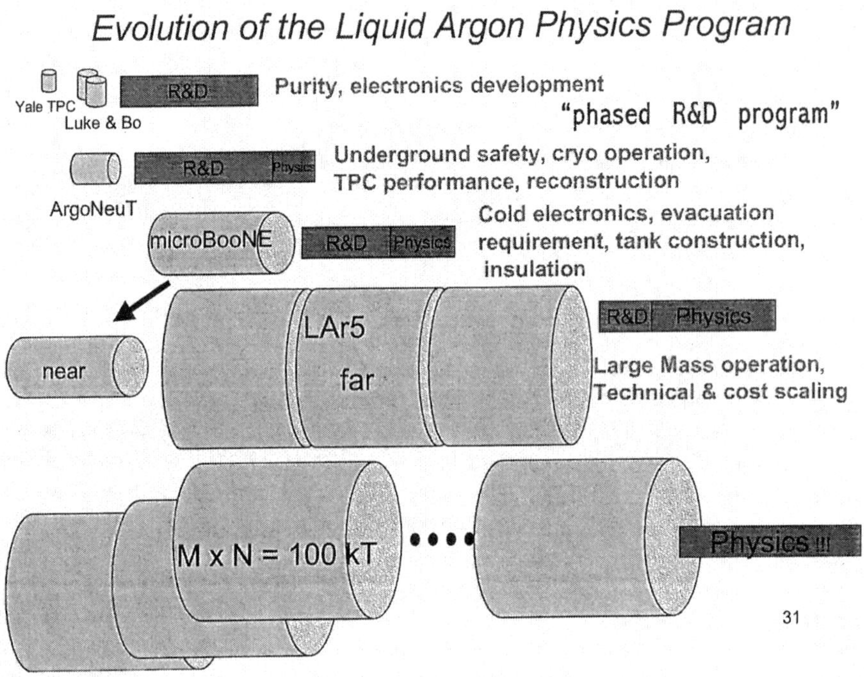

Figure 12: Possible evolution of a Liquid Argon detector program[4] in North America.

The reach for $\sin^2 2\theta_{13}$, the mass hierarchy and CP violating for the Fermilab Project X program is given in Fig.13. If $\sin^2 2\theta_{13}$ is significantly smaller than 10^{-3} or precision measurements of the oscillation parameters are required then protons from Project X could be used for a neutrino factory with detector(s) at DUSEL. With further technical developments these facilities could also be used for a future multi-TeV muon collider.

A brief outline of Project X is given and a possible neutrino program that could be performed. At this stage many options need to be studied in detail especially with regard to various funding profiles. However, it is clear that an intense neutrino source combined with very massive detectors is required to explore the size of θ_{13}, the

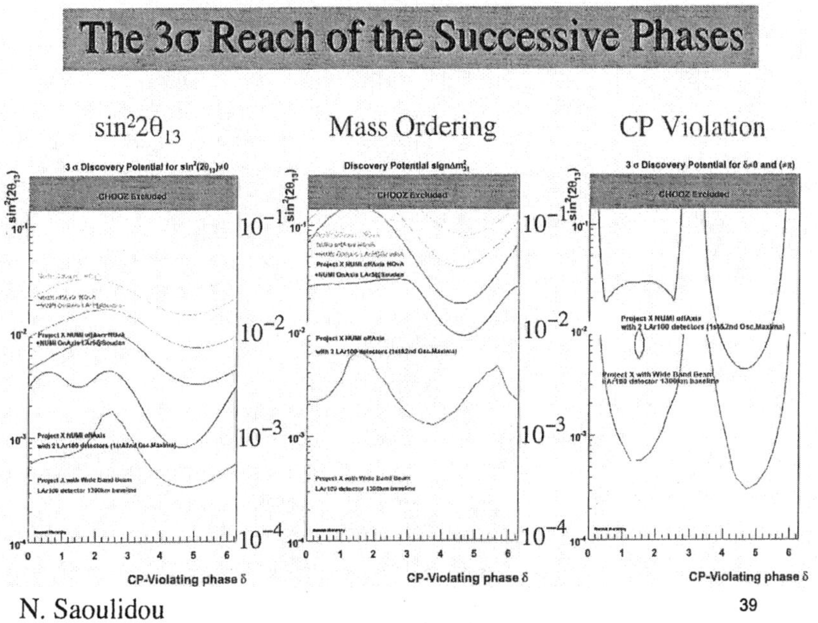

Figure 13: The evolution of the reach for $\sin^2 2\theta_{13}$, the mass hierarchy and CP violating for the Fermilab Project X program [7].

neutrino mass hierarchy and the CP violation in the neutrino sector.

2.6. Project X Summary

Project X with it's multi-megawatt high energy beams for neutrino physics and it's multi-megawatt low energy beams for rare Kaon physics, Muon physics and possibly other projects is a true Intensity Frontier Machine. The neutrino experiments have a clear goal to determine CP violation in the neutrino sector and determine the neutrino mass hierarchy. The Kaon program would look for 1000 events in both $K^+ \to \pi^+\nu\bar{\nu}$ and $K^0 \to \pi^0\nu\bar{\nu}$. The Muon program would look for $\mu + N \to e + N$ at unprecedented sensitivity as well as to push for a more precise measurement of (g-2)$_\mu$. Further in time is the possibility to use Project X to power a neutrino factory and then to regain the energy frontier with a multi-TeV muon collider. It is an exciting project, why don't you come an join us?

3. Dedication

I would like to dedicate this article to my thesis advisor, Sidney Coleman, who was a regular lecturer at this school and was always voted "best lecturer" whenever he appeared. I owe a great deal to Sidney. He demonstrated to me how theoretical physics research is performed by a true master. He also shared many other aspects of life as a theoretical physicist.

Figure 14: Sidney Coleman on the top of Mt. Monadnock, New Hampshire, USA in the fall of 1979. (Photo by S. Parke.)

Students[8] of Sidney Coleman

Robert Socolow 1964	Wu-Yang Tsai 1971	Stephen Parke 1980	Robert Mawhinney 1987
Jeffrey Mandula 1966	Erick Weinberg 1973	Fred Posner 1980	Alexander Safian 1987
Leonard Parker 1967	David Politzer* 1974	TheodorosTomaras 1980	Brian Hill 1988
Stephen Fels 1968	James Butler 1974	Bernard Grossman 1981	Paul Renteln 1988
Carl Bender 1969	Eldad Gildener 1975	Ahmed Abouelsaood 1984	Mark Alford 1990
Arnold Cantor 1970	Paul Steinhardt 1978	Philip Nelson 1984	John March-Russel 1990
Anthony Zee 1970	Ian Affleck 1979	Richard Woodard 1984	Katherine Benson 1991
John Mansfield 1970	Frank DeLuccia 1979	Gregory Moore 1985	Shane Hughes 1993
David Griffiths 1970	Lee Smolin 1979	MinosAxenides 1987	SteliosSmirnakis 1997
Lawrence Thebaud 1971	Gerald Sobelman 1979	Jacques Distler 1987	Nathan Salwen 2001

4. Acknowledgements

I would like to thank the Directors of this school, Prof. Zichichi and Prof. 't Hooft, for the opportunity to present the highlights from the Fermilab program at this school as well as to Fabiola Gianotti for showing me how to do this in an effective manner.

5. References

1) Lepton Photon 2009 especially the talks by Gregorio Bernardi and Florencia Canelli. See http://lp09.desy.de/
2) D. S. Ayres *et al.* [NOvA Collaboration], arXiv:hep-ex/0503053.
3) O. Mena and S. J. Parke, Phys. Rev. D **70**, 093011 (2004) [arXiv:hep-ph/0408070].
4) From the "Project X workshop Series", FNAL, 2007-2008.
5) A. Cervera, A. Donini, M. B. Gavela, J. J. Gomez Cadenas, P. Hernandez, O. Mena and S. Rigolin, Nucl. Phys. B **579**, 17 (2000) [Erratum-ibid. B **593**, 731 (2001)] [arXiv:hep-ph/0002108].
6) H. Nunokawa, S. J. Parke and J. W. F. Valle, Prog. Part. Nucl. Phys. **60**, 338 (2008) [arXiv:0710.0554 [hep-ph]].
7) "Fermilab Steering Group Report," CITATION = FERMILAB-PUB-07-672-DO;
8) "QFT & QCD: Past, Present and Future", Harvard University, Cambridge, Massachusetts March 18-19, 2005. See http://www.physics.harvard.edu/QFT/

CHAIRMAN: S. PARKE

Scientific Secretaries: G. Inguglia, T. Orimoto

DISCUSSION I

- *Preghenella:*

The Tevatron experiments and detectors are well known, and the analysis should be pretty well-tuned and defined. My question is would it be possible to run only one experiment to enhance the luminosity for free?

- *Parke:*

No, you would still have the same total number of collisions. The total number of collisions is controlled by how many antiprotons you can get into the machine, whether you burn them all in one experiment or half in CDF and half in D0. Also, we don't know how to squeeze the antiproton beam at CDF or D0 to make the instantaneous luminosity go up by a factor of two at that intersection point.

- *Ferrara:*

I want just to reformulate the question of this morning, about limits of the supersymmetric Higgs.

- *Parke:*

You can find the current limits in Bernardi's talk at Lepton-Photon 2009. Here is brief summary. With 10 fb^{-1} and 50% sensitivity improvements, using all the channels, the Tevatron will probe signal regions of the MSSM parameter space to 95% confidence level and all of the parameter space at 90% confidence level. So we should have some hints of supersymmetry, if MSSM is what nature has chosen.

- *Perrey:*

What are the advantages of using antiprotons compared to a proton-proton collider? Is it just the beam pipe or are there other advantages?

- *Parke:*

No, the Tevatron is a low energy machine, so that if you want to study top quarks, the cross section at proton-proton is about a fifth that of proton-antiproton, so that you'd have to have five times the luminosity. People studied this with the SSC and also the LHC in great deal, and they came to the conclusion that for a high energy of a machine you want proton-proton. But the Tevatron is a low energy machine.

- *Cerchiai:*

Could you tell us more about how you do the data analysis since there are more challenges due to the fact that there is a small signal to background ratio?

- *Parke:*

What I think is very, very important at this point is you need to have more than one type of analysis, since the signal to background is getting smaller and smaller. There are neural networks, boosted decision trees, matrix elements methods etc. All of these analyses should be done as independently as possible, and then compared at the end to decide whether they agree together or not.

- *Albrecht:*

I have a question to the di-vector boson decay, which decays completely hadronically. You said that the next step would be to measure the decay to b.

- *Parke:*

Yes, the figure I showed was the invariant mass of the dijets, where you have leptons recoiling against them, and we see an enhancement over the normal QCD background. So the obvious question you would like to ask is to do the same thing, but now tag the jets to find out what fraction of them are b-jets. Then make the same plot, and check the W-Z channel with the Z going to bbar, because that is a background to W-Higgs. If you can see that, then there is some hope that then you could start to see W-Higgs with Higgs going to bbar

- *Albrecht:*

How efficient is b-jet tagging?

- *Parke:*

Typically b-jet tagging efficiency is around 50-70 percent. You can apply a tight cut or a loose cut. If you apply a loose cut, you get more events, but you also have more fakes.

- *Corcella:*

Do you have any chance to see W-W scattering at the Tevatron? Or the statistics is too low?

- *Parke:*

Well, it's a combination of energy and luminosity. You need a fairly high energy machine and high luminosity to look for W-W. I don't remember the exact cross section at the Tevatron but it is very small.

- *Dixon:*

At least parts of a RunII-b detector upgrade did not happen; the worry was that the detectors would become radiation damaged. Is there any sign of that, or what are the expectations as you start to accumulate these really large integrated luminosities?

- *Parke:*

Yes, some of the inner detectors were supposed to be replaced at some point, and that got cancelled for various reasons. They do see damage to the zero-zero layers (the very inner-side layers). However, they are able to compensate for that by changing the voltage on the silicon vertex detectors. They don't expect to have any serious degradation until you get to about 15/fb, and that is sort of the natural scale that I've been talking about. We have 7/fb on tape now and doubling that will take you to that point, and doubling again will take a long time.

CHAIRMAN: S. PARKE

Scientific Secretaries: J. Albrecht, M. Williams

DISCUSSION II

- *Preghenella:*

Looking at the Argoneut event display you showed this morning, it seems that the tracks go straight. Don't they use a magnetic field in the TPC?

- *Parke:*

No. I don't believe anybody's ever built a liquid argon TPC with a magnetic field. Eventually you want to build a TPC with a magnetic field so you can see charged tracks, and see them curve. This has not been done. There are some issues with the materials you use if you have a magnetic field. This is a question people are thinking about for the future.

- *Epifanov:*

Are there any plans to use the muon collider as a source of muonic neutrinos and antineutrinos?

- *Parke:*

A neutrino factory is just a muon storage ring with long straight sections. Looking at slide 56, the collider is just the red circle – there are no straight sections on that. The straight sections on the storage ring are just for acceleration. The muons undergo two or three turns in the accelerator before they get put into the collider, which is deep underground. What you're talking about is an add-on to this. All the money you're spending is on the front section, in the muon cooling. For a neutrino factory you just need an extra storage ring with long straight sections, to send a neutrino beam to your favorite detector, whether it be in Dusel, or much further away. The way people have envisaged this at the moment is to go for the muon collider, and tack on a neutrino factory.

- *Perrey:*

How would the muon cooling work, and how do you avoid muon decay?

- *Parke:*

Muons decay very rapidly, so you have to accelerate them as quickly as possible, to get the gamma enhancement. To cool muons, you deaccelerate them somehow, and then reaccelerate them in one direction. You make the muons pass through material (like a gas), and that slows them down in all directions, and then you accelerate them in the z-direction. Now they are cooled in the x and y directions. Then you repeat this process. The problem is to do this quickly enough so that they don't decay. That's where the technology goes: into the cooling and accelerating of these muons: that's where the R&D will be concentrated.

- *Perrey:*

So the cooling is just for focusing the muons, and not to stop them decaying?

- *Parke:*

If you want to put muons into a storage ring, you must reduce the six-dimensional phase space, so you can put them in a bunch: that's what I mean by cooling. You do this by deaccelerating in all directions and reaccelerating in one. People have ideas about how to do this, but they involve things like 40 Tesla magnets, so you're pushing the limits of all sorts of technologies. I think a lot of R&D needs to be done before this becomes viable, but I do think that we need to do this R&D because this (a muon collider) may be the only option. Suppose that for some reason CLIC is not viable: you want to explore all the options and then at some point later on make a decision of which is the way to go.

- *Baldini:*

You mentioned a two sigma evidence for CPT violation. Can you please show the data?

- *Parke:*

Slide 65 shows the anti-neutrino data for the MINOS detector. The expectation, if there were no oscillations, would be the red curve. The CPT-conserving solution would be the blue curve, and the data is consistently below that. The observation is 42 events, 'CPT' predicts 58, and 'no oscillations' predicts 64, with the error-bars given here: it's not a lot of data. These are the anti-neutrino events in the neutrino running mode. The muon antineutrinos comprise 5-6% of the beam, so it's important to tune your Monte Carlo accurately to model this sub-dominant component. That's why they want to run with the horn switched to focus the negative pions, giving you anti-neutrinos. Slide 66 shows the allowed region in the ΔM^2-$\sin^2\theta$ space. The CPT conserving solution is included at around 1.9σ. It makes sense to run with anti-neutrinos and clear this up, to see whether neutrinos and anti-neutrinos are CPT-conserving. It's a low apple to pick, and you can make a statement fairly quickly: you only need three months of running to get a very good handle on ΔM^2. Of course it will not be as good as the neutrino running, but good enough to see if the results are in the same ball park.

- *Badziak:*

In the proposed muon collider, how many turns do the muons make before they decay (at maximal energy – 2 TeV)?

- *Parke:*

The muons make of the order of a thousand turns of this machine before they decay. You can calculate it: the circumference of the storage ring is around six kilometers, the muons are at 2 TeV and you know their lifetime. They're not around very long. At the Tevatron a typical p-pbar store last for twelve hours: colliding muons is a lot harder.

- *Davey:*

What kind of timescale is the muon collider working on?

- *Parke:*

I'm happy to discuss this in general terms. Looking at slide 47, you can see a list of projects. What I consider to be near-term is between now and 2015, mid-term is maybe 2015 to 2025, and long-term is after that. The uncertainties on these dates are +-5 years.

- *Chiu:*

How well does the LAR detector work as a proton decay detector?

- *Parke:*

Liquid argon is a very good detector to look for the K+ neutrino decay channel of proton decay, it's much better than water Cherenkov. The problem with water Cherenkov in this particular decay mode is that the K+ is below the Cherenkov threshold, so the efficiency is very low. Whereas LAR, the efficiency is quite high. In the standard decay mode (pi^0 e+), it will be very hard to beat SuperK, because SuperK is 50 kton that's been running for 10 years, maybe 20 years by the time this has been built. Things are changing very rapidly. I think the LAR is a much better idea for proton decay because you are exploring a new channel that hasn't been done very well before.

- *Chiu:*

How about as a Dark Matter detector?

- *Parke:*

You have to get your backgrounds to be exceptionally low for dark matter searches, and I think in such a big detector this will be very hard.

- *Albrecht:*

Can you sketch the differences between the rare Kaon program of the Tevatron and the CERN Na62 experiment?

- *Parke:*

The CERN Na62 experiment for K+ → pi+ v vbar is aiming for a 100 events, which gives you a statistical uncertainty of the branching fraction of 10%. This is good, but it's not down to the theoretical uncertainty. To get much above 100 events, you have to invent something new. Some of the technology of Na62 experiment was part of the CKM experiment that we were trying to do at Fermilab and that's basically been imported by the CERN group.

- *Albrecht:*

What is the main difference? The more intense Kaon beam?

- *Parke:*

That's part of it and you have to figure out ways to get your backgrounds way down. These are extremely difficult experiments; you are looking for processes that have a branching fraction of 10E-10, so you have to know incredible things about your detector.

The other decay mode, K0 → pi0 v vbar, is even harder to measure, because it is basically considered to be "nothing in, nothing out". You don't see the K0 coming in and you don't see the neutrinos going out, all you see is the pi0 decaying to two photons. The current limit on the decay mode is approximately 1000 times of what the Standard Model says it is, so there is a lot of room for improvement, but this is an exceptionally hard experiment.

- *Badziak:*

To what extent can we be sure that a difference in mass of top and anti-top, if discovered, would imply CPT violation?

- *Parke:*

CPT implies that particle and antiparticle have the same mass. So if you could do a measurement where you are convinced that it is correct, that the mass of the top quark is different than the mass of the anti-top quark, you would have evidence for CPT violation.

The error on this measurement by D0 is about (4+–4) MeV. This is a very good measurement of this difference.

- *Haidt:*

You mentioned the low energy neutrino beam. What is in the program foreseen?

- *Parke:*

The boosted neutrino beam has a low energy tail which goes down quite low. This is the beam that is used by miniBooNE and SciBooNE, but I think they probably don't see the neutrinos down below 100MeV. The miniBoone and SciBoone have a collection of data below 1GeV. In fact, the mini-Boone experiment, which was designed to either confirm or refute LSND, I think their great legacy will be in cross section measurements below 1GeV, both in the neutrino channel and the anti-neutrino channel. The estimate is that they will have in the next year nine papers on neutrino cross sections in different modes. This is very important for an experiment like T2K, the competitor of the Nova experiment because you need to know these cross sections as well as you can, especially if you are looking for CP violation.

Radiation Damage Studies for Silicon Sensors for the XFEL*

H. Perrey †

Abstract

For the study of radiation damage of silicon sensors by 12 keV X-rays for doses up to 1 GGy an irradiation facility has been set up at HASYLAB at DESY. Test structures (gate-controlled diodes) have been irradiated and the properties of the Si-SiO_2 interface under high irradiation have been studied using I/V, C/V, and TDRC measurements. In addition to a strong increase of the interface current and a large shift of the flat-band voltage, strong hysteresis effects have been found. The data can be qualitatively described by a model which includes interface traps, fixed and mobile oxide charges. It is found that above doses of several MGy the density of interface traps decreases, whereas the density of fixed and mobile oxide charges appears to saturate. The origin of these effects is not understood so far.

1 Introduction

Planed to be operational in 2014 the European X-ray free-electron laser (XFEL) will set new standards in brilliance for X-ray sources. With laser pulses of the order of 10 fs duration at wavelengths of about 0.1 nm the XFEL will open up new possibilities for many fields of research such as the analysis of single biomolecules, the recording of chemical reactions, and the processes inside a plasma.

Experiments at the XFEL pose unprecedented requirements to the detectors in terms of radiation tolerance: fluxes of up to 10^{16}[12 keV photons/ cm^2], corresponding to approximately 10^9 Gy in SiO_2, are expected. As the threshold for bulk damage is 300 keV, only surface damage is expected: build-up of

*Work partially supported by the European XFEL Company

†Institute for Experimental Physics, University of Hamburg, Germany

charges inside the oxide and of energy states ("traps") at the Si-SiO_2 interface. While both effects will change the field distribution, which is relevant for the device stability, the latter will also result in an increase of the dark current, which impacts the read-out electronics and may result in an increase of the noise.

The aim of this study is to determine the relevant parameters for device simulations with the help of test structures and thus allowing to reliably predict the sensor performance as function of dose. The work is part of the detector R&D for the European X-Ray Free-Electron-Laser, XFEL.

2 Test-Structures and Measurement Techniques

Gate-controlled diodes have been chosen as test-structures for this analysis. They are especially suited for this type of study as they allow the measurement of both diode and MOS (Metal-Oxide-Silicon) characteristics using one structure. Fig. 1 shows the gate-controlled diodes used for the study and sketches the measurement techniques used:

- C/V (Capacitance-Voltage) on the MOS capacitor realized by gate rings 2 and 3,
- I/V_{gate} (Current-Voltage) on gate ring 1 and the diode,
- TDRC (Thermally Depolarization Relaxation Current) on gate rings 2 and 3.

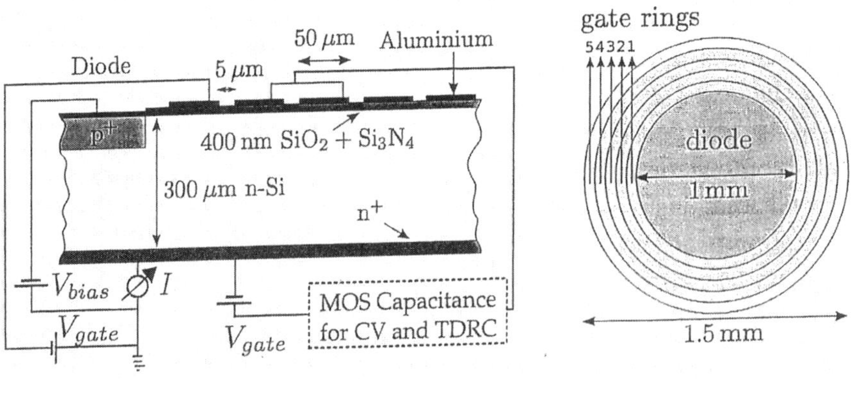

Figure 1: Schematic drawings of the gate-controlled diode: cross section with measurement techniques (*left*) and top view (*right*).

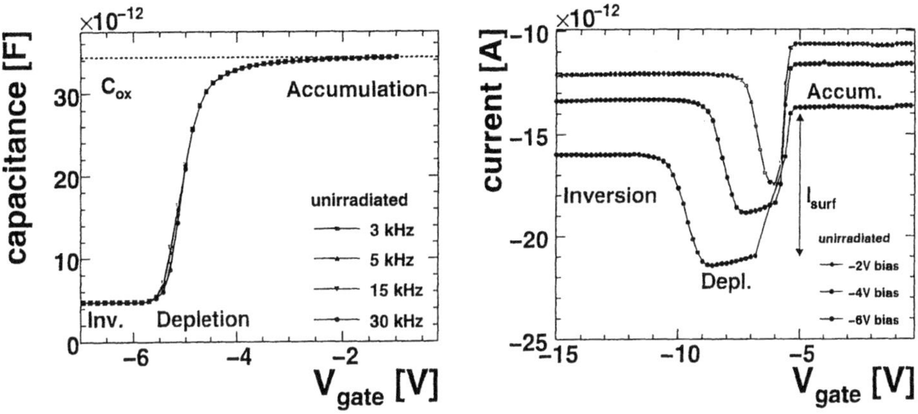

Figure 2: Examples for a C/V measurement (*left*) and an I/V measurement (*right*) of an unirradiated gate-controlled diode.

In *C/V measurements* the capacitance is determined as function of the voltage V_{gate} applied to the second and third gate ring while V_{gate} is modulated by a small-signal variation of frequency f_{AC} and amplitude V_{AC}. Figure 2 shows an example of a C/V measurement on an unirradiated sample at four different frequencies with $V_{AC} = 0.05\,\text{V}$.

With increasing negative voltage applied to the second and third gate rings the MOS-capacitor goes through three different conditions resulting in the characteristic shape of the C/V curve: *accumulation*, *depletion*, and *inversion*. These are illustrated in figure 3.

The MOS is said to be under accumulation conditions as long as the net electric field at the surface is positive, resulting in electrons accumulating at the surface. The measured capacitance in this case is equal to the oxide capacitance, C_{ox}, which is determined by the capacitor formed by the metal gates and the silicon with the oxide as dielectric.

The measured capacitance does not change with the gate voltage until the *flat-band voltage*, V_{fb}, is approached. When exceeded the net electric field at the surface becomes negative. It then pushes away electrons thus creating a region depleted of mobile charge carriers underneath the gates. This condition is called depletion. The depleted region acts as capacitor in series to the oxide capacitor therefore reducing the total measured capacitance.

The depletion region grows with increasing gate voltage until inversion is reached where the strong negative field causes holes to form a thin positively charged inversion layer underneath the gates. This charged layer effectively shields the depleted region from any further increase of the field. The mea-

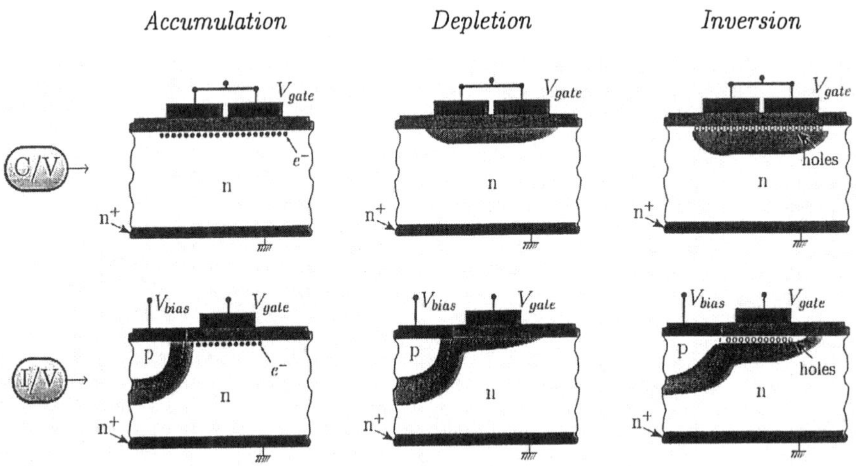

Figure 3: Charge distributions under accumulation, depletion, and inversion conditions for C/V measurements (*top*) and I/V measurements (*bottom*). The red colored areas indicate a region depleted of charge carriers.

sured capacitance therefore does not change anymore with increasing gate voltage.

Besides the bias applied to the gate the net electric field at the surface also depends on charges present in the oxide, N_{ox}, and charged energy states at the interface, N_{it}. Both cause a shift of the measured curve relative to the ideal C/V curve. Additionally, the interface states introduce a frequency dependence into the C/V measurements [1]: As the states differ in the time needed to charge and discharge, some are not "fast" enough to follow the applied small-signal voltage variation at frequency f_0, and therefore do not influence the measured capacitance $C(f_0)$. As shown in figure 2 for unirradiated samples the frequency dependence of the C/V curve is small indicating the presence of relatively few interface states.

Another measurement technique sensitive to the number of interface states is the *I/V measurement*: the current through the diode is measured versus the voltage V_{gate} applied to the first gate ring while a constant bias is applied between the diode and the backside contact. Figure 1 shows an illustration of the setup and figure 2 (right hand side) shows an example of the resulting I/V curve for different V_{bias} values.

Since in I/V measurements the voltage applied the gate ring is varied, the same biasing conditions as in C/V measurements occur: With increasing negative voltage on the first gate ring the area underneath the gate goes from accumulation to depletion and finally into inversion as illustrated in figure 3.

But only in the case of depletion the measured current changes significantly: When the depleted region underneath the first gate ring merges with the depleted part of the *pn*-junction of the diode, the accompanying electric field transfers free charge carriers generated at the Si-SiO_2 interface to the diode where they are measured. This current is referred to as *surface generation current*[1], I_{surf} [2]. When inversion is reached the interface is isolated from the field region through a thin inversion layer, therefore terminating I_{surf}. Consequently, the measured current drops to the value seen in accumulation plus a small generation current from the depleted region under the gate ring. Through the surface current the I/V measurement depends on the number of interface states, N_{it}.

The third measurement technique, *TDRC measurements*[3], allows to directly determine the density of interface states: the gate-controlled diode is first cooled to 30 K while biased at zero gate voltage (accumulation), thus filling all interface traps with electrons. Then V_{gate} is applied to the second and third gate rings and the sample is heated up with the constant rate of 0.183 K/s while the current from the discharge of the filled traps is measured as function of the temperature.

The measurement is performed for values of V_{gate} between -20 and -100 V. From the TDRC spectrum the density of interface traps, D_{it}, as function of the energy distance from the conduction band, $E_c - E_t$, is obtained. Fig. 6 shows the results of such a measurement for an irradiated sample[2] of 3.5 MGy.

The data show up to which energy in the band gap the interface traps are filled as function of the gate voltage V_{gate}. Up to approximately 0.45 eV the value derived for the interface trap density D_{it} is independent of the gate voltage, whereas at higher gate voltages a significant increase of D_{it} is observed. We interpret this as evidence for mobile oxide charges due to irradiation. These will be discussed in more detail in the next section.

3 Results

A total of 17 gate-controlled diodes have been irradiated in various steps to doses of 1 kGy–1 GGy. Immediately[3] after each irradiation step C/V, I/V, and later also TDRC measurements were performed. This allows to study the dose dependence of both the measured data and the resultant extracted

---

[1] There are different names in the literature: surface generation current, interface current, oxide current

[2] The TDRC signal for an unirradiated gate-controlled diode is below the noise-level.

[3] All measurements on irradiated structures have been performed within 4 h of the irradiation unless stated otherwise.

interface parameters.

Figure 4 (left) shows I/V measurements of the same gate-controlled diode for different irradiation doses. While the curve for the unirradiated sample is barely visible the current is increased considerably after irradiation. In fact, the surface generation current, I_{surf}, as determined from the curves, increased by two orders of magnitude after irradiation to 2 kGy and increases further for higher doses.

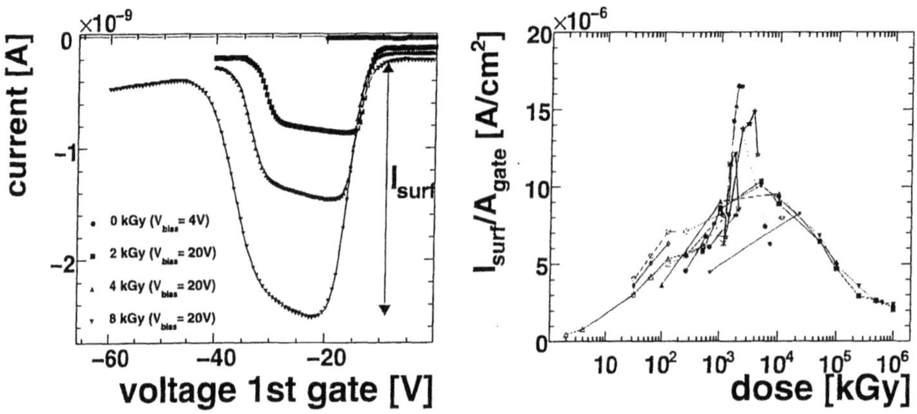

Figure 4: *Left:* I/V measurements for three different irradiation steps of the same sample. The curve for the unirradiated diode is included for comparison. *Right:* The surface current normalized to area versus dose. Shown are results for all diodes.

In figure 4 (right) the surface current normalized to gate area, I_{surf}/A_{gate}, is shown versus the dose for all 17 diodes and all irradiation steps. The plot shows that I_{surf}/A_{gate} reaches a maximum of $\approx 15\,\mu A/cm^2$ at approximately 2 MGy and then drops by almost a factor 5 when increasing the dose to 1 GGy. This behavior is independent of the dose rate during irradiation as the curve is reproduced by samples irradiated at less than 10% of the maximum dose rate. Monitoring of the temperature during irradiation indicates that temperature induced annealing is not the cause of this effect.

A similar behavior can be observed in the C/V measurements. The curves shown on the left in figure 5 exhibit both a change in slope and a shift in gate voltage depending on the irradiation dose. For doses below 5 MGy the curves are shifted toward more negative gate voltages with higher doses. For doses higher than 10 MGy the shift decreases with increasing dose.

As function of the dose, this shift[4] in gate voltage shows a similar behavior

[4]The shift can be parametrized e.g. by the gate voltage at which a specific capacitance

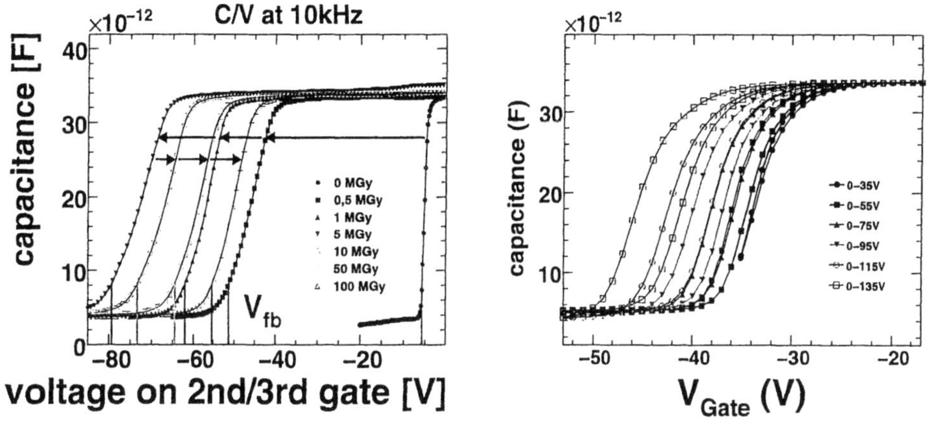

Figure 5: The capacitance versus the gate voltage measured at $f_{AC} = 10\,\text{kHz}$. *Left:* C/V for an unirradiated sample and six subsequent irradiation steps. The arrows highlight the shift of the curves. *Right:* C/V of a 1 MGy sample. The gate voltage is varied from 0 V to V_{gate}^{max} and back to 0 V for increasing maximum gate voltages V_{gate}^{max}.

as the I_{surf}/A_{gate} curve. The shift reaches a maximum at about 2 MGy and drops by a factor of two when increasing the dose to 1 GGy[4].

C/V measurements of irradiated samples also show strong "hysteresis" effects as shown in the right plot of figure 5: there is a significant shift between the "up" and "down" curve[5]. The shape of the C/V curve however is not affected. The voltage shift of the "up" curve depends on the history and storage condition of the diode. The voltage shift of the "down" curve depends on the maximum value of V_{gate} to which the MOS capacitor was biased and the time it remained at this voltage. We interpret this as evidence for mobile oxide charges close to the interface.

These mobile oxide charges slowly build up during biasing and are stable for up to several days. Thus they effectively prevent reproducible measurements of C/V curves. Only the very first "up" curve measured can be assumed not to be distorted by mobile oxide charges.

To further investigate the radiation induced reduction in I_{surf}/A_{gate}, the shift of the C/V curves, and D_{it} additional measurements were performed $\sim$ 8 months after the last irradiation campaign. The analysis, which aims at determining the interface charge trap density D_{it}, the fixed oxide charge

is measured or by the value of the flat-band voltage.

[5] "up" corresponds to the measurement when V_{gate} is ramped from zero to negative values; the "up" curve is always below the "down" curve

density N_{ox}^{fix}, and the mobile oxide charge density N_{ox}^{mob}, proceeds in the following steps:

- obtain D_{it} as function of $E_c - E_t$ from the TDRC spectrum [3],
- compare the calculated shape and frequency dependence of the C/V curve from model [1] with the measurements; from N_{it}, the integral over D_{it}, the voltage shift ΔV_{it} due to the interface traps is calculated,
- subtract from the voltage shift of the first "up" CV curve the voltage shift ΔV_{it} to obtain the fixed oxide charge density N_{ox}^{fix},
- obtain the mobile oxide charge density N_{ox}^{mob} from the difference of the voltage shift "down" minus first "up" CV curve.

It should be noted that in particular the values obtained for N_{ox}^{mob} depend on the maximum value of V_{gate} and on the time the MOS capacitance was biased at this value. The chosen measurement parameters have therefore been kept constant in order to obtain comparable results.

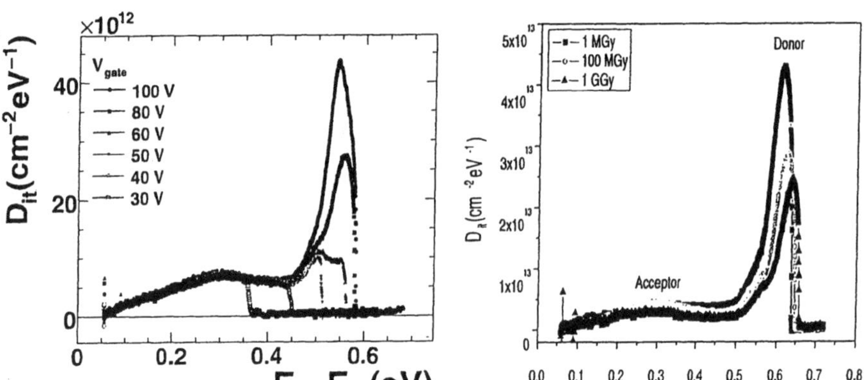

Figure 6: D_{it} versus position in band gap as function of the energy distance from the conduction band. *Left:* Measured on a 3.5 MGy sample with different values for V_{gate}. *Right:* Results for three different irradiated samples (measured ~ 8 months after irradiation) for $V_{gate} = -100\,\mathrm{V}$.

Figure 6 shows the results of TDRC measurements for three irradiated diodes that were analyzed as described above. We observe a significant decrease of D_{it} with increasing dose, in particular around the middle of the band gap.

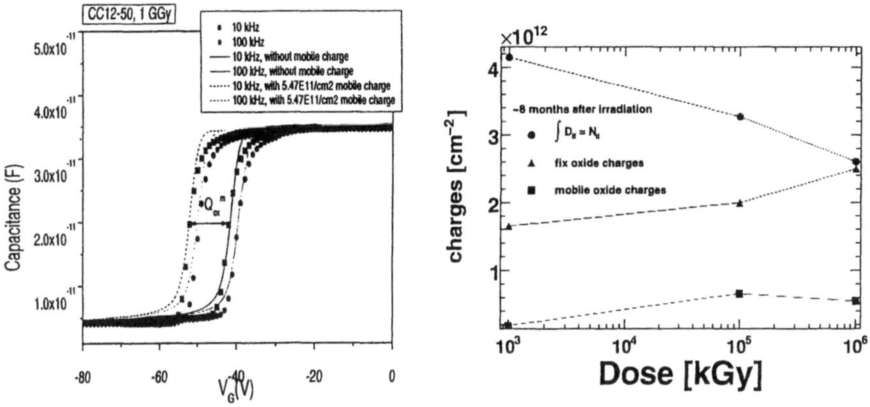

Figure 7: *Left:* Comparison of the measured C/V curves at 10 and 100 kHz with model calculations for a dose of 1 GGy. *Right:* The density of interface charges (N_{it}), fixed oxide charges (N_{ox}^{fix}) and mobile oxide charges (N_{ox}^{mob}) for different doses.

The results of a preliminary determination of N_{it}, N_{ox}^{fix} and N_{ox}^{mob} for doses of 1, 100 and 1000 MGy are shown in on the right in figure 7. On the left in figure 7 the comparison of the measured CV curves for 1 and 1000 MGy and frequencies of 10 and 100 kHz to the model calculations is shown. A qualitative agreement is achieved. It is found that, whereas N_{ox}^{fix} and N_{ox}^{mob} increase with dose, N_{it} decreases.

4 Summary and Conclusions

Measurements performed on gate-controlled diodes immediately after X-ray irradiation show that the surface current and the shift of C/V curves reach maxima at a dose of 1–5 MGy. Both parameters decrease for higher doses in the measured dose range up to 1 GGy.

From measurements at different dose rates we can exclude that the decrease at high doses is due to temperature effects during irradiation. So far the reason for the decrease is not understood.

Studies performed on the irradiated samples $\sim$ 8 months after the last irradiation step show a decrease of N_{it} between samples from 1 MGy $\rightarrow$ 1 GGy, whereas N_{ox}^{fix} and N_{ox}^{mob} increase with dose. A decrease of N_{it} reduces both the surface current and the shift of the C/V curves and is consistent with our observations.

It should be noted that the results presented are preliminary: improvements of the analysis techniques and the analysis of additional data taken for more dose values are under way. We also have started to implement the parameters found into a TCAD [5] simulation with the aim to reproduce the measurements and later use them for the design of radiation hard sensors for the AGIPD project [6].

Acknowledgments

The author would like to thank E. Fretwurst, R. Klanner, I. Pintilie, and T. Theedt for their strong support, insights, and valuable discussions.

Furthermore, I would like to thank the XFEL company for support. The work also profited from the infrastructure grant of the Helmholtz Alliance "Physics at the Terascale". The interest and support by the AGIPD collaboration, in particular by H. Graafsma, is very much appreciated. I would also like to thank André Rothkirch for his continuous help and assistance.

References

[1] E. H. Nicollian, J. R. Brews, MOS (Metal Oxide Semiconductor) Physics and Technology, Wiley, New York, 1982.

[2] A. S. Grove, Physics and Technology of Semiconductor Devices, Wiley, John & Sons, 1967.

[3] J. S. Uranwala, J. G. Simmons, H. A. Mar, Nonsteady-state techniques for determining the energy distribution of interface traps in MNOS (memory) devices, Appl. Phys. Lett. 26 (1975) 697.

[4] E. Fretwurst, F. Januschek, R. Klanner, H. Perrey, I. Pintilie, F. Renn, Study of the radiation hardness of silicon sensors for the XFEL, in: Nuclear Science Symposium Conference Record, IEEE, 2008, pp. 2535 – 2538.

[5] Synopsys, Inc., TCAD software.
URL `http://www.synopsys.com/products/tcad/tcad.html`

[6] Adaptive gain integrating pixel detector, web page (2009).
URL `http://hasylab.desy.de/science/developments/detectors/agipd`

Notes on Chern-Simons Theory in the Temporal Gauge

Andrey Smirnov [1], *ITEP, Russia*

Abstract

We analyze the perturbative series expansion of vacuum expectation values for Wilson loop operators in Chern-Simons gauge theory in the temporal gauge. Following to Labastida and Pérez we introduce the notion of the kernels of knot polynomial invariants. A method for exact calculation of the kernels of polynomial invariants is presented.

Contents

1 Introduction

In 1988, Edward Witten showed the connection between Chern-Simons (CS) field theory and the theory of knots in three-dimensional space [1]. Since that time the knot theory has been intensively studied using the standard quantum field theory methods. Different approaches inherent in quantum field theory established important connections between different type of knot invariants and provided a lot of new constructions for them.

The CS theory has been studied in different ways. Using series of non-perturbative methods Edward Witten in his original paper proved the equivalence of vacuum expectation values (vevs) for Wilson lines operators and known polynomial knot invariants [2, 3, 4, 5]. Perturbative studies performed in series of paper [6, 7, 8, 10, 11] established the connection between the coefficients of perturbative series expansion of vevs for Wilson loops and finite type (Vassiliev) knot invariants.

Perturbative series expansion has been studied for different gauge fixing. The study of CS theory in the covariant Landau gauge performed in [10, 11] showed the equivalence of coefficients for the perturbative series and the integral representation for invariants by Bott and Taubes [13]. The Kontsevich integral representation for Vassiliev invariants turned out to correspond to the coefficients of vevs of Wilson lines operators in the light-cone gauge [9, 14].

In this paper we concentrate our attention on the study of CS theory in the temporal gauge. Different aspects of this gauge was discussed in [8, 12]. The main ingredient for perturbative calculations in the temporal gauge is the gauge propagator. As it was noted in [8] the calculation of the gauge propagator in non-covariant gauges is plagued with ambiguities which should be solved by demanding additional properties for the correlation functions of gauge fields. As it was shown in [9] in non-covariant gauges we need to introduce some additional term to the propagator in order to obtain the knot invariants. Unfortunately, at present the correction term for

[1] E-mail: asmirnov@itep.ru

the propagator in the temporal gauge is not known yet. In this paper we work with the propagator (6) without the correction term that leads to non-invariant quantities for vevs of the Wilson loop operators. On the other hand side as it was argued in [8] the propagator (6) contains enough information about a knot to reconstruct the full knot invariants in any order. Following to definitions of work [8] we call these non-invariant vevs for Wilson loops *the kernels of polynomial invariants.*

The aim of this paper is to present the method for calculation of the kernels of the knot polynomial invariants in a very simple way. We show that there exists an operator $X : R_1 \otimes R_2 \to R_1 \otimes R_2$, where R_1 and R_2 are some representations of gauge group of CS action, such that the kernel of polynomial invariant for a knot with m intersection points can be obtained by appropriate contraction of indexes for m copies of the operator X.

The structure of the article is as follows: in section 2 we discuss the main ingredients of perturbative calculations in CS theory for the temporal gauge and explain in details the calculation of vevs for Wilson loops in the simplest abelian case. In section 3 we discuss the geometrical Labastida-Pérez formula for perturbative series expansion in non-abelian case and present some results of explicit calculations with this formula for the gauge group $GL(N)$. In section 4 we introduce a notion of the intersection operator, we calculate explicitly this operator for the case of $GL(2)$ gauge group. As an example, using this operator we calculate the exact answer for kernel of polynomial invariant for the trefoil knot in $GL(2)$ case.

2 Chern-Simons theory in the temporal gauge

2.1 The propagator of the gauge fields in the temporal gauge

Let G be a semi-simple Lie group and $A = A^a_\mu(x)dx^\mu F^a$ be a G-connection on $\mathbb{R}^3$, where F^a are the generators of Lie algebra of G in the fundamental representation. The Chern-Simons field theory is defined by the following action:

$$S[A] = \int_{\mathbb{R}^3} tr\left(A \wedge dA + g\frac{2}{3} A \wedge A \wedge A \right) \tag{1}$$

where tr denotes the trace over the fundamental representation of G and g is some parameter.

The CS theory is a particular example of a topological field theory [12], which means that all the observables in this theory are some topological invariants. In this way, the CS theory is a natural tool for study of three-dimensional topology, for example the topology of three-dimensional knots. Indeed, let us consider the following gauge invariant operator associated with a knot $c \in \mathbb{R}^3$:

$$W_R(c) = tr\left(P\exp\left(g\oint_c A^a_\mu(x)\, R^a\, dx^\mu \right)\right), \tag{2}$$

this operator is just a Wilson loop associated with knot c or a trace of the holonomy of A along the path c. The index R means that one-form A takes values in representation R of G. The natural class of the knot invariants provided by the CS theory is the vacuum expectation values of these operators:

$$< W_R(c) >= \int DA \exp(-S[A])\, W_R(c) \tag{3}$$

In [1] E.Witten using non-perturbative methods showed that these invariants are in fact well known to mathematicians polynomial knot invariants with respect to the argument $t = \exp(g^2)$. To compute (3) using standard perturbative methods (with respect to the parameter g) we need to chose a gauge fixing condition to make the associated functional integral well defined. Different choices of a gauge fixing lead to different descriptions of the same polynomial invariants [8, 9, 10]. The aim of this paper is to study the structure of the perturbative expansion of these polynomials in the so called temporal gauge. In the temporal gauge the condition imposed on the field A is:

$$A_0 = 0$$

This condition leads to the following action:

$$S[A]|_{A_0=0} = \int_{\mathbb{R}^3} d^3x\, A^a_\mu\, (\delta^{a\,b}\, \epsilon^{\mu\, 0\, \nu} \frac{\partial}{\partial x_0})\, A^b_\nu, \qquad \delta^{a\,b} = tr(F^a\, F^b) \tag{4}$$

We have the following equation for the Green function:

$$\delta^{a\,b}\, \epsilon^{\mu\, 0\, \nu}\, \partial_0\, \triangle^{b\,c,\,\nu\,\eta}(x) = \delta^{a\,c}\, \delta^{\mu\,\eta}\, \delta(x)$$

for the solution of this equation we have (for careful derivation of this formula see [8]):

$$\triangle^{bc,\nu\eta}(x) = -\epsilon^{\nu 0 \eta}\,\delta^{bc}\,\delta(x_1)\,\delta(x_2)\,\frac{1}{2}\text{sign}(x_0) + f^{bc\nu\eta}(x_1,x_2), \tag{5}$$

where $f^{bc\nu\eta}(x_1,x_2)$ is an arbitrary tensor that does not depend on x_0. Particular choice of $f^{bc\nu\eta}(x_1,x_2)$ depends on a prescription, some examples of different prescriptions can be found in works [8, 12]. In this work we will restrict our attention to the case $f^{bc\nu\eta}(x_1,x_2)=0$. In this case one obtains the following components of the propagator:

$$< A_0(x),\, A_\mu(y) >= 0, \mu = 0,1,2; \quad < A_\mu(x),\, A_\nu(y) >= \epsilon^{\mu\nu}\,\delta^{bc}\,\delta(x_1-y_1)\,\delta(x_2-y_2)\,\frac{1}{2}\text{sign}(x_0-y_0), \quad \mu,\nu=1,2 \tag{6}$$

As we will see below, perturbation theory based on the propagator (6) leads to the quantities for $< W_R(c) >$ that do not coincide with known knot invariants obtained using non-perturbative methods and depend on a projection of the knot to two-dimensional (x_1, x_2) plane. As it was shown in [8, 9] this deviation arise because we do not take into account the prescription depending term in (5).
In spite of the fact that the vacuum expectation values of Wilson lines in this description are not knot invariants, they contain a lot of information about knots and as it was conjectured in [8] they can be used for a reconstruction of the full invariants. Following to the notation of [8] we will denote these quantities by $< \hat{W}_R(c) >$ and call them *kernels of polynomial invariants*. The aim of this paper is to present the method that allows us to find exact expressions for the kernels of the polynomial invariants for every knot and for every representation in a very simple and elegant way.

2.2 The Abelian case as a basic example

Let us analyze in details the structure of the perturbative series expansion with propagator (6) in the Abelian case. In this case for the propagator we have:

$$\Pi_{\mu\nu}(x-y) =< A_\mu(x),\, A_\nu(y) >= \frac{1}{2}\epsilon^{\mu\nu}\,\delta(x_1-y_1)\,\delta(x_2-y_2)\,\text{sing}(x_0-y_0), \quad \mu,\nu=1,2, \quad A_0=0 \tag{7}$$

The Wilson lines are presented by a "simple" exponents:

$$W(c) = \exp\left(g \oint\limits_c A_\mu(x)\, dx^\mu\right) \tag{8}$$

We omit the label of representation because in the Abelian case the Wilson loop operators for different representations differ by a constant. For the perturbative expansion we get:

$$< \hat{W}(c) >=< \sum_{n=0}^{\infty} \frac{1}{n!}\left(g \oint\limits_c A_\mu(x)\, dx^\mu\right)^n >= \sum_{n=0}^{\infty} \frac{(2n-1)!!}{2n!} g^{2n} \left(\int_c\int_c dx_\mu\, dx_\nu < A_\mu(x), A_\nu(x) >\right)^{2n}$$

and after the summation we obtain:

$$< \hat{W}(c) >= \exp\left(g^2\, L_c\right) \tag{9}$$

where:

$$L_c = \frac{1}{2}\int_c\int_c dx_\mu\, dy_\nu < A_\mu(x), A_\nu(y) >= \frac{1}{2}\int_c\int_c dx^\mu\, dy^\nu \frac{1}{2}\epsilon^{\mu\nu}\,\delta(x_1-y_1)\,\delta(x_2-y_2)\,\text{sing}(x_0-y_0)$$

Let us parametrize the knot by a parameter t running from 0 to 1, then we can rewrite the last integral in the following form:

$$L_c = \frac{1}{4}\int_0^1\int_0^1 dt_1 dt_2 \left(\frac{dx_1}{dt_1}\frac{dy_2}{dt_2} - \frac{dx_2}{dt_1}\frac{dy_1}{dt_2}\right)\delta(x_1(t_1)-y_1(t_2))\,\delta(x_2(t_1)-y_2(t_2))\,\text{sing}(x_0(t_1)-y_0(t_2))$$

To perform the integration we need to solve the following equations:

$$\begin{cases} x_1(t_1) - y_1(t_2) = 0 \\ x_2(t_1) - y_2(t_2) = 0 \end{cases} \tag{10}$$

The solutions of these equations are the self-intersection points of two dimensional curve $(x_1(t), x_2(t))$ which is the projection of the knot c on the plane (x_1, x_2). Let us denote by $t_1^k < t_2^k$ the values of the parameter t in the intersection points, then the two-dimensional delta function in the integral can be presented in the form:

$$\delta(x_1(t_1) - y_1(t_2))\,\delta(x_2(t_1) - y_2(t_2)) = \sum_k \frac{(\delta(t_1 - t_1^k)\delta(t_2 - t_2^k) + \delta(t_1 - t_2^k)\delta(t_2 - t_1^k))}{|\frac{dx_1}{dt_1}\frac{dy_2}{dt_2} - \frac{dx_2}{dt_1}\frac{dy_1}{dt_2}|}$$

Substituting this expression into (10) and integrating over t_1 and t_2 we arrive to the following simple expression:

$$L_c = \sum_k \epsilon_k \tag{11}$$

where the quantities ϵ_k are the "sings" of the intersection points. They can take values ± 1 and are defined in the following way:

$$\epsilon_k = \frac{\frac{dx_1}{dt_1}\frac{dy_2}{dt_2} - \frac{dx_2}{dt_1}\frac{dy_1}{dt_2}}{|\frac{dx_1}{dt_1}\frac{dy_2}{dt_2} - \frac{dx_2}{dt_1}\frac{dy_1}{dt_2}|}\mathrm{sing}(x_0(t_1^k) - y_0(t_2^k)) \tag{12}$$

Note that the expectation value (9) for the knot can be expressed in terms of product over the intersection points:

$$< \hat{W}(c) >= \prod_k \exp(g^2\,\epsilon_k) \tag{13}$$

We see that the answer for vev in the abelian case has a form of a product over the intersection points of some quantity $\exp(g^2\,\epsilon_k)$ that depends only on the intersection point. As we will see below in non-abelian case this formula generalizes to the contraction of some tensors corresponding to intersection points.

3 The Non-Abelian Case

3.1 Labastida-Pérez formula

In the case of non-Abelian gauge group the Wilson line operator is more complicated:

$$W_R(c) = \mathrm{tr} P \exp\left(g \oint_c A_\mu(x)\, dx^\mu\right)$$

The vacuum expectation value has the form:

$$< W(c) >= \sum_{m=0}^{\infty} I_m^R(c) g^{2m}$$

where the expansion are presented in terms of ordered multidimensional integrals:

$$I_m^R(c) = tr(R^{a_1}\, R^{a_2} ... R^{a_m}) \int_0^1 dx_{\mu_1} \int_0^{x_{\mu_1}} dx_{\mu_2} ... \int_0^{x_{\mu_{m-1}}} dx_{\mu_m} < A_{\mu_1}^{a_1}\, A_{\mu_2}^{a_2} ... A_{\mu_m}^{a_m} >$$

where R^a are the generators of the gauge group in a representation R. Using Wick theorem, and the following facts:

$$\int_0^1 dx_{\mu_1} \int_0^{x_{\mu_1}} dx_{\mu_2} ... \int_0^{x_{\mu_{m-1}}} dx_{\mu_m} = \int dx_1 ... dx_m \prod_{k=1}^{m-1} \theta(x_k - x_{k+1}),$$

$$< A_0^a(x),\, A_\mu(y)^b >= 0, \mu = 0, 1, 2;$$

$$< A_\mu^a(x),\, A_\nu^b(y) >= \frac{1}{2}\epsilon^{\mu\nu}\,\delta^{bc}\,\delta(x_1 - y_1)\,\delta(x_2 - y_2)\,\mathrm{sign}(x_0 - y_0),\ \ \mu, \nu = 1, 2$$

we arrive to the following formula for I_m^R (J.M.F.Labastida and E.Pérez [8]):

$$
\begin{aligned}
I_m^R(c) = & \sum_{i_1<i_2<...<i_m} \epsilon_{i_1}\epsilon_{i_2}...\epsilon_{i_m} D(i_1, i_2, ..., i_m) + \\
& \frac{1}{(2!)^2} \sum_{\substack{\sigma\in S_2 \\ j\neq i_1,...,i_{m-2} \\ i_1,...,i_{m-2}}} \epsilon_j^2 \epsilon_{i_1}, ..., \epsilon_{i_{m-2}} D(j, \sigma, i_1, ..., i_{m-2}) + \\
& .... \\
& \frac{1}{(r!)^2} \sum_{\substack{\sigma\in S_r \\ j\neq i_1,...,i_{m-r} \\ i_1,...,i_{m-r}}} \epsilon_j^r \epsilon_{i_1}, ..., \epsilon_{i_{m-r}} D(j, \sigma, i_1, ..., i_{m-r}) + \sum_{\substack{\sigma\in S_m \\ j}} \epsilon_j^m D(j, \sigma)
\end{aligned}
\tag{14}
$$

The first term in this big sum comes from the contribution in which all the propagators are attached to different crossings. The second when two propagators are attached to the same crossing and rest to different crossings and so on. The factors $D(j, \sigma, i_1, ..., < i_{m-r})$ are group factors and they can be computed in the following way: we attach to every crossing i_k a group generator and r generators to the crossing j. Then travelling along the knot from some base point we multiply this generators in the order that they encounter. When we arrive to the crossing j first time one encounters product of r group generators and the second time the product is rearranged in accordance with the permutation $\sigma \in S_r$. After returning to the base point we should take a trace of obtained group factor. For more details and illustrative examples we send reader to [8].

Using modern computational tools such as `Maple` or `Mathematica` and formula (14) one can perform computations in perturbation theory up to order 5. In the tables below some exact results of these calculus for $gl(N)$-case in fundamental representation F are presented:

Knot	I_1^F	I_2^F	I_3^F	I_4^F
3_1	$3N^2$	$3/4\,N(5+N^2)$	$1/12\,N^2(53+N^2)$	$\frac{1}{192}N(284+363N^2+N^4)$
4_1	0	$3N(N-1)(N+1)$	0	$\frac{7}{144}N(N-1)(N+1)(5N^2+46)$
5_1	$5N^2$	$5/4\,N(9+N^2)$	$\frac{5}{36}N^2(149+N^2)$	$\frac{5}{576}N(1712+1287N^2+N^4)$
5_2	$4N^2$	$-1/2\,N(-17+N^2)$	$1/18\,N^2(181+11N^2)$	$\frac{1}{288}N(1494+1535N^2+43N^4)$
6_1	N^2	$1/4\,N(-29+31N^2)$	$1/36\,N^2(37N^2-31)$	$\frac{1}{576}N(-4228+2541N^2+1711N^4)$

In the case $N=1$ we get the following results:

$$
\begin{aligned}
&<\hat{W}_F(3_1)>= \sum_{m=0}^{\infty} I_m^F(3_1)\,g^{2m} = 1+3g^2+9/2\,g^4+9/2\,g^6+\tfrac{27}{8}g^8+\tfrac{81}{40}g^{10}+... = e^{3g^2}\\
&<\hat{W}_F(4_1)>= \sum_{m=0}^{\infty} I_m(4_1)\,g^{2m} = 1+0\,g^2+0\,g^4+0\,g^6+0\,g^8+0\,g^{10}+... = 1\\
&<\hat{W}(5_1)>= \sum_{m=0}^{\infty} I_m^F(5_1)\,g^{2m} = 1+5g^2+\tfrac{25}{2}g^4+\tfrac{125}{6}g^6+\tfrac{625}{24}g^8+\tfrac{625}{24}g^{10}+... = e^{5g^2}\\
&<\hat{W}_F(5_2)>= \sum_{m=0}^{\infty} I_m^F(5_2)\,g^{2m} = 1+4g^2+8g^4+\tfrac{32}{3}g^6+\tfrac{32}{3}g^8+\tfrac{128}{15}g^{10}+... = e^{4g^2}\\
&<\hat{W}_F(5_3)>= \sum_{m=0}^{\infty} I_m^F(5_3)\,g^{2m} = 1+g^2+1/2\,g^4+1/6\,g^6+1/24\,g^8+... = e^{g^2}
\end{aligned}
$$

We can see that in full agreement with the answer predicted by (13) the vevs of the Wilson loops in this cases present just a product of simple exponent operators over intersection points. In the first non-abelian case $N=2$ we get:

$$
\begin{aligned}
&<\hat{W}_F(3_1)>= \sum_{m=0}^{\infty} I_m^F(3_1)\,g^{2m} = 2+12g^2+\tfrac{27}{2}g^4+19g^6+\tfrac{73}{4}g^8+\tfrac{279}{20}g^{10}+... =?\\
&<\hat{W}_F(4_1)>= \sum_{m=0}^{\infty} I_m^F(4_1)\,g^{2m} = 2+18g^4+\tfrac{77}{4}g^8+... =?\\
&<\hat{W}_F(5_1)>= \sum_{m=0}^{\infty} I_m^F(5_1)\,g^{2m} = 2+20g^2+\tfrac{65}{2}g^4+85g^6+\tfrac{955}{8}g^8+\tfrac{871}{6}g^{10}+... =?\\
&<\hat{W}_F(5_2)>= \sum_{m=0}^{\infty} I_m^F(5_2)\,g^{2m} = 2+16g^2+13g^4+50g^6+\tfrac{1387}{24}g^8+\tfrac{2573}{30}g^{10}+... =?\\
&<\hat{W}_F(5_3)>= \sum_{m=0}^{\infty} I_m^F(5_3)\,g^{2m} = 2+4g^2+\tfrac{95}{2}g^4+13g^6+\tfrac{347}{3}g^8+... =?
\end{aligned}
$$

In this case it is very difficult to find any regularity in the coefficients of the expansions and summate them to the exact expressions of vevs for kernels of associated Wilson loops operators. It indicates that the answer for these vevs in non-abelian case has much more complicated structure. In section 4 we present the example of such vev for the simplest trefoil knot 3_1 in the case $N = 2$.

3.2 Two-Component Links

In order to derive the notion of intersection point operator let us analyze the vevs of two components links in CS theory. More precisely, we are interested in the following quantity:

$$I^{R_1\,R_2}(c_1,c_2) = \frac{<\hat{W}_{R_1}(c_1),\,\hat{W}_{R_2}(c_2)>}{<\hat{W}_{R_1}(c_1)><\hat{W}_{R_2}(c_2)>} = \sum_{k=0}^{\infty} I_k^{R_1\,R_2}(c_1,c_2)g^{2k} = 1 + \sum_{k=1}^{\infty} I_k^{R_1\,R_2}(c_1,c_2)g^{2k} \qquad (15)$$

where c_1 and c_2 are two contours in $\mathbb{R}^3$ and $W_{R_i}(c_i)$ are associated Wilson lines operators in representation R_i of the gauge group:

$$W_{R_i}(c_i) = tr P \exp\left(g \oint_{c_i} A_\mu\, dx^\mu\right) \qquad (16)$$

In (15) we divided vev $<\hat{W}_{R_1}(c_1),\,\hat{W}_{R_2}(c_2)>$ by the product od vevs $<\hat{W}_{R_1}(c_1)><\hat{W}_{R_2}(c_2)>$, which means that in perturbation series expansion we will only take into account the terms with different ends of the propagators attached to the different contours.

3.3 $I_1^{R_1\,R_2}(c_1,c_2)$ case

In the order g^2 we have the following integral:

$$I_1^{R_1\,R_2}(c_1,\,c_2) = g^2\, tr(R_1^a)\, tr(R_2^b) \oint_{c_1} dx^\mu \oint_{c_2} dy^\nu <A_\mu^a(x),\, A_\nu^b(y)>= 0$$

3.4 $I_2^{R_1\,R_2}(c_1,c_2)$ case

In the order g^4 we have two contributions:

$$I_2^{R_1\,R_2}(c_1,c_2) = M_1 + M_2 \qquad (17)$$

where

$$M_1 = tr(R_1^{a_1} R_1^{a_2})\, tr(R_2^{b_1} R_2^{b_2}) \oint_{c_1} dx_1^{\mu_1} \int_{c_1}^{x_1} dx_2^{\mu_2} \oint_{c_2} dy_1^{\nu_1} \int_{c_2}^{y_1} dy_2^{\nu_2} <A_{\mu_1}^{a_1}(x_1),\, A_{\nu_1}^{b_1}(y_1)><A_{\mu_2}^{a_2}(x_2),\, A_{\nu_2}^{b_2}(y_2)> \qquad (18)$$

$$M_2 = tr(R_1^{a_1} R_1^{a_2})\, tr(R_2^{b_1} R_2^{b_2}) \oint_{c_1} dx_1^{\mu_1} \int_{c_1}^{x_1} dx_2^{\mu_2} \oint_{c_2} dy_1^{\nu_1} \int_{c_2}^{y_1} dy_2^{\nu_2} <A_{\mu_1}^{a_1}(x_1),\, A_{\nu_2}^{b_2}(y_2)><A_{\mu_2}^{a_2}(x_2),\, A_{\nu_1}^{b_1}(y_1)> \qquad (19)$$

Introducing the parameterizations t and s for contours c_1 and c_2 respectively we get:

$$G(t,s) = \epsilon_{\mu\nu} \frac{dx^\mu}{dt}(t) \frac{dy^\nu}{ds}(s)\, \delta(x_1(t) - y_1(s))\, \delta(x_2(t) - y_2(s))\, \frac{1}{2}\mathrm{sign}(x_0(t) - y_0(s)) = \sum_k \epsilon_k \delta(t - t_k)\delta(s - s_k)$$

Where $\epsilon_k = \pm 1$ and sum runs over the intersection points of two contours lying in the plane (x_1, x_2) which are the projections of three dimensional contours on the plane. Using this notation we can rewrite the integrals for M_1 and M_2 in the form:

$$M_1 = tr(R_1^{a_1} R_1^{a_2})\, tr(R_2^{a_1} R_2^{a_2}) \int_0^1 dt_1 \int_0^{t_1} dt_2 \int_0^1 ds_1 \int_0^{s_1} ds_2\, G(t_1,\, s_1)\, G(t_2,\, s_2) \qquad (20)$$

$$M_2 = tr(R_1^{a_1} R_1^{a_2})\, tr(R_2^{a_2} R_2^{a_1}) \int_0^1 dt_1 \int_0^{t_1} dt_2 \int_0^1 ds_1 \int_0^{s_1} ds_2\, G(t_1,\, s_2)\, G(t_2,\, s_1) \qquad (21)$$

and using the following fact:

$$\int_0^1 dt_1 \int_0^{t_1} dt_2 \int_0^1 ds_1 \int_0^{s_1} ds_2 = \int_0^1 dt_1 \int_0^1 dt_2 \int_0^1 ds_1 \int_0^1 ds_2\, \theta(t_1 - t_2)\theta(s_1 - s_2) \tag{22}$$

we have:

$$M_1 = tr(R_1^{a_1} R_1^{a_2})\, tr(R_2^{a_1} R_2^{a_2}) \sum_{k_1\, k_2} \epsilon_{k_1} \epsilon_{k_2}\, \theta(t_{k_1} - t_{k_2})\, \theta(s_{k_1} - s_{k_2})$$
$$M_2 = tr(R_1^{a_1} R_1^{a_2})\, tr(R_2^{a_2} R_2^{a_1}) \sum_{k_1\, k_2} \epsilon_{k_1} \epsilon_{k_2}\, \theta(t_{k_1} - t_{k_2})\, \theta(s_{k_2} - s_{k_1}) \tag{23}$$

and finally we arrive to the following expression:

$$I_2^{R_1\, R_2}(c_1, c_2) = M_1 + M_2 = \sum_{\sigma \in S_2} tr(R_1^{a_1} R_1^{a_2})\, tr(R_2^{a_{\sigma(1)}} R_2^{a_{\sigma(2)}}) \sum_{k_1\, k_2} \epsilon_{k_1} \epsilon_{k_2}\, \theta(t_{k_1} - t_{k_2})\, \theta(s_{k_{\sigma(1)}} - s_{k_{\sigma(2)}}) \tag{24}$$

where S_2 is the permutation group of two elements.

3.5 $I_m^{R_1\, R_2}(c_1\, c_2)$ Case

The last formula (24) has obvious straightforward generalization for $I_m^{R_1\, R_2}(c_1\, c_2)$:

$$I_m^{R_1\, R_2}(c_1\, c_2) = \sum_{\sigma \in S_m} M_\sigma \tag{25}$$

where S_m is the permutation group for a set with m elements.

$$M_\sigma = tr(R_1^{a_1} ... R_1^{a_m}) tr(R_2^{a_{\sigma(1)}} ... R_2^{a_{\sigma(m)}}) \sum_{k_1\, k_2 ... k_m} \epsilon_{k_1} \epsilon_{k_2} ... \epsilon_{k_m} \left(\prod_{i=1}^{m-1} \theta(t_{k_i} - t_{k_{i+1}}) \right) \left(\prod_{j=1}^{m-1} \theta\left(s_{k_{\sigma(j)}} - s_{k_{\sigma(j+1)}}\right) \right) \tag{26}$$

Let us note that in agreement with Abelian case the sum of M_σ without the part containing non-commuting lie-algebra structure is expressed through the linking number of two contours:

$$\sum_{\sigma \in S_m} \left(\sum_{k_1\, k_2 ... k_m} \epsilon_{k_1} \epsilon_{k_2} ... \epsilon_{k_m} \left(\prod_{i=1}^{m-1} \theta(t_{k_i} - t_{k_{i+1}}) \right) \left(\prod_{j=1}^{m-1} \theta\left(s_{k_{\sigma(j)}} - s_{k_{\sigma(j+1)}}\right) \right) \right) = \frac{L_{12}^m}{m!} \tag{27}$$

where the quantity L_{12} is just a sum of crossing sings over two dimensional intersection points:

$$L_{12} = \sum_k \epsilon_k$$

The definition of L_{12} coincides precisely with definition of linking number of two knots. The linking number of two two-component link known to be the topological invariant of the link, and we arrive to the following result: in abelian case the vev (15) is just the exponent of linking number of the link:

$$I(c_1, c_2) = \sum_{m=0}^{\infty} g^{2k} \frac{L_{12}^m}{m!} = \exp(g^2\, L_{12})$$

4 The intersection point operator method

4.1 The intersection point operator

Using the following property of the tensor product of operators:

$$tr(R_1^{a_1} R_1^{a_2} ... R_1^{a_m}) tr(R_2^{a_{\sigma(1)}} R_2^{a_{\sigma(2)}} ... R_2^{a_{\sigma(m)}}) = tr(R_1^{a_1} R_1^{a_2} ... R_1^{a_m} \otimes R_2^{a_{\sigma(1)}} R_2^{a_{\sigma(2)}} ... R_2^{a_{\sigma(m)}})$$

we can rewrite expression for $I_m^{R_1\,R_2}(c_1, c_2)$ in the form:

$$M_\sigma = tr(\hat{M}_\sigma)$$

$$\hat{M}_\sigma = R_1^{a_1}...R_1^{a_m} \otimes R_2^{a_{\sigma(1)}}...R_2^{a_{\sigma(m)}} \sum_{k_1\,k_2...k_m} \epsilon_{k_1}\epsilon_{k_2}...\epsilon_{k_m} \left(\prod_{i=1}^{m-1} \theta(t_{k_i} - t_{k_{i+1}})\right) \left(\prod_{j=1}^{m-1} \theta\left(s_{k_{\sigma(j)}} - s_{k_{\sigma(j+1)}}\right)\right) \qquad (28)$$

Let us consider in spite of the sums (15) and (25) the following operator:

$$M = \sum_{m=0}^{\infty} g^{2m} \sum_{\sigma\in S_m} R_1^{a_1} R_1^{a_2}...R_1^{a_m} \otimes R_2^{a_{\sigma(1)}} R_2^{a_{\sigma(2)}}...R_1^{a_{\sigma(m)}} \sum_{k_1\,k_2...k_m} \epsilon_{k_1}\epsilon_{k_2}...\epsilon_{k_m} \prod_{i=1}^{m-1} \theta(t_{k_i} - t_{k_{i+1}})\, \theta\left(s_{k_{\sigma(j)}} - s_{k_{\sigma(j+1)}}\right) =$$

$$= 1 + g^2 R_1^{a_1} \otimes R_2^{a_1} \sum_{k_1} \epsilon_{k_1} + g^4 \sum_{\sigma\in S_2} R_1^{a_1} R_1^{a_2} \otimes R_2^{a_{\sigma(1)}} R_2^{a_{\sigma(2)}} \theta(t_{k_1} - t_{k_2})\theta(s_{k_{\sigma(1)}} - s_{k_{\sigma(2)}}) + ... \qquad (29)$$

and let us calculate the contribution coming from one point. To find this contribution we need to imply $k_1 = k_2 = ... = k_m$ for g^{2m} order in expansion, and take into account the following fact:

$$\left(\prod_{k=1}^{m-1} \theta(t_{k_i} - t_{k_{i+1}})\right) |_{k_1=k_2=...k_m} = \frac{1}{m!}, \qquad \left(\prod_{k=1}^{m-1} \theta(s_{k_{\sigma(i)}} - s_{k_{\sigma(i+1)}})\right) |_{k_1=k_2=...k_m} = \frac{1}{m!}$$

Let us denote:

$$\{R_i^{a_1} R_i^{a_2}...R_i^{a_m}\} = \frac{1}{m!} \sum_{\sigma\in S_m} R_i^{a_{\sigma(1)}} R_i^{a_{\sigma(2)}}...R_i^{a_{\sigma(m)}}$$

then for single point contribution we get:

$$X(\epsilon_j g^2) = \sum_{k=0}^{\infty} \frac{\epsilon_j^k g^{2k}}{k!} \{R_1^{a_1} R_1^{a_2}... R_1^{a_k}\} \otimes \{R_2^{a_1} R_2^{a_2}... R_2^{a_k}\} \qquad (30)$$

We can treat $X(\pm g^2)$ as an operator acting in the space $R_1 \otimes R_2$. It is convenient to think about the operator X as about a tensor with four indexes X_{km}^{ij} such that each index is associated with some leg of intersection vertex (fig. 1).

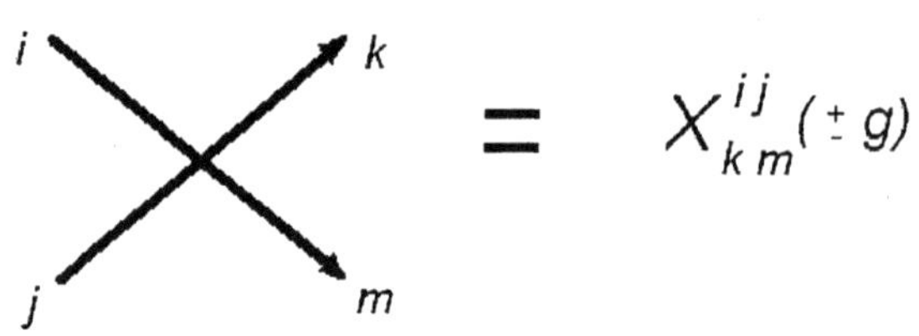

Figure 1: Representation of intersection operator X

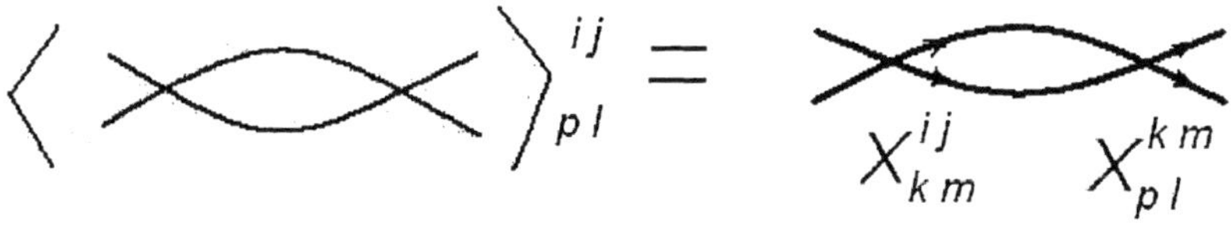

Figure 2: The contribution coming from two adjacent points of a link

The main feature of this operator is that two point contribution can be simply expressed as corresponding product of two such operators (fig.2).

4.2 The vev of the trefoil knot in $GL(2)$ case

As an example let us consider how the intersection point operator (30) can be used for calculation of kernels of polynomial invariants. Let us summate the series (30) for for operators F^a in fundamental representation of $gl(2)$. In this case we have:

$$X(g^2) = X(g^2) = \sum_{k=0}^{\infty} \frac{g^{2k}}{k!} \{F^{a_1} F^{a_2} ... F^{a_k}\}^{\otimes 2} = A(g^2)\, \hat{Id} + B(g^2)\, \hat{P} \qquad (31)$$

where $\hat{Id}$, $\hat{P}$ are the identity and interchange operators in $\mathbb{C}^2 \otimes \mathbb{C}^2$ and:

$$A(g^2) = 1/6\, e^{g^2} + 1/6\, e^{g^2} g^2 + 5/6 - 1/3\, g^2,$$
$$B(g^2) = 2/3\, e^{g^2} - 2/3 + 1/6\, g^2 + 1/6\, e^{g^2} g^2$$
$$\hat{Id} = \begin{bmatrix} 1 & 0 & 0 & 0 \\ 0 & 1 & 0 & 0 \\ 0 & 0 & 1 & 0 \\ 0 & 0 & 0 & 1 \end{bmatrix}, \quad \hat{P} = \begin{bmatrix} 1 & 0 & 0 & 0 \\ 0 & 0 & 1 & 0 \\ 0 & 1 & 0 & 0 \\ 0 & 0 & 0 & 1 \end{bmatrix}$$

To find the kernel of polynomial invariant for example of trefoil knot, we need to chose some projection of

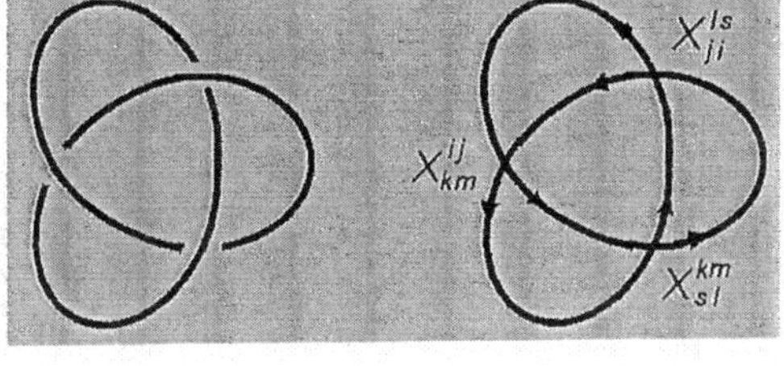

Figure 3: Two dimensional projection of the trefoil knot and contraction of intersection operators.

the knot to two-dimensional plane. For instance as in the fig.3. Then we attach to every intersection point of obtained two dimensional curve the tensor $X(\epsilon g^2)$, where ϵ is the sign of the intersection point defined by (12). Finally, to find the kernel of polynomial invariant in this case, we contract the indexes of the tensors in accordance with two dimensional diagram fig.3:

$$< \hat{W}(3_1) >= \sum_{ij\,kmsl} X^{ij}_{km}\, X^{km}_{sl}\, X^{ls}_{jk} =$$

$$2\,(5/6\,e^{g^2}+1/3\,e^{g^2}g^2+1/6-1/6\,g^2)^3+$$
$$+6\,(1/6\,e^{g^2}+1/6\,e^{g^2}g^2+5/6-1/3\,g^2)^2(2/3\,e^{g^2}-2/3+1/6\,g^2+1/6\,e^{g^2}g^2)+$$
$$+2\,(2/3\,e^{g^2}-2/3+1/6\,g^2+1/6\,e^{g^2}g^2)^3$$

The expansion in g^2 coincide precisely with result obtained in section 3.1 for trefoil knot 3_1 in the $N=2$ case by means of Labastida-Pérez formula:

$$<\hat{W}(3_1)>=2+12\,g^2+\frac{27}{2}g^4+19\,g^6+\frac{73}{4}g^8+\frac{279}{20}g^{10}+O(g^{12})$$

Acknowledgments

The author is grateful to A.Morozov for fruitful discussions and interest to this work. The work was partly supported by RFBR grant 09-02-00393, RFBR grant 06-01-92054-KE_a, RFBR-CNRS grant 09-01-93106, and grant for support of scientific schools NSh-3036.2008.2. The work was also supported in part by the "Dynasty" Foundation.

References

[1] E.Witten, Commun.Math.Phys. 121, 351-399 (1981)

[2] V.F.R. Jones, Bull. AMS 12 (1985) 103; Ann. of Math. 126 (1987) 335.

[3] P.Freyd, D.Yetter, J.Hoste, W.B.R.Lickorish, K.Millet, and A.Ocneanu, Bull. AMS 12 (1985) 239.

[4] L.H.Kauffman, Trans, Am. Math. Soc. 318 (1990) 417

[5] Y.Akutsu and M. Wadati, J.Phus.Soc.Jap. 56 (1987) 839 and 3039

[6] M.Alvarez, and J.M.F.Labastida, Nucl. Phys. B395,(1993) 198; B433 (1995) 555.

[7] J.M.F.Labastida and E.Pérez, Nucl.Phys. B527, (1998) 499;

[8] J.M.F.Labastida and E.Pérez, CERN-TH/98-193, US-FT-11/98 (1998) 499;

[9] J.M.F.Labastida and E.Pérez, Journ.Math.Phys. 39, (1998) 5183

[10] D.Bar-Natan, Ph.D. Thesis, Princeton Univericity (1991).

[11] E.Guadagnini, M.Martellini and Mitchev, Phys.Lett.B227 (1989) 111; B228 (1989) 489; Nucl.Phys. B330 (1990) 575.

[12] A.S.Cattaneo, P.Cotta-Ramusino, J.Frohlich and M. Martellini, Jour.Math.Phys. 36 (1995) 6137.

[13] R.Bott and C.Taubes, Jour.Math.Phys. 35 (1994) 5247.

[14] M.Kontsevich, Advances in Soviet Math. 16, part 2 (1993) 137

Dark Matter via Many Copies of the Standard Model

Alexander Vikman
Center for Cosmology and Particle Physics,
NYU, New York, NY 10003, USA

Abstract

In this proceedings I review our results from [1] where we have demonstrated that baryons in many possible hidden copies of the standard model can naturally account for the dark matter. The right abundance of the hidden-sector baryons and the correct spectrum of density perturbations are simultaneously generated during modulated reheating. We have also shown that for natural values of inflaton coupling constants, dictated by unitarity, the dark-matter abundance is predicted to be proportional to the ratio of observed cosmological parameters: the square of the amplitude of cosmological perturbations and the baryon-to-photon number ratio.

Introduction- One of the major unsolved problems in cosmology is the understanding of the nature of dark matter (DM). Astrophysical data imply that if dark matter interacts in any other way than through gravity, it does so extremely weakly, leaving no observational signature of this fact large enough to be detectable to date (see, however, the recent PAMELA [2] and ATIC [3] results). Indeed, some of the most stringent constraints on the collider physics of supersymmetry or extra dimensions arise as a result of the requirement to predict the appropriate dark-matter abundance. These extensions of the standard model (SM) are invoked, among others, to account for the hierarchy problem and contain within them natural candidates for dark matter.

Recently, a different solution to the hierarchy problem was proposed in [4]. Namely, it was realized that in any effective field theory with a large number N of elementary particle species, the fundamental scale at which gravity becomes strong is lower than the Planck mass $M_{\rm Pl}$ and is given by

$$M_* \equiv \frac{M_{\rm Pl}}{\sqrt{N}}. \tag{1}$$

This scale serves as the cut-off for the low-energy effective theory. The generalization of the bound (1) to quasi-de-Sitter inflationary vacua was given in [5]. Notice, that the scale M_* also puts the upper bound on the mass of the species of number N. Therefore, if there were to exist $N = 10^{32}$ different species in our universe, this would place the scale beyond which gravity is strong at 1 TeV. In this way, the strong-gravity effects can potentially provide a mechanism which regulates the ultraviolet-sensitive radiative corrections to the mass of the Higgs. The existence of this strong-coupling scale was initially proved by considering the implications of models with a large number of species on the consistency of black-hole (BH) evaporation [4]. Moreover, the same result has been justified by considering BH entanglement entropy [6], quantum information theory and holography [7]. The possibility of the perturbative generation of the hierarchy (1) between the Planck mass and fundamental gravity scale as a result of the renormalization of the gravitational coupling by the loops of N-species was suggested even earlier [8]. This bound can be also used to restrict possible number of species. In particular, if future experiments measure (e.g. through B-Mode of CMB) the gravitational waves from inflation [9] on the level predicted in simplest chaotic inflation [10], then $N \ll 10^{12}$. Moreover, the detection of gravitational waves from inflation would put strong constraints on all gravity modifications with a "low" strong-coupling scale.

In [1] we proposed that there may exist an additional, discrete permutation symmetry, $P(N)$, under which the standard model is charged, and therefore, there exist N copies of the SM augmented by N copies of the inflaton.

Our idea is that the matter we usually consider to be baryonic all belongs to the same species, while the dark matter consists of baryons of all the other copies of the standard model. Current astrophysical observations constrain DM to be effectively a collisionless fluid that behaves as if it only interacted gravitationally. In our scenario, this property results from: (a) Low partial densities of each species except ours, which are achieved through the proposed inflationary and reheating mechanisms; (b) The suppression of any cross-couplings between different species compared to couplings within one sector, which—we demonstrate— follows directly from unitarity.

Consistent inflation, without strong-gravity effects, can occur only below the cutoff M_*, given by Eq. (1). In this case, it is natural to implement a modulated-reheating scenario [11] to generate the appropriate amplitude for the power spectrum of cosmological perturbations. In order to populate one of the species more than any other, the final stage of inflation should occur in one species only, so that reheating happens mostly within this preferred sector. However, a cross-coupling of the inflaton with the other species, which naturally appears at least owing to gravity, ensures that some of the inflaton decays occur into all the other copies. The cosmological perturbations generated in this scenario are guaranteed to be adiabatic if the modulator is common to all species.

It is important that the partial decay rates of the inflaton field into different species are strongly restricted by unitarity considerations. As a result, dark-matter abundance is predicted in terms of known cosmological parameters—the dimensionless amplitude of the power spectrum of cosmological perturbations $\Delta^2_{\mathcal{R}}$ and the baryon-to-photon number ratio $\eta_{\rm b}$:

$$\frac{\Omega_{\rm DM}}{\Omega_{\rm b}} \sim \frac{\Delta^2_{\mathcal{R}}}{\eta_{\rm b}}, \tag{2}$$

with the only additional requirement being that baryon-antibaryon annihilation be suppressed in the dark sectors.

In our species, the plasma is dense and thermal equilibrium is achieved: it is key that the majority of baryons annihilate with antibaryons, creating the observed small baryon-to-photon number ratio. In the other species, given a large enough N, the individual partial densities are extremely low. We show that all annihilation processes are frozen out at all times, which ensures that all the dark baryons survive.

Thus the combination of: couplings bounded by unitarity, constraints on the modulator vacuum expectation value (VEV) resulting from the observed cosmological-perturbation amplitude, and baryon asymmetry as given by the observed baryon-to-photon number ratio in our model predicts the correct dark-matter abundance when inflation occurs around the natural cut-off scale.

Inflationary Background- We consider a theory with a large number, N, of standard model copies, labeled by an index $j = 1, 2, \ldots N$, where N is a free parameter which can be as huge as $N \sim 10^{32}$. Taking in mind the benefits of inflation we endow each j-th SM copy with its own inflaton field [1] Φ_j which is able to realize chaotic inflation/slow-roll. For maximal predictivity and simplicity, we assume that all sectors are exact replicas of SM, and that all of them are related by an exact permutation symmetry $P(N)$. It is clear from our experience of the world that this $P(N)$ symmetry must be broken by the cosmological background since we are coupled to a fraction larger than one in N of the matter in the universe. We posit that the seed of this symmetry breaking was prepared by inflation. All we require is that the evolution of the last few e-foldings of inflation be driven by a single inflaton species. As we have shown in [1] this scenario can be naturally realized provided there are cross-couplings between the species in the potential:

$$V = \sum_j V(\Phi_j) + \lambda_{22} \sum_{j_1 \neq j_2} \Phi_{j_1}^2 \Phi_{j_2}^2 + \frac{\lambda_{24}}{M_*^2} \sum_{j_1 \neq j_2} \Phi_{j_1}^2 \Phi_{j_2}^4 + \text{higher terms}. \quad (3)$$

Here it is important that the cross couplings λ_{22} etc are suppressed by powers of N. In particular λ_{22} scales as N^{-1}, see Fig. 1. Similar bounds on cross-couplings follow from many other considerations including the renormalization of the kinetic terms of the species due to loops of other species. This scaling is also fully compatible with the typical values of the cross-couplings induced via the exchange of virtual black holes [4]: the dominant contribution comes from black holes of mass M_* and each exchange by a virtual black hole between the different species requires a suppression by $1/N$ in the amplitude.

Let us assume that all the couplings are positive-definite. In that case, despite the required suppression of the cross couplings , putting one of the inflatons on the inflationary slope at some VEV, say $\Phi_1 \gtrsim M_{\rm Pl}$, forces all the other inflatons to remain at zero VEV or, if they had a non-zero VEV before, to occupy zero VEV during the time of order H^{-1}. Indeed, the effective mass term for all inflatons with index $j \neq 1$ in the background of Φ_1 is given by

$$M_{\Phi_{j\neq 1}}^2 = \frac{\partial^2 V(\Phi_j)}{\partial \Phi_j^2} + 2 \sum_\alpha \frac{\lambda_{2,2\alpha}}{M_*^{2\alpha-2}} \Phi_1^{2\alpha} \quad (4)$$

This effective mass is at least of order M_*, and thus all the other inflaton fields strongly violate the slow-roll condition: the effective friction cannot prevent them from fast roll towards the minimum. At the same time, quantum perturbations for the degrees of freedom with mass $M_{\Phi_{j\neq 1}} > H_*$ are not

[1] It would be interesting to consider whether this role could be played by the Higgs field

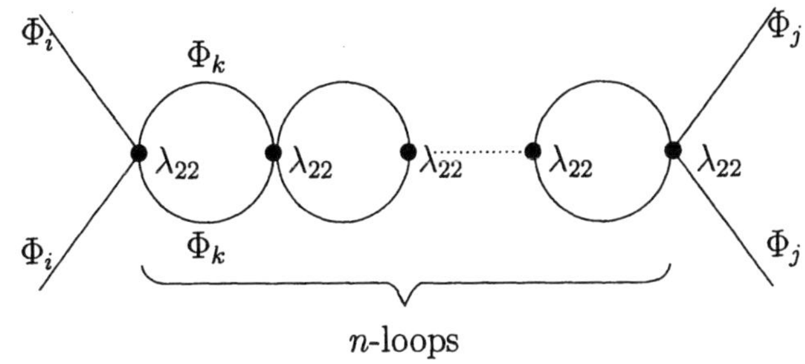

Figure 1: Introducing a cross-species coupling to the inflaton potential involves diagrams which couple species through loops. Since the species index on the loops is free, these amplitudes are relatively enhanced by a factor N for every loop, $\mathcal{M} \sim \lambda_{22}(\lambda_{22}N)^n$. Perturbative unitarity of the theory therefore requires that the coupling λ_{22} be suppressed by at least a factor N^{-1} in order to prevent the amplitudes from diverging.

enhanced, leaving no way to produce a VEV for them. Thus all other inflatons with $j \neq 1$ drop out from the effective theory consideration similarly to disappearance of states with masses larger than $M_{\rm Pl}$ in low energy theories originating form e.g. String Theory.

Moreover, non-perturbative black-hole arguments can be used to derive a bound in (quasi) de-Sitter spaces on the masses of particles which have N species [5], $m < M_{\rm Pl}/\sqrt{N}$. Any inflationary vacuum in which this bound is not satisfied must destabilize and cannot support slow-roll. If the coupling λ_{22} exceeded $1/N$, the inflationary vacuum with $\Phi_1 \gg M_{\rm Pl}$ would result in the presence of order N species with masses $\gg M_{\rm Pl}/\sqrt{N}$, violating the bound above. This argument is fully non-perturbative and is independent of the unitarity considerations, but results in the same constraint on the cross-species couplings as those we found through unitarity argument above.

Thus the cross couplings of the form Eq. (3) exclude the possibility of inflating collectively with $\Phi_k \gtrsim M_{\rm Pl}$ for more than one given sector. Using the further generalization of the cross-coupling terms in Eq. (3):

$$V = \sum_i V(\Phi_i) + \sum_{\alpha_1,\ldots\alpha_n} \frac{\lambda_{\alpha_1,\ldots\alpha_n}}{M_*^{\alpha_1+\ldots\alpha_n-4}} \sum_{j_1,\ldots j_n} \Phi_{j_1}^{\alpha_1} \ldots \Phi_{j_n}^{\alpha_n}.$$

and similar arguments one can eliminate the undesired possibility of inflating collectively with all inflatons at small VEVs, but with $\sum_j \Phi_j^2 \sim M_{\rm Pl}^2$. Such a situation could also provide sufficient inflation but reheats to all the species comparably. However, with the cross-couplings present, the cross-terms add up in such a way that the mass of all the inflatons is too high and no inflation can occur.

Perturbations via Modulated Reheating- The main benefit of chaotic inflation is a natural quantum mechanical generation [12] of primordial cosmological perturbations. However, the amplitude of cosmological perturbations produced in this standard way in the setup described above is much too low. It is natural to generate the cosmological perturbations though modulated reheating [11], which enables the generation of sufficient, red-tilted cosmological perturbations for the small values of H_* required by the bound Eq. (1). In this mechanism, one introduces a new scalar field χ, the *modulator*, which controls the decay rate, Γ, of the inflaton field into SM particles. This modulator χ is assumed to be light, $M_\chi \ll H_*$, so that substantial quantum fluctuations $\delta\chi \sim H_*$ are generated during inflation. These fluctuations become frozen (classical) as they leave the horizon. Moreover, it is assumed that χ does not contribute significant energy density during inflation and does not affect inflationary dynamics: $H_*^2 \gg M_\chi^2 \chi^2 / M_{\rm Pl}^2$. After inflation, during the oscillatory stage the Hubble constant decreases quickly, and eventually becomes less than the decay rate Γ. Around that time reheating occurs: inflaton decay becomes efficient enough to populate the universe with SM particles. The local VEV $\langle\chi\rangle$ controls the decay rate of the inflaton, Γ, and therefore controls the position of the reheating surface. Perturbations in the field value of the modulator lead to perturbations in the reheating surface. The couplings which provide the dominant decay channels are linear in the modulator field. Assuming that the universe was dust-dominated prior to reheating, radiation-dominated afterwards and that reheating occurs instantaneously at $H \simeq \Gamma$, one obtains [11] for dimensionless curvature power spectrum

$$\Delta_{\mathcal{R}}^2 = \frac{H_*^2}{(6\pi)^2 \langle\chi\rangle^2}. \tag{5}$$

For the correct amplitude of the curvature perturbations, the vacuum expectation value for the modulator during inflation is then required to be

$$\langle\chi\rangle \sim \frac{H_*}{\Delta_{\mathcal{R}}} \sim 10^5 \, H_* \,. \tag{6}$$

Inflaton Decay, Reheating and Freeze-Out in Dark Sectors- We require that the modulator χ be common for all species. Thus all the perturbations in the decay products of the inflaton are generated by a single source: perturbations in the modulator field. This ensures that, even if the decay products are not thermally coupled after reheating, their curvature perturbations will be identical and will not contain an isocurvature mode. The relative contributions to the energy density in different particle types will be the same everywhere, only the time of reheating will have changed. It is natural (for details see [1]) to parameterize the dominant couplings between the inflatons and the SM species (call them Q_j) in the following way

$$\Phi_i \frac{\chi}{M_*} \left(g\, \bar{Q}_i Q_i + \tilde{g}\, \bar{Q}_j Q_j \right) , \tag{7}$$

where the two coupling constants g and $\tilde{g}$ are responsible for the diagonal and non-diagonal decays respectively.

In our analysis, we shall assume that the values of these coupling constants are the most natural (the largest possible) compatible with the requirement of unitarity at the energies below the cutoff scale. The most stringent unitarity bound on g comes from the consideration of the processes involving inter-species interactions via the exchange of a virtual χ see e.g. Fig. 2(a):

$$g \lesssim N^{-1/2} . \tag{8}$$

The same constraint can be also obtained by requiring that the decay rate of the modulator χ in the presence of the inflaton VEV $\langle \Phi_1 \rangle \gtrsim M_{\rm Pl}$ be less than the fundamental scale M_*.

The modulated-reheating mechanism requires the presence of a background field VEV, $\langle \chi \rangle$. On such a background, we can consider processes involving the exchange of virtual-inflaton species, e.g. shown in Fig. 2(b) between different species. The constraint on the coupling then becomes

$$\tilde{g} \lesssim \Delta_{\mathcal{R}} N^{-1} \left(\frac{M_*}{H_*} \right) \sim 10^{-5} N^{-1} , \tag{9}$$

where we have used (6).

The decay products after reheating all come from a single inflaton species, Φ_1. The coupling g is relatively large, Eq. (8), and controls the diagonal decay into matter species Q_1. However, particles in the other copies are also produced by virtue of the much smaller off-diagonal coupling $\tilde{g}$, Eq. (9). The individual energy densities in all the other sectors are negligible relative to our SM sector; however, the integrated energy density is significant. The

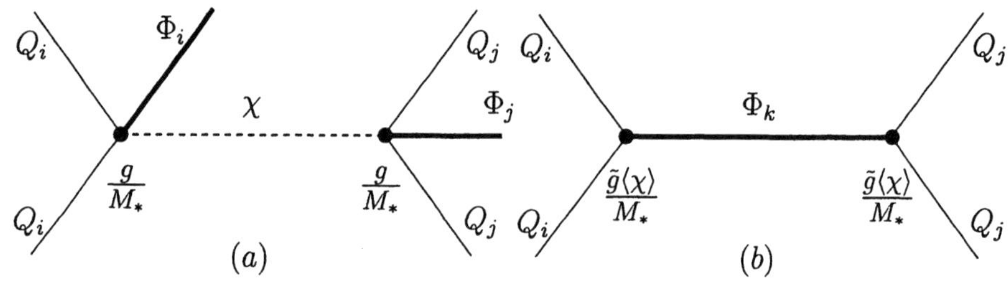

Figure 2: Processes which lead to strongest constraints on the coupling constants g and $\tilde{g}$ arising as a result of interaction terms Eq. (7). Bounds are governed by the requirement for perturbative unitarity. (a) Exchange of a virtual χ. (b) Exchange of a virtual inflaton Φ_k; total amplitude involves a sum over all inflatons k.

ratio of the energy densities between the SM and any other given j-th copy is set by the ratio of the inflaton decay rates into the two sectors

$$\frac{\rho_j}{\rho_1} = \frac{\Gamma_j}{\Gamma_1} = \frac{\tilde{g}^2}{g^2} \sim \frac{M_*^2}{N\,\langle\chi\rangle^2} \sim \left(\frac{M_*}{H_*}\right)^2 \Delta_{\mathcal{R}}^2\, N^{-1}\,. \tag{10}$$

where we have used Eq. (6). Or, equivalently, the total energy density in the dark sector right after the reheating is

$$\sum_{j\neq 1} \rho_j \sim \rho_1 \left(\frac{M_*}{H_*}\right)^2 \Delta_{\mathcal{R}}^2 \sim 10^{-10}\rho_1, \tag{11}$$

and is strongly suppressed compared to the energy density in our sector. Note that this result does not depend on the number of species N. Since interactions between the sectors are suppressed, the other species are never in thermal equilibrium with our sector. This means that they are not included in the naive tally of energy density in radiation and therefore contribute an energy density that is unaccounted for during Big-Bang nucleosynthesis. The smallness of Eq. (11) guarantees that the nucleosynthesis bound on the amount of extra radiation in the Universe [13] is satisfied.

In [1] we have estimated the thermalization rate as

$$\frac{\Gamma_{\rm T}}{H_{\rm RH}} \gtrsim \frac{\alpha^2}{\Delta_{\mathcal{R}}^2} \gg 1, \tag{12}$$

where α is a typical SM coupling constant. Therefore the thermalization happens well within a Hubble time. Consequently for the reheating temperature we obtained

$$T_{\rm RH} \lesssim \frac{M_{\rm Pl}}{\Delta_{\mathcal{R}}} N^{-3/4} \sim 10^5 M_{\rm Pl} N^{-3/4}. \tag{13}$$

For $N \sim 10^{32}$ the reheating temperature is around or less than 1 GeV and it is only higher for smaller N. The number of species N clearly cannot exceed 10^{32} by more than a few orders of magnitude, since otherwise the strong-gravity effects would have already been seen by accelerators. Hence, in our scenario $T_{\rm RH} \gtrsim 1$ GeV for all interesting N, which ensures that BBN proceeds in a standard way. The thermal history for our SM sector is unaltered: The high density of SM baryons and anti-baryons ensures that they remain thermally coupled, converting the baryons and antibaryons into photons extremely efficiently. Only as a result of some primordial baryon/anti-baryon asymmetry does a significant number of baryons survive. Nonetheless, these represent a tiny fraction of the original number, which is known to be

$$\eta_{\rm b} = \frac{n_{\rm b}}{n_\gamma} = 2.7 \cdot 10^{-8}\, \Omega_{\rm b} h^2 = 6.2 \cdot 10^{-10}. \tag{14}$$

This number is also known to not have been significantly different as early as at the time of nucleosynthesis [14]. Therefore, for every baryon in our species that exists today, approximately $\eta_{\rm b}^{-1}$ baryons in our copy of the SM were in existence after reheating.

For the baryons not in our species, the story is very different: given a large enough N, their partial densities following reheating are extremely low. In our model, we require that they be so rare that all annihilation processes are always frozen out and the other species never reach thermal equilibrium. We define freeze-out as the time at which the annihilation rate for the j-th copy, $\Gamma_j^{\rm A}$, falls below the Hubble parameter. In [1] we found that the strongest bound comes from annihilation to two photons $j\bar{j} \to 2\gamma$. Namely for given N the annihilation rate $\Gamma_{(j\bar{j}\to 2\gamma)}/H$ is maximal at $T_{\rm RH}$ and consequently N will be most strongly constrained at reheating where

$$\frac{\Gamma_{(j\bar{j}\to 2\gamma)}}{H_{\rm RH}} \sim \frac{\alpha^2}{N^{3/2}} \left(\frac{M_*}{M_{\Phi_1}}\right) \left(\frac{M_{\rm Pl}}{m_{\rm p}}\right) \ln\left(\frac{2M_{\Phi_1}}{M_*} \frac{M_{\rm Pl}}{m_{\rm p}\sqrt{N}}\right).$$

Throughout the paper we assume that $M_{\Phi_1} \lesssim M_*$, therefore requiring that this process be frozen out at reheating, constrains the number of species to be $N > 10^{11}$.

Thus, shortly after reheating, the universe is left with our sector at temperature $T_{\rm RH}$, while the other sectors—with much lower particle densities—remain unthermalized and consist of a collection of particles with energies $\sim M_{\Phi_1}$.

It is obvious that DM does not form bound objects like dark stars etc. in this scenario. Indeed the total energy of DM in the galaxy is of the same order as the energy of baryons. However, the energy stored in a given dark copy of SM is 10^{-32} of this total energy. Thus, in the galaxy, each dark copy of the SM possess 10^{-21} of the energy of a typical star. On the other hand the unitarity bounds cross-specie couplings to be extremely small and the baryons become nearly collisionless, interacting effectively only through gravity.

The limits on the mass for such dark-matter candidates are weak: constraints arising from structure formation limit the mass of a sterile neutrino to be $m_\nu \gtrsim 2 - 8$ keV [15]. More generally, DM must transition to non-relativistic at temperatures above a few keV so as not to erase the power spectrum at observable scales. In our proposal, this transition occurs above MeV even for the largest N and would not affect the density perturbations in this manner at observable scales.

Finally, we can calculate the contribution of such dark matter to the total energy budget of the Universe today. Taking the result Eq. (11) for the ratio of the density in our sector to the dark-matter density and adjusting it for the fact that only a fraction given in Eq. (14) of baryons in our sector will fail to annihilate to radiation, we obtain,

$$\frac{\Omega_{\rm DM}}{\Omega_{\rm b}} \sim \eta_{\rm b}^{-1} \frac{M_*^2}{\langle \chi \rangle^2} \sim \frac{\Delta_{\mathcal{R}}^2}{\eta_{\rm b}} \left(\frac{M_*}{H_*}\right)^2 \gtrsim 4 \tag{15}$$

This is rather remarkable: taking the inflationary Hubble parameter and inflaton mass to their most natural value, around the cutoff, together with the saturation of the unitarity constrains on coupling constants gives the correct dark-mater abundance. It would be very interesting to find a baryogenesis mechanism which would be able to relate baryon asymmetry to the amplitude of cosmological perturbations, or other models which could connect these to the DM abundance. This model also represents DM particles as practically an ideal dust which most certainly is undetectable in direct detection experiments. The general falsifiable prediction of this scenario is an emergence of strong gravitational effects at the LHC. For more details about the phenomenology of this setup see [16].

Acknowledgments- It is a pleasure to thank Gia Dvali and Iggy Sawicki for a fruitful collaboration on this topic. Moreover, I am grateful to Iggy Sawicki for a careful reading of the first version of this manuscript. Our original work [1] has benefited a lot from the discussions with A. Boyarsky, G. Gabadadze, A. Gruzinov, S. Hofmann, T. Levi, D. Malyshev, M. Manera, M. Redi, J. Roberts and N. Weiner. This work is supported by the James Arthur Fellowship. Last but not least I would like to thank the organizers of the 47th Course of the International School of Subnuclear Physics in beautiful Erice-Sicily.

References

[1] G. Dvali, I. Sawicki, and A. Vikman, *JCAP* **0908** (2009) 009, [arXiv:0903.0660].

[2] **PAMELA** Collaboration, O. Adriani *et. al.*, *Nature* **458** (2009) 607–609, [arXiv:0810.4995];

[3] J. Chang *et. al.*, *Nature* **456** (2008) 362–365.

[4] G. Dvali, arXiv:0706.2050; G. Dvali and M. Redi, *Phys. Rev.* **D77** (2008) 045027, [arXiv:0710.4344].

[5] G. Dvali and D. Lust, *JHEP* **06** (2008) 047, [arXiv:0801.1287].

[6] G. Dvali and S. N. Solodukhin, arXiv:0806.3976.

[7] G. Dvali and C. Gomez, *Phys. Lett.* **B674** (2009) 303–307, [arXiv:0812.1940]; R. Horvat, *Phys. Lett.* **B674** (2009) 1–3, [arXiv:0812.2768].

[8] G. R. Dvali and G. Gabadadze, *Phys. Rev.* **D63** (2001) 065007, [hep-th/0008054]; G. Veneziano, *JHEP* **06** (2002) 051, [hep-th/0110129].

[9] A. A. Starobinsky, *JETP Lett.* **30** (1979) 682–685; *Phys. Lett.* **B91** (1980) 99–102.

[10] A. D. Linde, *Phys. Lett.* **B129** (1983) 177–181.

[11] G. Dvali, A. Gruzinov, and M. Zaldarriaga, *Phys. Rev.* **D69** (2004) 023505, [astro-ph/0303591]; *Phys. Rev.* **D69** (2004) 083505, [astro-ph/0305548].

[12] V. F. Mukhanov and G. V. Chibisov, *JETP Lett.* **33** (1981) 532–535.

[13] V. Barger, J. P. Kneller, H.-S. Lee, D. Marfatia, and G. Steigman, *Phys. Lett.* **B566** (2003) 8–18, [hep-ph/0305075].

[14] S. Weinberg, *Cosmology.* Oxford University Press, first ed., 2008.

[15] U. Seljak, A. Makarov, P. McDonald, and H. Trac, *Phys. Rev. Lett.* **97** (2006) 191303, [astro-ph/0602430];A. Boyarsky, J. Lesgourgues, O. Ruchayskiy, and M. Viel, arXiv:0812.0010; arXiv:0812.3256.

[16] G. Dvali and M. Redi, *Phys. Rev.* **D80** (2009) 055001, [arXiv:0905.1709].

DIPLOMAS:

Twenty three Diplomas were open for competition among the participants. They have been awarded as follows:

- John S. BELL Diploma to:

  Pierpaolo MASTROLIA
  CERN, Geneva, Switzerland

- Patrick M.S. BLACKETT Diploma to:

  Mickey Guotai CHIU
  Brookhaven National Laboratory (BNL), Upton, NY, USA

- James CHADWICK Diploma to:

  Yvonne PETERS
  University of Manchester, UK and FNAL, Batavia, IL, USA

- Sidney COLEMAN Diploma to:

  Alexander VIKMAN
  New York University, NY, USA

- Richard H. DALITZ Diploma to:

  Gianluca INGUGLIA
  University of Groningen, The Netherlands

- Paul A.M. DIRAC Diploma to:

  Ely D. KOVETZ
  Tel Aviv University, Israel

- Vladimir N. GRIBOV Diploma to:

  Philipp BURDA
  ITEP, Moscow, Russia

- Robert HOFSTADTER Diploma to:

  Andrea ALICI
  INFN – University of Bologna, Italy

- Gunnar KÄLLEN Diploma to:

  Oleksandr GAMAYUN
  Bogolyubov Institute for Theoretical Physics, Kiev, Ukraine

- Seymour J. LINDENBAUM Diploma to:

  Andràs AGOCS
  RMKI Research Institute for Particle and Nuclear Physics of the Hungarian Academy of Sciences, Budapest, Hungary

- Yuval NE'EMAN Diploma to:

  Bianca Letizia CERCHIAI
  LBNL, Berkeley, CA, USA

- Giuseppe P.S. OCCHIALINI Diploma to:

  Kerstin PEREZ
  Columbia University of New York / CALTECH, New York, NY, USA

- Oreste PICCIONI Diploma to:

  Johannes ALBRECHT
  Ruprecht-Karls Universität, Heidelberg, Germany

- Bruno PONTECORVO Diploma to:

  Gennaro CORCELLA
  Enrico Fermi Centre, Rome, Italy and Scuola Normale Superiore, Pisa, Italy

- Giampietro PUPPI Diploma to:

  Raphaelle ICHOU
  SUBATECH, Nantes, France

- Isidor I. RABI Diploma to:

  Floriana GIANNUZZI
  University of Bari, Italy

- Giulio RACAH Diploma to:

  John DAVEY
  Imperial College, London, UK

- Julian S. SCHWINGER Diploma to:

  Marcin BADZIAK
  Warsaw University, Poland

- Bruno ROSSI Diploma to:

  Roberto PREGHENELLA
  Enrico Fermi Centre, Rome, Italy and INFN – University of Bologna, Italy

- Victor F. WEISSKOPF Diploma to:

  Daniel TAPIA TAKAKI
  IPN – Université de Paris Sud, Orsay, France

- Eugene P. WIGNER Diploma to:

  Andrey SMIRNOV
  ITEP, Moscow, Russia

- Bjorn H. WIIK Diploma to:

  Hanno Christopher PERREY
  Universität Hamburg, Germany

- Chien Shiung WU Diploma to:

  Denis A. EPIFANOV
  Budker Institute for Nuclear Physics – INP Novosibirsk, Russia

AWARDS

The Awards for the Best 'New Talent – Speakers' have been attributed to:

- **Andrey SMIRNOV**
  ITEP, Moscow, Russia
  for his original presentation in
  MATHEMATICAL PHYSICS

- **Hanno Christopher PERREY**
  Universität Hamburg, Germany
  for his original presentation in
  DETECTOR PHYSICS

- **Alexander VIKMAN**
  New York University, NY, USA
  for his original presentation in
  THEORETICAL PHYSICS

- **Mickey Guotai CHIU**
  Brookhaven National Laboratory (BNL), Upton, NY, USA
  for his original presentation in
  EXPERIMENTAL PHYSICS

Award for Best Question given to:
Maxim KURKOV
St. Petersburg Nuclear Physics Institute, Russia

Award for Best Student given to:
Roberto PREGHENELLA
Enrico Fermi Centre, Rome, Italy and INFN – University of Bologna, Italy

Award for Best Scientific Secretary given to:
Mohammad Ayaz AHMAD
Aligarh Muslim University, India

PARTICIPANTS

Andrea ACHILLI Italy	University of Perugia PERUGIA, Italy
András AGOCS Hungary	RMKI Research Institute for Particle and Nuclear Physics of the Hungarian Academy of Sciences BUDAPEST, Hungary
Mohammad Ayaz AHMAD India	Aligarh Muslim University ALIGARH, India
Muhammad AJAZ Pakistan	COMSATS Institute of Information Technology ISLAMABAD, Pakistan
Johannes ALBRECHT Germany	Ruprecht-Karls Universität HEIDELBERG, Germany
Andrea ALICI Italy	INFN and University of Bologna BOLOGNA, Italy
Richard ARNOWITT USA	Texas A&M University COLLEGE STATION, TX, US
Paolo ASCHIERI Italy	University of Alessandria ALESSANDRIA, Italy
Marcin BADZIAK Poland	Warsaw University WARSZAWA, Poland
Rinaldo BALDINI Italy	Enrico Fermi Centre ROME, Italy
Alessandro BETTINI Italy	INFN - University of Padova PADOVA, Italy
Philipp BURDA Russia	ITEP MOSCOW, Russia

Alberto CALIVA
Italy

University of Palermo
PALERMO, Italy

Bianca Letizia CERCHIAI
Italy

Lawrence Berkeley National Laboratory
BERKELEY, CA, US

Mickey Guotai CHIU
USA

Brookhaven National Laboratory
UPTON, NY, US

Juansher CHKAREULI
Georgia

F. Andronikashvili Institute of Physics
TBILISI, Georgia

Luisa CIFARELLI
Italy

University of Bologna
BOLOGNA, Italy

Fabrizio COCCETTI
Italy

Enrico Fermi Centre
ROME, Italy

Eugenio COCCIA
Italy

LNGS - INFN
L'AQUILA, Italy

Gennaro CORCELLA
Italy

Enrico Fermi Centre
ROME, Italy

John DAVEY
UK

Imperial College London
LONDON, UK

Carmelo DI GIGLIO
Italy

University of Bari
BARI, Italy

Lance DIXON
USA

SLAC National Accelerator Laboratory
MENLO PARK, CA, US

Michael R. DOUGLAS
USA

NHETC - The State University of NJ
PISCATAWAY, NJ, US

Georgi DVALI
Georgia

CERN
CH 1211 GENEVA 23, Switzerland

Denis A. EPIFANOV
Russia

Budker Institute for Nuclear Physics - INP
NOVOSIBIRSK, Russia

Sergio FERRARA
Italy
CERN
CH 1211 GENEVA 23, Switzerland

Ioannis FLORAKIS
Greece
École Normale Supérieure
PARIS, France

Oleksandr GAMAYUN
Ukraine
Bogolyubov Institute for Theoretical Physics
KIEV, Ukraine

Floriana GIANNUZZI
Italy
University of Bari
BARI, Italy

Barbara GUERZONI
Italy
INFN - University of Bologna
BOLOGNA, Italy

Dieter HAIDT
Germany
Hadermannsweg 41A
D-22459 HAMBURG (GERMANY)

Rolf-Dieter HEUER
Germany
CERN
CH 1211 GENEVA 23, Switzerland

Raphaelle ICHOU
France
SUBATECH
NANTES, France

Vladimir IGLOVIKOV
Russia
St. Petersburg Nuclear Physics Institute
ST. PETERSBURG, Russia

Gianluca INGUGLIA
Italy
University of Groningen
GRONINGEN, The Netherlands

Tomasz JELINSKI
Poland
Warsaw University
WARSZAWA, Poland

Yaroslav KLOPOT
Ukraine
Bogoliubov Theoretical Physics Laboratory
DUBNA, Russia

Christian P. KORTHALS-ALTES
The Netherlands
CNRS-Luminy
MARSEILLE, France

Ely D. KOVETZ
Israel
Tel Aviv University
TEL AVIV, Israel

Alekxandra KUREK Poland	Astronomical Observatory of the Jagiellonian University KRAKÓW, Poland
Maxim KURKOV Russia	St. Petersburg Nuclear Physics Institute ST. PETERSBURG, Russia
Dieter LÜEST Germany	Ludwig-Maximilians-Universitaet MÜNCHEN, Germany
Markus MARIENFELD Germany	DESY HAMBURG, Germany
Alessio MARRANI Italy	Stanford University STANFORD, CA, US
Pierpaolo MASTROLIA Italy	CERN GENEVA, Switzerland
Larry MCLERRAN USA	Brookhaven National Laboratory UPTON, NY, US
Alexander MONIN Russia	University of Minnesota MINNEAPOLIS, MN, US
Dmitry NESTEROV Russia	Lebedev Physical Institute MOSCOW, Russia
Francesco NOFERINI Italy	INFN - CNAF and University of Bologna BOLOGNA, Italy
Toyoko ORIMOTO Japan	CALTECH PASADENA, CA, US
Simone PACETTI Italy	Enrico Fermi Centre ROME, Italy
Luca PARISI Italy	University of Salerno SALERNO, Italy
Stephen PARKE USA	FNAL BATAVIA, IL, US

Bakur PARSAMYAN
Armenia
University of Torino
TORINO, Italy

Vladyslav P. PAUK
Ukraine
Taras Shevchenko National Kiev University
KIEV, Ukraine

Kerstin PEREZ
USA
Columbia University of the New York / CALTECH
NEW YORK, NY, US

Hanno Christopher PERREY
Germany
Universität Hamburg
HAMBURG, Germany

Yvonne PETERS
Germany
University of Manchester
MANCHESTER, UK

Roberto PREGHENELLA
Italy
Enrico Fermi Centre
ROME, Italy

Anastasiia REBESH
Ukraine
Bogolyubov Institute for Theoretical Physics
KIEV, Ukraine

Mikhail SHAPOSHNIKOV
Russia
École Polytechnique Fédérale de Lausanne
LAUSANNE, Switzerland

Andriy SHCHERBAKOV
Ukraine
INFN - Laboratori Nazionali di Frascati
FRASCATI, Italy

Andrey SMIRNOV
Russia
ITEP
MOSCOW, Russia

Paul SORENSEN
USA
Brookhaven National Laboratory
UPTON, NY, US

Daniel TAPIA TAKAKI
Mexico
IPN Orsay - Université Paris Sud
ORSAY, France

Gerardus 't HOOFT
The Netherlands
Utrecht University
UTRECHT, The *@phys.uu.nl*

Michael J. TANNENBAUM
USA
Brookhaven National Laboratory
UPTON, NY, *@bnl.gov*

Valiya Neerilakkal UVAIS
Aligarh Muslim University

India — ALIGARH, India

Naomi VAN DER KOLK
Germany
CERN
CH 1211 GENEVA 23, Switzerland

Alexander VIKMAN
Germany/Ukraine
New York University
NEW YORK, NY, US

Horst WENNINGER
Germany
CERN
CH 1211 GENEVA 23, Switzerland

Mark WILLIAMS
UK
Lancaster University
LANCASTER, UK

Günther WOLF
Germany
DESY
HAMBURG, Germany

Stanislav YASNOPOLSKIY
Russia
RRC I.V. Kurchatov Institute
MOSCOW, Russia

Ivan YELETSKIKH
Ukraine
National Academy of Sciences of Ukraine
KHARKOV, Ukraine

Chiara ZAMPOLLI
Italy
CERN
CH 1211 GENEVA 23, Switzerland

Antonino ZICHICHI
Italy
Dept. of Physics,
University of Bologna
BOLOGNA, Italy